Land Use Changes in Europe

The GeoJournal Library

Volume 18

The titles published in this series are listed at the end of this volume.

Land Use Changes in Europe

Processes of Change, Environmental Transformations and Future Patterns

A study initiated and sponsored by the
International Institute for Applied Systems Analysis
with the support and co-ordination of
the Stockholm Environment Institute

edited by

F. M. BROUWER
*Agricultural Economics Research Institute,
The Hague, The Netherlands*

and

A. J. THOMAS
M. J. CHADWICK
*Stockholm Environment Institute at York,
York, United Kingdom*

Published for the Stockholm Environment Institute
and the International Institute for Applied Systems Analysis

KLUWER ACADEMIC PUBLISHERS
DORDRECHT / BOSTON / LONDON

Library of Congress Cataloging-in-Publication Data

Land use changes in Europe : processes of change, environmental
  transformations, and future patterns / edited by F.M. Brouwer, A.J.
  Thomas, M.J. Chadwick.
        p.   cm. -- (GeoJournal library ; v. 18)
    Revised papers from the Workshop on Land Use Changes in Europe
  held in Poland in Sept. 1988.
    "Published for the Stockholm Environment Institute and the
  International Institute for Applied Systems Analysis."
    Includes index.
    ISBN 0-7923-1099-3 (hb : acid-free paper)
    1. Land use--Europe--Congresses. 2. Land use--Environmental
  aspects--Europe--Congresses.   I. Brouwer, Floor.   II. Thomas, A. J.
  III. Chadwick, M. J. (Michael J.) IV. International Institute for
  Applied Systems Analysis. V. Stockholm Environment Institute.
  VI. Series.
  HD586.LS36   1991
  333.73'13'094--dc20   90-26080

ISBN 0-7923-1099-3

Published by Kluwer Academic Publishers,
P.O. Box 17, 3300 AA Dordrecht, The Netherlands.

Kluwer Academic Publishers incorporates
the publishing programmes of
D. Reidel, Martinus Nijhoff, Dr W. Junk and MTP Press.

Sold and distributed in the U.S.A. and Canada
by Kluwer Academic Publishers,
101 Philip Drive, Norwell, MA 02061, U.S.A.

In all other countries, sold and distributed
by Kluwer Academic Publishers Group,
P.O. Box 322, 3300 AH Dordrecht, The Netherlands.

Printed on acid-free paper

Preface

The patterns of land use that have evolved in Europe reflect the boundaries set by the natural environment and socio-economic responses to the needs of the population. Over the centuries man has been able to overcome increasingly the constraints placed on land use by the natural environment through the development of new technologies and innovations, driven by an increasing population and rising material expectations. However, activities are still ultimately constrained by natural limitations such as climatic characteristics and associated edaphic and vegetational features.

A major problem for land management, in its broadest sense, can be a reluctance to foresee the consequent ecological changes. This means that mitigating strategies will not be implemented in time to prevent environmental degradation and social hardship, although in many parts of Europe, over some centuries, demands have been met in a sustainable way, by sound, prudent and temperate expectations that have dictated management regimes.

The management of land in Europe has always been a complex challenge: land is the primary, though finite resource. Decisions regarding the use of land and manipulation of ecosystem dynamics today may affect the long-term primary productivity of the resource. Decisions to change land use may be virtually irreversible; urbanization is an illustration of the influence of population density on the land resource. The International Institute for Applied Systems Analysis (IIASA), in co-operation with the Institute of Geography and Spatial Organization of the Polish Academy of Sciences, organized a Workshop on *Land Use Changes in Europe* in Poland in September, 1988 to consider all these interacting features of European land use against the background of possible global climatic change. The Workshop was the source of draft papers and these, in an edited form, constitute the majority of Chapters in this book. The meeting focused on six discrete but interrelated topics including: i) major land use determinants; ii) present land use patterns; iii) the main processes of change of major importance for future land use; iv) historical land use changes in Europe; v) likely future land use patterns; and vi) policy implications and the identification of management strategies. Thus, this book covers a wide spectrum of issues. Most are related to the potential impact of climatic change and how this must be considered in the long-term and on a broad-scale. Some of the questions that need to be addressed include the topics listed. How will the characteristics of the land resource change and what are the implications of those changes on the environment and its capacity to supply these? What policies need to be introduced to encourage sensitivity to environmental supply limitations? What scale of response is required?

Following the workshop sponsored by IIASA, the editing tasks were undertaken by F.M. Brouwer (formerly of IIASA), and Alison Thomas and M.J. Chadwick of the Stockholm Environment Institute at York (SEIY). The Stockholm Environment Institute (SEI), Stockholm, Sweden, financed this phase of the work. Isobel Devane prepared the manuscript for publication along with Susan Sparrow who provided the final version of many of the figures. Andrew Lees assisted in the preparation of the Index.

During the Workshop in Poland many of the arrangements were made by Dr. R. Kulikowski of the Institute of Geography and Spatial Organization of the Polish Academy of Sciences. The Stockholm Environment Institute and IIASA express their warm appreciation of his efforts and those of the Director of the Institute, Professor J. Kostrowicki.

F.M. Brouwer
A.J. Thomas
M.J. Chadwick

<u>September, 1990, York.</u>

List of Contributors

S. Anderberg, Department of Social and Economic Geography, Lund University, Lund, Sweden.

I. Aselmann, Max-Planck Institute for Chemistry, Mainz, Federal Republic of Germany.

N. Brink, Division of Water Management, Department of Soil Sciences, Swedish University of Agricultural Sciences, Uppsala, Sweden.

F.M. Brouwer, Agricultural Economics Research Institute, LEI, s'-Gravenhage, The Netherlands.

T. Carter, Finnish Meteorological Institute, Helsinki, Finland.

M.J. Chadwick, Stockholm Environment Institute at York, York, UK.

H. Coccossis, University of the Agean, Athens, Greece.

M. Falkenmark, Committee for Hydrology, Natural Science Research Council, Stockholm, Sweden.

A. Friend, Institute for Research on Environment and Economy, University of Ottawa, Ottawa, Canada.

G.P. Hekstra, Ministry of Housing, Physical Planning and Environment, Leidschendam, The Netherlands.

A.C. Imeson, Landscape and Environmental Research Group, University of Amsterdam, Amsterdam, The Netherlands.

N. Karavayeva, Institute of Geography, USSR Academy of Sciences, Moscow, USSR.

L. Kauppi, Water and Environment Research Institute, National Board of Waters and the Environment, Helsinki, Finland.

A. Kedziora, Department of Agrobiology and Forestry, Polish Academy of Sciences, Poznan, Poland.

J. Kostrowicki, Institute of Geography and Spatial Organization, Polish Academy of Sciences, Warsaw, Poland.

J. Lee, The Agricultural Institute, Johnston Castle Research Centre, Wexford, Ireland.

H.N. van Lier, Department of Physical Planning and Rural Development, Agricultural University of Wageningen, Wageningen, The Netherlands.

E.W. Manning, Sustainable Development Branch, Environment Canada, Ottawa, Canada.

T.G. Nefedova, Institute of Geography, USSR Academy of Sciences, Moscow, USSR.

P. Nijkamp, Department of Economics, Free University, Amsterdam, The Netherlands.

J. Okuniewski, Institute of Rural and Agricultural Development, Warsaw, Poland.

L.R. Oldeman, International Soil Reference and Information Centre, Wageningen, The Netherlands.

J. Olejnik, Department of Agrobiology and Forestry, Polish Academy of Sciences, Poznan, Poland.

T. Paces, Department of Geochemistry, Geological Survey, Prague, Czechoslovakia.

M. Parry, Atmospheric Impacts Research Group, University of Birmingham, Birmingham, UK.

J. de Ploey, Laboratory for Experimental Geomorphology, Leuven, Belgium.

M. Posch, International Institute for Applied systems Analysis, Laxenburg, Austria.

L. Ryszkowski, Department of Agrobiology and Forestry, Polish Academy of Sciences, Poznan, Poland.

F. Soeteman, Department of Economics, Free University, Amsterdam, The Netherlands.

I. Szabolcs, Research Institute of Soil Science and Agricultural Chemistry, Hungarian Academy of Sciences, Budapest, Hungary.

V.O. Targulian, International Institute for Applied Systems Analysis, Laxenburg, Austria.

W.H. Verheye, Institute of Geology, State University of Gent, Gent, Belgium.

Table of Contents

Chapter 1

LAND RESOURCES, LAND USE AND PROJECTED LAND AVAILABILITY FOR ALTERNATIVE USES IN THE EC

J. Lee

1.1. Introduction

The European Economic Community has a wide range of soils, climates, vegetation, topography, and a corresponding wide range in land use suitability. The soil resource is subject to limitations and may suffer destruction or deterioration, particularly through erosion effects, mismanagement and pollution. The intensity of land use can vary considerably, with land under arable use accounting for 90 per cent of the utilized agricultural area (UAA) in Denmark and as little as 10 per cent in Ireland. The agro-ecological environment of north-west Europe, which is characterized by relatively high precipitation, is conducive to pasture production, whereas in the Mediterranean and much of the Continental zones, moisture stress is a limiting factor and agricultural land use has a large arable component. Livestock densities and the intensity of fertilizer application may also vary considerably.

1.2. Land Use Structure

1.2.1. Agricultural and forestry use

Agriculture or utilized agricultural area (UAA) is the largest user of land in the European Community (Table 1.1) taking up well over half of the total land area in most countries. The proportion of UAA ranges from 43.5 in Greece to as much as 80.8 per cent in Ireland. The high UAA in countries such as United Kingdom and Ireland is explained partly by the inclusion of rough grazing land in the UAA category.

Whereas 52.6 per cent of the UAA in the EEC-12 is classified in the arable use category (Table 1.1), the range is from 18.7 per cent in Ireland to 91.9 per cent in Denmark. Similarly, permanent grassland shows significant variation. Woodland ranges from 44.7 per cent of the land in Greece (Table 1.1) to as little as 4.7 per cent in Ireland, compared with the EEC-12 average of 24.1 per cent.

Germany and France account for 60 per cent of the EEC-10 wooded area. Of the 31 million ha of forest in the EEC-9, 19 million

1

F. M. Brouwer et al. (eds.), Land Use Changes in Europe, 1–20.
© 1991 *Kluwer Academic Publishers. Printed in the Netherlands.*

Table 1.1. Principal categories of land use in EEC countries. (Area in 1,000 ha. In brackets: UAA and wooded area as a percentage of total land area; arable land and permanent grassland as a percentage of the total amount of land that is utilized for agricultural purposes.)

Country	UAA	Arable land	Permanent grassland	Wooded area
Belgium	1,412 (46.3)	743 (52.6)	632 (44.8)	617 (20.4)
Denmark	2,818 (65.4)	2,592 (91.9)	214 (7.6)	493 (11.6)
France	31,388 (57.2)	17,737 (56.5)	12,094 (38.5)	14,635 (26.9)
Germany	12,000 (48.3)	7,244 (60.4)	4,537 (37.8)	7,360 (30.1)
Greece	5,741 (43.5)	2,925 (50.9)	1,789 (31.2)	5,755 (44.7)
Ireland	5,676 (80.8)	1,062 (18.7)	4,612 (81.3)	327 (4.7)
Italy	17,445 (57.9)	9,061 (51.9)	4,944 (28.3)	6,097 (20.7)
Luxembourg	128 (49.4)	56 (43.8)	70 (54.7)	82 (31.8)
Netherlands	2,024 (51.0)	876 (43.3)	1,108 (54.7)	330 (9.1)
Portugal	4,532 (49.2)	2,906 (64.1)	761 (16.8)	2,968 (32.4)
Spain	27,222 (53.9)	15,651 (57.5)	6,645 (24.4)	12,511 (25.1)
UK	18,612 (76.2)	6,952 (37.4)	11,583 (62.2)	2,297 (9.5)
EEC-12	128,999 (57.1)	67,805 (52.6)	48,990 (37.9)	53,473 (24.1)

Source: Eurostat Crop Production, 1988.

hectares were classified as productive in 1978 (EEC, 1978). Of the remaining 12 million ha, at least 4 million ha were considered suited for conversion to productive forest, the remaining 8 million ha being considered less suitable for timber production, but serving a vital environmental role, for example, in the prevention of erosion. The 12 million ha include certain Alpine forests near the upper limit of tree growth, coppice and shrub areas on poor sites, special vegetation types like the Mediterranean maquis, as well as potentially productive forests managed as nature reserves. France and Italy account for 10 million of the 12 million ha.

1.2.2. **Changes in land use**

Table 1.2 describes the distribution of utilized agricultural area in Europe (not including the European part of the USSR). Over the period 1968-71 to 1983 the UAA of Europe declined by 8 million ha. For comparison, the extent of decline was 6.7 million ha in western Europe, 4.5 million ha in EEC-10 and 6 million ha in EEC-12. The decline

Table 1.2. Utilized agricultural area in Europe in millions of hectares (per cent decrease in brackets).

	1969-71	1983	Decrease
Europe	233.45	225.37	8.08 (3.46)
Western Europe	158.11	151.42	6.69 (4.23)
EEC-10	105.63	101.08	4.55 (4.31)
EEC-12	142.38	136.35	6.03 (4.24)

Source: FAO Production Yearbooks, 1981 and 1984.

occurred in almost all countries with the highest percentage occurring in Italy, Belgium, The Netherlands and Luxembourg. Agricultural area comprises arable land and permanent pasture.   Tables 1.3 and 1.4 indicate that both the arable and grassland components declined by 4.5 per cent in the EEC-12.

Table 1.3. Arable land area in Europe in millions of hectares (per cent decrease in brackets).

	1969-71	1983	Decrease
Europe	130.63	125.86	4.77 (3.65)
Western Europe	79.15	75.71	3.44 (4.35)
EEC-10	51.59	49.08	2.51 (4.86)

Source: FAO Production Yearbooks, 1981 and 1984.

Table 1.5 indicates that forestry increased by 2.6 million ha (5 per cent) in the EEC-12 over the 1969-71 to 1983 period. Figure 1.1 depicts the major land use trends over the 1961-65 to 1983 period.

1.3. Climate and Soils

In his classification of world climates Papadakis (1966) distinguishes ten major categories of which five are represented in Europe (Figure 1.2).   Each zone is designated by a number representing the climatic type.   The dominant European climate is of Marine type, which is subdivided into Warm, Cool, Cold Marine, Warm and Cool Temperate climatic units.   Southern Europe is characterised by a Temperate Mediterranean climate.

Table 1.4.   Permanent pasture in Europe in millions of hectares (per cent decrease in brackets).

	1969-71	1983	Decrease
Europe	88.50	85.55	2.95 (3.33)
Western Europe	67.57	64.43	3.14 (4.65)
EEC-10	48.16	46.34	1.82 (3.78)
EEC-12	60.29	57.56	2.73 (4.53)

Source: FAO Production yearbooks, 1981 and 1984.

Table 1.5.   Wooded area in Europe in millions of hectares (per cent increase in brackets).

	1969-71	1983	Increase
Europe	148.8	154.0	5.2 (3.49)
Western Europe	112.5	116.6	4.1 (3.64)
EEC-10	33.7	34.9	1.2 (3.56)
EEC-12	51.5	54.1	2.6 (5.04)

Source: FAO Production Yearbooks, 1981 and 1984.

1.3.1.   Soils and land use

Europe comprises several major soil regions (EEC, 1985).   These soil regions are depicted in Figure 1.2, which indicates that the EEC-12 is characterized by nine major soil regions.   It is evident from this figure that Europe has a wide range of soils occurring over a wide climatic range.   However, soil units with Cambisols and Luvisols are dominant, and account for over 60 per cent of the total area.   Their land use varies according to climate, topography, and degree of stoniness.   The Eutric Cambisols which are commonly stony or lithic in character are found most extensively in the Mediterranean zone, whereas the Dystric Cambisols are extensive under Cool Temperate conditions.   In the Mediterranean zone, the Eutric Cambisols are subject to severe drought limitations.   They are devoted to grazing from low yielding pastures, with arable cultivation and viticulture being important locally.

The Eutric Cambisols are also highly important arable and grassland soils in the UK (see cambisols complex in Figure 1.2) and eastern Denmark and are capable of yielding 7.5 t ha^{-1} from cereals and

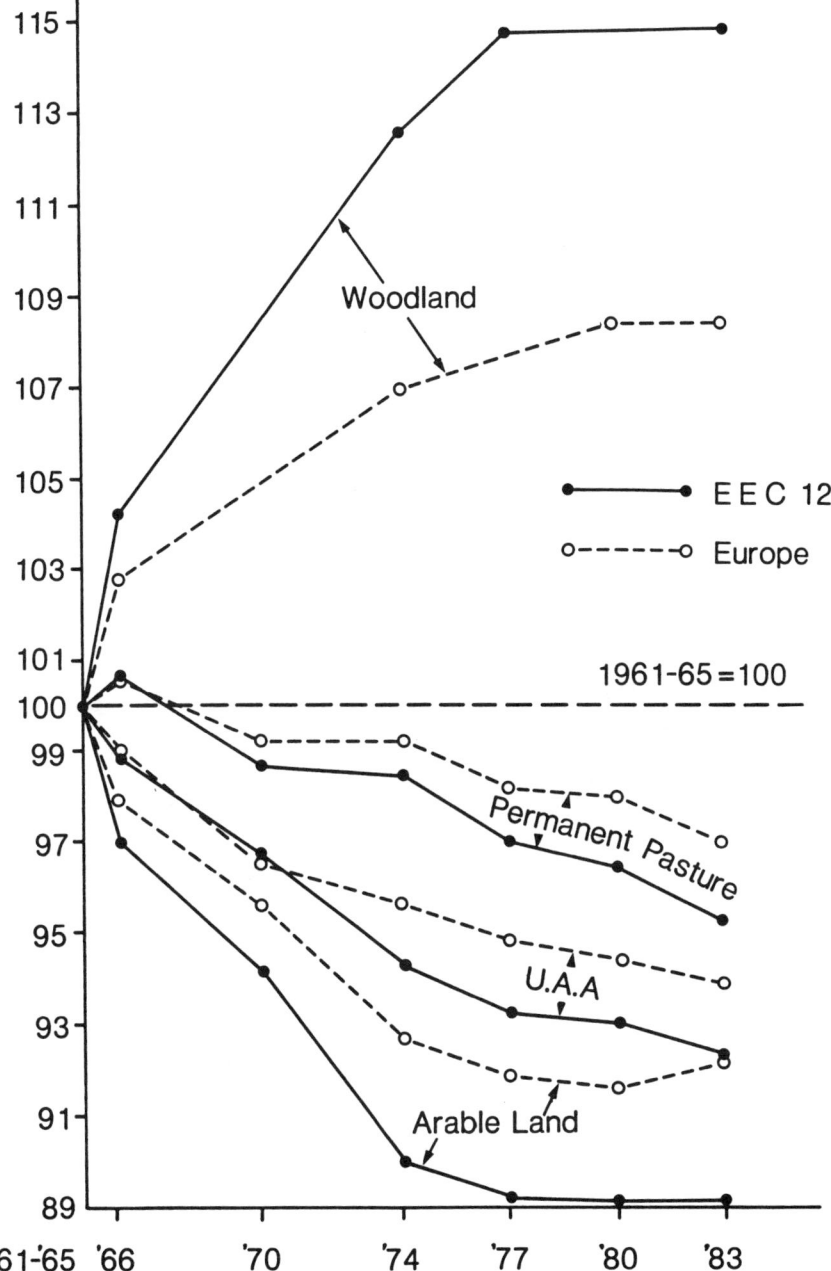

Figure 1.1. Land use trends in Europe and EEC-12.

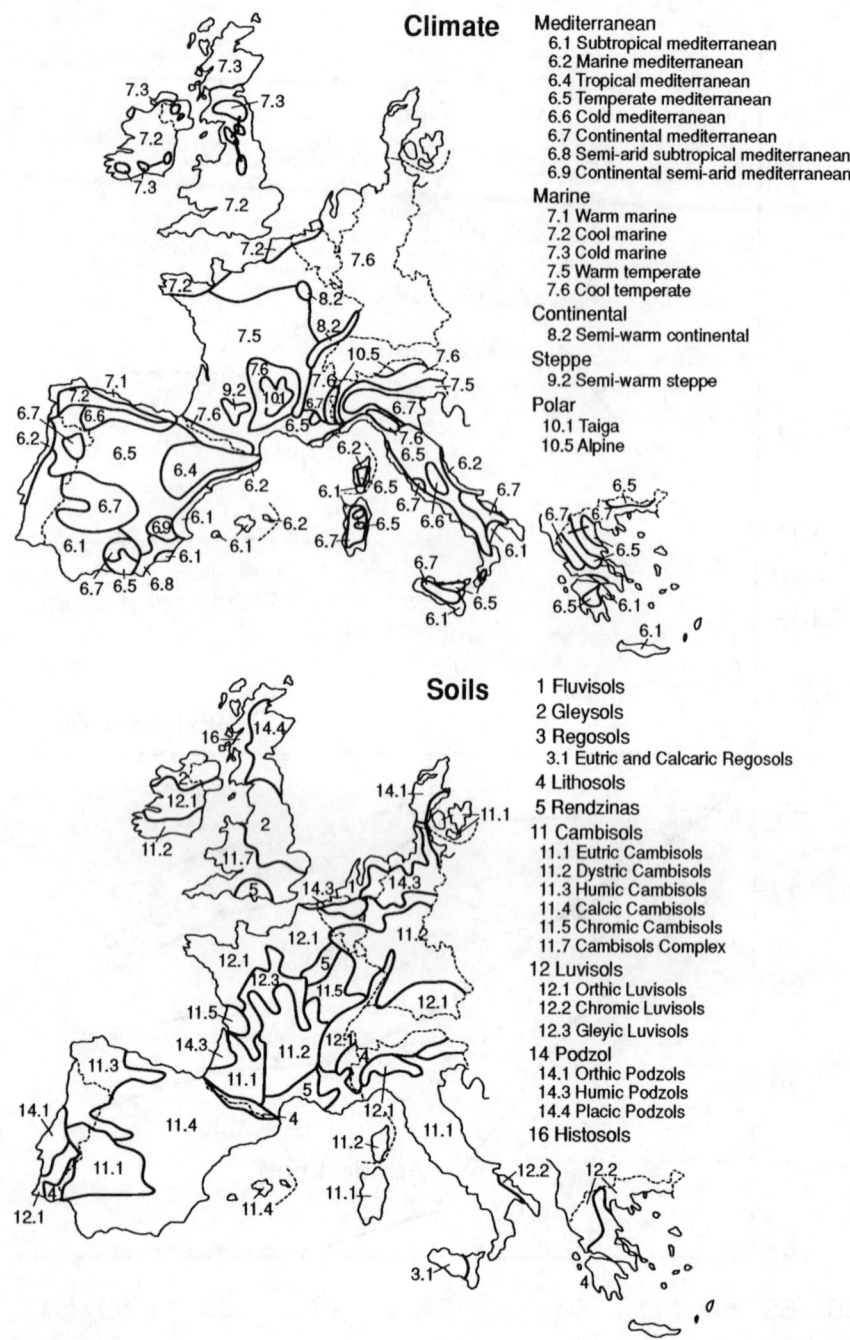

Climate

Mediterranean
  6.1 Subtropical mediterranean
  6.2 Marine mediterranean
  6.4 Tropical mediterranean
  6.5 Temperate mediterranean
  6.6 Cold mediterranean
  6.7 Continental mediterranean
  6.8 Semi-arid subtropical mediterranean
  6.9 Continental semi-arid mediterranean

Marine
  7.1 Warm marine
  7.2 Cool marine
  7.3 Cold marine
  7.5 Warm temperate
  7.6 Cool temperate

Continental
  8.2 Semi-warm continental

Steppe
  9.2 Semi-warm steppe

Polar
  10.1 Taiga
  10.5 Alpine

Soils

  1 Fluvisols
  2 Gleysols
  3 Regosols
    3.1 Eutric and Calcaric Regosols
  4 Lithosols
  5 Rendzinas
  11 Cambisols
    11.1 Eutric Cambisols
    11.2 Dystric Cambisols
    11.3 Humic Cambisols
    11.4 Calcic Cambisols
    11.5 Chromic Cambisols
    11.7 Cambisols Complex
  12 Luvisols
    12.1 Orthic Luvisols
    12.2 Chromic Luvisols
    12.3 Gleyic Luvisols
  14 Podzol
    14.1 Orthic Podzols
    14.3 Humic Podzols
    14.4 Placic Podzols
  16 Histosols

Figure 1.2. Climate and soil regions of EEC-12.

10-12 DM t ha^{-1} from pastures. The Dystric Cambisols associated with mountainous regions such as the Massif Central, Bavarian Woods, Famenne, and the Italian Alps are largely devoted to forestry or grazing, whereas under lowland conditions in the UK, Ireland and France they are highly important agricultural soils capable of pasture yields of 11-15 DM t ha^{-1} and cereal yields of 7.5 t ha^{-1}.

The Orthic Luvisols have a moderately high nutrient status. These soils are found most extensively in France and Germany and are highly suited to intensive arable and grassland production where they occur under favorable climatic conditions on level topography. In Ireland pasture yields of 13 DM t ha^{-1} are attainable and in the Rhine Valley cereal yields of 6 t ha^{-1} are achievable. In contrast, Chromic Luvisols are subject to severe drought limitation in Greece, Spain and Italy in addition to susceptibility to erosion.

Fluvisols occupy 5.7 per cent of the land area and are among the most intensively farmed soils in Europe. They are particularly significant in the Mediterranean zone, where the level topography is well suited to irrigation technology. They are devoted to intensive arable cropping, viticulture and market gardening in the Mediterranean zone, whereas under Cool Temperate conditions in The Netherlands they are devoted to intensive grassland and arable cropping. Similarly, the Rendzinas which occur on suitable terrain are important arable and grassland soils in France, Germany and the UK. They are devoted to forestry and grassland at high elevation occupying 5 per cent of the land area.

Although the Vertisols which are confined mainly to Spain and Italy are subject to structure and texture limitations, they are nevertheless important arable soils. They occur on level terrain suited to irrigation technology and despite their limitations they are intensively cultivated.

The Gleysols occupy over 3 per cent of the land area and are subject to drainage and permeability limitations. They are most extensive in the UK and Ireland where they occur under Cool marine conditions. They are largely limited to pasture use except in lower-rainfall zones such as eastern England where they are important arable soils.

The Podzols and Rankers which are commonly subject to depth, drainage, stoniness, elevation, accessibility or climatic constraints, occupy over 9 per cent of the land area. In mountainous regions, they are devoted largely to forestry or rough grazing. However, the Podzols on suitable terrain, which occur in west Denmark and north-east Scotland constitute important arable and grassland soils in these regions. Similarly, the Lithosols which occupy 5 per cent of the land area are largely associated with mountainous/hilly terrain and are subject to serious limitations of stoniness, rockiness, slope and elevation. They are largely afforested or devoted to rough grazing of garrigue type as in Greece or in the Cévennes. Locally, they may be devoted to vine, olive or almond culture in the Mediterranean zone. Although Histosols in their natural state are subject to severe use constraints, in their reclaimed state in The Netherlands, Germany and Ireland, they are capable of exceptionally high arable, horticultural and grassland yields.

1.3.2. Evaluating land base for agriculture

The publication of the Soil Map of the EC (1:1,000,000) (EEC, 1985) has enabled an exploratory evaluation of land suitability for grassland and arable uses to be carried out at Community level (Lee, 1984 and 1986).

The *grassland suitability assessment* (Table 1.6 and in Figure 1.3) shows that:

i)   43.04 million ha (28 per cent) of the EEC-10
        has very high - high productivity (Suitability
        Classes A and B);

ii)   56.33 million ha (36 per cent) of the EEC-10 has
        moderate/low - moderate/high productivity
        (Suitability Classes C and D);

iii)   56.33 million ha (36 per cent) of the EEC-10 has
        low - very low productivity (Suitability Classes D
        and E);

The *arable suitability assessment* (Table 1.7 and Figure 1.3) shows that:

i)   30.64 million ha (20 per cent) of the EEC-10 is
        well suited (Class 1);

ii)   26.68 million ha (17 per cent) of the EEC-10 is
        suited (Class 2);

iii)   48.54 million ha (32 per cent) of the EEC-10 is
        moderately/poorly suited (Class 3);

iv)   48.14 million ha (31 per cent) of the EEC-10 is
        unsuited (Class 4).

It is significant that France alone accounts for over 50 per cent of total EEC-10 Class 1 land. Class 2 arable land (having slight or slight/moderate limitations) occupies 26.68 million ha or 17 per cent of the land area with France accounting for about 30 per cent of the total.

Table 1.6. Extent of grassland suitability classes in EEC-10 (percentage of total area). Five suitability classes were defined as A, B, C, D and E, with degree of limitation increasing from A to E.

| | Suitability class | | | | |
	A	B	C	D	E
Belgium	4	50	45	-	1
Denmark	-	38	43	17	3
France	-	21	29	37	13
Germany	13	34	43	5	5
Greece	11	1	-	19	69
Ireland	32	13	21	-	34
Italy	17	6	4	35	38
Luxembourg	-	12	80	-	8
Netherlands	22	52	19	4	3
UK	8	17	42	2	30
EEC-10	9	19	26	22	24

Source: Lee, 1984.

In the approach adopted by Lee, three land qualities have assumed major importance:

1) moisture availability;
2) poaching susceptibility;
3) accessibility to machinery
(irrigation is assumed for suitability classes A and B in Italy and Greece which under rainfed conditions would mainly correspond to class D).

It is notable that 31 per cent of the EEC-10 land area is unsuited to arable cropping, and the remaining 32 per cent of land is in the Class 3 category. While cultivation may be feasible on Class 3 land, in practice it is subject to moderate or moderate/severe limitations of soil/topography and must largely be excluded from consideration for mechanized arable systems. Therefore, the land pool for viable arable use options is limited to about 57 million ha with France accounting for 40 per cent of the total.

Although it would be misleading to assume that the existing arable areas are synonymous with Classes 1 and 2 land, the indications are that the land quality of the arable areas is high. For the EEC-10, the

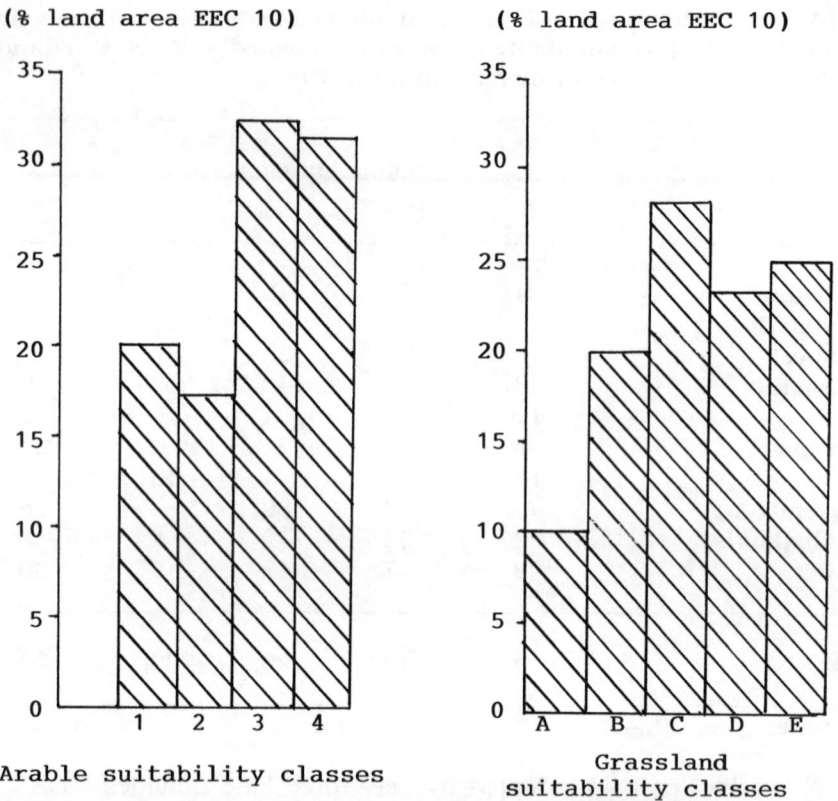

Figure 1.3.   Extent of arable and grassland suitability classes in the EEC-10.

extent of Classes 1 and 2 is 8 per cent greater than the existing arable area. The extent of Classes 1 and 2 land as compared with arable area tends to be high throughout the EEC-10 with the exception of the Mediterranean zone.

1.4. Slope Characteristics of Land

The slope characteristics of land are significant not only in terms of utilization but also in terms of susceptibility to water erosion. The legend of the Soil Map of the European Community (EEC, 1985) includes four categories of slope as follows:

$$
\begin{array}{lll}
a = & 0 \text{ - } 8\% & \text{level} \\
b = & 8 \text{ - } 15\% & \text{sloping} \\
c = & 15 \text{ - } 25\% & \text{moderately steep} \\
d = & > 25\% & \text{steep}
\end{array}
$$

Table 1.7.   Extent of arable land suitability classes in the EEC-10 countries in 1,000s of hectares (percentage of total area in brackets).

| | Suitability class | | | |
	1	2	3	4
Belgium	760 (25)	1,140 (37)	1,060 (35)	93 (3)
Denmark	765 (18)	3,355 (79)	- -	13 (3)
France	15,440 (29)	7,700 (14)	18,700 (35)	12,100 (22)
Germany	6,300 (30)	2,810 (13)	11,150 (53)	930 (4)
Greece	550 (4)	950 (7)	1,810 (14)	9,810 (75)
Ireland	1,010 (12)	1,380 (16)	1,100 (13)	4,940 (59)
Italy	3,350 (13)	3,640 (14)	8,100 (31)	10,870 (42)
Luxembourg	38 (15)	- -	198 (77)	21 (8)
Netherlands	920 (27)	1,060 (31)	1,100 (33)	300 (9)
UK	1,510 (7)	5,650 (26)	5,330 (25)	8,950 (42)
EEC-10	30,643 (20)	26,685 (17)	48,548 (32)	48,144 (31)

Source: Lee, 1986.

The classification scheme includes four Suitability Classes as follows:

Class 1 -	well suited land, with no or slight limitations;
Class 2 -	suited land, with slight or slight/moderate limitations;
Class 3 -	moderately suited land, with moderate or moderate/severe poorly suited limitations;
Class 4 -	unsuitable land with severe or very severe limitations.

Degree of limitation increases from the lowest to the highest.

The four slope classes indicate the slope which dominates the Soil Association area. Through an analysis of the national legends provided for the Soil Map of Europe project, a slope inventory of European land has been derived. The results are summarized in Table 1.8. Only 25 per cent of the EEC-12 is categorized as level land, with an additional 14 per cent in the level/sloping class. The extent of accessible land from a mechanization suitability viewpoint ranges from over 90 per cent

Table 1.8.   Percentage distribution of slope classes by country in the EEC-12.

Country	a	a/b	a/d	b	b/c	b/d	c	c/d	d
Belgium	25	2	-	45	-	-	28	-	-
Denmark	45	-	-	55	-	-	-	-	-
France	37	23	-	13	1	-	9	17	-
Germany	47	8	-	16	9	-	19	-	1
Greece	12	-	-	4	8	11	-	3	62
Ireland	8	25	-	44	23	-	-	-	-
Italy	15	10	-	0	17	3	0	44	11
Luxembourg	7	-	16	-	-	-	76	-	-
Netherlands	91	-	-	9	-	-	-	-	-
Portugal	2	2	-	18	30	24	1	23	-
Spain	18	4	-	14	22	-	1	40	1
UK	14	40	2	15	23	4	-	2	1
EEC-12	25	14	0	14	13	3	5	20	6

```
a =   0 -  8%  level
b =   8 - 15%  sloping
c = 15 - 25%  moderately steep
d =      > 25%  steep
```

in The Netherlands to about 16 per cent in Greece. Almost 31 per cent of EEC-12 is moderately steep/steep ranging from about 65 per cent in Greece to zero in Denmark, Ireland, and The Netherlands.   The Mediterranean region is characterized by a high proportion of moderately steep/steep land. The percentage areas of Greece, Italy and Spain in this category are >65 per cent, 56 per cent and 42 per cent, respectively. For Portugal about 24 per cent is in this category, with an additional 54 per cent in the sloping/moderately steep and sloping/steep categories. In the Mediterranean zone, where sloping land areas are intensively cultivated or where overgrazing of sparsely vegetated areas occurs, soil degradation is a major hazard.   For instance, the intensive cultivation of sloping/moderately steep Chromic Cambisols in Spain under subtropical Mediterranean conditions results in a high vulnerability to erosion (FAO/UNESCO, 1977).

1.5. Assessing Future Land Requirements for Major Uses in the EEC

1.5.1. Grassland

Milk and dairy product consumption is projected to remain static around present levels. Consumption of beef and veal is expected to make a small increase. Development in sheep and goat production will depend strongly on policies for the less favored areas.

The implications of these projections for grass production is that if recent trends in fertilizer usage persist, there will be a surplus of grassland for the projected ruminant requirements. This surplus is put at 5.3 million ha (van Dijk and Hoogervorst, 1982).

1.5.2. Cereals

The area under cereals in 1982 was 28 million ha. On the basis of a comprehensive analysis and taking all factors into consideration, a cereal surplus of 58 million tons is forecast for the year 2000, providing present trends prevail (Rexen and Munck, 1984). A total restriction on the importation of cereal substitutes and maize would reduce the surplus to 38 million tons. Assuming a projected cereal yield of 5.65 t ha^{-1} and the total curtailment of cereal substitutes, there would still be a surplus of 6.7 million ha under cereals by the year 2000. Assuming no restriction, the surplus is equivalent to 10.3 million ha under cereals (Lee, 1986).

1.5.3. Sugar beet and potatoes

These crops currently occupy 2.8 million ha. At present the prospects for the world sugar market are not encouraging, and when this and various other factors such as competition from isoglucose and a projected European population increase of 0.3 per cent year^{-1} are considered, the prognosis is that the sugar beet area will decline by a minimum of 0.2 million ha. The area under potatoes will decline by 0.15 million ha (Lee, 1986).

1.5.4. Oil and protein crops

Area under oilseed is projected to increase by 0.42 million ha from an existing (1981) level of 1.7 million ha. Research indicates that it is technically feasible to substitute 66 per cent of imported protein for pig/poultry production by the combined use of peas, beans and lupines. If there is a deliberate policy by EEC to achieve this substitution, an area of 4-5 million ha would be required (Lee, 1986).

1.5.5. **Forestry**

National forest projections are not (readily) available. However, detailed analysis of future land requirements for forestry in UK to the year 2000 concluded that the expected additional demand for forest land represented between 1.6 and 3.7 per cent of the total productive agricultural land area (CAS, 1976).   Since forestry is likely to be confined to the poorer quality farmland and unfarmed land, a correction was made for this in arriving at the above estimate. In Ireland, the area under forest is projected to increase from 437,000 ha in 1983 to 539,000 ha (8 per cent of land area) by 2000 (Anonymous, 1984). This is an increase of 102,000 ha or about 5,000 ha annually.   The land involved is most likely to be submarginal for either arable land use or mechanized grassland.

The EEC-9 woodland area increased by 3.9 million ha over the decade 1961-71.   Over the next decade 1970-80 the increase was only 1.2 million ha. If this rate of planting continued to 2000, an additional 8 million ha would be diverted to forest between 1985-2000.

In 1978 the Commission of the EC pointed out that the afforestation of bare land offered great scope for adding to timber production (EEC, 1978).   They estimated that there were at least 4 million ha of land which had become submarginal for farming, but were well suited to timber production.   A conversion of 4 million ha to forestry may be viewed as an upper limit.

1.5.6. **Urban**

Trends in urbanization suggest that an additional 0.8 million ha will be diverted to this use, assuming a projected EEC-10 population increase of 16 million by the year 2000 (Lee, 1986).

1.6. **Projected Land Use Changes**

Land use changes projected for the year 2000 are set out in Figure 1.4. For grassland a decline of 5.3 million ha is projected, and it is most likely that the areas involved would be the lower grassland suitability Classes D and E. Areas capable of moderate intensification are unlikely to be diverted to other uses. In France, however, Class D grassland may correspond with Class 2 arable.   For arable cropping it is most likely that land in surplus would be in the lower suitability Classes 2 and 3 in southern Europe.

It is logical that the highest yielding cereal area will be the most competitive. The northern region of Europe is the most productive for cereals with southern Europe (southern France, Italy and Greece) having a yield disadvantage. Any expansion in oil crops which takes place will be directed to Class 1 and 2 arable suitability areas and similarly for

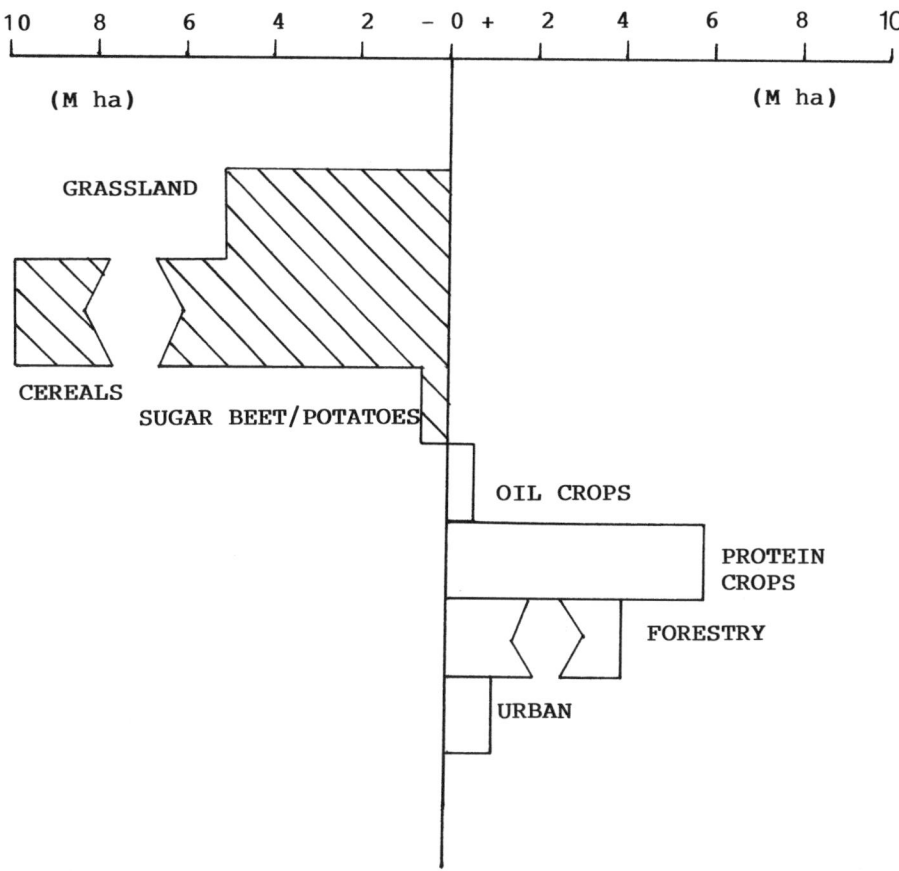

Figure 1.4. Projected land use changes in the EEC for the year 2000 in millions of hectares. (Source: Lee, 1986.)

protein crops. An expansion of 4 million ha in protein crops assumes significant intervention by the EEC in the pricing mechanism for these crops, and whether in fact these crops will become attractive economically is at present debatable. Similarly, if afforestation is to expand at a greater rate than indicated, significant EEC intervention will also be required in the production economics of forestry. Forestry expansion will be diverted to grassland suitability Classes D and E with perhaps some areas of Class C_3 land which have a high forestry potential, (for example, Gleysols in Ireland).

1.6.1. Land of low agricultural potential

There are 48 million ha in Arable Suitability Class 4 corresponding to Grassland Suitability Class E. This land is largely unsuited to mechanized (conventional machinery) arable or grassland farming. The land use options would be restricted to woodland or production from natural vegetation (for example, bracken, heather, agrostis in northern Europe or grassland, shrublands and forest range lands in Mediterranean Europe). The latter areas are often dominated by dwarf shrubs of maquis type. Woodland production levels would be considerably higher in northern Europe compared with the Mediterranean zone.

1.6.2. Use of arable land

Scenario 1: availability of 10.2 million ha of land in Arable Suitability Classes 2 and 3 (10.3 million ha for cereals + 0.35 million ha for sugar beet and potatoes - 0.42 million ha oil crops). (This scenario assumes existing policy on the import of cereal substitutes.)

Scenario 2: availability of 6.3 million ha of land in Arable Suitability Classes 2 and 3 (6.7 million ha for cereals - 0.42 million ha for oil crops). (This scenario assumes the total curtailment on the import of cereal substitutes.)

Scenario 3: availability of 2.3 million ha of land in Arable Suitability Classes 2 and 3 (6.7 million ha for cereals - 4.4 million ha for protein and oil crops). (This scenario assumes 4.4 million ha under protein and oil crops and total curtailment on the import of cereal substitutes.)

1.6.3. Substituting technology for land

Under improved management systems it is technically possible to release sizeable land areas from ruminant production. Research indicates that Ireland's current total ruminant livestock production of 6.2 million livestock units could technically be maintained on less than 2.8 million ha, thus releasing 1.4 million ha of lowland grass for alternative use. Some of this land would be in Arable Suitability Classes 2 and 3. Similarly, UK studies (Wilkins *et al.*, 1981) suggest that through intensive use of grassland it would be technically possible to release 4 million ha from grassland to other uses. A study of grassland production in Europe (Lee, 1984) concluded, that through improved fertilizer use it is possible to significantly raise production levels of much of the EEC-10 grassland area of 53 million ha. This would include the entire range of land suitability classes with the exception of The Netherlands.

There is also considerable scope for substituting known technology for land with respect to arable systems. Table 1.9 is a comparison of actual average yields of the major EEC arable crops with yield levels that are technically achievable.

Table 1.9.   Comparison of actual and potential arable crop yields   (in t ha^{-1} as average for the EEC-10).

Crop	Actual	Potential
Wheat	4.8	12.0
Barley	4.2	11.1
Grain maize	6.3	12.0
Sugar beet	6.8	13.8
Potatoes	28.5	90.0

It is evident that there is scope for doubling yields above the average levels recorded.   However, in practice the achievement of potential yield levels is constrained by limiting agro-ecological conditions such as disease and weed incidence.   In addition, severe environmental problems are likely to be associated with this increased intensification; the environmental risks are discussed in other Chapters in this book. Nevertheless, it must be concluded that, with better application of technology to the existing arable areas of the EEC-10, there would be scope for a sizeable land transfer to alternative uses.

Table 1.10 indicates that a yield increase of 20 per cent from the listed crops could in time release 16 per cent (4.2 million ha) of arable land for alternative uses.

1.7. **Concluding Remarks**

Future EEC land use may be considered in the context of the following alternatives:

- increased food exports
- import substitution of animal feedstock
- increased forestry
- biofuel production
- biomass for chemical feedstock
- nature protection

The subsidizing of agriculture in the EC is a major constraint to land use diversification.   The level of this subsidization will determine to a major extent future land use balance.   This has been addressed in a recent EEC sponsored project conducted under the FAST programme (Conrad, 1987).   Table 1.11 shows projected land use changes under three levels of subsidization as follows:

Table 1.10. Areas of arable land released at two postulated levels of
yield increase in the EEC-10 (20% and 50% increase) in 1,000 tons and
area in 1,000 ha.

	Wheat	Barley	Grain maize	Sugar beet (white sugar)	Potato
Total production	55,437	36,352	18,924	11,968	28,270
Area (1983)	11,012	8,863	2,957	1,721	1,073
Area required (120% average yield)	9,184	7,388	2,464	1,473	895
Area released	1,828	1,475	493	248	178
Area required (150% average yield)	7,352	5,911	1,971	1,178	716
Area released	3,660	2,952	986	543	357

i)   agriculture subsidized at present level and also
     subsidization of alternative uses;

ii) agriculture subsidized only for self-sufficiency in
     food production and little subsidization of
     alternative uses;

iii) minimal subsidization.

Under the scenario of subsidization of agriculture and crop
subsidization for energy and chemical production, there would be a land
area deficit of 6 to 23 million ha.   If subsidies were withdrawn,
agricultural land would be in surplus to the extent of 35-37 million ha
corresponding to about 33 per cent of total UAA.   If agricultural
subsidization was restricted to attainment of self-sufficiency in food
production, probably through the implementation of quota systems, land
surplus would be 9-12 million ha. The overall conclusion is that land
scarcity would be limiting only if biomass production for energy,
chemicals and fibers were to be subsidized to a degree to make these
options economically feasible.

Table 1.11. Changes in land use for the EEC-10 under different conditions up to the year 2000 (in millions of ha).

Type of land use	Fully sub-sidized (food and non-food)	Food only subsidized c	Minimally subsidized
Arable land	-12	-14	-21
Grassland	-7	-8	-20
Permanent crops	-1	-1	-1
Wooded area[a]	+3	+3	+3
Urban settlement and transportation net-works	+1	+1	+1
Increased food export	0	0	0
Feedstuff import substitution	+9	+5[d]	0
Biomass for energy	+3 to +10[f]	0	0
Biomass for chemicals and fibres (non-wood)	+4 to +14	+1 to +4	+1 to +3
Nature conservation areas	+6[b]	+1[e]	0
Fallow land (net balance of zero)	-6 to -23	+9 to +12	+35 to +37

Source: Conrad, 1987.

a - figures do not take forestry for energy and fiber production into account;
b - figures assume 4 million ha already devoted to nature conservation;
c - assumes reduction of surpluses;
d - source: Lee (1986);
e - partially subsidized;
f - difficult to assess.

In the longer term, the production of crops for both industrial and food/feed purposes is indicated. There will be a need for integrated planning and evolution of production systems for food, feed, timber, chemicals and fiber and the concept of agro-industrial farming may well be the instrument for future land use efficiency.

REFERENCES

Anonymous, 1984, *Forestry and Wildlife Service*, Dublin.

Centre for Agricultural Strategy (CAS), 1976, *Land for Agriculture*. Report 1. University of Reading, UK.

Conrad, J., 1987, Alternative uses for land and the new farmworker; segregation versus integration. IIUG Report 87-1, Internationales Institut für Umwelt und Gesellschaft, Berlin.

van Dijk, G. and Hoogervorst, N., 1982, The demand for grassland in Europe towards 2000. Some implications for a possible scenario. Proc. 9th General Meeting of the European Grassland Federation. 21-32 Reading.EEC, 1978, Forestry policy in the European Community. COM (78)   62/final.

EEC, 1985, Soil Map of the EC (1:1,000,000). Directorate General for Agriculture - Coordination of Agricultural Research, Brussels.

FAO/UNESCO, 1977, Explanatory note, U.N. Conference on Desertification A/Conf. 74/2.

Lee, J., 1984, Suitability and productivity of land resources of EEC-10 for grassland use. A study carried out on behalf of the EC, Johnstown Castle, Wexford.

Lee, J., 1986, The impact of technology on the alternative uses for land, Report FOP 85, Commission of the European Communities, Brussels.

Papadakis, J., 1966, *Climates of the World and their Agricultural Potentialities*. Buenos Aires, Argentina.

Rexen, R. and Munck, L., 1984, Cereal crops for industrial use in Europe. Report prepared for the Commission of the European Communities.

Wilkins R., Newton J. E., James P. J. and Walshingham J. M., 1981, Possibilities for change in the output of grassland: an overall view. In J. L. Jollans (ed.), *Grassland in the British Economy*, CAS Report 10, Reading.

Chapter 2

TRENDS IN THE TRANSFORMATION OF EUROPEAN AGRICULTURE

J. Kostrowicki

2.1. Introduction

The concept of an agricultural typology was developed during the period 1964-1976 by the Commission on Agricultural Typology of the International Geographical Union (Kostrowicki, 1964, 1975, 1977 and 1980b; Kostrowicki and Szyrmer, 1988). The concept formed the basis for the preparation of a European map of agricultural typology. The objective of this Chapter is to discuss the major trends responsible for transforming European agriculture, based on this map and several related national studies.

Agricultural typology can be defined as a classification procedure where classes (or types) of agriculture are identified and aggregated according to their similarities based on characteristics that are recognized as most representative of such groups.

The individual farm or agricultural holding is the basic unit for agricultural typology. Various levels of aggregation (such as administrative or other spatial units) can be used, irrespective of their deficiencies, in particular when a large number of small holdings are investigated.

Agricultural typology analysis has three basic methodological difficulties:

1) selection and identification of attributes that characterize the nature of agriculture;

2) selection and identification of variables which characterize individual attributes;

3) selection of methods and tools for comparison of the observations (including spatial detail).

There are four significant attributes or characteristics of agriculture which are universal in nature and therefore would be suitable to characterize trends in agriculture on a global or continental

21

F. M. Brouwer et al. (eds.), Land Use Changes in Europe, 21–47.
© 1991 Kluwer Academic Publishers. Printed in the Netherlands.

scale:

   i)   *social attributes*, based on the social structure
         of agriculture, including variables such as land
         ownership, and the scale of agricultural practice;

   ii)   *operational attributes*, these characteristics reflect
         the use of input factors, such as labor intensity,
         machinery and fertilizer use;

   iii) *production attributes*, include the production per
         hectare, per person employed, as well as the
         level of local subsistence production for local
         use or for markets;

   iv)   *structural attributes*, the relative importance of
         various aspects of agricultural practice in terms
         of land use, gross production and commercial
         output.

   A limited number of variables have been selected which, together,
should express the major attributes of agriculture on a continental scale.
   The agricultural typology used for the European map is based on
28 variables. They were derived as a result of a series of experiments
in different parts of the World. Each of the four groups of attributes is
represented by seven variables. The 28 variables are identified in Table
2.1 (Kostrowicki, 1982).

Table 2.1. The variables used as the basis for constructing an
agricultural typology.

A. **Social attributes**
1)   Percentage rate of land held in common in the total agricultural
     land (arable land, also including fallow, perennial crops and
     permanent grassland).
2)   Percentage rate of land in labor and share tenancy (share
     cropping) in the total agricultural land.
3)   Percentage rate of land owned by private persons (irrespective of
     the land tenure system) in the total agricultural land.
4)   Percentage rate of land operated by the consciously planned
     collective or state enterprises in the total agricultural land.
5)   Number of people actively employed in agriculture per agricultural
     holding.
6)   Amount of land per agricultural holding in hectares.

Table 2.1. (continued)

7)   Amount of agricultural gross production in conventional units per agricultural holding.

B. Operational attributes

8)   Number of people actively employed in agriculture (100 ha^{-1} of agricultural land).

9)   Number of drawing animals (for example, horses, mules, asses) used in agricultural practice 100 ha^{-1} of cultivated land (arable land without fallow, perennial crops, and cultivated grassland without uncultivated meadows and pasture).

10)   Number of tractors and other self-propelling machinery in horse power 100 ha^{-1} of cultivated land.

11)   Amount of chemical fertilizers in pure content (NPK) ha^{-1} of cultivated land.

12)   Percentage rate of irrigated land of the total cultivated land.

13)   Percentage rate of harvested land in the total arable land.

14)   Number of farm animals 100 ha^{-1} of agricultural land.

C. Production attributes

15)   Gross production in conventional units ha^{-1} of agricultural land.

16)   Gross production ha^{-1} of cultivated land.

17)   Gross production per person actively employed in agriculture.

18)   Commercial production per person actively employed in agriculture.

19)   Percentage rate of commercial agricultural production in gross agricultural production.

20)   Commercial production in conventional units ha^{-1} of agricultural land.

21)   Degree of specialization in commercial agricultural production.

D. Structural attributes

22)   Percentage rate of perennial crops (for example, trees, shrubs, vines) and semi-perennial crops (covering land without rotation for some years such as hops, cotton, sugar cane) in the total agricultural land.

23)   Percentage rate of permanent grassland (including leys within field-grass system and current fallow if used for grazing) in the total agricultural land.

24)   Percentage of land under food crops in the total agricultural land.

25)   Percentage rate of animal products in gross agricultural production.

26)   Percentage rate of animal products in commercial agricultural production.

Table 2.1. (continued)

27) Percentage rate of industrial crops (for example, fiber, oil, sugar
     crops and tobacco) in gross agricultural crops.
28) Percentage rate of herd (herbivorous) animals in the total number
     of farm animals.

The variables need to be normalized since they are expressed in
different units of measurement. This normalization is based on a series
of global experiments for each of the variables. They are subdivided
into five categories and ranked from very low to very high (Table 2.2).

A European map was published in 1984 (Kostrowicki *et al.*, 1984;
Kostrowicki, 1982 and 1988; Aitchison, 1986). This map is neither a
land use map nor a map of crop and/or animal distribution. It is
rather a synthetic map that represents all significant characteristics or
attributes of European agriculture, based on information from the period
1975-1980.

The approach adopted for the agricultural typology map of Europe
has also been applied to various individual countries in Europe, for
example:

-    Poland (Kostrowicki, 1970 and 1974; Stola, 1970
     and 1983; Beigajlo, 1968 and 1973; Szczęsny,
     1978, 1979b, 1981, 1983 and 1988; Tyszkiewicz,
     1975).
-    France (Bonnamour, 1972, 1973 and 1975;
     Bonnamour *et al.*, 1971a and b; Bonnamour
     and Gillette, 1980; Bonnamour, 1984).
-    Belgium (Christians, 1975; Stola, 1975 and
     1983).
-    Austria (Szczęsny, 1979a, 1986).
-    Bulgaria (Tyszkiewicz, 1979; Galczyńska, 1984).
-    USSR (Gorbunova *et al.*, 1979).
-    Yugoslavia (Tyszkiewicz, 1980).
-    Switzerland (Szczęsny, 1982 and 1986).
-    Sweden (Tyszkiewicz, 1982).
-    Finland (Rikkinen and Vapaoksa, 1983).
-    Norway (Tschudi and Johanson, 1983).
-    United Kingdom (Kostrowicki, forthcoming).

The European agricultural typology map is static reflecting the
situation for the period 1975-1980. However, several of the national
studies have focused on changing trends in individual countries based
on the same agricultural typology. Together these allow certain
conclusions to be drawn regarding the trends responsible for the

Table 2.2. Classes of world ranges of the variables used to produce the agricultural typography.

No. of the variable	1 very low	2 low	Classes 3 medium	4 high	5 very high
1	- 20	20-40	40-60	60-80	80-
2	- 20	20-40	40-60	60-80	80-
3	- 20	20-40	40-60	60-80	80-
4	- 20	20-40	40-60	60-80	80-
5	- 2	2-8	8-50	50-200	200-
6	- 5	5-20	20-100	100-1,000	1,000-
7	-100	100-1,000	1,000-10,000	10,000-100,000	100,000-
8	- 3	3-15	15-40	40-150	150-
9	- 2	2-8	8-15	15-30	30-
10	- 6	6-15	15-35	35-90	90-
11	-10	10-30	30-80	80-200	200-
12	-10	10-25	25-50	50-80	80-
13	-10	10-30	30-70	70-130	130-
14	-10	10-30	30-80	80-160	160-
15	- 5	5-20	20-45	45-100	100-
16	- 5	5-20	20-45	45-100	100-
17	-40	40-100	100-250	250-800	800-
18	-20	20-60	60-180	180-600	600-
19	-20	20-40	40-60	60-80	80-
20	- 3	3-12	12-30	30-80	80-
21	- 0.1	0.1-0.2	0.2-0.4	0.4-0.8	0.8-
22	-10	10-20	20-40	40-60	60-
23	-20	20-40	40-60	60-80	80-
24	-20	20-40	40-60	60-80	80-
25	-20	20-40	40-60	60-80	80-
26	-20	20-40	40-60	60-80	80-
27	-20	20-40	40-60	60-80	80-

transformation of European agriculture.

Six types of world agriculture have been described to date and have been identified in the transformation process of European agriculture:

E - Traditional extensive (primary) agriculture.
L - Traditional large-scale (Latifundia) agriculture.
T - Traditional small-scale (peasant) agriculture.
M - Market-oriented agriculture.
S - Socialized agriculture.
A - Large-scale, highly specialized livestock breeding.

Within these six first order types, 22 types of the second order and 64 types of the third order have been observed in Europe to date (Kostrowicki, 1984, 1985 and 1986). They are described in Table 2.3.

2.2. Types of Agriculture in Europe

2.2.1. Traditional extensive (primary) agriculture (E)

Traditional extensive agriculture is characterized by a traditional system of land tenure (with land held in common), with service or labor tenancy or share-cropping. It is also characterized by a low labor input, very low level of capital investments, low intensity of land use (fallowing) and/or animal breeding, very low land productivity, low labor productivity as well as poor or no linkages with the market sector. It can generally be characterized as semi-subsistence agriculture.

Crop cultivation and livestock breeding were separate practices as livestock was usually grazed on grassland held in common around the arable land, as well as in woodlands, and only occasionally fed on the stubble after harvest. Livestock, therefore, did not contribute much to the maintenance of soil productivity (Duby, 1962; Le Roy Ladurie, 1975; Slicher van Bath, 1968).

The original form of this type of agriculture is no longer found in Europe but is still characteristic for many developing countries. A more advanced and transitional form existed in parts of Europe until the nineteenth century, although now it is only observed locally, for example, in some less developed areas of southern Europe, such as in central Spain, southern Italy, southern Yugoslavia, north-eastern Poland and part of northern Scandinavia. An extensive system of crop rotation with fallow was practised in these areas until recently (Faucher, 1949; Montelius, 1953; Bodvall, 1957; Petrov, 1968, Andreae, 1964; Boesch and Bronhofer, 1956, Biegajlo, 1969; Tyszkiewicz, 1969; Cabo Alonso, 1981).

In some parts of north-eastern Poland the traditional extensive type of agriculture was common until the 1960s comprising a three field

Table 2.3. Types of world agriculture.

E.   TRADITIONAL EXTENSIVE (PRIMEVAL) AGRICULTURE

    Et   Current fallow, subsistence to semi-subsistence

        Etm -   mixed

L.   TRADITIONAL LARGE-SCALE (LATIFUNDIA) AGRICULTURE

    Ll   Low intensive low to medium productive

        Llm -   mixed

T.   TRADITIONAL SMALL-SCALE (PEASANT) AGRICULTURE

    Ti   Very small-scale, intensive crop growing

        Tir -   low productive, semi-subsistence

        Tiu -   irrigated, medium productive, semi-subsistence to semi-commercial

    Tm   Mixed

        Tmh -   labor intensive, semi-irrigated, low pro- ductive, subsistence

        Tmb -   medium intensive, medium productive, semi-subsistence

        Tme -   very small-scale, medium intensive, low productive, subsistence

        Tmm -   medium intensive, low to medium productive, semi-commercial

        Tmo -   very small-scale, low to medium intensive, with high land productivity and low labor productivity, subsistence to semi-subsistence

        Tmy -   very small-scale, labor intensive, low to medium productive, semi-subsistence with crop growing prevalent

        Tmc -   labor intensive, medium productive, semi-commercial with crop growing prevalent

Table 2.3. (continued)

Tmg - medium intensive, low productive, semi-commercial with crop growing prevalent, perennial fruit crops important

Tmf - very small-scale, intensive, medium productive, semi-commercial with perennial crops dominant

Tmk - labor intensive, productive, semi-commercial with livestock breeding prevalent

M. MARKET-ORIENTED AGRICULTURE

Ms Specialized perennial industrial crop

Mso - small-scale, low labor intensive, medium productive

Mi Small-scale, intensive crop

Min - very small-scale, very high land productivity, medium labor productivity with fruit crops dominant

Mif - productive, specialized in perennial fruit crops growing

Miv - very high land productivity, medium to high labor productivity with vegetable growing dominant

Mig - industrialized high intensive, highly productive horticulture

Mm Small-scale, mixed

Mmr - capital intensive, medium productive with crop growing prevalent

Mmc - intensive, productive with crop growing prevalent

Mmf - medium intensive, medium productive with crop growing prevalent, perennial fruit crops important

Mmv - highly capital intensive, highly productive with temporary crops dominant

Table 2.3. (continued)

Mmz - low labor intensive, medium to high capital intensive, low to medium productive, semi-commercial

Mmp - very small-scale, intensive, with high land productivity, low labor productivity, semi-commercial with perennial fruit crops important

Mmt - medium labor intensive, capital intensive, productive, semi-commercial with perennial fruit crops important

Mmm - medium labor intensive, highly capital intensive, productive

Mmh - very small-scale, intensive, very high land productivity, medium labor productivity, with livestock breeding prevalent

Mmg - low intensive, medium productive, semi-commercial with livestock breeding prevalent

Mmw - medium labor intensive, highly capital intensive, medium productive, semi-commercial with live-stock breeding prevalent

Mma - low labor intensive, capital intensive, productive with livestock breeding prevalent

Mmi - medium labor intensive, highly capital intensive, highly productive with livestock breeding prevalent

Ml   LARGE-SCALE, INTENSIVE

Mlm - medium labor intensive, capital intensive, mixed

Mlc - low labor intensive, capital intensive, productive with crop growing dominant

Me   LARGE-SCALE LABOR EXTENSIVE

Mel - medium capital intensive, with low land-high labor productivity with crop growing prevalent

Table 2.3. (continued)

Mem - medium capital intensive, low land-high capital intensive, with low land, high capital productivity, mixed

Meb - medium to high capital intensive, with low land and very high labor productivity, with livestock breeding dominant

Mea - highly capital intensive, with very low land-high labor productivity, with livestock breeding dominant

S.   SOCIALIZED AGRICULTURE

Se   INCIPIENT

Sec - low to medium intensive, low productive, with crop growing dominant

Sep - low labor intensive, medium capital intensive, low to medium productive with crop growing prevalent

Sem - medium intensive, low to medium productive, mixed

Sm   MIXED

Smw - low labor intensive, medium capital intensive, low productive

Smm - low labor intensive, capital intensive, medium productive

Smg - very low labor intensive, capital intensive, with high productivity of arable land and very low productivity of grassland

Smy - low labor intensive, capital intensive, low productive with crop growing prevalent

Smc - medium intensive, medium productive, with crop growing prevalent

Smu - medium labor intensive, capital intensive, productive with crop growing prevalent

Table 2.3. (continued)

Smi - low labor intensive, capital intensive, irrigated, productive to highly productive, with crop growing prevalent

Sma - low labor intensive, medium capital intensive, low productive with livestock breeding prevalent

Smd - medium labor-highly capital intensive, productive with livestock breeding prevalent

Sg INTENSIVE CROP GROWING AND EXTENSIVE LIVESTOCK BREEDING

Sgt - intensive, temporary industrial crop growing with extensive grazing

Sh INTENSIVE SPECIALIZED CROP GROWING

Shy - medium labor intensive, highly capital intensive medium productive crop growing with perennial fruit crops important

Shv - productive, highly specialized in vegetable growing

Shh - industrialized, highly intensive, highly productive, highly specialized horticulture

Ss SPECIALIZED PERENNIAL CROPS GROWING

Ssf - medium intensive, medium productive fruit crop growing

Ssc - low labor intensive, capital intensive, medium productive, industrial crop growing

Sc LABOR EXTENSIVE

Scm - medium capital intensive, with very low land productivity, medium labor productivity, mixed

Scc - medium capital intensive, with very low land productivity, medium labor productivity, highly specialized in temporary crop growing

Table 2.3. (continued)

A. SPECIALIZED LIVESTOCK BREEDING

   Ar   ANIMAL GRAZING

         Arr -   market-oriented (reindeer)

         Aro -   socialized, a - reindeer

   Ad   INDUSTRIALIZED LIVESTOCK BREEDING

         Add -   market-oriented

         Ado -   socialized

---

system with fallow (Beigajlo, 1960, 1964 and 1968). This traditional extensive type of agriculture has become more intensive and is presently a transitional form between the types E and T.

Similar processes have taken place in Yugoslavia where different types of agriculture have been identified such as a transition between the traditional extensive type (E) and the traditional small-scale type of agriculture (T). In addition, labor intensive types of traditional small-scale agriculture are identified (Kostrowicki, 1984, 1985 and 1986).

This traditional extensive type of agriculture could still be found during the 1960s and 1970s in parts of Macedonia on the southern Balkan Peninsula and could be considered as a reflection of the historical forms of European agriculture.

2.2.2. **Traditional large-scale agriculture (L)**

Traditional large-scale agriculture (Latifundia) differs from the former extensive type in terms of the scale of operation because of its stronger dependency on a market. It is characterized by private or institutional ownership of land, low labor requirements but a medium level of capital, as well as low levels of land and labor productivity.

In medieval Europe, Latifundia formed a dual economy with the former type and was characterized by strong interrelationships between powerful landowners and peasants based on a villein system or various forms of share-cropping (Liversage, 1945; Milhau, 1953).

Today, this type of agriculture in its original form is observed only locally in Europe as it has been transformed into either i) small-scale farming after land reallocation programs, or ii) an extensive form of market-oriented agriculture as a result of agricultural advancements. It is found mainly in parts of southern Europe, including parts of the Iberian Peninsula, and, to a smaller extent, in southern Italy (Sicily, Calabria) where it exists as a type in which both share tenancy and extensive forms of crop rotation are applied. More extensive types are practised in these areas, for example, "al tercio" where two out of three fields are fallow each year. There remain some even more extensive types of agriculture (Artola, 1978; Cabo Alonso, 1981) that have survived mainly in those areas where fields could not be sown each year due to insufficient precipitation.

Traditional large-scale agriculture was long practised in eastern Europe. It has been transformed rapidly in the USSR into the socialized type of agriculture (S), either directly into state farms (Sovkhozes) or first subdivided into small private farms, and then collectivized during the 1930s (kolkhozes) with the former village agriculture (types E and F).

Similar processes have taken place in the eastern part of central Europe since the Second World War. As a result of the land reforms a proportion of the large-scale farms was transformed into State farms and some land was subdivided among peasants. A few years later the remaining holdings, together with original peasant farms, were collectivized with the exception of Poland and Yugoslavia where small farms still play an important role. In fact, Poland and Yugoslavia are the only countries in Europe where three types of the first order (that is, T, M and S) are observed, together with remnants of the extensive and traditional type of agriculture (type E).

2.2.3. Traditional small-scale agriculture (T)

The extensive dual system of agriculture described above has changed in several parts of Europe; the transformation took place between the sixteenth century and the middle of the nineteenth century and occurred in response to the increasing population and the development of a market-oriented economy. The villein system of labor service was substituted by rent and long-term fallow with short fallow (one to two years). This was possible because of the closer integration between crop cultivation and livestock breeding. Livestock increasingly grazed on fallow because the size of arable land used for growing fodder crops (both legumes and roots) had increased. The introduction of "structure forming" crops that enriched soils with nitrogen and the "intensifying crops" that required a more careful cultivation method and a higher concentration of manure (Kostrowicki, 1964) gradually led to the reduction and eventual elimination of fallow.

European agriculture also changed with the introduction of intensive crops such as maize and potatoes, originating from America. These crops contributed largely to the increase in agricultural productivity, although they require careful soil cultivation and fertilization. Their introduction has contributed to the development of livestock breeding which, in turn, made it possible not only to maintain but also to improve soil productivity using manure (Duby, 1962; Kerridge, 1967; Kostrowicki, 1980a; Slicher van Bath, 1968, 1977;

Chorley, 1981; Grigg, 1980, 1982 and 1987).

The Mediterranean two-year crop rotation with fallow was gradually transformed into a system without fallow. Similarly, the three-field system with fallow (winter crops - spring crops - fallow) in central Europe was transformed into the three-year crop rotation without fallow and later on into an even more rotational four-year crop rotation such as i) root crops (manured), ii) spring crops, iii) structure forming crops (legumes) and iv) winter grains (Kostrowicki, 1964).

Fallow was transformed in most of northern Europe by the introduction of sown grass (ley) system. This system is characterized by a rotation between crop cultivation and pasture.

The trend toward intensified agriculture has been enhanced by the reclamation of wetland (in The Netherlands, the UK and the Po Valley of Italy). Irrigation has also had a significant influence on the level of intensity. Irrigation was introduced in the Mediterranean countries during the Middle Ages but has never covered extensive areas.

The traditional small-scale type of agriculture (T) still covers major parts of southern Europe and the eastern part of central Europe. It is based either on private ownership or advanced forms of land tenancy usually for a long period of time with fixed rent. It is also characterized by limited connections to a market, that is, by sale of whatever is left after feeding the farmer, the local population and farm animals, and results in limited investment for more advanced means of production such as chemical fertilizers and machinery).

Within the traditional type of small-scale agriculture in Europe three types of the second order and twelve types of the third order have been identified (Table 2.3).

First, an intensive form of small-scale agriculture based on semi-subsistence agriculture with very high labor requirements and an emphasis on crop growth. It is characteristic of the traditional and irrigated type of agriculture in several parts of Asia and has been observed in Europe only in the Macedonia region of Yugoslavia.

Secondly, a mixed type of agriculture which is slightly more capital intensive, with low levels of productivity, has been observed in the Macedonia region of Yugoslavia. This type of agriculture, however, is less intensive in terms of land and labor requirements.

Other types of traditional mixed agriculture, especially those with an arable focus, are more frequent in southern Europe while those in which animal production is in balance with or prevails over crop production are mainly characteristic of the eastern part of central Europe. Agricultural practices with medium labor requirements, low capital input, medium productivity, low commercialization and a prevalence of animal production in commercial production are closest to the traditional extensive type of agriculture. This has been observed in the north-eastern part of Poland (Biegajlo 1960 and 1968) where the three-field system with fallow was employed until the early 1960s. It has also been identified in some parts of Yugoslavia and on some private farms in Romania. This is very small-scale agriculture with low- or medium-capital input, high land but low labor productivity and a very low level of commercialization dominated by animal production.

The traditional mixed type of agriculture with a medium level of labor intensity, a low capital intensity, medium land productivity but low labor productivity and medium commercialization is much more widespread in Europe. It is characteristic of large areas in the eastern and central part of Poland, parts of Yugoslavia, as well as some parts of Austria, Italy, northern Spain and Portugal.

A special type of traditional mixed agriculture is based on part-time farming and is found in some suburban areas of Poland and Yugoslavia.

Three third order types have been identified in Europe within the traditional mixed type of agriculture with an emphasis on crop growing (Table 2.3).

The first type, which is nearest to the traditional extensive type of agriculture, is characterized by medium labor and capital inputs, medium intensity of using crop land (still some short fallow), low productivity of total agricultural land but high productivity of cultivated land, low labor productivity, medium level of commercialization and a considerable proportion of both perennial crops and permanent grassland. Originally described from southern Greece, it has also been identified in some parts of southern Italy, in Spain and in Portugal.

The second type, which is more characteristic of the Mediterranean part of Europe, is characterized by small-scale agriculture with high inputs of labor and animal power, medium to high level of land productivity, but low labor productivity, a medium level of commercialization and an emphasis on crop growing with a high proportion of perennial crops. Although it was originally described in various parts of Italy, it has also been identified in Spain, Portugal and Yugoslavia. This type is now gradually disappearing and is inevitably being transformed into its market-oriented equivalent.

The third type, which was originally observed in Apulia on the south-eastern coast of Italy, has also been identified in other parts of Italy and in southern Spain. It is a very small-scale, highly labor intensive but also capital intensive type of agriculture with medium land and labor productivity, a medium level of commercialization, but high specialization in perennial crops, namely vines and olives.

2.2.4. **Market-oriented agriculture (M)**

Food requirements in Europe have increased with the growing rates of urbanization. The productivity of the traditional semi-subsistence types of agriculture was undoubtedly insufficient for the rising population.

Land and labor productivity remained relatively low until the twentieth century. This meant that "few societies could support a large non-agricultural population. Thus, towns were comparatively few in number, very large towns were rare and only a small part of the population was engaged in activities other than agriculture. The urban population was therefore a small part of the total population. The smallness of Europe's urban population was a function reflecting not only the low productivity of agriculture but also the high costs of transporting food. Not surprisingly, most European towns were located on rivers and before the nineteenth century the major cities were ports since the cost of shipping food by sea was approximately one-eighth of the cost of overland transportation. The growth of towns, either clusters of towns, as in the Low Countries, or very large towns, such as Paris, London, Naples or Rome, had a profound influence on the agriculture of their hinterlands. The demand for food meant that there was a ready market and the commercialization of agriculture was first associated with the major cities. The high value of land meant that small farms predominated and the fallow period was soon extinguished since soil fertility could be maintained by using horse manure and other organic wastes from the town. Farming was intensive, and the products generally high value goods. These regions benefited from capital investment from the town and generally represented the advance guard of agricultural improvement" (Grigg, 1987).

The process of industrialization since 1800 has had major impacts on agriculture, these can be summarized as follows:

i)   rapid increase of population - the subsequent increase in the level of commercialization in agriculture and the decline of subsistence farming;

ii)   increase of income of nearly all groups in society; the demand for more expensive food, and particularly livestock products increased with rising income; livestock products were introduced and cereals were used to feed livestock (Grigg, 1987; Aymard, 1979; Teuteberg, 1975);

iii) the demand for raw materials in the textile industry increased; these were either imported (cotton) or produced by domestic agriculture (wool and flax); vegetable oils were used for cooking and for many industrial purposes (Grigg, 1987);

iv)   an increased proportion of agricultural input requirements was obtained through the market, including pesticides, fertilizers and machinery (Grigg, 1987).

These trends led to the development of the market-oriented type of agriculture. Today, farmers are almost entirely tied to the market structure for agricultural inputs as well as produce distribution and the agricultural activity is at the mercy of fluctuations in supply and demand as well as costs and prices. Such market forces may be controlled by government intervention or by some international agreement (such as the Common Agricultural Policy in the EC). Market influence and intervention have not developed at the same rate universally. This uneven development, as well as differences in the natural and other external conditions of agricultural development, have led to a considerable differentiation in the pattern of market-oriented agriculture in Europe, particularly in relation to labor and capital inputs, productivity, commercialization and specialization levels. On the other hand, in this type of agriculture the size of operation plays only a secondary role.

Market-oriented agriculture is presently the most differentiated type of agriculture in Europe (Table 2.3). Various subcategories have been identified within the mixed type of market-oriented agriculture. Crop production and animal production are of equal importance. The categories show a wide variation in terms of intensity, productivity and degree of commercialization. Most common is a small-scale, mixed type of agriculture which replaced the traditional mixed type of agriculture over large regions of the lowlands in central Europe, ranging from western Poland, across Germany, the northern part of Austria and Switzerland, to the north-eastern part of France, Belgium, parts of The Netherlands, south-eastern part of England, Denmark and southern regions in northern Europe.

The mixed type of market-oriented agriculture has also evolved with an increasing rate of intensification, with increasing farm size coupled with increasing specialization, either to grow crops or for livestock breeding.

A mixed type of agriculture with an arable emphasis also appears in several forms with varying degrees of intensity and technological advancement. Predictably the least intensive (and the least productive) is characteristic of less developed areas in Europe where perennial crops do not play a major role (for example, in central France as well as in parts of Italy and Spain).

Systems where perennial crops (mainly fruits) are most important are being displaced by a commercial type of agriculture and transformed into intensive farms, fully specialized for growing fruit. This has developed along the Mediterranean coasts of Italy, France and Spain, and has recently shown a considerable growth, particularly in southern France.

A special type, transitional between the traditional large-scale type and the small-scale intensive type of agriculture, is both labor and capital intensive, highly productive without a major degree of specialization. It is found in Campania (in southern Italy around Naples) and also on the Catania Plain.

The only type of the traditional mixed agriculture in Europe, with livestock breeding being prevalent (originally a characteristic for most of the European mountain areas) has now been transformed into various subtypes of market-oriented agriculture. It has become a more intensive practice in northern Spain, but less so elsewhere in Europe. These, in turn, with further development, tend to be transformed into more intensive and still more commercial types and later on into more productive and more specialized systems.

Changes have taken place most rapidly in Scandinavian countries. In Norway, for example, the **Mmg** type, which was dominant in the 1960s, was almost entirely substituted in the 1970s by the **Mmw** and then **Mma** types (Table 2.3). The transformation has been slower in Austria (Szczęsny, 1979a), particularly in the higher alpine areas.

The second order types represent large-scale, market-oriented agriculture, either more intensive (**Mi**) or extensive (**Me**). Although more characteristic of North America and Australia, they have been developed in Europe either by the evolution of former landed estates (Latifundia) or by concentration of small-scale agriculture.

Concentration of small-scale agriculture continued in western Europe, particularly after the Second World War, with the exception of the UK, where it had occurred much earlier. On the other hand, the transformation process of the traditional Latifundia into modern, market-oriented, large-scale agriculture is now complete in most European countries, but is still going on in Spain and Portugal (Kostrowicki, 1984 and 1986).

Some types of the specialized intensive crop agriculture are rather common in Europe such as specialized tree crop, fruit growing in parts of Italy and southern France. Highly commercial agriculture, specializing in market gardening, and highly industrialized horticulture, specializing in vegetable or flower production in more or less artificial environments (greenhouses or hot beds), are only observed in small areas.

2.2.5. Socialized agriculture (S)

Socialized agriculture is primarily characterized by the social ownership of land that is being held in the form of collective or State farms. It is a large-scale type of agriculture with high capital inputs and a high degree of commercialization.

It was first developed in the USSR during the 1920s with the nationalization of the former privately owned estates (which were mainly of the type L) and converted into State farms or with the collectivization of small-scale peasant farms (mainly of the T type).

An example in Europe of a spontaneous organization of a socialized type of agriculture is the formation of collective farms in Portugal during the so-called "carnation revolution" and other sporadic attempts elsewhere to abolish land ownership by individuals.

Socialized agriculture has many lower order types in Europe. One particular type is characteristic of the initial stages of development of socialized agriculture immediately after nationalization or collectivization in the 1920s. The size of farms varied greatly in relation to the size of the former privately owned estates or village territories. Manpower was still an important resource while capital was scarce. This mixed type of agriculture was dominant in the USSR at the beginning of the collectivization and in most of the eastern part of central Europe after the land reforms of the 1950s. During the late 1970s it was only evident in some parts of Romania and Portugal. In other parts of Europe this initial stage of socialized agriculture was transformed into a mixed type of socialized agriculture.

Socialized, mixed agriculture is now found in the form of several low order types with varying levels of intensity, productivity or orientation, that in some cases also reflect their development path. The most extensive types of agriculture cover most of the northern part of the USSR. Towards the south they are substituted by more intensive types **Smc** and **Smm** (Table 2.3). Farther south, the crop growing oriented type is gradually transformed into the **Smu** type characteristic for most of the Ukraine, other parts of the chernozem zone of the USSR, as well as parts of Romania and Bulgaria. On the other hand, the **Smm** type, quite common in the central part of European Russia, as well as parts of the Baltic Republics, is evolving towards the more intensive and more productive type with livestock breeding prevalent. This type is predominant in the German Democratic Republic but is also appearing in some parts of Czechoslovakia, Hungary and Poland.

On the contrary, more extensive but more specialized types of socialized agriculture are prevailing in the East such as the **Sc** type where specialization in grain crops occurs.

A much more intensive type of agriculture is socialized horticulture, scattered throughout the countries mentioned above, including the **Shg** type parallel to **Mig** type of the market-oriented agriculture.

A peculiar second order type has been described from southern Bulgaria (Galczyńska, 1984). This type of agriculture is characterized by high labor and capital inputs, intensive use of crop land, mainly for tobacco, in the mountain valleys, but at the same time very extensive use of large tracts of mountain pastures for livestock grazing. Similar types are found in Albania and some Transcaucasian Soviet Republics.

2.2.6. **Highly specialized livestock breeding (A)**

This type of agriculture is represented by two second order types including extensive commercial herding and highly industrial livestock breeding. The first has been identified in Europe as third order types only, either as the very extensive reindeer grazing practice in the northernmost outskirts of Europe (northern Scandinavia and northern parts of the USSR) or very extensive livestock grazing in the southeastern margins of Europe, around the Caspian Sea. The second one is very labor and capital intensive, very productive and highly commercial and was identified, at least locally over the continent.

2.3. **Agricultural Regions in Europe**

Six agricultural regions can be distinguished in modern-day Europe (Figure 2.1).

I   The *northern region* is mainly characterized by market-oriented agriculture with livestock breeding as well as transitional forms between traditional and market-oriented agriculture. It is located mainly in lowlands or close to urban markets. This region covers the Scandinavian countries including Denmark, the United Kingdom and Ireland, and some areas of continental Europe with the northern parts of the Federal Republic of Germany, parts of The Netherlands and France (Normandy).

II   The *western part of central Europe* is mainly characterized by market-oriented agriculture, either of a small-scale, mixed nature or a large-scale, intensive type. Market-oriented types, with livestock breeding, and transitional forms from traditional to market-oriented agriculture are identified in the mountain areas (for example, the Alps and the Massif Central), and highly intensive types of mixed and market-oriented farming in The Netherlands and Lombardy in northern Italy. This type of agriculture extends from the Elbe River and Upper Danube to the Atlantic Ocean and covers most of the Federal Republic of Germany, Austria, Switzerland, parts of northern Italy and the central part of France.

III   The *southern region* shows a diverse pattern of agricultural types because of the wide variety of natural conditions and different levels of development. Market-oriented agriculture with crop cultivation, together with agriculture specializing in crop cultivation (mainly fruit crops), are dominant types

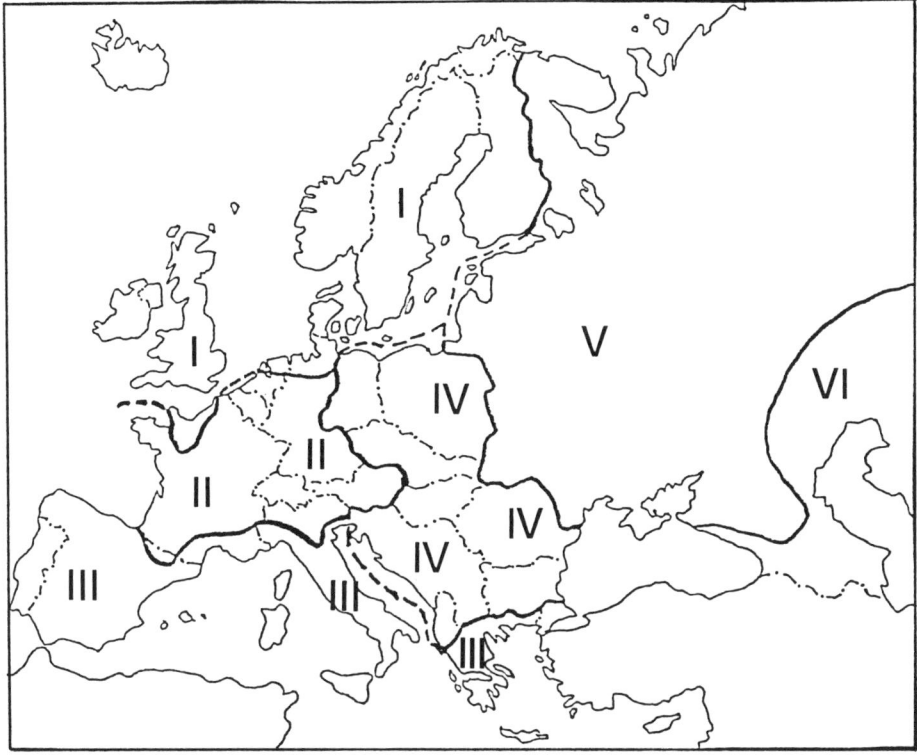

Figure 2.1. Agricultural regions in Europe. (Source: Kostrowicki, 1984.)

in this region. Traditional agriculture is also still found in this region. It includes the northern part of Portugal and Spain, southern France and most of the areas from Italy to Greece.

IV   The *eastern part of the central region* exhibits three main types of agriculture; traditional agriculture (T), market-oriented agriculture (M), and socialized agriculture (S). This region covers all the socialist countries of east-central and the south-eastern part of Europe.

V   The *eastern region* is dominated by large-scale socialized agriculture, including the European part of the USSR.

VI   The *south-eastern region* which is of marginal importance in
       a European context (covering the Caucasus region). This
       region includes the central Asian republics of the USSR with
       socialized, intensive, irrigated agriculture predominating, as
       well as more extensive types of agriculture on non-irrigated
       land which includes livestock breeding.

2.4. Conclusions

Agriculture is an economic activity which is still dependent on
environmental conditions although it has developed rapidly during the
last decades.

Based on investigations of the dynamic nature of the classification
of agriculture applied here, it is foreseen that the traditional type of
agriculture will disappear in Europe within the next few decades; and it
will be substituted by a market-oriented type of agriculture. The same
trend can be predicted, but with less certainty, about transformations
within socialized agriculture (Kostrowicki 1985, 1986). These are,
however, only very general predictions. A much more detailed
assessment could be made were the typology of European agriculture
elaborated for another period using the same approach and techniques.
The practical use of such trend identification could be increased if not
only agricultural trends but also changes in the natural, social,
economic, technological and other external conditions of agricultural
development were taken into account. This task remains to be
undertaken.

REFERENCES

Aitchison, J.W., 1986, Classification of agricultural systems. In M.
     Pacione (ed.), *Progress in Agricultural Geography.* London. 38-
     69.
Andreae, B., 1964, Bietriebsformen in der Landwirtschaft. Entstehung
     und Wandlung von Bodennutzungs-, Viehhaltungs- und
     Bietriebssystemen in Europa und Übersee sowie neuer Methoden
     ihrer Abgrenzung. Stuttgart.
Artola, M., *et al.*, 1978, El latifundio. Propiedad y exploatacion, ss. XVII
     - XX (Madrid), 197 p. (+ graphs and maps).
Aymard, M., 1979, Toward the history of nutrition: some methodological
     remarks. In R. Foster and O. Ranum (eds.), *Food and Drink in
     History: Selections from the Annales.* John Hopkins University
     Press, Baltimore. 1-16.
Biegajlo, W., 1960, Recherches geographiques sur le système
     d'assolement triennal avec jachere en Pologne. Ergon II. Fasc.
     suppl. 370-374.

Biegajlo, W., 1964, The ways of transition from the three-field system to modern farming as currently observed in Poland's underdeveloped region of Bialystok. *Geographia Polonica*, **2**, 153-158.

Biegajlo, W., 1968, Types of agriculture in north-eastern Poland (Bialystok voivodship). *Geographia Polonica*, **14**, 275-282.

Biegajlo, W., 1969, Uzytkowanie ziemi i gospodarka nolma w środkowej Macedonii: wieś Elovec (Utilisation du sol et agriculture en Macedonia Centrale: village Elovec). *Dokumentacja Geograficzna*, **5**, 5-32.

Biegajlo, W., 1973, Typologia rolnictwa na przykladzie Województwa Bialostockiego (Agricultural Typology. A Study made on the example of the Bialystok voivodship). Report published by the Institute of Geography and Spatial Organization, Polish Academy of Sciences, Warsaw (mimeographed).

Bodvall, A., 1957, Periodic settlement, land clearing and cultivation with special reference to Boothlands of North Hälsingland. *Geografiska Annaler*, **39**(4), 213-256.

Boesch, H. and Bronhofer, M., 1956, The decline of the Dreizelgen-System in north-eastern Switzerland. Resumé du XVIII Congres International de Geographie 4, Rio de Janeiro. 21-25.

Bonnamour, J., 1972, Essai de typologie économique des systèmes d'exploitation en France. In C. Vanzetti (ed.), *Agricultural Typology and Land Utilization, Verona.* 223-229.

Bonnamour, J., 1973, Typologie des systèmes d'exploitation en France. In L.G. Reeds (ed.), *Agricultural Typology and Land Use.* Hamilton, Ontario. 73-86.

Bonnamour, J., 1975, Methodes d'analyse utilisées pour l'evaluation des systèmes d'exploitation par regions agricoles. In C. Vanzetti (ed.), *Aricultural Typology and Land Utilization, Verona.* 71-85.

Bonnamour, J. (ed.), 1984, Les types d'agriculture en France 1970 et 1980. Recueil des cartes. Centre de Géographie Rurale de l'E.N.S. de Fontenay aux Roses.

Bonnamour, J. and Gillette, Ch., 1980, Les types d'agriculture en France 1970. Essai méthodologique. CNRS, Paris.

Bonnamour, J., Gillette, Ch. and Guermond, Y., 1971a, Les systèmes regionaux d'exploitation agricole en France. Methode d'analyse typologique. *Etudes Rurales*, **43-44**, 78-168.

Bonnamour, J., Gillette, Ch. and Guermond, Y., 1971b, Typologie des systèmes d'exploitation agricole utilisee en France. Essai methodologique. *Annales de Géographie*, **80** (438), 144-166.

Cabo Alonso, A., 1981, Distribucion de sistemas de cultivo en los secanos herbaceos espanoles. In E. Bustos (ed.), *I Coloquio iberico de Geografia.* Salamanca. 79-88.

Chorley, G.P.H., 1981, The agricultural revolution in northern Europe 1750-1880; nitrogen, legumes and crop productivity. *Economic History Review*, **34**, 71-93.

Christians, Ch., 1975, La typologie de l'agriculture en Belgique: methodes, problèmes, resultats. In C. Vanzetti (ed.), *Agricultural Typology and Land Utilization*. Verona. 93-110.

Duby, G., 1962, *Rural Economy and Country Life in the Medieval West*. Edward Arnold, London.

Faucher, D., 1949, Geographie agraire, types de cultures. Paris.

Galczyńska, B., 1984, Agricultural typology of Bulgaria. *Geographia Polonica*, **50**, 169-177.

Gorbunova, L.I., Komleva, M.V. and Shishkina, L.V., 1979, Agricultural typology of the USSR. *Geographia Polonica*, **40**, 83-92.

Grigg, D.B., 1980, *Population Growth and Agricultural Change: an Historical Perspective*. Cambridge University Press, London.

Grigg, D.B., 1982, *The Dynamics of Agricultural Change: the Historical Experience*. Hutchinson, London.

Grigg, D.B., 1987, The industrial revolution and land transformation. In M.G. Wolman and F.G.A. Fournier (eds.), *Land Transformation in Agriculture*. SCOPE 32. Chichester. 79-110.

Kerridge, E., 1967, The Agricultural Revolution. Allen and Unwin, London.

Kostrowicki, J., 1964, Geographical typology of agriculture in Poland. Methods and Problems. *Geographia Polonica*, **1**, 111-146.

Kostrowicki, J., 1970, Types of agriculture in Poland. A preliminary attempt at a typological classification. *Geographia Polonica*, **19**, 99-110.

Kostrowicki, J., 1974, Les transformations dans la repartition spatiale de l'agriculture en Pologne et essai du pronostic de l'evaluation ulterieure. *Geographia Polonica*, **32**, 331-342.

Kostrowicki, J., 1975, The typology of world agriculture. Principles, methods and model types. In C. Vanzetti (ed.), *Agricultural Typology and Land Utilization*, Verona. 429-479.

Kostrowicki, J., 1977, Agricultural typology. Concept and method. *Agricultural Systems*, **2**, 33-45.

Kostrowicki, J., 1980a, Geografia dell'agricolutra. Ambiente Società, Sistemi dell-Agricoltura. Milano. 711.

Kostrowicki, J., 1980b, A hierarchy of world types of agriculture. *Geographia Polonica*, **43**, 126-148.

Kostrowicki, J., 1982, The types of agriculture map of Europe. *Geographia Polonica*, **48**, 79-91.

Kostrowicki, J., 1984, Types of agriculture in Europe. A preliminary outline. *Geographia Polonica*, **50**, 131-149.

Kostrowicki, J., 1985, The transformations from the traditional to the market-oriented or socialized agriculture as seen from the types of agriculture map of Europe. In M. Shafi and M. Raza (eds.), *Spectrum of Modern Geography. Essays in Memory of Prof. Mohammad Anas.* New Delhi. 379-394.

Kostrowicki, J., 1986, Transformations de l'agriculture europeenne à la lumière de la carte des types agricoles de l'Europe. *Geographia Polonica*, **52**, 191-208.

Kostrowicki, J., 1988, Types of agriculture map of Europe. Concept, Method Techniques. In F. Chiffelle (ed.), Aménagement rural - Rural Planning. *Geo-Regards*, **15**, 115-137.

Kostrowicki, J. (forthcoming), Types of agriculture in Britain in the light of the Types of Agriculture Map of Europe. *Geographia Polonica.*

Kostrowicki, J. *et al.*, 1984, Types of Agriculture Map of Europe. 1:15 million. 12 sheets.

Kostrowicki, J. and Szyrmer, J., 1988, Agricultural Typology, Guidelines. Institute of Geography and Spatial Organization, Polish Academy of Sciences, Warsaw (mimeographed).

Le Roy Ladurie, E. (ed.), 1975, *Histoire de la France Rurale.* Paris.

Liversage, 1945, Land tenure in the colonies. Cambridge. 151.

Milhau, J., 1953, Les diverses formes de métayage. *Etudes et Travaux du Conseil Economique.* Vol. 23. Paris.

Montelius, S., 1953, The burning of forest land for cultivation of crops. *Geografiska Annaler*, **35**, 41-54.

Petrov, V.P., 1968, Podsechnoye zemliediceliye. Kiev. 221.

Rikkinen, K. and Vapaoksa, T., 1983, The application of world agricultural typology to Finland. *Geographia Polonica*, **46**, 93-106.

Slicher van Bath, B.H., 1968, *The Agrarian History of Western Europe, A.D. 500-1850.* Edward Arnold, London.

Slicher van Bath, B.H., 1977, Agriculture in the rural evolution. In E.E. Rich and C.H. Wilson (eds.), *The Cambridge Economic History of Europe*, vol. 5, The Economic organization of Early Modern Europe. Cambridge University Press, Cambridge. 42-133.

Stola, W., 1970, Procedure of agricultural typology. The case of Ponidzie, Central Poland. *Geographia Polonica*, **19**, 111-118.

Stola, W., 1975, Changements dans les types de l'agriculture belge dans les années 1950-1970. In C. Vanzetti (ed.), *Agricultural Typology and Land Utilization.* Verona. 339-356.

Stola, W., 1983, Essai d'application des méthodes typologiques à l'étudee comparée sur le développement des agricultures Belge et Polonaise. *Geographia Polonica*, **46**, 159-173.

Szczęsny, R., 1978, Changes and trends in the spatial pattern of types of indivdual agriculture in Poland. 1960-1970. In J. Kostrowicki and W. Tyszkiewicz (eds.), *Transformations of Rural Areas. Proceedings of the lst Polish-Yugoslav Seminar.* Polish Academy of Sciences, Institute of Geography and Spatial Organization, Warsaw. 155-164.

Szczęsny, R., 1979a, Changing types of Austrian agriculture 1960-1970. *Geographia Polonica*, **40**, 161-169.

Szczęsny, R., 1979b, Transformation of different types of agriculture in Poland between 1970-1976. In G. Barta (ed.), *Rural Transformation in Hungary and Poland, Polish-Hungarian Seminar.* September 27-October 1, 1978. Bozsok. Budapest. 40-61.

Szczęsny, R., 1981, Transformations of agricultural types in Poland 1970-1976. In Noor Mohammad (ed.), *Perspectives in Agricultural Geography*, vol. 1, New Delhi. 209-228.

Szczęsny, R., 1982, Typy rolnictwa Szwajcarri. (Agricultural types of Switzerland.) *Przeglad Geograficzny*, **54** (4), 511-532.

Szczęsny, R., 1983, Changes in the spatial structure of Polish agriculture. In M. Klemencic and M. Pak (eds.), *Geographical Transformation of Rural Areas. Proceedings of the 3rd Yugoslav-Polish Geographical Seminar.* Ljubljana, October 19-21, 1983. Ljubljana. 151-160.

Szczęsny, R., 1986, Agricultural typology of the Alpine areas: Austria and Switzerland. *Geographia Polonica*, **52**, 209-219.

Szczesny, R., 1988, Przemiany struktury przestrzennej rolnictwa Polski w latach 1970-1980. Przestrzenne zróznicowanie typów rolnictwa (Changes of the Spatial Structure of Polish Agriculture in the years 1970-1980). Wroclaw. 171.

Teuteberg, H.J., 1975, The general relationship between diet and industrialization. In E.R. Forster (ed.), *European Diet from Pre-industrial to Modern Times.* Harper and Row, London. 61-110.

Tschudi, B. and Johanson, H., 1983, Types of agriculture in Norway by the typogram method. *Geographia Polonica*, **46**, 83-91.

Tyszkiewicz, W., 1969, Z badań nad uzytkowaniem ziemi w poludniowej Macedonii, wieś Asamati (Recherches sur l'utilisation du sol en Macedoine centrale: village Asamati. *Dokumentacja Geograficzna*, **5**, 39-49.

Tyszkiewicz, W., 1975, Types of agriculture in Poland as a sample of the typology of world agriculture. In C. Vanzetti (ed.), *Agricultural Typology and Land Utilization.* Verona. 391-408.

Tyszkiewicz, W., 1979, Agricultural typology of the Thracian basin, Bulgaria, as a case of the typology of world agriculture. *Geographia Polonica*, **40**, 171-186.

Tyszkiewicz, W., 1980, Types of agriculture in Macedonia as a sample of the typology of world agriculture. *Geographia Polonica*, **43**, 163-185.

Tyszkiewicz, W., 1982, Zastosowanie metod typologicznych do badań rolnictwa Szwecji (The application of the typological methods to the studies of Swedish agriculture), *Przeglad Geograficzny*, **53** (4), 551-570.

Chapter 3

FUTURE LAND USE PATTERNS IN EUROPE

F.M. Brouwer and M.J. Chadwick

3.1. Introduction

Land is a key resource for most socio-economic activities (agriculture, wood production, industry, recreation) and infrastructure (settlements, transportation and communication networks), and a vital component of natural ecosystems (such as forests). The use of land is characterized by large transformations over time (for example, Wolman and Fournier, 1987 present a state-of-the-art view on the availability of land for agriculture, as well as the quality of the land that is required to produce food and fiber, including the major types of land transformations in agriculture). The transformation of European land results from a complex set of interactions. The most important include:

i)   *socio-economic and historical changes*, such as in land ownership and tenure, population growth, urbanization, industrialization, development of technology, the establishment of transportation and communication networks. Grigg (1987) describes the constant modification of the environment by human interference to facilitate the production of crops and livestock;

ii) *political decisions*, such as subsidies and taxes for using the land; in addition, decisions like the Corn Laws of Britain in 1846 also determined land use patterns, since they made it easier for overseas producers to sell wheat in that country; a more recent example of policy decisions being important for land use patterns is the Common Agricultural Policy (CAP) in the European Economic Communities (EEC), established in 1957 to achieve, among other things, an increasing level of self-sufficiency in food consumption and an equilibrium in the markets of agricultural products (EC Commission, 1985); and

iii) *environmental conditions*, particularly climatic factors and soil quality.

Large areas of European agricultural land have been transformed over the past few decades to land used for non-agricultural purposes,

F. M. Brouwer et al. (eds.), Land Use Changes in Europe, 49–78.

mainly for the development of infrastructure and communication, urban areas, industries and recreational facilities. In consequence in Europe arable land has decreased over the past 30 years by about 6 million ha. The amount of land transformed over this period was greater than at any previous time and it has been commented that "at the prevailing rates of land conversion, the whole of Britain would be covered by bricks, mortar and asphalt by the twenty-first century" (Hill, 1986). While the changes in European infrastructure and land use were greater than in any previous century, society has also shown a flexibility in adapting to and incorporating such transformations. This flexibility is reflected by the fact that the total European population living in urban areas increased both in absolute and relative terms, while agricultural production also rose. However, human development now requires an alternative view of land management. In the past man was able to cultivate rural areas, establish new settlements and industries, and move to other places when these areas became degraded. But land is now becoming a critical factor in the establishment of new activities, as a more closed and interdependent system of socio-economic activities and environmental constraints has developed. A long-term and broad-scale perspective on the interface between socio-economic and technological changes, environmental transformations and land use patterns is required. A need for an integrated approach to long-term land use planning is evident as conflicts arise due to the constraints imposed by natural environmental features, limiting our social responses and the degrees of freedom for action.

Clark and Munn (1986) have recently outlined the extent to which long-term and extensive societal transformations may predicate significant changes in many environmental spheres. An investigation of future land use patterns, notably one that examines possible strategies for future sustainable development, involves the following considerations:

- *time perspective*: this should be long enough to foresee interactions between environmental transformations and socio-economic development;

- *spatial scale*: this should i) be sufficiently broad to incorporate large-scale transformations in the environment, and ii) enable a linkage between causes, consequences and management of human development at the local, regional and global scale;

- *flexibility and resilience* - such that options should be kept open for long-term alternative development and for adaptation to environmental transformations;

- *significant change*: the use of land cannot be sustained under static conditions, but is "...rather a process of change in which the exploitation of resources, the direction of investments, the orientation of technological development and institutional change are made consistent with future as well as present needs..." (WCED, 1987);

- *technological innovation*: characterized with a focus on how it would modify the environmental consequences of socio-economic development;

- *allocation and renewability* of resources.

This paper focuses on "not-impossible" future land use patterns in Europe associated with a few plausible long-term and large-scale transformations in technological innovations and in the environment over the next century.

The present land use patterns in Europe and the changing role of anthropogenic and environmental factors which determine changes in such patterns will be discussed in Section 3.2. A so-called Conventional Wisdom Scenario for agricultural and land use development from 1980 to 2030 will be discussed. The Conventional Wisdom Scenario considers extrapolation of present trends in agricultural productivity, but the consequences of the application of new technologies are not incorporated. Development in technology, notably information technology and biotechnology, and transformations in the environment will be discussed in Section 3.3, and some scenarios for agricultural and land use development are presented. Some "not-impossible" future changes in climate and soils and how these might influence future land use patterns, are discussed in Section 3.4.

Such "not-impossible" environmental transformations and technological trends might have a low probability of occurrence, although they would certainly have a large impact on society. This will be further explored in Section 3.5 in the framework of future land use changes in Europe.

3.2. Land Use Patterns in Europe

3.2.1. The present situation

The current land use distribution in Europe is summarized in Table 3.1. Europe has been subdivided into five regions. The Table has four main land use categories: i) arable land for growing annual crops and land under permanent crops, ii) permanent meadows and pastures for growing forage crops, iii) forests and woodland (land under natural or planted stands of trees), and iv) "other" land. The last category includes land that is being used for settlements, transportation and communication networks, recreation, industry, as well as unproductive land (for example due to contamination in former times by opencast mining, waste disposal or intensive agricultural use).

The percentage distribution of the four land use categories is depicted in Figure 3.1.

The Nordic countries cover about 10 per cent of the European land area and are characterized by large wooded areas; 20 per cent of European forests are found in the Nordic countries. About 60 per cent of the EEC-9 is covered with cultivated land that is used for agricultural purposes. The central part of Europe includes much of the mountain areas which are largely covered with pastures and forests. Eastern Europe includes the centrally planned economies with arable land comprising about 40 per cent of the total land area.

A substantial part of the total land area in Europe is devoted to agriculture and forestry. The total coverage of arable land, pasture land and forests is around 90 per cent (Figure 3.1). A wide variation exists

Table 3.1. Land use distribution in Europe during the 1980s in million ha.

Region	Arable	Pasture	Forest	Other	Total
Nordic (1)	6.2	0.9	58.1	37.3	102.5
EEC-9 (2)	50.7	40.9	32.3	26.6	150.5
Central (3)	1.9	3.6	4.3	2.5	12.3
South (4)	36.5	23.4	32.5	8.1	100.5
East (5)	237.0	117.0	168.0	39.0	561.0
Europe	332.3	185.5	295.3	113.6	926.7

Source: FAO Production Yearbook, 1984; data for the European part of the USSR are from Alayev et al., 1987.

1)   Finland, Norway and Sweden;
2)   Belgium/Luxembourg, Denmark, France, Germany, FR, Ireland, Italy, The Netherlands and United Kingdom;
3)   Austria and Switzerland;
4)   Albania, Greece, Portugal, Spain and Yugoslavia;
5)   Bulgaria, Czechoslovakia, GDR, Hungary, Poland, Romania and the European part of the USSR.

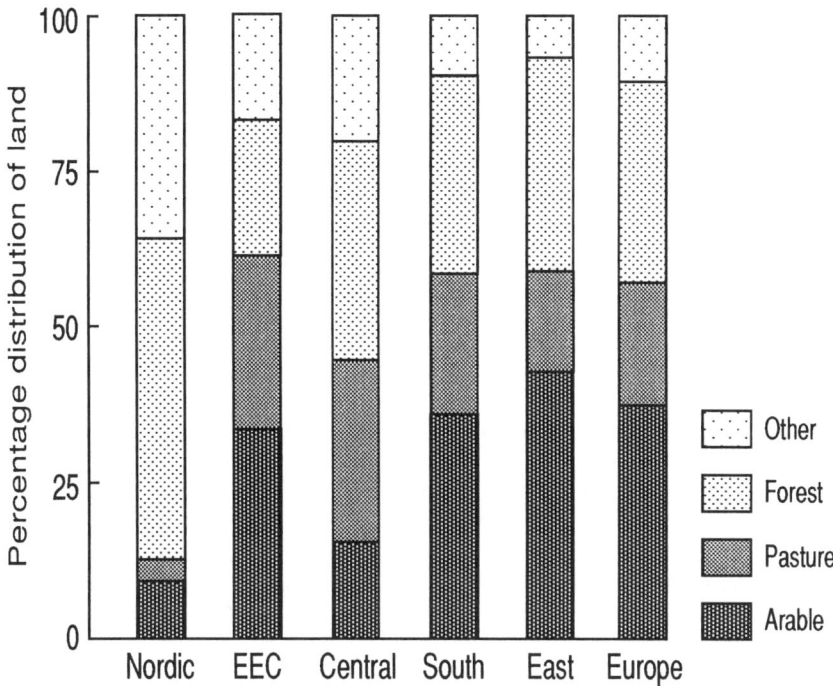

Figure 3.1.   Percentage distribution of land use categories in Europe.

over Europe as a result of many contributing factors.   Environmental factors such as climatic and soil conditions are one set of determinants; another set of factors is the combination of socio-economic, political, historical and technological factors (these might change over a relatively short period of time).   These factors not only determine the way in which the land is used but also the productivity of cultivated and forest land.

The post-war period of agricultural development in Europe is mainly characterized by steadily increasing yields (productivity per unit area), both in absolute and relative terms, improved varieties, a better understanding of crop physiology, larger farms, an increasing use of chemical fertilizers and pesticides and also by the replacement of labor with machinery and capital.   The relative increases in agricultural production were large in Europe during this period amounting to about 70 kg ha^{-1} yr^{-1} for wheat in the original nine countries of the EEC (de Wit *et al.*, 1987), and about 50 kg ha^{-1} yr^{-1} for wheat in the Eastern part of Europe (Wong, 1986).

Some of the major changes since 1950 of inputs of machinery, chemical fertilizers and irrigation in Europe are summarized in Table 3.2.

Table 3.2. Some major input characteristics of European agriculture between 1950 and 1980 (not including the European part of the USSR).

Type	Unit	1950	1960	1970	1980
Machinery	1,000 tractors	1,050	3,764	6,169	8,465
Fertilizer					
(NPK)	1,000 tonne	8,000	13,000	24,900	32,800
Irrigation	1,000 ha	na	6,000	10,794	14,452

Sources: various FAO Production Yearbooks.

Machinery increased to replace labor. The application of chemical fertilizers (mainly nitrogen, phosphorus and potassium) has contributed to a large extent to the increased productivity of arable land (Olson, 1987). The total extent of irrigated agricultural land more than doubled (especially in Bulgaria, France, Italy, Romania and Spain). Irrigation may transform the land through the construction of canal networks and through the change in the water and salt balance of the cultivated land (Shannon, 1987).

The present situation of agricultural productivity in Europe can be characterized by i) overproduction in relation to markets, ii) considerable differences in productivity over the continent, and iii) different potentials for increases in productivity under present technological conditions. These may be detailed as follows:

i)   one of the original purposes of the Common Agricultural Policy (CAP), namely to increase the level of self-sufficiency in the EEC, was a driving force for various transformations in European agriculture (larger farms, improvements in efficiency), which have resulted, since 1980, in overproduction of many products (cereals, milk);   an example may be found in the total domestic consumption of cereals around 1965, which was about 20 million tonnes larger than the total production levels; this trend has changed such that since 1980 the domestic consumption has increasingly exceeded production (Porceddu, 1986);

ii)   agricultural land productivity also shows a wide variation in Europe; in the EEC, for example, about 50 per cent of the total wheat production is produced on about 25 per cent of the total wheat acreage (Strijker and de Veer, 1986); the average wheat yields during the early 1980s were about 3 tonne ha^{-1} in Greece, Italy and Spain, while they were over 7 tonne ha^{-1} in Denmark, The Netherlands and United Kingdom;

iii)   the productivity of agricultural land could further increase since
         the present levels of agricultural production in Europe are still
         below potential (de Wit *et al.*, 1987); the production potential can
         be constrained by biophysical factors (climate and soils), but also
         by socio-economic factors (location, transportation facilities, level
         of return on investments) or social conditions (availability of
         educational facilities, age structure and managerial skills of the
         farming population); the different productivity levels of agricultural
         land can be evaluated by distinguishing between core and
         marginal land, such that marginality of land is based on land that
         has achieved absolute and relative limits with present technologies
         (Beattie *et al.*, 1981).

Large areas of European agricultural land are likely to be taken
out of production over the next decades for the reasons described, this
trend will be emphasized when technological improvements in society
and environmental transformations are also taken into account. In a
country like Britain, 6 million ha of agricultural land could be taken out
of production, when all currently available techniques to improve yields
are implemented (Milne, 1987). The present agricultural surpluses in
the EEC correspond to an area of some 9 million ha (Lewis, 1987). The
landscape of Europe could therefore change drastically in a period
between the 1980s and the middle of the next century.

The key questions to be considered with respect to such land use
policies are, among others, what land will be set aside from agriculture
and to what alternative use will this land be put (recreation, nature
conservation, forests to grow biomass, wood production and cottage
industry).

3.2.2. **Conventional Wisdom Scenario**

Possible changes to future land use patterns will be described here
in terms of scenarios. A Conventional Wisdom Scenario for agricultural
development from 1980 to 2030 is presented in Table 3.3, as an
extrapolation of present trends. The Scenario up to 2000 considers an
annual increase in productivity of 1 per cent for cereal crops and an
annual increase in productivity of 0.5 per cent for the period 2000-
2030. Table 3.3 shows the actual and projected land (in million ha),
total production (in million tonnes) and yield (in tonne ha^{-1}) for various
regions under a Conventional Wisdom Scenario.

In total some 40 million ha of arable land might be taken out of
production if the present trends of increasing land productivity
continue. This corresponds to about 35 per cent of the present land
used for growing cereal crops. The largest decrease in land used for
agriculture would be for the EEC-9 region where about 50 per cent of
the land might be set aside for non-agricultural purposes.

Table 3.3. Conventional Wisdom Scenario for agriculture for the period 1980-2030.

Region	1980			2000			2030		
	land	productivity	yield	land	productivity	yield	land	productivity	yield
Nordic	3	10	3.3	2	9	4.5	2	10	5.0
EEC-9	27	120	4.5	18	98	5.4	13	82	6.3
Central	1	5	4.5	1	6	5.5	1	6	5.8
South	15	42	2.8	14	46	3.3	11	44	4.0
East	84	189	2.3	68	200	3.0	60	216	3.6
Europe	130	366	2.8	103	359	3.5	87	378	4.2

In the following two Sections we shall describe a "not-impossible" transformation in technology and how this might affect the use of land, and then assess the possible linkages between environmental transformations and land use changes, notably a change in climate and soils.

3.3. New Technologies and Changing Land Use Patterns

It has been pointed out that it is not only environmental changes but also changes in social, economic and technological factors that result in the transformation of land use patterns. The present land use distribution is the result of structural changes in the European agricultural industry, changing technologies and the demographic transitions resulting from urbanization. Rapid advances in new technologies, notably biotechnology and information technology, may particularly lead to further unexpected transformations in European land use over the next 70 years. A recent report (OTA, 1985) gives projections for production growth for the period until the year 2000, both under present technological conditions and under a wide application of new technologies for agriculture in the USA. The annual growth rate of wheat production, now about 1 per cent, might increase to over 2 per cent when further new technologies are applied in agriculture. Although this report summarizes the agricultural position of the USA, the future trends for technology described may also be applied to European agricultural development in the next 10 to 20 years. Such developments may, for example, result in conservation of land because of land being set aside from agriculture owing to increased productivity.

Biotechnology research will lead to the emergence of new products i) to control disease by breeding crops resistant to pests and diseases or implementing widespread biological control strategies, ii) to produce cultivars for harsh environments (for example, salty soils and poor climatic conditions), and iii) to produce cultivars with altered nutritional requirements. The breeding of plants that are resistant to diseases or certain pests could result in a considerable reduction in the use of insecticides and herbicides. In addition, the improvements in biotechnology may result in growing crops that are able to use atmospheric nitrogen and therefore require less chemical fertilizers and so diminish nitrate pollution (Arntzen, 1984; Nielsen, 1986). However, when pest-resistant crops are applied on a large scale, and cheap herbicides are available, these may transform the cultivated land in such a way that it is not possible to grow anything else (von Weizsäcker, 1986).

The role of biotechnology in producing biomass for energy purposes and the alteration of land use patterns in western Europe are assessed by Lewis (1987).   By the year 2000 a maximum of 1.2

Exajoule net biomass energy could be produced each year in the EEC-9, requiring about 7 million ha of land previously used for agriculture or forests. The total production of biomass energy in Europe might double by the year 2000 when methanol fuel will be produced from cultivated crop land in Norway, Sweden, Portugal, Spain and Greece.

Information technology is the application and integration of computers and electronics into farm management. A wide application of available information technology (microcomputers and software) at the farm level may improve the management of pests. The application of information technology may decentralize production and also conserve energy and material through an improvement in the efficiency of using agricultural inputs (von Weizsäcker, 1986).

The potential impacts of the application of new technologies for future land use patterns may be summarized as:

i) *an increase in productivity*, wheat production in countries such as The Netherlands, France and United Kingdom, which is presently between 6 and 8 tonne ha^{-1}, could increase to some 12 tonne ha^{-1}; such an increase in productivity may result in a further reduction of agricultural land when compared to the Conventional Wisdom Scenario;

ii) *adaptation to poor local conditions*, the productivity of marginal land could be improved if crops suited to harsh environments became available (for example, land affected by salts or drought); this can be incorporated in a land use policy to improve the marginal land and to maintain the socio-economic structure of rural areas, but it would further aggravate the current problem of agricultural surpluses;

iii) *improvement of the environment*, through the application of, among others, nitrogen fixing species, integrated pest management or recycling of agricultural waste; the application of integrated farming to control pests and diseases by the development of biological control methods and the reduction of pesticide use may result in a decrease of physical productivity, although the economic yields may still increase owing to a decrease in input of fertilizers and pesticides (Bal and van Lenteren, 1987), this approach could be promising for the highly productive agricultural land of Europe; about 90 per cent of the present total use of pesticides is obsolete and its abandonment would require a transformation of chemical farming to a more biological-oriented farming practice;

iv) *unexpected risks of biotechnology*, the land may be transformed so that it becomes impossible to grow anything else other than man-made, high-technology crops;

v)   *increasing monoculture*, the wide application of clones will affect
        the diversity of landscape in rural areas.

        Figures 3.2 and 3.3 successively show some trends for land use
and yield of cereal crops, considering different developments in the
application of new technologies in agriculture.   Two scenarios of
technological development are presented, one is based on an *increase in
productivity*, and the other is based on the *adaptation to poor local
conditions*.   The results of the Conventional Wisdom Scenario are also
given to show the order and magnitude of changes.
        The scenario which is based on an increase in productivity due
to the application of new technologies may show an increase in yield
about 50 per cent above that of the Conventional Wisdom Scenario.   In
addition, it may well result in a decreased use of agricultural land as
compared to the Conventional Wisdom Scenario.   On the other hand, a

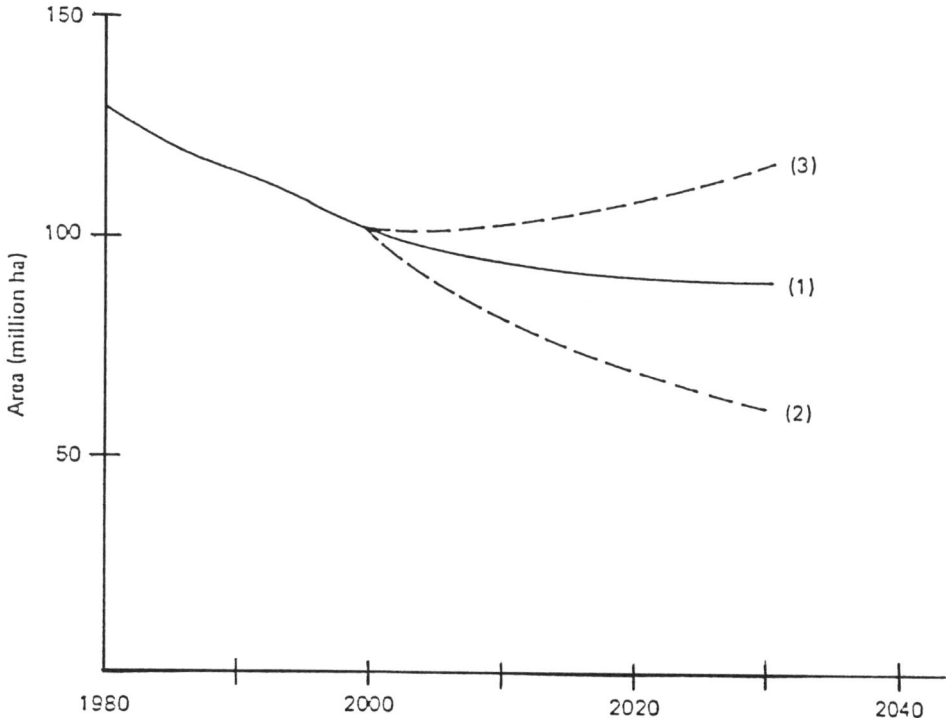

Figure 3.2.   Agricultural scenarios for Europe to the year 2030: land
utilized for agriculture (1 = Conventional Wisdom; 2 = increase in
productivity due to technology; 3 = emphasis on adaptation to poor local
conditions).

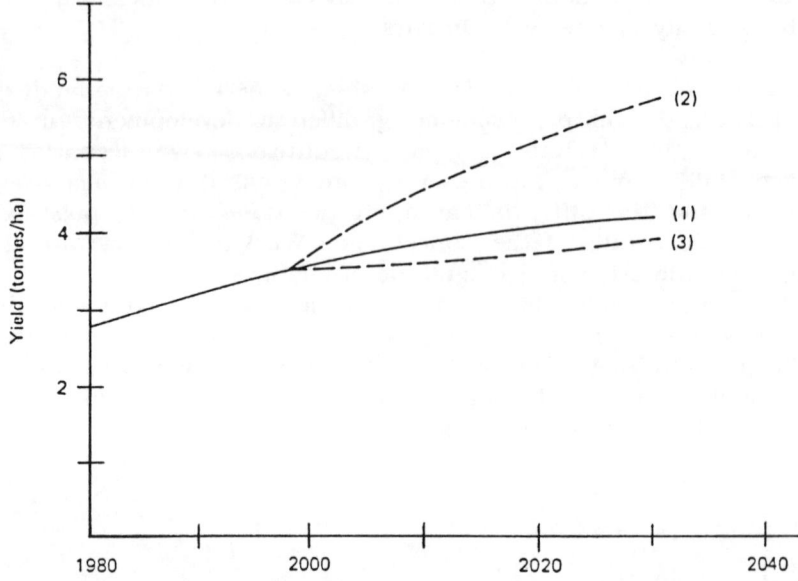

Figure 3.3. Agricultural scenarios for Europe to the year 2030: yields of cereal crops (1 = Conventional Wisdom; 2 = increase in productivity due to technology; 3 = emphasis on adaptation to poor local conditions).

scenario which is based on an adaptation to poor local environmental conditions alone would show an increase in yields which would be less than the increment under the Conventional Wisdom Scenario, while the total land cover that is set aside from agriculture could be reduced.

To summarize, the application of new technologies could result in different trends of European land use patterns from those described by the Conventional Wisdom Scenario for the period 1980-2030. A further increase in productivity may well result in a further increase in the amount of land set aside from agriculture, but a land use policy focusing on an adaptation to poor local environmental conditions could result in an increase in the land devoted to agriculture compared to the Conventional Wisdom Scenario. In the following Section the possible linkages will be assessed between environmental transformations over the next decades and land use changes.

3.4. Climate Change and Soil Degradation in Relation to Changing Land Use Patterns

3.4.1. Introduction

Temperature and precipitation are important factors determining natural as well as managed vegetation. Water availability, for example, is a major limiting factor for crop growth and forest production and about 80 per cent of the fresh water consumed in the World is used agriculturally (Pimentel, 1986).

The present climatic conditions in Europe are summarized in Table 3.4. Large parts of Europe (including western Europe, Nordic countries and the central part of Europe) are characterized by relatively large amounts of rainfall during summer or autumn, and minimum rainfall during the spring and winter period. On the other hand, the Mediterranean climate is mainly characterized by maximum rainfall in the autumn or winter period, with minimum amounts during the summer period. The average January temperature in this part of Europe is over 5°C with only the occasional occurrence of frost, while the average monthly July temperature is over 20°C. Figures 3.4 and 3.5 show the distribution of mean annual temperature and precipitation over Europe.

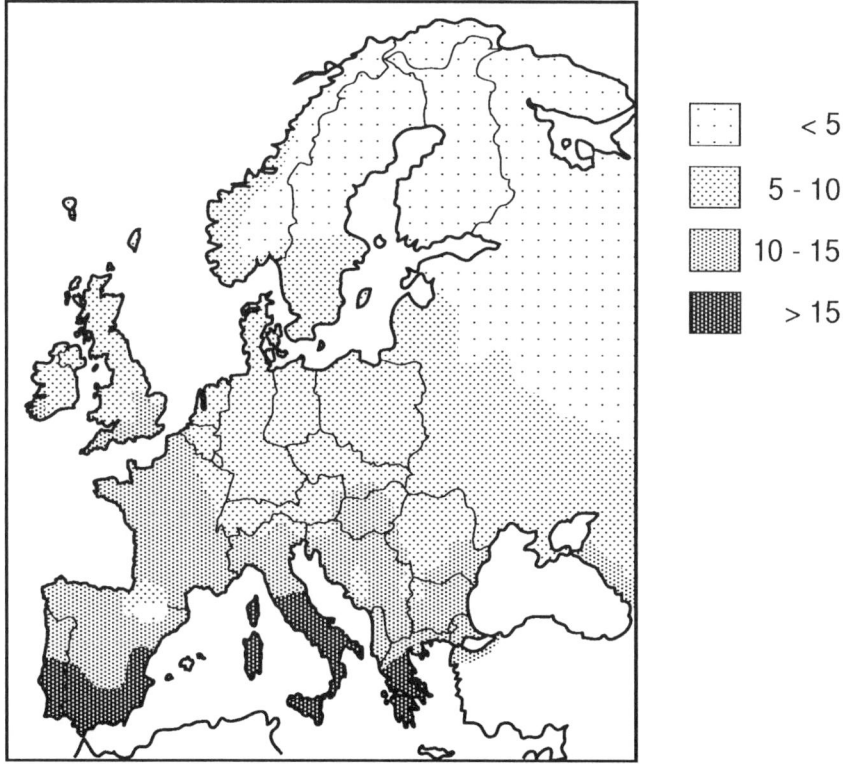

Figure 3.4. Mean annual temperature in Europe (in °C). (Source: Müller, 1982.)

Table 3.4. Climatic conditions in Europe (data from weather stations are based on averages for a thirty-year period from 1931-1960, and are from Müller, 1982).

Country	Precipitation (mm)			Temperature (°C)		
	Mean annual	Maximum	Minimum	Mean annual	January	July
Albania	1050	Winter	Summer	15	5	25
Austria	1150	Summer	Winter	7	-4	16
Belgium/Luxembourg	925	Summer	Spring	9	1	16
Bulgaria	625	Summer	Winter	12	-1	23
Czechoslovakia	675	Summer	Winter	9	-2	19
Denmark	675	Summer	Spring	8	0	17
Finland	550	Summer	Spring	2	-10	16
France	775	Autumn	Spring	11	3	19
Germany, FR	750	Summer	Winter	8	0	17
Germany, DR	625	Summer	Winter	8	-1	18
Greece	625	Winter	Summer	16	6	26
Hungary	675	Summer	Winter	11	-2	22
Ireland	1,000	Autumn	Spring	10	5	15
Italy	825	Autumn	Summer	14	5	24
Netherlands	750	Autumn	Spring	9	1	17
Norway	1,000	Summer	Spring	4	-5	14
Poland	600	Summer	Winter	8	-3	19
Portugal	800	Winter	Summer	15	9	22
Romania	725	Summer	Winter	10	-3	21
Spain	650	Winter	Summer	14	7	23
Sweden	700	Summer	Spring	4	-7	16
Switzerland	1,275	Summer	Winter	8	-1	17
United Kingdom	1,050	Autumn	Spring	9	4	15
USSR	550	Summer	Winter	5	-8	18
Yugoslavia	950	Autumn	Winter	12	0	22

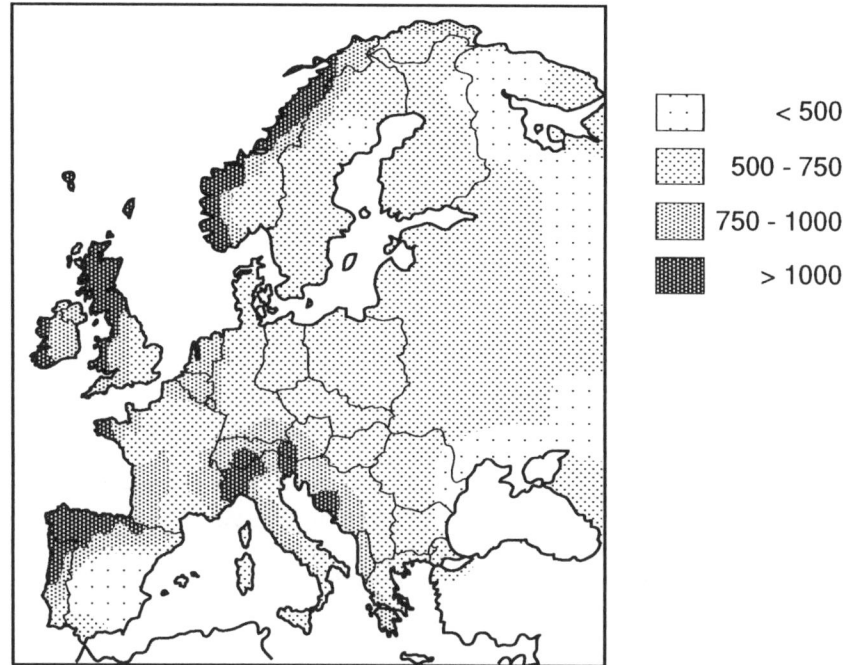

☐	< 500
▨	500 - 750
▧	750 - 1000
■	> 1000

Figure 3.5.   Mean annual precipitation in Europe (in mm).   (Source: Müller, 1982.)

A shift of climate is expected to occur on a global scale due to increased atmospheric concentrations of carbon dioxide and other greenhouse gases (chlorofluorocarbons, nitrous oxide, methane and ozone).   Pre-industrial concentrations of CO_2 were about 280 ppmv (ppmv = parts per million volume) and present concentrations are around 340 ppmv.   The pre-industrial concentration of carbon dioxide may be doubled by the year 2030 (Lough et al., 1983).   Such a scenario of increasing atmospheric CO_2 largely depends on (uncertain) projections for fossil fuel consumption and rates of deforestation.   The last hundred years has already seen an increase in global mean temperature of between 0.3 and 0.7°C.   This increase is consistent with the observed increasing concentrations of carbon dioxide and other greenhouse gases. Recent estimates from simulation experiments with general circulation models and empirical studies on a doubling of atmospheric CO_2 suggest a global mean temperature increase of between 1.5 and 4.5°C.   However, the magnitude and order of variation as well as change over the seasons will vary with latitude.

It is possible to obtain some detail of a climatic scenario for Europe in a warmer world by reference to historical analogues.   An historical approach utilizes information on "warm" and "cold" climatic periods between 1880 and 1980. Lough *et al.* (1983) pointed out the utility of such an approach although it should be borne in mind that measurement disparities inherent in the historical data and features such as the development of very large cities may be the cause of some of the differences between "warm" and "cold" climatic periods over this time. However, future climatic features based on historical analogues may be appropriate for *shorter term* (to 2030) climatic indications but not in the *longer term* (from 2030 onwards to 2075) as the increase in greenhouse gas concentrations becomes more pronounced and more extreme climatic events become better established. Thus, although it is possible to sketch out a shorter-term climatic scenario based on past climatic analogues such an approach has to give way to results from general circulation models, such as the BMO model described by Mitchell (1983), to obtain details of the longer-term climatic scenario.

There is no suggestion that future climatic changes will be uniform with time but similarly the use of different methods of gauging climatic change scenarios does not imply any element of "phasic" change. Climatic change is very much a "moving target" phenomenon. Furthermore, both methods of displaying climatic change scenarios show that seasonal variation of changes in temperature and precipitation in Europe are likely to occur. Such seasonal characteristics and the order of variation with time are important for the suitability of using land for agriculture and other purposes, since they affect water availability for agricultural crops or forests and the length of the growing season.

3.4.2. Shorter-term European climate changes

Shorter-term climate changes based on the analysis of historical analogues from Lough *et al.* (1983) can be characterized as follows (Figures 3.6a and 3.6b show the mean annual changes in temperature and precipitation).

Mean annual temperature shows an increase over all of Europe, with the largest increases in the northern part (over 1°C) and the south-eastern part of the continent (over 0.5°C). *Mean annual precipitation* decreases over large areas of Europe, with the exception of Norway, Sweden, Finland, parts of the United Kingdom, France, Spain and central and eastern Europe, for which small increases may occur.

Figures 3.7a and 3.7b show the changes in temperature in summer and winter. The main features of the temperature scenarios illustrated are the slight to moderate warming over much of Europe in summer but slight cooling in winter. The largest increase in temperature (over 1°C) is found in the summer period, with the exception of the south-eastern part of Spain which shows a slight cooling. The majority of Europe shows a slight cooling in winter (less than 0.5°C), with the exception, for the most part, of the Nordic countries, United Kingdom, Ireland and southern Spain which would

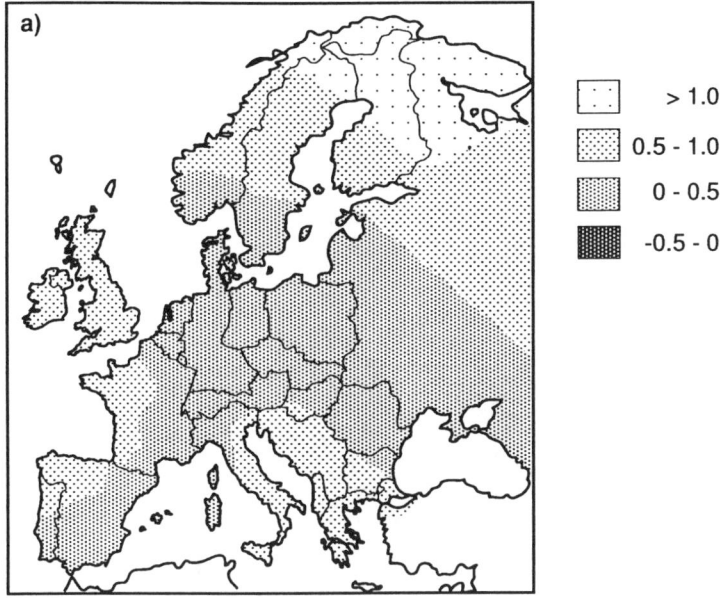

Figure 3.6a. Changes in mean annual temperature (in °C).*

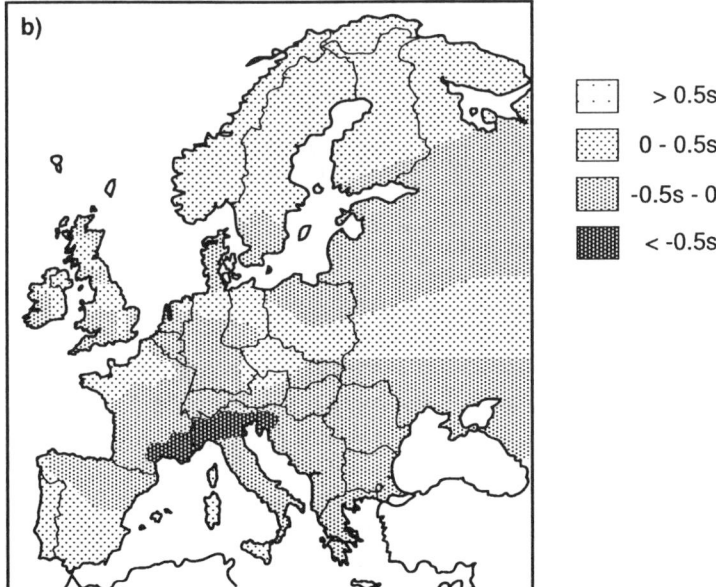

Figure 3.6b. Changes in mean annual precipitation (as multiples of the standard deviation).

*Figures 3.6, 3.7 and 3.8 have been produced by reference to LOUGH et al. (1983), Journal of Climate and Applied Meteorology(American Meteorological Society).

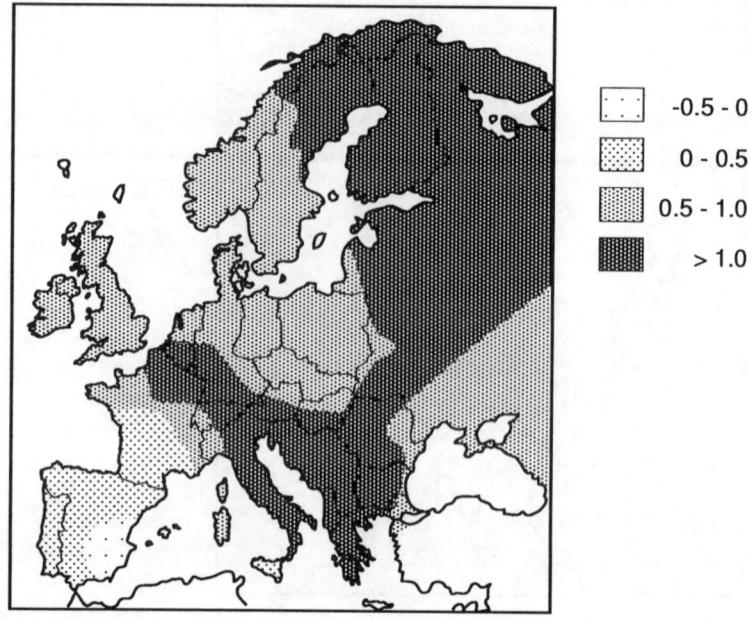

Figure 3.7a. Temperature changes in summer (in °C).

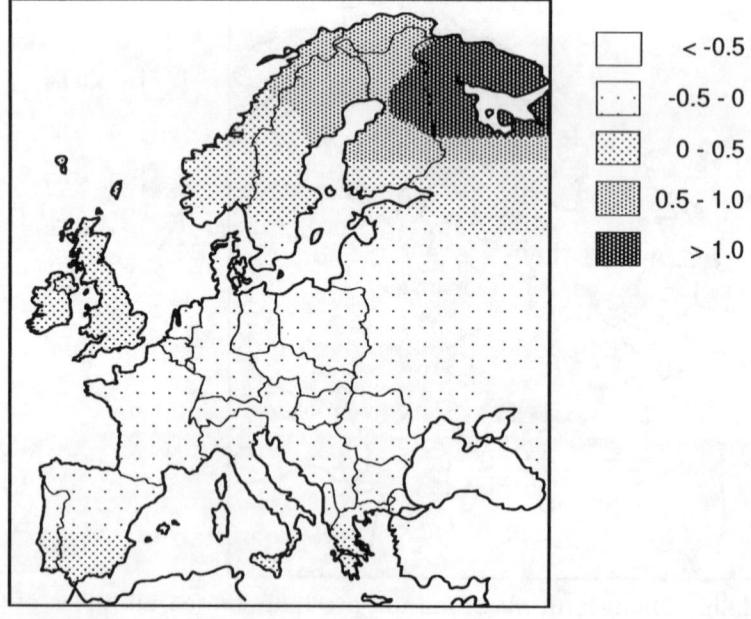

Figure 3.7b. Temperature changes in winter (in °C).

experience a slight temperature increase. Spring and autumn patterns (not shown) are similar to those in summer.

Figures 3.8a and Figure 3.8b show the changes of precipitation in summer and winter. There is a general tendency for drier summers and wetter winters. Major decreases during the summer period are found in areas which already experience minima during that season (Italy, Portugal, Spain and the south-eastern part of Europe).

3.4.3. **Longer-term changes in European climate**

Predictions regarding the climatic conditions that would result from a doubling of atmospheric carbon dioxide globally before the middle of the next century and an equilibrium response in climate can be obtained from general circulation models. Results show the seasonal variation of climate in a warmer Europe. The increase in mean annual temperature in Europe is expected to be in the region of 3 to 4°C. The mean annual precipitation pattern shows an increase roughly north of the 50° latitude and a decrease south of it. The annual increase in northern Europe might be as large as 150 mm, while the decrease in the Mediterranean area might be some 300 mm. Figure 3.9 shows the changes in mean annual temperature in Europe for this scenario, compared to present climatic conditions.

The scenario considers an increase in temperature during all seasons in Europe. The regional distribution of the seasonal variation in temperature shows the largest increase during the winter season; this might be as large as 7°C in northern Europe and around 2°C in the Mediterranean region. Figure 3.10 shows the changes in mean annual precipitation in Europe for this scenario, relative to the present climatic conditions.

The climate change scenario also considers a resulting sea level rise over the next 70 years of between 40 and 160 cm (UNEP/WMO/ICSU, 1988). This is especially important for the coastal lowlands of western Europe (France, Belgium and The Netherlands) and the Mediterranean coastal areas (Italy).

3.4.4. **Soil degradation factors and the change in climate**

The kind of broad-scale transformation in climate discussed in the previous subsections would have profound effects on all types of land use. Cultivated land and forest are vulnerable to a change in climate. It has been emphasized that "some regions now marginal for crop production because of climate may become even more so" (Crosson, 1986). This is illustrated in the Mediterranean area which is characterized by high soil moisture deficit in the summer period. However, other types of land use might also be highly vulnerable to rapid climatic shifts. Weinberg (1985) considers the adaptation or land

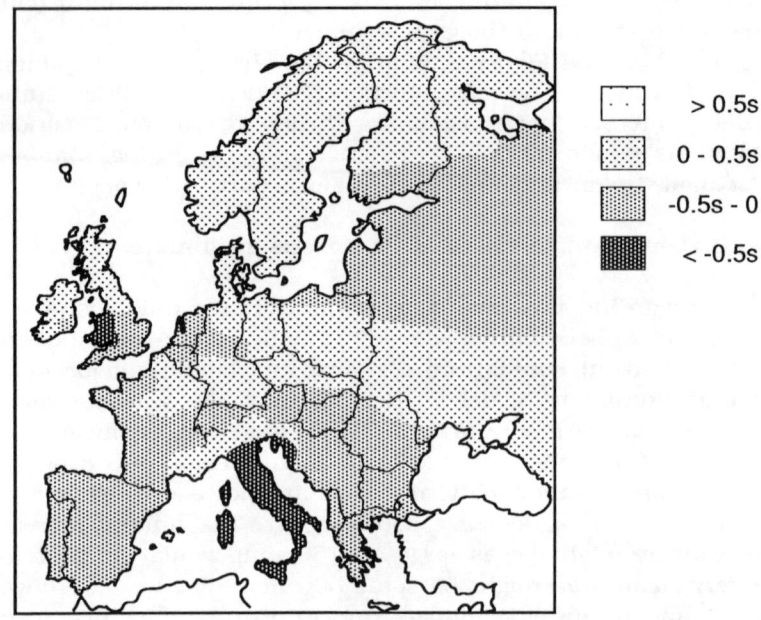

Figure 3.8a. Changes in precipitation in summer (as multiples of standard deviation).

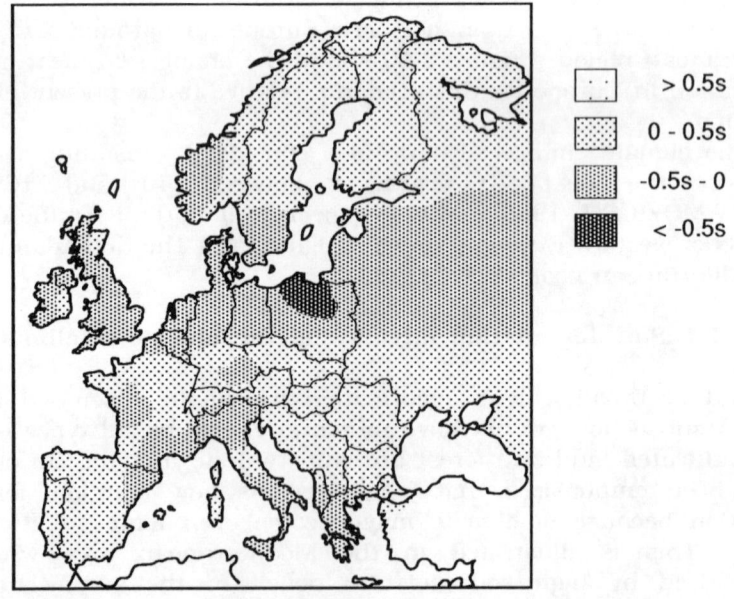

Figure 3.8b. Changes in precipitation in winter (as multiples of standard deviation).

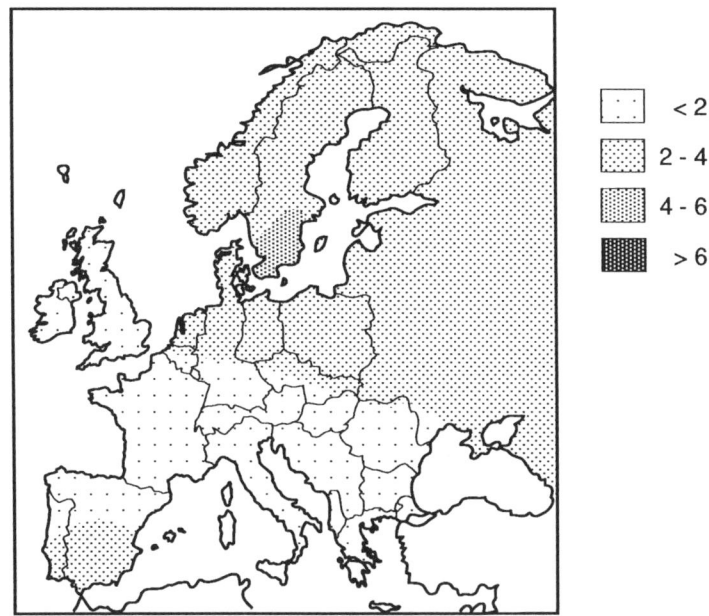

Figure 3.9.   Changes in mean annual temperature (in °C) (BMO scenario).

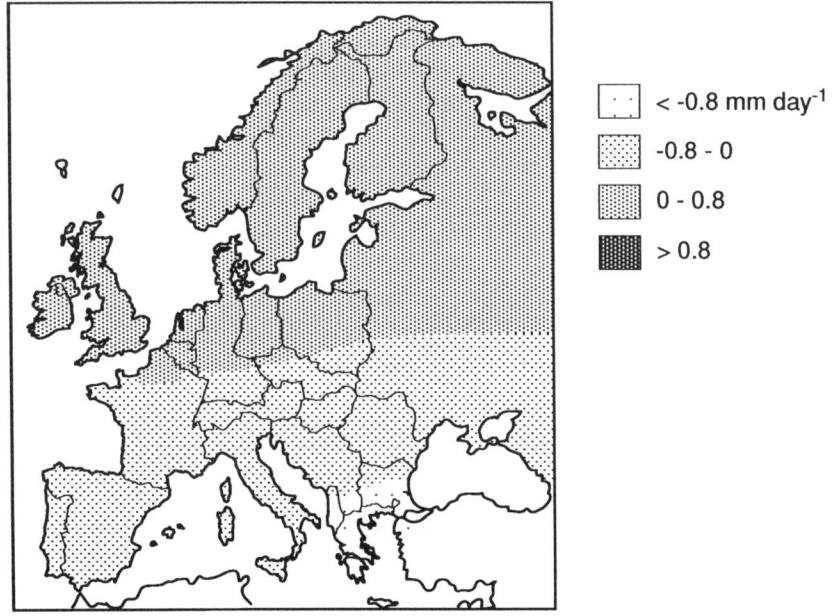

Figure 3.10. Changes in mean annual precipitation (in mm day⁻¹) (BMO scenario).

use change that might be required for urban centers and infrastructure due to a climate change (for example, change in transportation and communication networks due to a sea level rise and the protection of land against the increasing risk of flooding). Soil degradation factors will be considered first. Important degradation factors that relate to changes in agricultural land use and climatic factors are: erosion, either by wind or water; depletion of nutrients and organic matter; and salinization.

Soil erosion causes problems for agriculture not only through the removal and transport of the fertile soil, but it may also cause problems for settlements in lowlands which can be damaged by flooding and siltation.

Depletion of nutrients and organic matter occurs when more nutrients are regularly removed from the soils than are replaced during the year. This type of land degradation may accelerate over the next decades in areas of increased mean annual precipitation due to leaching, or where increased mean annual temperature results in increased oxidation of soil organic matter. A change in climate may result in an increasing soil moisture deficit in the Mediterranean area, especially important during the growing season for crops. This increasing soil moisture deficit might result in an increased requirement for the application of fertilizers to retain productivity of that land which, again, may result in an increase of leaching problems.

Salinization may occur as a result of salt accumulation from surface evaporation of groundwater or irrigation water in the Mediterranean area and where the intrusion of salts in groundwater occurs due to a sea level rise in the north-western part of Europe. A change in salt accumulation in the future may result from:

i)   increased irrigation in agriculture;

ii)   increased soil moisture deficit owing to a change
        in climate;

iii)   sea level rise, mainly in the coastal lowlands of
        western Europe.

The land degradation factors that are described here in relation to a change in climate during the period between the present day and the middle of the next century will be further elaborated in the next Section in terms of future land use changes.

3.5. Future Land Use Changes in Europe

The Conventional Wisdom Scenario considers that over an approximate 100-year period global population will stabilize (at about 10 billion), energy use will increase by about a factor of 6 and a steady growth in the scale of agricultural production (an increase of a factor of 4) will occur. It is essentially a "continuing trend" scenario although means whereby the trends are perpetuated are not taken to be necessarily those prevailing at present. It is possible to conceive of "surprise rich" scenarios in place of this relatively "surprise free" one. The value of doing this is emphasized by those who point out that "if we know anything about the future, it is that projections will not hold forever" (Svedin and Aniansson, 1987). Surprises are sure to occur. Therefore, the argument goes, let there be an attempt to build an element of surprise into certain scenarios. This has been attempted in some studies (Svedin and Aniansson, 1987) so that changes occur in population density (away from Europe to south and east Asia, for example) or population stabilization does not occur. These scenarios, known as the Big Shift and the Big Load respectively, are described along with other scenarios by Svedin and Aniansson (1987). Such changes would imply changes in various aspects of the agricultural system and some details are summarized in Table 3.5.

Table 3.5. Changes associated with increases in agricultural production in Europe for three scenarios described by Svedin and Aniansson (1987). 1975 = 100.

	Conventional Wisdom	Big Shift	Big Load
Fertilizer applied	200	250	100
Irrigation	125	100	145
Land acreage	95	130	70
Yield	250	200	290

It can be inferred from this information that a number of other changes are occurring that are only peripherally related to agriculture (such as use of land for biomass production for energy purposes in the Big Shift scenario and the advent of food shortages in Europe under the Big Load scenario).

It is important to distinguish between end points predicted by these scenarios and the means by which these may be attained. For example, a Conventional Wisdom Scenario that considers population, energy use and agricultural production changes with time does not expose the possible land use changes in Europe or allow these to be assessed realistically as it does not explicitly explore the impact of technological, socio-economic and physico-environmental (particularly climatic) factors that are likely to change over the next 50-100 years.

Cropping and other land use boundaries are determined by the interplay of all these factors. Andreae (1981) distinguishes a number of boundaries to land use including those based on profitability (where returns tend to zero), technological boundaries (the boundary up to which a particular land use could be undertaken at a certain stage of technology if economic considerations are waived) and effective boundaries that represent the actual limit of a certain form of land use.

Some technological boundaries are well documented for Europe: grain corn (*Zea mays*) has shifted its northern boundary in Europe by approximately 5° latitude in the last 25 years due to selection by plant breeders, developing varieties that can produce economic yields under a shorter growing season and mature at lower mean summer temperatures. Other technological advances, especially those associated with genetic engineering activities, are likely to result in the potential for large shifts in the critical limits for a whole range of crops currently used in European agriculture and horticulture and, to a lesser extent, for forest crops over the next hundred years. This could lead to the clearance of large areas of land not presently devoted to agriculture to enable these new varieties to be exploited. In addition, land use changes will undoubtedly result from intervention policies that evolve from the mismatch of production and demand. The set aside policies at present under implementation in the EEC may result in considerable changes in land use although it is not clear at present whether these changes will involve taking out of production large areas of "marginal" farm land or the application of "production quotas" to the more productive land. This would have the effect of reducing the more extensive effects of such a policy and would allow the implementation of more conservation-oriented farming systems on land hitherto farmed extremely intensively. It is also necessary to consider limitations on productivity resulting from soil degradation (due to erosion, salinization and loss of fertility) and adverse pollution loads.

In the light of this it might be envisaged, in evolving a "not impossibe" scenario applicable to Europe, that production surpluses will marginally limit crop extension; that new crop varieties will allow yields to continue to increase; that major pollution sources will be, on the whole, abated as it is seen that such measures represent cost-effective responses to pollution damage in the longer term. However, interacting with these factors, ecological land use boundaries are likely to shift in response to an overall climatic change that imposes new geographical limits on plant growth and land use systems.

Land use boundaries are imposed by a range of ecological limitations: altitude, temperature, moisture availability (aridity and waterlogging) and even slope and exposure. In view of the climatic changes postulated by general circulation models that explore the effects of increased concentration of greenhouse gases, it seems necessary to consider as a minimum activity limitations related to changes in moisture availability and mean annual temperature and the closely related feature of length of growing season, as a "back cloth" to the other factors stimulating change.

In northern Europe the length of growing season is defined by the period over which mean daily temperatures above 5°C prevail. Here, plant growth is not, in general, limited by lack of soil moisture, although areas do exist where crop yields could be enhanced in some years, particularly in May-August by an increased water availability. Soil moisture deficits, however, are much more frequent limitations to growth in southern Europe, particularly in Mediterranean regions. Although soil moisture deficits are recharged during winter and soil moisture storage can meet moisture demand during short periods when evapo-transpiration exceeds precipitation, soil moisture deficits soon restrict transpiration and gas exchange and the rate of dry weight increment is reduced.

Figures 3.11 and 3.12 indicate the length of the growing season. This is defined here by the period of the year, in months, that the mean temperature is above 5°C and the period during which precipitation exceeds 0.5 of the potential evapotranspiration (Verheye, 1986). When the growing season in Europe under present climatic conditions (Figure 3.11) is compared with the period based on the BMO scenario of climatic conditions, it can be seen that the greatest increases are in northern Europe (due to an overall rise in mean annual temperature) and the largest decreases are in the Mediterranean area where increasing soil moisture deficit will limit crop production.

The result of the changes in the length of growing season outlined above is the background against which other interacting factors must be placed (such as technological, pollution and socio-economic changes) to arrive at an indication of land use changes. In general, there is a potential for the northern crop boundaries to shift 5-7° in a northerly direction and an associated limitation on crop growth in parts of the Mediterranean with the southerly boundary shifting 3-5° N. Figures 3.13 and 3.14 give some indication of this for selected crops.

The potential northern shift of the boundary for sugar beet, Winter wheat, Spring barley and potatoes could mean that large areas of land hitherto devoted to forestry (the northern boreal forest) could be cleared for cultivation of some of these crops. Even some soil constraints could be overcome by the application of well tried farming methods practised previously on areas of difficult soils. Approximately 18-20 million ha of present day forest land in northern Europe would be potentially exploitable in this way. In southern Europe (Mediterranean and bordering the Black Sea) it would probably be no longer feasible to grow perennial tree crop products (citrus fruits and olives) successfully and the potential would switch to such crops as cotton (and possibly rice) under irrigation. In between these two extremes significant shifts in cropping regimes would be possible.

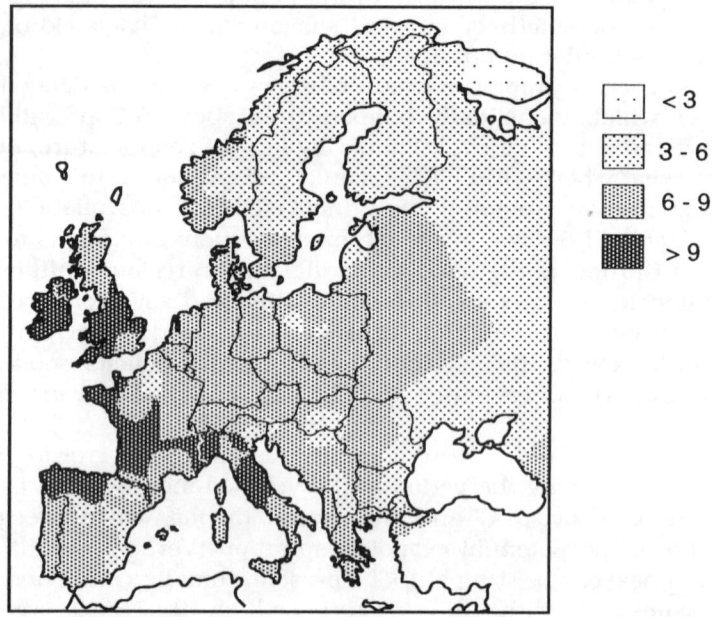

Figure 3.11.   Growing season in Europe (in months) (present climatic conditions).

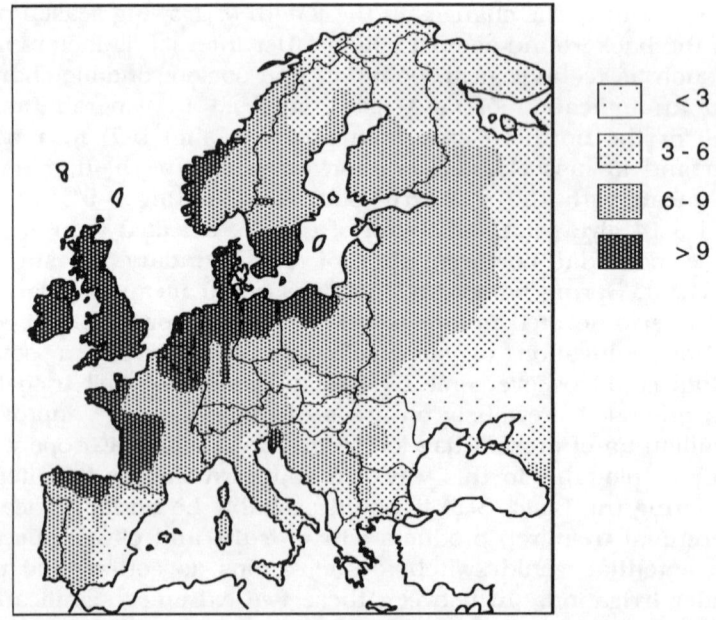

Figure 3.12.   Growing season in Europe (in months) (BMO scenario).

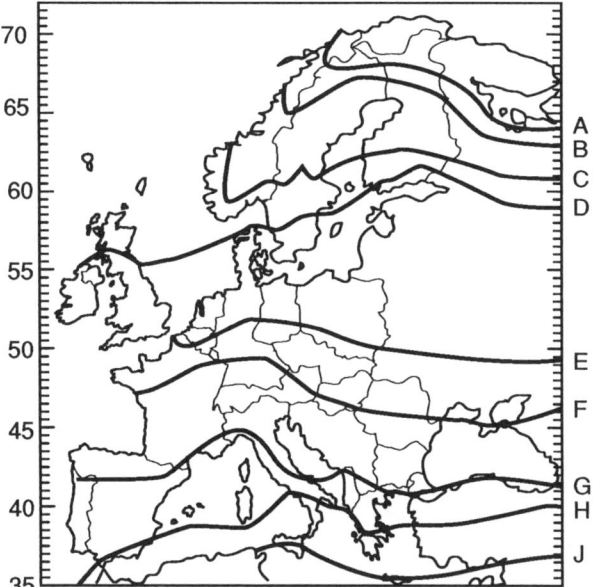

Figure 3.13. Northern cropping boundaries in Europe for 9 crops in 1975 (Andreae, 1981). Letters indicating crops given below.

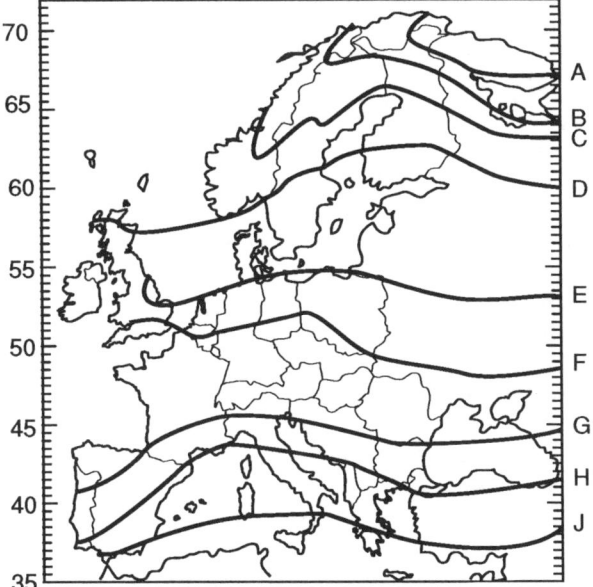

Figure 3.14. Northern cropping boundaries for 9 crops in 2075 (A - potatoes; B - Spring barley; C - Winter wheat; D - sugar beet; E - maize; F - grapevines; G - olives; H - citrus fruits and J - cotton).

3.6. **Concluding Remarks**

It has been stressed that changes in technology and socio-economic conditions will combine with any environmental transformations to determine future patterns of land use in Europe by the middle of the next century. That large potentials for a shift of present land use boundaries exist does not mean that such changes will occur extensively. But, equally, it is inconceivable that no change in land use patterns will take place in response to trends in climatic modification and associated soil features, catchment characteristics and other ecological conditions. The factors we have discussed serve to focus on the considerations that need to be weighed in reaching an assessment of the plausible land use patterns that will develop by the middle of the next century.

REFERENCES

Alayev, E.B., Yu. B. Badenkov and Karavayeva, N.A., 1987, *The Russian Plain: Regional View*, paper presented at the conference Earth as Transformed by Human Activities, Clark University, 24-31 October, 1987.

Andreae, B., 1981, *Farming, Development and Space*, Gruyter, Berlin.

Arntzen, C.J., 1984, Biotechnology and agricultural research for crop improvement, in: *Cutting Edge Technologies*, National Academy of Engineering, National Academy Press, Washington, D.C. 52-61.

Bal, A. and van Lenteren, J.C., 1987, Geitegreerde bestrijding van plagen (integrated pest management), Series: Ecological Pest Management, Ministry of Housing, Physical Planning and Environment, The Hague (in Dutch).

Beattie, K.G., Bond and W.K., Manning, E.W., 1981, The agricultural use of marginal land: a review and bibliography. Working Paper 13. Lands Directorate, Environment, Canada, Ottawa.

Clark, W.C. and Munn, R.E. (eds.), 1986, *Sustainable Development of the Biosphere*, Cambridge University Press, Cambridge.

Crosson, P., 1986, Agricultural development - looking into the future, in: W.C. Clark and R.E. Munn (eds.), *Sustainable Development of the Biosphere*, Cambridge University Press, Cambridge. 104-136.

EC Commission, 1985, *The Green Paper: Perspectives for the Common Agricultural Policy*, The Agricultural Information Service of the EC Commission, Brussels.

Food and Agriculture Organization of the United Nations (FAO), 1984, FAO Production Yearbook, volume 38, Rome.

Grigg, D., 1987, The industrial revolution and land transformation, in: M.G. Wolman and F.G.A. Fournier (eds.), *Land Transformation in Agriculture*, SCOPE 32, John Wiley, New York. 79-109.

Hill, R.D., 1986, Land use changes on the urban fringe, *Nature and Resources*, **22**(1), 24-33.

Lewis, C., 1987, The role of biotechnology in assessing future land use within Western Europe, Report FOP 87, *Commission of the European Communities*, Brussels, 125 pp.

Lough, J.M., Wigley, T.M.L. and Palutikof, J.P., 1983, Climate and climate impact scenarios for Europe in a warmer world, *Journal of Climate and Applied Meteorology*, volume 22. 1673-1684.

Milne, R., 1987, Putting the land out to grass, *New Scientist*, 17 December 1987. 10-11.

Mitchell, J.F.B., 1983, The seasonal response of a general circulation model to changes in CO_2 and sea temperature, *Quarterly Journal Royal Meteorological Society*, **109**, 113-152.

Müller, M.J., 1982, *Selected Climatic Data for a Global Set of Standard Stations for Vegetation Science*, Series: Tasks for Vegetation Science, volume 5, Dr. W. Junk (Publishers), The Hague.

Nielsen, M.H., 1986, Industrial biotechnology in Europe: the overall context, in: D. Davies (ed.), *Industrial Biotechnology in Europe: Issues for Public Policy*, Frances Pinter Publishers, London. 28-34.

Olson, R.A., 1987, The use of fertilizers and soil amendment, in: M.G. Wolman and F.G.A. Fournier (eds.), *Land Transformation in Agriculture*, SCOPE 32, John Wiley, New York. 2O3-226.

OTA (Office of Technology Assessment), 1985, Technology, public policy and the changing structure of American agriculture, OTA Report, Washington, D.C.

Pimentel, D., 1986, Water resources for food, fiber and forest production, *Ambio*, **15**(6), 335-340.

Porceddu, E. (1986), Agriculture as a customer and supplier of biotechnology, in: D. Davies (ed.), *Industrial Biotechnology in Europe: Issues for Public Policy*, Frances Pinter Publishers, London. 83-96.

Shannon, L., 1987, The impact of irrigation, in: M.G. Wolman and F.G.A. Fournier (eds.), *Land Transformation in Agriculture*, SCOPE 32, John Wiley, New York. 115-131.

Strijker, D. and de Veer, J., 1986, Regional impacts of the common agricultural policy of the EC, *Agricultural Economics Research Institute*, Internal Report 328, The Hague.

Svedin, U. and Aniansson, B. (eds.), 1987, *Surprising Futures*, Notes from an International Workshop on Long-term World Development, Swedish Council for Planning and Coordination of Research, Stockholm.

UNEP/WMO/ICSU, 1988, Briefing paper on policies for responding to climatic changes, Policy Issues Workshop, Bellagio, Italy, November 1987.

Verheye, W.H., 1986, Principles of land appraisal and land use planning within the European Community, *Soil Use and Management*, **2**(4), 120-124.

WCED (World Commission on Environment and Development), 1987, *Our Common Future*, Oxford University Press, Oxford/New York.

Weinberg, A.M., 1985, 'Immortal' energy systems and intergenerational justice, *Energy Policy*, February 1985. 51-59.

von Weizsäcker, E., 1986, The environmental dimension of biotechnology, in: D. Davies (ed.), *Industrial Biotechnology in Europe: Issues for Public Policy*, Frances Pinter Publishers, London. 35-45.

de Wit, C.T., Huisman, H. and Rabbinge, R., 1987, Agriculture and the environment: are there other ways?, *Agricultural Systems*, **23**, 211-236.Wolman, M. and Fournier, F.G.A. (eds.), 1987, *Land Transformation in Agriculture*, SCOPE 32, John Wiley, New York.

Wolman, M.G. and Fournier, F.G.A. (eds.), 1987, *Land Transformation in Agriculture*, SCOPE 32, John Wiley, New York. 79-109.

Wong, L-F., 1986, *Agricultural Productivity in the Socialist Countries*, Westview Special Studies on Agriculture Science and Policy, Westview Press, Boulder and London.

Chapter 4

THE ROLE AND IMPACT OF BIOPHYSICAL DETERMINANTS ON PRESENT AND FUTURE LAND USE PATTERNS IN EUROPE

W.H. Verheye

4.1. Introduction

Natural vegetation patterns result from a continuous interaction between local conditions of climate, soils and topography. The original plant cover has been largely modified by human interference and it has been subsequently replaced by crops providing the high yields that are required for commercial food supply.

As a result of increasing demographic pressures and changes in socio-economic conditions, agricultural activities have gradually been intensified in Europe over the past 2,000 years. Not only has the total acreage of cultivated land increased, but also yields and production methods have improved due, in part, to the wide application of chemical fertilizers, pest control, the introduction of new genotypes and the application of modern farming technology.

The changes in society that have taken place over the past centuries, from an almost exclusive rural lifestyle towards more urban-based societies, has created a situation where the role of farmers is no longer limited to supplying of a variety of fresh food crops for the small, local and dominantly rural community. The role has been transformed to producing bulk products for consumers in urban areas, or for the agro-industry. Hence, the nature and intensity of agricultural activities in a certain area, cropping patterns and the importance of cultivated versus uncultivated land are no longer a true reflection of the basic food requirements, nor of the activities and/or density of the local population.

The evolution of land use patterns in Europe can be characterized by five major stages.

1) In prehistoric times, agriculture was limited and most of the food requirements for the population were met through gathering and picking of natural products. Land was almost completely covered by natural vegetation, and environmental deterioration effects due to land use activities were not apparent.

F. M. Brouwer et al. (eds.), Land Use Changes in Europe, 79–97.
© 1991 *Kluwer Academic Publishers. Printed in the Netherlands.*

2)   Agriculture first started to evolve around rural settlements in early
        historical periods, where it developed into a shifting cultivation
        system - a small scale farming system where most of the land
        remains under natural or semi-natural vegetation.

3)   Population increases in the Middle Ages (500-1500 A.D.) began the
        shift of many human activities from rural areas to urban centers,
        consequently food supply requirements for the cities also increased.
        This initiated, for example, the clearance of forests around the
        major urban centers and the release of land for crop growth.

4)   The steady population increase in Europe, and the development of
        large industrial centers, further emphasized the differences in
        lifestyle between the rural and urban communities, with the farmer
        becoming a professional food producer. Increased food demand
        was met initially by the extension of arable land and,
        subsequently, by intensification of agricultural practices and a
        decrease in the average field size. The first signs of soil
        deterioration were observed because of continuous cultivation and
        the lack of proper fertilizers.

5)   A technology-oriented agriculture has been introduced during the
        second half of the twentieth century, which makes full use of
        sophisticated plant materials, agrochemicals and modern farming
        techniques. Crop production is now achieved by fewer farm
        workers supplying larger markets. Land reclamation and re-
        allocation schemes are widely applied, and the size of farms shows
        an increasing trend. Marginally productive land has been taken
        into production, facilitated by an increased application of
        agrochemicals. Environmental deterioration effects in relation to
        the land use pattern are observed, manifested as an increased
        erosion and pollution hazard to soils and groundwater.

        Under present conditions of continuously increasing yields,
saturated consumer markets and growing concern about the protection
of the environment, it is foreseen that large areas of land in western
Europe presently used for growing crops will be taken out of production
in the near future. Part of this land might be devoted to urban centers,
used for industrial and/or leisure purposes, while other zones may be
turned into (agro) forestry or be transformed into more extensive
agricultural systems with a reduced fertilizer input. A scenario of
projected land use pattern will largely, if not exclusively, be determined
by financial incentives (including both subsidies and penalties) proposed
to the farmers.
        In central and eastern Europe where production has not yet
reached a level that saturates the consumer markets, present conditions

are projected to move towards a maximum land occupation and/or an increase in production. However, development trends largely depend on the status which is given to the farmer and on financial expectations.

4.2. Major Determinants of Land Use Patterns in Europe

Land use patterns in Europe have been affected mainly by the interaction, to various degrees of influence, of three major factors, including i) *biophysical* or environmental conditions, related to climate, soils and landform; ii) *socio-economic* conditions of farmers and consumers; and iii) *technological* achievements and progress. The dominant role of each of these factors, and their impact on agricultural, industrial, urban or other activities, have largely determined the extent and geographical distribution of natural land versus cultivated land, field size and crop pattern.

4.2.1. Biophysical conditions

The regional biophysical or environmental conditions determine almost exclusively the nature and extension of natural biotopes. Such conditions mainly refer to climatic parameters (including rainfall, temperature, frost hazard and sunshine), and to soil parameters (including root depth, moisture-holding capacity, permeability and nutrient status).

Soil and climate are both of paramount importance in an intensive cropping system in terms of i) the selection of crops, ii) the nature of the management system, and iii) the type of farming operations. Indeed, they provide the framework within which the system may operate and define the natural constraints which have to be overcome (such as water shortage requiring irrigation and nutrient deficiencies requiring fertilizer applications) and indicate the tillage procedures which have to be followed and the harvesting constraints (such as optimal workability conditions during particular periods of the year).

4.2.2. Socio-economic conditions

The socio-economic context determines primarily the incentives for farmers in terms of financial benefits and/or social status. Benefits are not only related to yields, but also depend on demand for agricultural products, fluctuations in market prices, or national export policies.

Other factors which may interfere at this level deal with the land tenure and ownership rules, the government policy to support agriculture or agricultural goods, market protectionism and/or the introduction of a preferential tax system.

Economic conditions change over time, and may even change overnight as a result of sudden political decisions. The implication for land use patterns therefore requires a continuous monitoring of world market conditions, as well as of the local or regional political decisions.

In the Middle Ages agriculture was largely oriented towards cereal production for human and animal consumption. Horses disappeared from the farmyard with the gradual introduction of machines and the barley acreage decreased accordingly. Food production in western Europe has generally been sufficient to cover most of the common regional needs, but only 100 years ago even the slightest fluctuation in yield could have important consequences for the urban proletariat. Yields have largely increased with the introduction of chemical fertilizers and modern agro-technics, and in recent years agricultural overproduction has increased in the European Community. However, growing concern about the protection of the environment may inevitably lead to the implementation of policies for reducing cropping areas, substantial changes in cropping patterns, or a return to a more natural vegetation cover.

4.2.3. **Technological progress**

Modern technological evolution in the field of agriculture can not be completely separated from recent socio-economic trends. As a result of technological progress in farming techniques and crop production, present day agriculture in western Europe has rapidly shifted from labor-intensive practices to a system supported by capital and technology. Such a shift has been slower in eastern Europe, but is nevertheless evolving in the same direction. This evolution has an obvious impact on the social living conditions of the rural population and has influenced the land use patterns; for example, the growing dominance of monocultures and the increase in average field sizes.

The number of people engaged in European agriculture has decreased over the past 20 years by 45 per cent, whereas total crop production increased by some 60 per cent between 1965 and 1973 and by another 10 per cent between 1974 and 1985. The area of utilized agricultural land decreased by 10 per cent coincident with an increase of land for urban and industrial centers, and an increase in wooded areas (Chapter 1). Yields are projected to increase further, and there is increasing evidence that over the next decades more agricultural land will return to natural or semi-natural vegetation in Europe. The present environmental policies to preserve rural areas in Scandinavia and the EC may also have major implications for future land use patterns. The imposition of a tax on the use of fertilizers, for example, would increase the operating costs for farmers and, hence, locally affect land use towards non-food production (Neville-Rolfe and Caspari, 1987). In the same context, agricultural land may be converted to (agro) forestry or to any form of diversified land use without direct production of food.

Whatever the future evolution in the land use patterns and

technological progress might be, man still operates within the opportunities and/or constraints of natural environmental conditions. Hence, the role and impact of socio-economic and technological determinants will always be limited within this biophysical framework. For proper land use planning major attention must therefore be given to the climatic, soil and topographic parameters that influence vegetation, including managed vegetation, and the land management activities linked to it. A proper knowledge of such biophysical determinants and their impact on crop production would not only provide a better understanding of the actual land use pattern and its limitations, but would also contribute to a realistic development of possible future scenarios which may modify that pattern.

4.3. Effect of Biophysical Parameters on Crop Growth

The natural vegetation of an area is a direct expression of the combined effect of soil and climatic parameters. The plants require an adequate moisture and nutrient supply for a proper vegetative growth, while minimum levels of temperature and radiation are also required for efficient photosynthesis and biomass production.

Crop growth requirements vary from species to species. Hence, under given environmental conditions one species will perform better than others or will be ruled out completely, either through competition with other species, or because the combined effects of soil and climate do not allow successful growth.

The poor heath vegetation, for example, in various parts of Britain, Belgium, The Netherlands and Germany, is the result of environmental conditions dominated by permeable, acid and nutrient-deficient soils developing under high rainfall regimes. In these areas only a narrow range of species enjoys optimum growing conditions. This is in strong contrast with Mediterranean type vegetation where high temperature and summer moisture stress are major growth determinants. Similar considerations hold for Atlantic vegetation and Alpine biotopes.

As for natural vegetation a similar case applies for cultivated crops. Major factors that play a decisive role may be regrouped into climatic and soil parameters on one hand, and on workability conditions on the other hand. The major impacts of such biophysical parameters on grassland and Winter wheat production will be discussed in more detail in the following Sections.

4.3.1. Climatic conditions

The climatic parameters that influence plant growth and crop production mainly involve temperature and moisture conditions and the correlated air humidity, sunshine duration and day length conditions. The period of the year when both moisture and energy conditions are favorable for germination and plant development corresponds with the growing season. In Mediterranean areas this period is almost exclusively determined by the summer rainfall deficit; in northern latitudes the limiting factor is generally the winter temperature.

The length and position in the year of the growing season varies over the continent and, hence, defines the crop calendar. In areas where this period is longer than the time which is normally required for crop production, it may be considered that, apart from other requirements, there are no growth constraints. If the period is shorter than the time required for normal vegetative or reproductive cycle, then optimal crop development and production will be constrained.

A relatively accurate definition of the growing season can be obtained through a simple mathematical model that is based on water balance and long-term average temperature (FAO, 1978). The growing season is evaluated as the time (in days) that i) mean rainfall exceeds half the potential evapotranspiration (PET), extended by the time that a theoretical amount of 100 mm moisture stored in the soil is evaporated, and that ii) the mean temperature exceeds 5°C. In this approach PET is calculated from the Penman formula, slightly modified by FAO (Doorenbos and Pruitt, 1977; Frere and Popov, 1979). If the initial input data to this model are provided on a monthly basis the growing season will be defined in relatively broad terms. For a more detailed assessment, the use of 5- or 10-day averages are recommended.

Figure 4.1 shows how the growing season is calculated at two European stations, using monthly precipitation (P), potential evapotranspiration (PET), and temperature (T). Nice in France is representative of the summer-drought affected Mediterranean areas with no constraints imposed by low winter temperatures; Ukkel-Brussels, in Belgium, represents climatic conditions of more northern latitudes where, due to low winter temperatures, only plants that pass through a dormancy season within the time of full moisture availability are suitable.

In terms of crop growth requirements, however, the optimal and marginal conditions may vary within the different substages of the growing season. For a perennial crop such as grass, for example, the number of parameters to be taken into account is relatively small, but for most arable crops such as wheat, barley or sugar beet, which pass through various phenological sub-stages during a growth cycle, the specific requirements in terms of temperature, radiation or moisture supply are much more complex.

Grass growth (and ultimately grassland production) is mainly determined by temperature and moisture conditions. Growth is restricted below 5°C and above 25°C, in the latter case moisture stress is induced by high evaporative demands. In terms of water requirements, grasses need an almost continuous moisture supply in the root zone, which may be provided through rainfall, capillary rise from groundwater or soil moisture storage.

Nice,
France

T	8.3	8.9	10.8	13.1	16.5	19.9	22.4	22.3	20.5	16.4	12.2	9.1	°C
P	68	61	73	73	68	35	20	27	77	124	129	107	mm
PET	25	37	56	80	105	129	147	126	89	53	30	20	mm

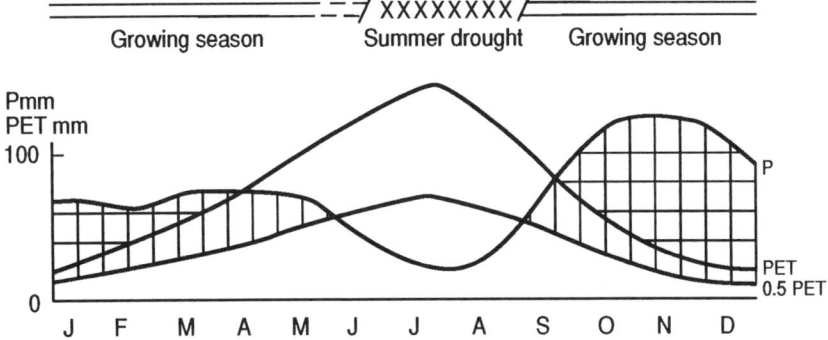

Ukkel - Brussels,
Belgium

T	2.7	3.1	5.5	8.2	12.8	14.9	16.8	16.4	14.0	10.0	5.2	3.4	°C
P	67	50	60	63	63	65	90	75	66	76	78	82	mm
PET	7	16	36	60	94	99	101	83	50	18	5	5	mm

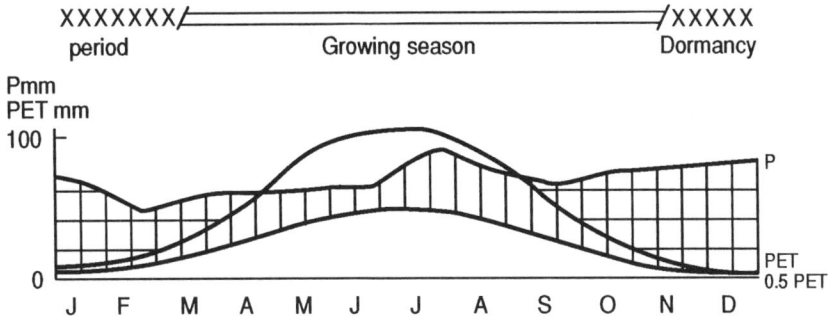

Figure 4.1. Position of the growing season based on mean monthly precipitation (P, in mm), potential evapotranspiration (PET, in mm, and calculated by a modified Penman method) and temperature (T, in °C) in a Mediterranean and a northern latitude station.

Climatic growth requirements for cereals and most other annual crops are much more complex than they are for perennial crops. Winter wheat varieties in Europe require an average growing season of 180-210 days, depending on the length of the winter dormancy, the average day length and the daily temperature. Different development stages can be recognized within the growing season, corresponding to the germination (15-20 days), the initial vegetative (20-30 days) and full vegetative (50-60 days) stages and the flowering (30-40 days) and ripening stage. The overall water consumptive needs are estimated to be 500-800 mm for the whole growing season, but crops have specific needs at each stage, in terms of temperature, radiation and moisture.

The main climatic requirements for growing Winter wheat are summarized in Table 4.1. The establishment of requirement tables to evaluate growth and productivity has mainly focused on optimal and marginal yield levels in quantitative terms. In future, another additional aspect, mainly dealing with product quality, has to be emphasized. It is foreseen that this could lead to contrasting figures when compared to those dealing with the quantitative evaluation. Indeed, for many plants there is an improvement in quality if they pass through a period of moisture stress or are exposed to low temperatures. Such climatic stresses may enhance fruit flavor, alter the biological or chemical composition of the plant oil, and/or improve the overall quality of the product.

4.3.2. Edaphic conditions

As for climate, soil conditions also have to be assessed in relation to the specific land utilization type and/or cropping pattern. Within the range of parameters that affect plant growth and biomass production, either directly or indirectly, a differentiation can be made between a soil's physical and drainage properties, and its chemical and fertility status.

Grass growth is largely dependent on the soil moisture regime and is therefore directly related to the drainage, moisture retention and permeability of the profile. External soil drainage is generally related to aspect at least to the extent that location governs the depth to the groundwater table and the likelihood of flooding. Soil moisture retention and permeability are affected by particle size, packing density and structure, depth and stoniness (Lee, 1984). For mechanized grassland farming, additional attention has to be paid to "poaching" susceptibility and land accessibility (Section 4.3.3), which are controlled by the period and extent of topsoil saturation, and by slope.

A tentative assessment of the major soil requirements for mechanized Winter wheat production is given in Table 4.2. This shows that medium textured soils are most suitable for wheat production, while sandy soils and heavy clays induce poor growth or result in serious yield reduction. The texture ratings may be shifted as a function of the

Table 4.1.   Main climatic requirements for Winter wheat growth in Europe.

| Climatic parameter | None | Degree of constraint | | Severe |
		Slight	Moderate	
Length of growing season (days)	180-210	150-180 210-250	100-150 >250	<100
Rainfall in growing season (mm)	600-800	500-600	800-1,200 400-500	>1,200 <400
Mean monthly temperature in growing season (°C)	15-20	12-15 20-25	<12 >25	
Mean monthly rainfall in vegetative stage (mm)	60-90	30-60 90-120	<30 >120	
Mean monthly temperature in vegetative stage (°C)	8-12	6-8 12-18	<6 >18	
Daily minimum temperature in vegetative stage (°C)	<8	8-10	>10	
Monthly rainfall in flowering stage (mm)	30-45	45-60	>60	
Mean monthly temperature in flowering stage (°C)	14-20	10-14 20-24	<10 >24	
Mean monthly temperature in ripening stage (°C)	16-22	12-16 22-30	<12 >30	

relative importance of the soil water holding capacity in extremely dry or moist environments.

Soil drainage affects the air-moisture equilibrium and, hence, governs the development of the rooting pattern. Wheat, like most deep-rooting crops, does not tolerate waterlogging for a long period, although the crop can stand occasional floods. Optimal cultivation conditions therefore occur on non-flooded soils with good drainage and ample aeration. Imperfectly and poorly drained lands have a definite marginal growth potential on medium and fine-textured materials, but are more suitable where there are coarse-textured substrates.

In terms of soil fertility a distinction should be made between the relatively stable and natural fertility status (related to cation retention capacity) and the man-induced nutrient level, expressed in terms of N, P and K, and the availability of micro-elements. The sensitivity of crops to increasing concentrations of salts, sodium compounds and heavy metals is becoming increasingly important with altered environmental conditions.

Although wheat is a relatively exacting crop, it can grow on a wide variety of soils with different nutrient levels. A base saturation below 50 per cent in both the topsoil and the subsurface horizons, or a low organic material content in the surface layer, can easily be reclaimed, and may therefore only be considered as a slight to moderate limitation. This type of crop is moderately tolerant to salinity and alkalinity and will develop without major limitations under salinity conditions in the root zone less than 4 mmhos per cm, and a level of sodium saturation which is less than 15 per cent.

4.3.3. **Workability conditions**

The agricultural potential of land is not only determined by the effective biophysical production capacity of the soil, but also by the physical possibilities for proper land preparation and efficient crop harvesting. Those characteristics, commonly identified as workability in the case of arable cropping systems, and as sensitivity to poaching and accessibility in the case of mechanized grassland farming (Section 4.3.2) depend on the soil-water status and on slope. Those particular land qualities can be quantitatively assessed by considering the following factors:

1)   climatic conditions, mainly related to the amount, frequency and intensity of rainfall during particular periods of the crop calendar; that is when i) fields have to be prepared for sowing and planting, ii) additional fertilizers have to be applied and/or iii) the mature crop has to be harvested;

2)   soil permeability, related to structure, texture and surface compaction;

3)   slope, which affects soil-water intake and runoff processes, and sets the critical limits above which machinery can no longer be used effectively.

Even if grass growth in terms of biomass production is not directly affected by any problem of access or poaching, this will not necessarily be true once advanced production methods are adopted. Hence, assessing the suitability for mechanized grassland farming also requires evaluation of the impact of slope and the extent of soil saturation (expressed as a combination of rainfall excess and texture).

Table 4.2. Main soil requirements for Winter wheat growth and production under high management level in Europe.

| Soil parameter | None | Degree of limitation | | Severe |
		Slight	Moderate	
Soil physical properties:				
Soil texture in root zone	Medium	Medium to heavy	Medium to coarse or very heavy	Coarse or heavy + massive
Soil structure development in root zone	Good	Medium	Medium to poor	Poor (massive or single grain)
Stoniness in root zone (%)	None	<15%	15-30%	>30%
Soil wetness	Good	Moderate	Imperfect	Poor
Flooding hazard	None	Slight	Moderate and short	Frequent or long
Soil chemical and fertility properties:				
Cation exchange capacity (meq/100 g soil)	>25	15-25	8-15	<8
Base saturation root zone (%)	80-100	50-80	<50	
N reserve [a]	...	...	...	...
P reserve	...	...	...	...
K reserve	...	...	...	...
Salinity root zone (mmhos/cm)	0.2	2-4	4-8	>8
Alkalinity in root zone (ESP)	0-15	15-25	>25	

Table 4.2. (continued)

Workability conditions:

Monthly P/PET during
field preparation
period in:

- clay textures	<0.6	0.6-0.8	0.8-1.0	>1.0
- sandy textures	<0.8	0.8-1.0	1.0-1.2	>1.2
Slope (%)	0-3	3-5	5-8	>8

[a]Figures to be defined in relation to specific varieties.

In the case of Winter wheat cultivation in western Europe, field preparation generally takes place in October-November, which is a relatively critical period for water excess on impermeable and heavy-clay soils. Under these conditions some areas are difficult to prepare properly for the cultivation of winter cereals and, therefore, have to be planted with Spring barley. Such conditions are observed, for example, in Ireland.

Another example refers to the harvest problems for sugar beet, a crop which is generally taken from the field as late as possible (October-November) when soils are almost saturated with water. As the harvesting involves heavy machinery, it may result in a serious deterioration of soil structure particularly in soils with low organic matter content, and ultimately in local waterlogged field conditions during a winter period with heavy rainfall.

4.4.   Land Assessment Based on the Physical Potential of the Land

4.4.1. Assessment procedure

Once the role and impact of the different biophysical parameters on crop growth and yield production are understood, this basic knowledge may be applied to an assessment that will quantify land use potential (for a given crop under given management conditions). Many attempts to produce "land use capability" assessments have been undertaken nationally in Europe (Bibby and Mackney, 1972; Mackney, 1974). For example, a provisional land use capability classification was produced by the Soil Surveys of England and Wales, and of Scotland. This was a modified form of the Land Capability Classification of Klingebiel and Montgomery (1961), resulting in an agricultural land classification (MAFF, 1970). Seven classes of land were identified from "land with minor limitations" (Class 1) to "land with extremely severe limitations that cannot be rectified" (Class 7). Limitations such as wetness, gradient,

susceptibility to erosion or climate were recognized, the last being subdivided into three groups based on average rainfall, average potential
transpiration and long-term average mean daily maximum temperature.
Land capability maps were produced (Corbett and Tatler, 1970; Green
and Fordham, 1973). More recently, very detailed assessments of soil
qualities that limit land use and crop production have been made in
connection with detailed soil surveys (Matthews, 1975). Factors assessed
include:

- susceptibility to poaching and compaction;
- clay (or clay loam) surface texture;
- seasonal surface wetness;
- high groundwater;
- susceptibility to drought;
- susceptibility to wind erosion;
- susceptibility to flooding;
- ground frosts;
- stoniness;
- shallow soils;
- indurated layer and humus-iron pan;
- steepness and landslipping;
- natural activity;
- blocking of drains.

Detailed matrices are produced of these factors against the soil series
that are mapped on a three-tier scale: limiting factor not serious; limiting
factor present; limiting factor potentially serious (Matthews, 1975). This
allows land use limitations of areas whose soils have been mapped to be
identified and corrective actions to be prescribed. Recently a revised
land classification has incorporated new criteria using climatic limitations
(MAFF, 1988). The new Agricultural Land Classification System has five
grades from "excellent quality agricultural land" (Grade 1) to "very poor
quality agricultural land" (Grade 5). The new Grade 5 incorporates,
largely, the previous Classes 5, 6 and 7.

The principles of such a land evaluation procedure to be described
here, however, are based on a matching exercise between specific growth
requirements and particular environmental conditions. The more
divergent the environmental conditions are from the ideal situation for
a particular crop the less suitable is the match and the lower the
potential productivity.

Because growth requirements are different for each crop, it may
happen that land is suitable for one crop and unsuitable for another.
It is therefore necessary to make an evaluation for different specific
crops or cropping systems and compare the different alternative
scenarios.

This procedure is based on the concepts developed by FAO in the

"Framework for Land Evaluation" (FAO, 1976) and on a series of follow-up studies, carried out in agroecological zones projects (FAO, 1976), or by other individuals or institutions (Sys, 1976; Verheye, 1980; Higgins and Kassam, 1981; FAO, 1983, Dumanski and Stewart, 1983). The initial FAO work was conceived mainly for the intertropical belt but has been adapted for temperate areas (Dumanski and Stewart, 1983, *op. cit.*) and has been accepted as a working methodology for land evaluation at EC-level (Verheye, 1984, 1986a and 1986b).

Using this concept, land assessment is considered a step-by-step approach, which includes the following stages:

a)   definition of the land utilization type for which the assessment will be made;

b)   definition of the specific crop growth and production management requirements in terms of climatic, soil and topographic parameters;

c)   evaluation of the environmental conditions dealing with the climate, soils and topography of the area concerned;

d)   matching crop and management needs with the environmental conditions and detection of the nature and degree of eventual constraints, hence permitting definition of preliminary suitability classes;

e)   field checking and validation of the preliminary classification;

f)   definition of final suitability classes, identification of growth limitations and a proposal for reclamation and improvement.

4.4.2. **Potential for mechanized grassland farming**

The major biophysical growth determinants for grass and the optimal production criteria for mechanized grassland farming were outlined in Section 4.3. The suitability assessment criteria may be summarized as follows:

1)   the mean air temperature should be in the range 5-25°C;

2)   the soil moisture regime should ensure sufficient plant-available water in the root zone for a minimum of 6 months (moderate constraint), 8 months (slight constraint) or 10 months (no constraint);

3)   fields should be easily accessible for mechanized field operations linked to maximum slopes of 8 per cent (no constraint), 15 per

cent (moderate constraint) or 25 per cent (severe constraint);

4)   the risk of poaching and reduced soil aeration should be limited in time, as influenced by a continuous soil moisture saturation within the time period defined by 1) and 2), and should not exceed two (slight limitation) or three months (moderate limitation).

A matching of growth and production requirements with the soil and climatic data base derived from the EC soil map at scale 1:1,000,000 (EC, 1986) and agroclimatic map of the EC has led to the definition of the following suitability classes:

Class 1:   land highly suitable for mechanized grassland farming, with no or only slight limitations.

Class 2:   land suitable for mechanized grassland farming, referring to areas with a maximum of two moderate limitations; three subclasses can be identified at this level depending on the nature of the constraints:

Class 2a: areas with climatic restrictions, mainly affecting the growing period of grass and, ultimately, the total biomass production;

Class 2b: areas with slope restrictions, mainly affecting the farm management operations;

Class 2c: areas with both climatic and slope constraints;

Class 3:   marginally suitable land for mechanized grassland farming with more than 2 moderate limitations and/or at least 1 severe limitation.

The geographical distribution of these different suitability classes in France is schematically represented in Figure 4.2. In interpreting Figures 4.2a and 4.2b it should be emphasized that, given the generalized nature of the basic formation, this map does not pretend to localize exactly the potential grassland farming areas in the country. It should nevertheless be considered as an example for an overall assessment at a European level, which identifies the broad grassland zones and which, if required, may be used as a working tool for additional and more detailed investigations in areas which have a proper potential.

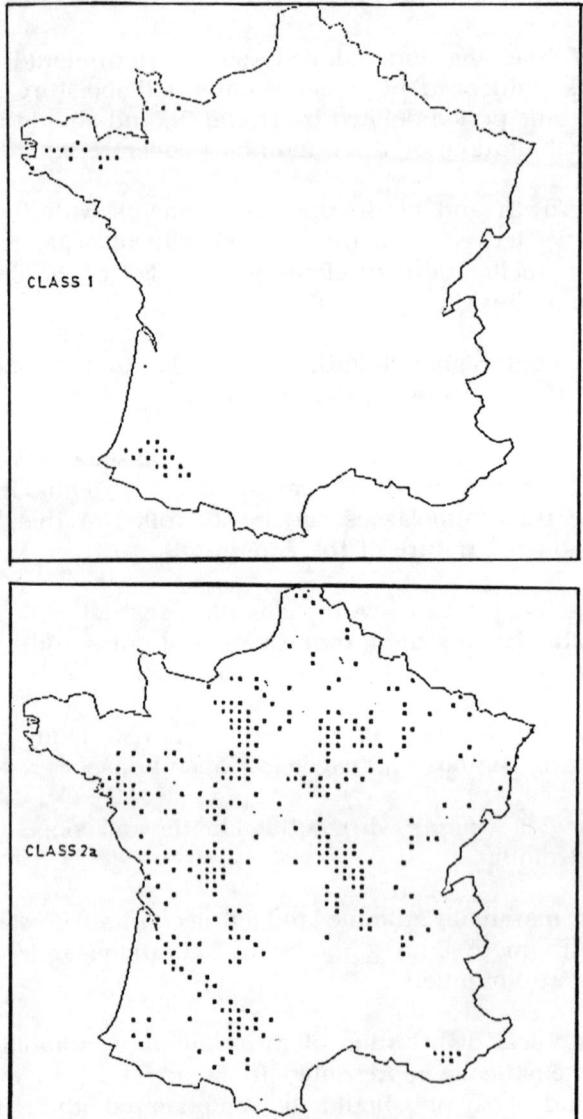

Figure 4.2a. Suitability evaluation for mechanized grassland farming in France.

      Class 1 :   Highly suitable.

      Class 2a: Suitable, but with climatic constraints.

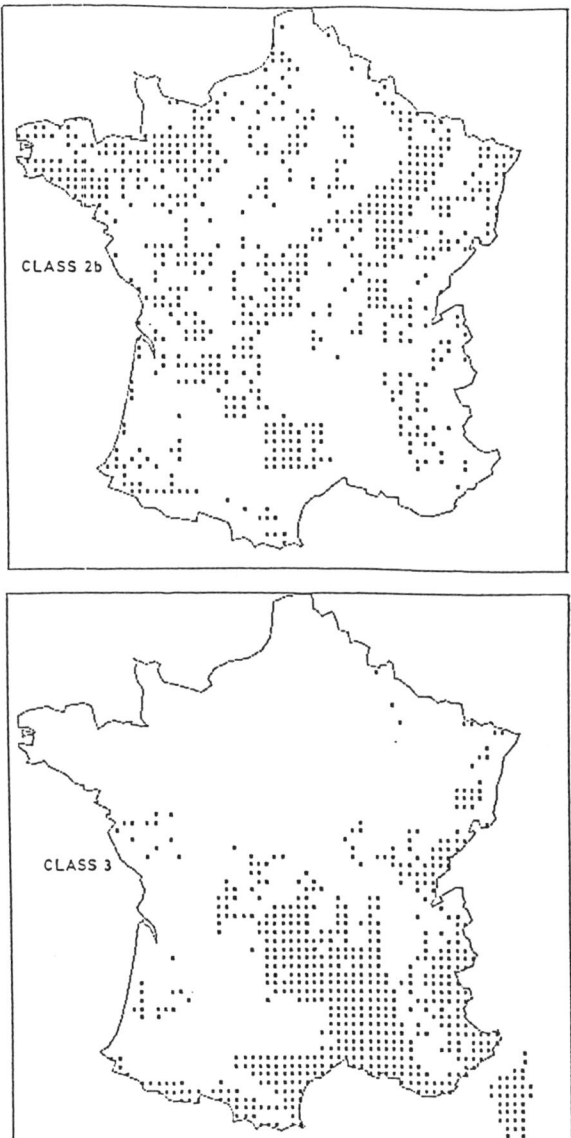

Figure 4.2b. Suitability evaluation for mechanized grassland farming in France.

    Class 2b: Suitable, but with slope constraints.

    Class 3 :   Marginally suitable, with major climatic and/or slope constraints.

4.5. Concluding Remarks

The classification of land in terms of its biophysical potential does not necessarily provide the most optimal land use pattern since such a classification is not a comprehensive study of all influencing elements. Indeed, other factors, mainly social and/or economic in nature, may have a much larger impact than biophysical parameters. In remote areas where, due to long distances and the absence of a proper market, there is no incentive for the farmer to grow commercial crops, the major effort will be concentrated on the production of basic food requirements, regardless of the best suitability of the land. Conversely, in the hinterland of densely populated centers, most of the agricultural activities will be oriented towards the direct market needs at the particular time of the year, even if this is not the best temporal organization in terms of yield expectations. However, the distance to market influence is lessened as transport networks are improved.

The approach outlined above provides a scientifically sound framework for future applied systems analyses. The procedure has two major advantages which may be of paramount importance for rural land use planning. Primarily, it permits the identification and direct quantification of the constraints which hamper optimal crop production and, as such, highlights the possibilities for reclamation or amelioration of land for future uses. Secondly, it allows comparison of the effects of different input scenarios on land use changes, giving the full range of alternatives which may be of relevance and/or taken into consideration by decision makers. Such a procedure should therefore become an excellent working tool in studies which focus on the possibe implications of anticipated changes of future environments in Europe.

REFERENCES

Bibby, J.S. and Mackney, D., 1972, *Land Use Capability Classification.* The Soil Survey, Harpenden.
Corbett, W.M. and Tatler, W., 1970, Soils in Norfolk: Sheet TM49 (Beccles North). *Soil Surv. Record.* Soil Survey, Harpenden.
Doorenbos, J. and Pruitt, W.O., 1977, Crop water requirements for irrigated agriculture. *FAO Irrigation and Drainage Bulletin,* vol. 24. FAO, Rome.
Dumanski, J. and Stewart, R.B., 1983, *Crop Production Potentials for Land Evaluation in Canada.* Land Resources Research Institute, Agriculture Canada, Ottawa.
EC, 1986, Soil map of the European Communities at scale 1:1,000,000, with explanatory text. Directorate General for Agricultural Research, Brussels.
FAO, 1976, A framework for land evaluation. *FAO Soils Bulletin,* vol. 32. FAO, Rome.

FAO, 1978, Report on the agroecological zones project. Volume 1: Methodology and results for Africa. *World Soil Resources Report*, vol. 48. FAO, Rome.

FAO, 1983, Guidelines for land evaluation for rainfed agriculture. *FAO Soils Bulletin*, vol. 52. FAO, Rome.

Frere, M. and Popov, G.F., 1979, Agrometeorological crop monitoring and forecasting. *FAO Plant Production and Protection Paper*, vol. 17. FAO, Rome.

Green, R.D. and Fordham, S.J., 1973, Soils in Kent I. *Soil Surv. Record*, **14**. Soil Survey, Harpenden.

Higgins, G.M. and Kassam, A.H., 1981, The FAO agroecological zone approach to determination of land potential. *Pedologie*, **31**, 167-168.

Klingebiel, A.A. and Montgomery, P.H., 1966, Land capability classification. *Agricultural Handbook No. 210*. Soil Conservation Service, USDA, Washington, D.C.

Lee, J., 1984, Suitability and productivity of land resources of EEC-10 for grassland use. I. Grassland suitability and productivity. EEC, Brussels and An Foras Taluntais, Wexford, Ireland.

Mackney, D. (ed.), 1974, *Soil Type and Land Capability*. Soil Survey, Harpenden.

MAAF, 1970, *Agricultural Land Classification*. Agricultural Land Service Report, **11**. Ministry of Agriculture, Fisheries and Food, London.

MAFF, 1988, *Agricultural Land Classification of England and Wales*. Ministry of Agriculture, Fisheries and Food, London.

Matthews, B., 1975, Soils in North Yorkshire I. *Soil Surv. Record*, **23**. Soil Survey Harpenden.

Neville-Rolfe, D. and Caspari, C., 1987, Potential for change in the use of the land in the European Community for non-food purposes up to the year 2000. FAST-Paper 178, Directorate General Scientific Research and Development, Brussels.

Sys, C., 1978, Evaluation of land limitations in the humid tropics. *Pedologie*, **28**, 307-335.

Verheye, W., 1980, Soils and soil suitabilities for the Rufiji Basin, Tanzania. Gent State University, Soils Department (mimeographed).

Verheye, W., 1984, Land evaluation methodology and interpretation of the EC-Soil Map of Europe. In J.C.F. Haans, G.G. Steur and G. Heide (eds.), *Progress in Land Evaluation*. A.A. Balkema Publishers, Rotterdam/Boston. 67-75.

Verheye, W., 1986a, Land evaluation: Chapter 6, Explanatory text EC Soil Map European Communities. Directorate General for Agricultural Research, EEC, Brussels. 76-79.

Verheye, W., 1986b, Principles of land appraisal and land use planning within the European Community. *Soil Use and Management*, **2**, 120-124.

Chapter 5

ATMOSPHERIC METHANE: ESTIMATES OF ITS PAST, PRESENT AND FUTURE AND ITS ROLE IN EFFECTING CHANGES IN ATMOSPHERIC CHEMISTRY

I. Aselmann

5.1. Introduction

Land use and industrialization in the past have led to local or regional environmental degradation; now our activities have reached a level at which we are facing serious global consequences. This is particularly true for the atmosphere, in which anthropogenically released gases have led to global changes in the chemical composition of the air and changes in the Earth's radiation budget, with likely consequences for the climate in the near future. Expected temperature rises due to increasing concentrations of greenhouse gases such as carbon dioxide (CO_2), methane (CH_4), nitrous oxide (N_2O), ozone (O_3) and chlorofluorocarbons (CFCs), have the potential to shift the climate regions within the coming decades (Bolin et al., 1986). This may be beneficial for some regions, but catastrophic for others. The regional climate shifts predicted are still subject to a high degree of uncertainty since the response of the climate system to perturbations is not fully understood. Model calculations agree to the extent that radiative forcing of the atmosphere due to increased concentrations of greenhouse gases globally leads to a warming of the Earth's surface, but differ considerably in their predictions on a regional level.

Considerable work has been devoted to the global carbon cycle, its perturbation by man and the effect of rising CO_2 levels on future climate (Bolin, 1981; Clark, 1982). However, within the last two decades insights into atmospheric chemistry and improved understanding of the climate system have shown that the combined effects of other trace gases are of similar magnitude. This Chapter focuses on methane, which has a two-fold effect. Each methane molecule is about 30 times more efficient in greenhouse forcing than CO_2. Rising methane levels influence the global distribution of ozone (O_3) and the hydroxyl radical (OH) in the atmosphere. Both species are key factors in the chemistry of the atmosphere, and ozone itself is a very efficient greenhouse gas. Our understanding of the global methane budget, however, still suffers

99

F. M. Brouwer et al. (eds.), Land Use Changes in Europe, 99–126.

from poor quantification of the various sources which puts constraints on present chemistry models and predictions of climate change.

Some likely consequences of increasing trace gas levels for the atmosphere are briefly described in Sections 5.2 to 5.4. This will be followed with a critical discussion of methane, with emphasis on its sources and sinks in the present and past, and a tentative projection into the future.

5.2. The Greenhouse Effect

Solar radiation penetrates the atmosphere and heats the Earth's surface. The energy is re-radiated in the infrared (IR) range of the spectrum. This energy is partly absorbed by the atmosphere and emitted back to the surface, primarily by water vapor (H_2O), carbon dioxide (CO_2) and clouds. H_2O and CO_2 have strong absorption bands in the spectral region of 3-8 µm and for wavelengths larger than about 13 µm (Figure 5.1).

The effect of IR absorption by these gases is reduced radiation back to space. The result felt at the Earth's surface is an increase in the temperature by some 34K referred to as the natural greenhouse warming. Without the natural greenhouse gases the global mean temperature at the Earth's surface would be around -18K. The energy

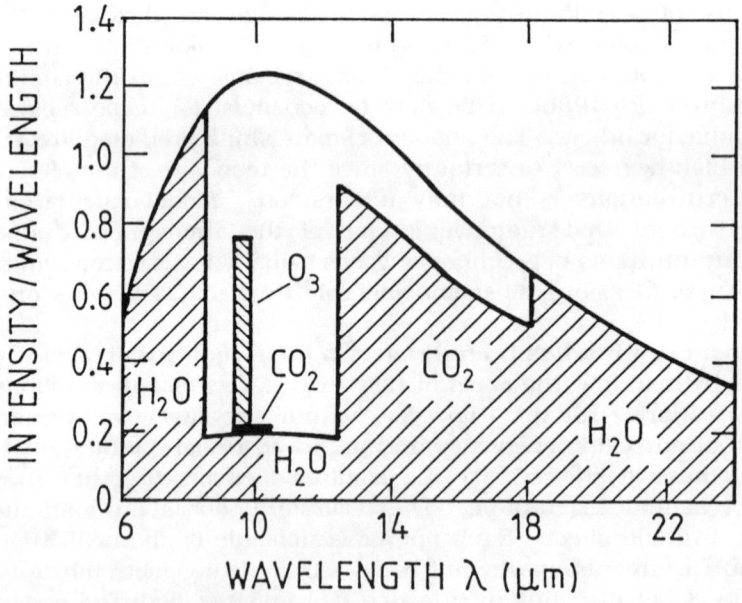

Figure 5.1. Infrared emission of the Earth's surface at T-288K with relative values of the overhead absorption by natural greenhouse gases (after Chamberlain *et al.*, 1982).

trap works because of the temperature decrease with increasing altitude in the troposphere. Greenhouse molecules at higher levels absorb IR radiation from the warmer surface but emit it at their lower ambient temperature with less energy. The net effect is a reduced radiation to space. About 90 per cent of the greenhouse warming is due to water vapor, CO_2 and clouds. The remaining part is contributed by O_3, CH_4 and N_2O (Ramanathan et al., 1987). There is a relatively transparent region in the IR spectrum between 7-13 μm commonly referred to as the atmospheric window, where about 70-90 per cent of the IR emission is lost to space (Figure 5.1). Trace gases with strong IR absorption in this region, such as CO_2, O_3, CH_4, N_2O, and in particular CFCs, can therefore effectively reduce the IR transmission to space. Rising concentrations of these gases can trap more energy in the atmosphere and lead, after a transient time, to a new equilibrium between surface radiation, incoming solar radiation and outgoing IR radiation at the boundary of the atmosphere.

Figures 5.2a, 5.2b, 5.2c and 5.2d display the observed increases of the important greenhouse gases which have ground sources. The data are based on recent measurements and analyses of air trapped in polar ice. Although the causes are not fully explained in every case, it is certain that the observed rises are primarily man-induced (Logan et al., 1981; Ramanathan et al., 1985; WMO, 1985). Rising levels of carbon dioxide are primarily due to fossil fuel burning. It has been suggested that the fate of the World forests and a possible reduction in the humus content of soils have contributed large quantities of biogenic CO_2 to the observed increase (Woodwell et al., 1978; Houghton et al., 1983). However, recent refinements of biogenic sources, studies on the isotopic composition of pre-industrial and recent carbon dioxide, as well as comparison of theoretical carbon uptake by oceans with historic CO_2 records derived from ice core samples imply no significant net biogenic contribution to the overall CO_2 increase (Pearman, 1988). Sources of methane are the main topic of this Chapter and will be discussed later. Sources of nitrous oxide are nitrification and denitrification processes on natural soils and fertilized land. It is also produced in combustion processes and is released from the oceans. The contribution of the various sources are only rough estimations, and the small increase of some 0.2 per cent per year may only be explained qualitatively in terms of an increasing use of nitrogenous fertilizers and by combustion of biomass and fossil fuels (WMO, 1985). Chlorinated fluorocarbons (CFCs) are entirely man-made. Production started in the late 1930s for purposes such as refrigerants, spray propellants, foaming of plastics and solvents. Due to their chemical inertness they have long life times (65 years for F11 and 110 years for F12). The increasing production of CFCs has therefore led to drastic accumulation in the atmosphere. The only sink for CFCs is photolysis in the stratosphere where they constitute a major threat to the ozone layer (Molina and Rowland, 1974).

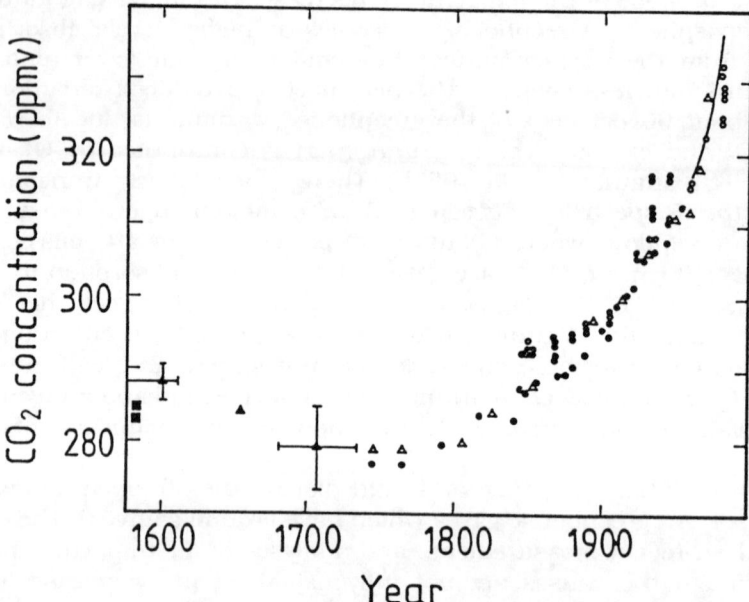

Figure 5.2a.   Changes of atmospheric CO_2 concentrations over the last 400 years deduced from Antarctic ice core samples. Data are compiled from various sources.   Horizontal bars represent the uncertainty in the age of the gas trapped in the ice; vertical bars are the standard deviations of concentrations.   The full line represents recent air measurements from the Antarctic. (Source: Pearman, 1988.) (ppmV = parts per million by volume (10^{-6}).)

5.3.   Chemical Interferences

Ozone plays a key role in the chemistry of the atmosphere and the radiation budget of the planet.   About 90 per cent of atmospheric O_3 occurs in the stratosphere with the highest concentrations between 15 and 35 km, commonly referred to as the ozone layer.   Stratospheric ozone is produced by photolysis of oxygen molecules (O_2) which produces oxygen atoms that recombine with O_2 molecules to form ozone (O_3). This reaction is confined to UV-radiation with wavelengths of less than 245 nm.   Ozone itself absorbs the UV-C $(\lambda < 290$ nm) and most of the UV-B radiation $(\lambda < 320$ nm) very effectively in its Hartley and Huggins bands and prevents harmful short wave radiation reaching the surface (Brasseur and Solomon, 1984).   Ozone destruction takes place through photolytic decay with less energetic radiation in catalytic cycles involving

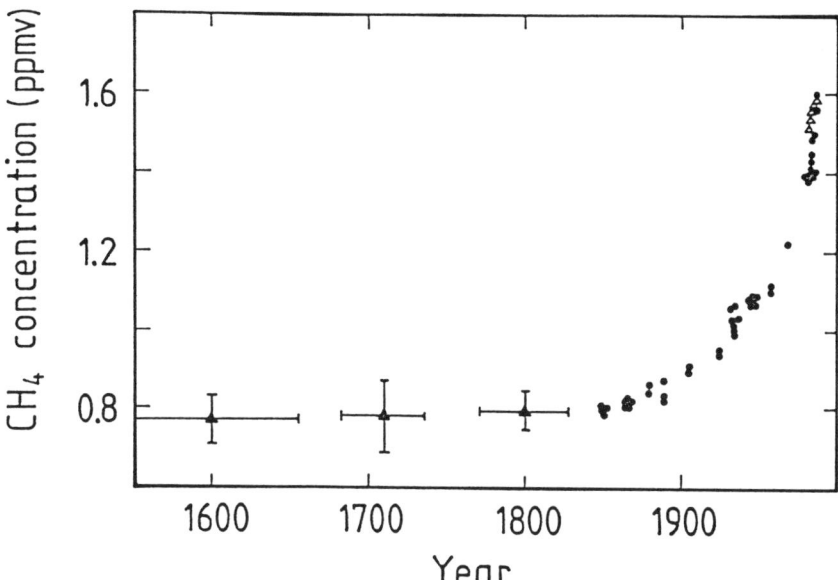

Figure 5.2b. As Figure 5.2a for methane. Triangles denote recent air measurements from Cape Grim, Tasmania. (Source: Pearman, 1988.)

other trace gases. These catalysts are decay products of tropospheric gases such as water vapor (H_2O), methane (CH_4), nitrous oxide (N_2O) and chlorinated hydrocarbons such as methylchloride and man-made chloro-fluorocarbons (CFCs). These are relatively stable in the troposphere and through mixing processes can reach the stratosphere. There they are destroyed by UV radiation (photolysis) and by reaction with excited oxygen atoms or hydroxyl radicals. Products of the destruction are highly reactive species, so-called radicals, that initiate the catalytic reaction cycles in which, for example, ozone is destroyed. Increasing concentrations of N_2O, CH_4 and CFCs therefore have the potential to lower stratospheric ozone concentrations. However, since the catalytic cycles interfere with each other and since the reaction rates are mostly temperature and pressure dependent, the effect of rising trace gas concentrations is not linear and may be quite different at different stratospheric levels. The net effect of NO and NO_2 (also written as NO_x, products of stratospheric N_2O decay) above 25 km is to lower ozone concentrations. Below this altitude NO_x has a protective effect on ozone by interfering with HO_2 and chlorine catalytic O_3 destruction. Likewise, methane has a protective effect on ozone in the middle stratosphere as it reacts with chlorine radicals produced from photolytic destruction of CFCs, thus reducing ozone destruction by chlorofluorocarbons. In the

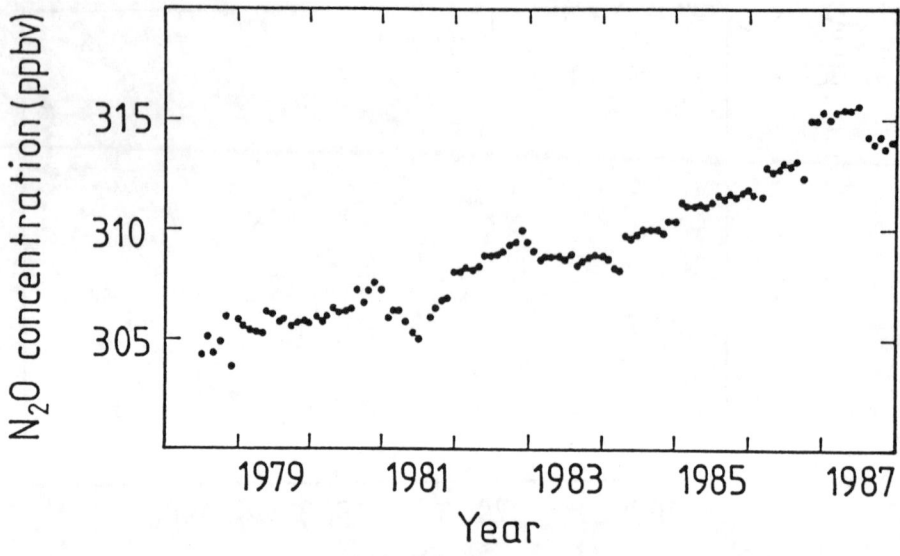

Figure 5.2c.   Air measurements of N₂O from Cape Grim, Tasmania. (Source: Pearman, 1988.) (ppbV = parts per billion by volume (10^{-9}).)

upper stratosphere, however, methane destruction leads to the formation of water vapor which enhances O_3 decay by odd hydrogen catalytic cycles (Crutzen and Schmailzl, 1983).

Part of the ozone is transported down to the troposphere by air mixing from the stratosphere.   Ozone production in the troposphere occurs by photochemical smog reactions with nitrogen oxides, methane and carbon monoxide as precursors.   Part of the tropospheric ozone is photolytically destroyed through UV radiation, which produces excited oxygen atoms (written as O(^{1}D) or O*).   These react in a second step with water vapor to form hydroxyl (OH) radicals.   The OH radical is very reactive, which explains its minute mixing ratio of 2×10^{-14} and its short life time of a few seconds.   It is, however, the major cleansing agent of the atmosphere, responsible for the removal of most anthropogenically released gases (Levy, 1971; McConnell et al., 1971).   Primary reaction partners for OH radicals are carbon monoxide and methane, which are relatively abundant trace gases in the troposphere.   Methane is oxidized via various reactions to carbon monoxide.   Oxidation of carbon monoxide may lead to a net gain or loss of ozone and hydroxyl radicals, depending on the presence of odd nitrogen (NO_x) (Logan et al., 1981).   Natural sources of NO are microbial activity in soils and lightning, but major emissions stem from combustion processes, which in most cases are

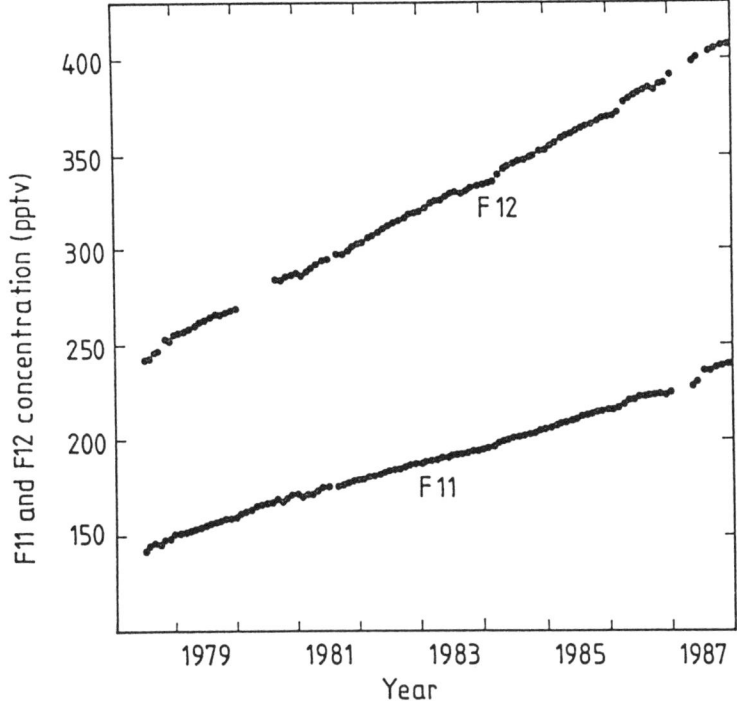

Figure 5.2d.   Increases of Trichlorofluoromethane (F11) and
Dichlorodifluoromethane (F12) as observed at Cape Grim, Tasmania, for
the period 1978-1987. (Source: Pearman, 1988.) (pptV = parts per
trillion by volume (10^{-12}).)

anthropogenic. In unpolluted NO_x poor environments such as remote
marine regions oxidation of one methane molecule may destroy on
average 1.7 ozone molecules and 3.5 OH radicals. At the same time in
NO_x polluted areas such as the industrial regions of the Northern
Hemisphere and in tropical regions, in which biomass burning takes
place, the oxidation of one molecule of CH_4 leads to a net gain of 3.7
molecules of O_3 and 0.5 molecules of OH (Crutzen, 1986 and 1988). It
is hypothesized that due to increasing levels of methane, the global
distribution of OH may decrease (Khalil and Rasmussen, 1987), which
implies a decrease in the cleansing power of the atmosphere. However,
there is no direct proof as the technical ability to measure OH in the
atmosphere is still problematic (Perner *et al.*, 1987). Tropospheric ozone,
on the other hand, has been observed to be rising at various stations in
the mid-northern latitudes (Logan, 1985), but the few measurements in
the Southern Hemisphere show steady concentrations (Pearman, 1988).
The global trend is still not clear since ozone shows a rather high spatial

and temporal variability in concentration, which results from its relatively short tropospheric life time of a few weeks.

5.4. Climatic Implications

Estimating the transient effect of rising trace gases on climate suffers from many feedback mechanisms within the climate system that are not fully understood. Climate sensitivity to rising levels of greenhouse gases depends to a large degree on the heat uptake by the World's oceans, on evapotranspiration from the Earth's surface and the subsequent changes in the water content of the atmosphere, on changes in the cloud cover and on changes in the albedo of land surfaces (Dickinson, 1986). For example, higher levels of tropospheric water vapor enhance the radiative forcing of the troposphere and may lead to different cloud cover. Clouds are a key parameter in the radiation budget, but their role in a warmer climate either as a negative or positive feedback has not yet been established. An increase in low clouds acts as negative feedback because of the increase in the planetary albedo; whereas, an increase in high clouds enhances the greenhouse effect. Most climate models assume an enhanced greenhouse effect of an increased high cloud cover due to rising CO_2 levels, but the actual response of clouds to increased radiative forcing of the atmosphere is still not known (Dickinson and Cicerone, 1986; Ramanathan *et al.*, 1987). Likewise, the role of the oceans in modulating the transient response of the climate to rising greenhouse forcing is not well established.

Calculations of the radiative forcing due to increased greenhouse gases are shown in Figure 5.3 (Ramanathan *et al.*, 1987). The decadal equilibrium change in the surface air temperature between 1850 and the 1980s is given. Equilibrium surface air temperature denotes the change (after a transient time to reach steady-state in the radiation budget) solely due to radiative forcing with no climatic feedback considered. The important result of this computation is the relative contribution of the various gases to the greenhouse effect. Between 1850 and 1960 CO_2 was the dominant cause of radiative forcing, whereas at present (the 1990s) the contribution from the other gases is of similar magnitude. Radiative forcing due to methane is 40 per cent that of CO_2 and similar to the combined contribution from F11, F12 and N_2O. The estimated contribution from tropospheric ozone, other CFCs and stratospheric water vapor is only tentatively indicated, since it includes uncertainties mainly due to the high spatial variability of tropospheric ozone, its precursors and lack of data for the other constituents.

Time dependent simulations of trace gas increments from 1860 to 2050 with a one-dimensional climate-photochemical model, coupled with a simple ocean model to account for heat transfer by sea water, predict a temperature increase of some 1.8K during this period (Brühl and Crutzen, 1988). This scenario assumes for the year 2050 a CO_2 mixing

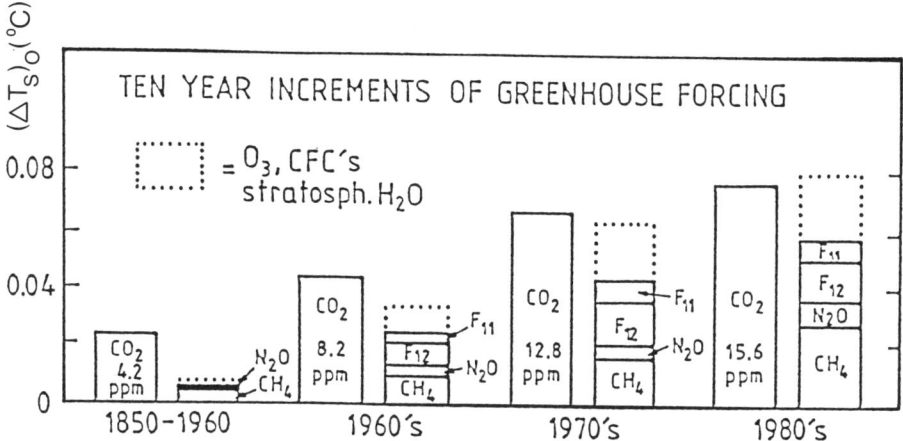

Figure 5.3.   Ten year increments of greenhouse forcing of the climate system. ($T_s)_o$ is the calculated surface temperature at equilibrium in the absence of any cimate feedback due to decadal rises in trace gas concentrations. The effect of ozone, CFCs other than F11 and F12, and stratospheric water is tentatively estimated to be equal to F11 and F12. (Source: Ramanathan *et al.*, 1987.)

ratio of 525 ppmV and 2.5 ppmV for methane, starting with 280 ppmV and 0.8 ppmV in 1860, respectively.   The mixing ratio for N_2O is projected using a formula that describes both the pre-industrial values and recent growth rates.   For anthropogenic induced releases of CO and NO_x emissions are linked to the expected population growth and CO_2 emissions.   Emissions for F11 and F12 were assumed constant at the rate of 1974, the year of maximum F11 and F12 production prior to 1985. This implies a small increase in the total CFC production until 2000 and a constant emission thereafter. Chemical reactions of all other important gases have been considered. The distribution of ozone is of particular importance as it may both ameliorate or enhance radiative forcing caused by its property to absorb solar radiation and to absorb and emit IR radiation. As an efficient greenhouse gas its increase in the troposphere enhances tropospheric warming. Predicted changes in the ozone and temperature profiles during the 1860-2050 period are shown in Figure 5.4. The model calculates quite different ozone trends for the two hemispheres. Northern Hemisphere O_3 concentrations should have doubled during the last 125 years, and are likely to increase by a further 50 per cent by 2050. No significant change is predicted for the Southern Hemisphere since the major sources of tropospheric ozone precursors (NO_x, CH_4 and CO) are in the Northern Hemisphere.

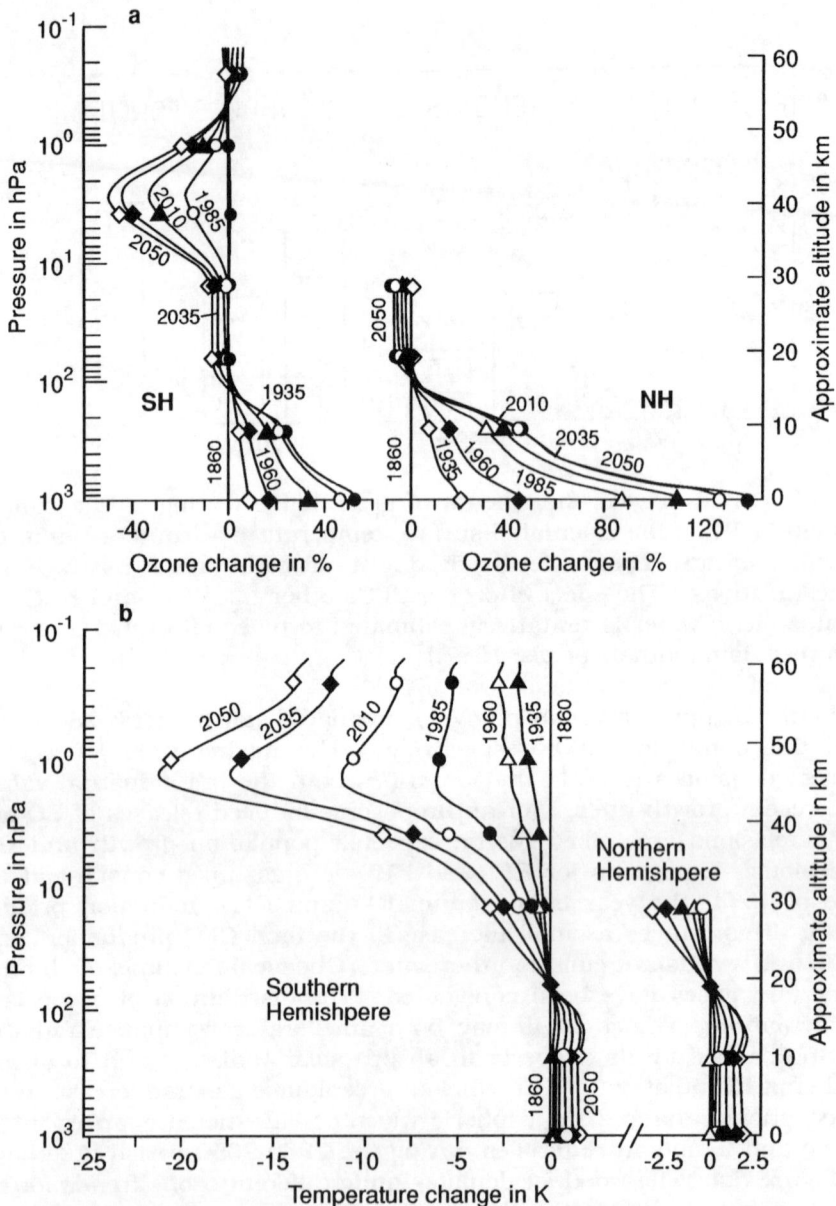

Figure 5.4. Computed ozone(a) and temperature(b) changes between 1860 and 2050 due to increases in CO_2, CH_4, N_2O, CO, NO_x and CFCs with a one-dimensional coupled climate chemistry model. (Source: Brühl and Crutzen, 1988.)

Hemispheric differences are insignificant in the stratosphere. Maximum ozone depletion of 50 per cent by 2050 is calculated for altitudes between 35-45 km. This would be reduced to some 30 per cent if the Montreal Regulations on the reduction of CFC production are effective and if no increase in methylchloroform (CH_3CCl_3, a widely used non-toxic degreasing solvent) takes place. Stratospheric temperature profiles will strongly be effected primarily by ozone depletion and by rising levels of CO_2, which may lead to stratospheric cooling of almost 25K during the period 1860-2050 in the middle and upper stratosphere. The calculated temperature change at surface until 1985 through increases in greenhouse forcing is about 0.6K. This is in the range of 0.4-0.7K predicted by Ramanathan *et al.* (1987) and agrees qualitatively with the observed temperature trend between 1860 and the present (Jones *et al.*, 1986; Hansen and Lebedeff, 1987). Although the model predicted a clear temperature trend over the last 100 years, it is still within the natural noise of climatic variability. Enhanced radiative forcing will most probably lead to substantial warming in the foreseeable future. All models predict that the global mean surface temperature will rise by at least 1K within the coming decades (Tucker, 1988). The actual rise will depend on future rates of increase of trace gases and the actual response of the climate system. Methane will increase in importance if the present rapid increase in emissions continues in the coming decades. Consequently, understanding the global methane cycle is important both for describing changes in the chemistry of the atmosphere and climate perturbation.

5.5. The Global Methane Budget

The present concentration of methane in the atmosphere is equal to about 1.7 ppmV (parts per million by volume) which is equivalent to an atmospheric burden of some 4,700 Tg (Teragram = 10^{12}g) (WMO, 1985). Investigations of air trapped in polar ice show that methane has been slowly increasing from about 0.75 ppmV some 300 years ago with a current growth rate of more than 1 per cent (50-60 Tg) annually (Figure 5.2b). The atmospheric concentration of methane is determined by its sinks and sources which are shown in Table 5.1. The presently known sinks of CH_4 are the oxidation by OH radicals in the troposphere and the stratosphere, and uptake by dry soils. Estimates of the OH oxidation are based on photochemical models and may be certain within 30 per cent (WMO, 1985). Uptake by dry soils has been observed at various sites and measured loss rates have been extrapolated to a global scale (Harriss *et al.*, 1982; Keller *et al.*, 1983; Seiler, 1984). These sinks, plus the yearly increase, account for 500 ± 150 Tg per year which must be balanced by sources. The methane sources are anaerobic decay of organic matter in natural wetlands, rice paddies and landfills, enteric fermentation in animals, biomass burning and fossil carbon sources such

Table 5.1. The global methane budget.

Present Concentration:
ca. 1.7 ppmV equal to some
4.700×10^{12} g
(WMO, 1985)

	Tg/year	
Sinks: Reaction with OH	315 - 425 - 550	Crutzen and Gidel, 1983 Crutzen and Schmailzl, 1983 WMO, 1985
Uptake by aerobic soils	5 - 20 - 30	Seiler et al., 1984 Keller et al., 1983 Harriss et al., 1982 WMO, 1985
Accumulation (1.1% year^{-1})	50 - 55 - 60	Rasmussen and Khalil, 1984
Total Sinks:	370 - 500 - 640	
Sources: Natural:		
Swamps, marshes, bogs	30 - 170	Aselmann and Crutzen, 1989
Wild herbivores	2 - 6	Crutzen et al., 1986
Other fauna	≤30	Crutzen et al., 1986
Anthropogenic:		
Rice paddies	60 - 140	Aselmann and Crutzen, 1989
Domestic animals	70 - 80	Crutzen et al., 1986
Landfills	30 - 70	Bingemer and Crutzen, 1987
Biomass burning	30 - 50 - 110	Seiler, 1984 Crutzen et al., 1979 Crutzen et al., 1985
Total non-fossil sources:	222 - 394 - 606	
Fossil sources:	100 - 190 - 290	(assumed as 32% of the over- all source)
Total Sources:	322 - 584 - 896	

as pipeline leaks of natural gas and coal mining operations. Although these sources have been identified as major CH_4 emitters, their individual source strengths are not well established and the overall range imperfectly matches the range given for the sinks (Table 5.1). These sources will be discussed in more detail in the following Sections.

5.6. Biogenic Sources

Biogenic methane is the end product of biological decay under anaerobic conditions accomplished through a diverse population of micro-organisms. Complex organic substances are broken down or transformed to simpler structures, of which hydrogen, carbon dioxide, acetate, formate, methanol and some methylamines form the substrates for methanogenesis (Wolin and Miller, 1987). However, only acetate, formate and hydrogen together with carbon dioxide are globally important in the major methanogenic environments; the intestinal tract of animals, the forestomach of ruminants, the waterlogged soils of wetlands such as swamps, marshes, bogs and landfills.

Methanogenesis in ruminants, such as cattle, sheep, buffalo, goats and camels, is an integral part of their food digestion and has been studied intensively for its economic significance. Diverse micro-organic populations in the forestomach (rumen) break down the cellulosic compounds of the feed, which would otherwise not be metabolizable for the animal. Methane is a byproduct of digestion and cannot be used by the animal. Methane yields, taken as the fraction of total energy intake (energy in feed) that is lost as CH_4, typically range between 4 per cent and 10 per cent for ruminants, but are less for non-ruminating animals like horses (2-3 per cent) or pigs (0.4-2 per cent) (Crutzen et al., 1986). Methane yields not only depend on the type of animal but also on the quantity and quality of the feed, and to some extent on the individual performance of animals. Under consideration of different management practices in various regions of the World, which take into account the type of animal, quality and quantity of feed intake, Crutzen et al. (1986) have calculated mean emission rates for different livestock categories and used these for global extrapolation. Table 5.2 summarizes the computation for the major livestock species separately for the developed and developing parts of the World. Cattle and buffalo, due to their large population and high individual production rates of 35-55 kg and 50 kg CH_4 per head per year (respectively), are the most important source accounting for more than 80 per cent of the total emission of 74 Tg from livestock.

A similar approach for the global wildlife population is less reliable, since statistics are mostly non-existent and since only a few studies report on the metabolism of wild animals. However, there are estimates of the global population of wild ranging ruminants (McDowell, 1976) and census data from reservations such as the Serengeti, which have been

Table 5.2. Methane production by domestic livestock.

Animal type and regions	Population (x 10^6)	CH$_4$ production per individual (kg year^{-1})	CH$_4$ production by total population (Tg year^{-1})	Production total (Tg year^{-1})
Cattle:				
Developed countries, Brazil and Argentina	572.6	55	31.5	
Developing countries	652.8	35	22.8	
Total	1,225.4			54.3
Buffalo:	124.1	50		6.2
Sheep:				
Developed countries	399.7	8	3.2	
Developing countries and Australia	737.6	5	3.7	
Total	1,137.3			6.9
Goats:	476.1	5	2.4	2.4
Camels:	17.0	58	1.0	1.0
Pigs:				
Developed countries	328.8	1.5	0.5	
Developing countries	444.8	1.0	0.4	
Total	773.6			0.9
Horses:	64.2	18	1.2	1.2
Mules, Asses:	53.9	10	0.5	0.5
Humans:	4,669.7	0.05	0.2	0.3
Total:				73.7

Source: Crutzen et al., 1986.

used to calculate the methane contribution from wild herbivores such as deer, elk, wildebeest, elephants, hares and rabbits (Crutzen *et al.*, 1986). Their contribution is rather small accounting for some 2-6 Tg annually.

A vivid controversy exists about the role of termites as a source of methane. Highly contradicting estimates of 2-5 Tg (Seiler *et al.*, 1984), 15 Tg (Fraser *et al.*, 1986), 50 Tg (Khalil and Rasmussen, 1983) and 150 Tg (Zimmerman *et al.*, 1982) were drawn from measured flux rates from various genera of termites and subsequent extrapolation to the global scale. As the latter three publications were based on flux rates measured in laboratories under conditions which are criticized for being unrealistic in respect to the proper ecology of social insects, the high emissions of 50 Tg and 150 Tg from termites are likely to be overestimates. Further research on methane production from invertebrates is warranted.

Previous estimates of the global methane emission from natural wetlands relied on a relatively small number of flux rates measured in a few wetland types and were further handicapped by rather conflicting values regarding the global extent and distribution of swamps, marshes and bogs. Commonly, the extent of wetland areas was taken to be between $2 - 3.4 \times 10^6$ km^2 (Ehhalt and Schmidt, 1978; Seiler, 1984) which, according to recent studies, is an underestimation. Two new, independently derived datasets by Matthews and Fung (1987) and Aselmann and Crutzen (1989) show that the actual extent of natural wetlands is between $5 - 6 \times 10^6$ km^2. Matthews and Fung (1987) derived their data by comparing three sets of maps on soils, vegetation and inundation, and distinguished between five broad categories; forested and non-forested bogs and swamps, and alluvial formations. Aselmann and Crutzen (1989) used monographs of the major wetland regions together with physical maps to derive a global wetland dataset which distinguished between bogs, fens, swamps, marshes, floodplains and shallow lakes. It also includes information on the seasonal and permanent status of these wetlands, which is of particular significance for methane studies since anaerobic conditions in soils only prevail under waterlogged conditions. Their dataset basically confirms the high global value derived by Matthews and Fung, but partial disagreement exists for certain regions, especially in the tropics such as the Amazon lowlands, for which available information is weak.

More complete information is available for rice paddies through the FAO Production yearbook series and other agricultural statistics publications. The present global area of wet rice harvested each year is about 1.3×10^6 km^2. This includes double and in certain cases even triple cropping of irrigated fields. However, due to limited water supply, only one rice crop is grown per year in most cases, this commonly takes 4-5 months to mature defining the methane production period (Aselmann and Crutzen, 1989). Almost 90 per cent of the global paddy area is cultivated in Asia.

Considerable effort in recent years has been devoted to flux measurements of methane in wetlands in order to extend the data base for source strength estimation. The presently available data for natural wetlands cover major wetland types from the tropics to the Arctic and long-term measurements in major rice regions are currently being carried out (W. Seiler, personal communication). They have revealed some of the ecology of methanogenesis in these environments and several factors have been identified that influence methane fluxes. Unfortunately, *in situ*, these combine in unpredictable ways and observed fluxes typically range over several orders of magnitude, even within the same sample site. Consequently, interpretation and extrapolation of the data can still only be very tentative as long as ecophysiological models are lacking (Aselmann, 1989). Nevertheless, averaged mean emission rates for various wetland categories do indicate some differences, but ranges are very large and overlap greatly between the categories (Table 5.3). Rice fields and marshes show highest emission rates while peat bogs show the lowest. The ensuing global emission is 75 Tg CH_4 per year from natural wetlands and 90 Tg CH_4 per year from rice paddies. Figure 5.5 displays the latitudinal distribution of these emissions. The major portion is emitted between April and September when, in the Northern Hemisphere, large boreal and arctic wetland areas thaw and when the monsoon in tropical Asia allows rice cultivation.

The importance of decaying solid wastes from landfills was first recognized by Sheppard *et al.* (1982), who tentatively estimated an annual upper limit of 170 Tg from this source by extrapolating from US data. A more detailed study by Bingemer and Crutzen (1987) shows that the global methane production from solid wastes is more likely to be around 30-70 Tg annually. The potential methane emission from landfills is primarily a function of the composition of solid wastes, or more precisely, the amount of biodegradable carbon in various compounds. High waste generation of paper and paperboard are typical for industrialized countries while vegetable refuse dominates the wastes in developing economies (Bingemer and Crutzen, 1987). On the basis of major socio-economic regions that have distinct per capita waste production and composition, these authors estimate that 85 Tg of biodegradable carbon from municipal solid wastes are dumped in landfills annually. This does not include wastes from industry similar in nature to municipal refuse. A tentative extrapolation from West German data yields another 25 Tg of biodegradable carbon from industry. Altogether 113 Tg (30 per cent) of organic carbon in landfills may be biodegradable. With an average methane production of 0.5 kg per kilo of biodegradable carbon and considering the variability inherent in the conversion of organic carbon to methane in landfills and in the diffusion of CH_4 to the atmosphere, the contribution from landfills is estimated by Bingemer and Crutzen (1987) to be between 30-70 Tg per year.

Table 5.3.   Current global methane emissions from wetlands.   (Values in parenthesis give the possible ranges.)

Wetland categories	Emission rate (mg CH$_4$ m^{-2} day^{-1})	Area (10^{12} m^2)	Emission (Tg year^{-1})
Bogs	13 (1-40)	2.05	5 (0.4-16)
Fens	72 (20-280)	1.65	20 (4-77)
Swamps	105 (70-150)	1.14	20 (13-28)
Marshes	238 (140-330)	0.29	18 (11-25)
Floodplains	100 (50-200)	0.82	10 (5-19)
Lakes	47 (20-80)	0.12	2 (1-3)
Natural Wetlands		6.07	75 (30-170)
Rice Fields	300-1,000	1.31	92 (60-140)
Total		7.38	90 - 310

Source: Aselmann and Crutzen (1989).

5.7.   **Methane of Non-biological Origin**

Crutzen *et al.* (1979) have shown that burning of biomass globally can release substantial amounts of atmospheric trace gases, including methane. Based on various studies, mostly from fires under controlled conditions in the USA, they derived methane yields of 1-2.2 per cent, which is the ratio of CH$_4$ to CO$_2$. More recent studies of plumes from

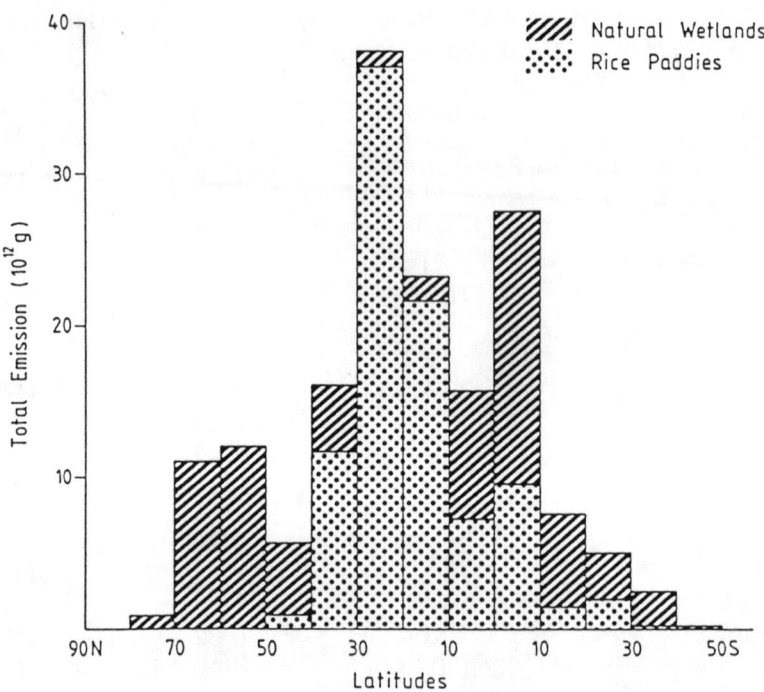

Figure 5.5. Current emissions from natural wetlands and rice paddies along 10° latitude belts, based on a dataset on global wetland distribution (Aselmann and Crutzen, 1989) and the mean emission rates given in Table 5.3.

burning savannas and forest fires in Brazil during the dry season give somewhat lower values. Mean values of 0.8-1.6 per cent have been found (Crutzen et al., 1986). However, in order to estimate the global methane flux from this source, the amount of biomass burnt annually around the globe has to be known. Based on the extrapolation from demographic and land use data, Seiler and Crutzen (1980) estimated that, due to slash and burn practices in shifting cultivation, burning of savannas, wild fires and combustion of agricultural wastes and fuel wood, 4.8-8.8 x 10^{15} grams of dry matter may be burned per year. This gives, together with the above methane yields, a range of 30-110 Tg CH_4 from burning biomass (Crutzen et al., 1979; Seiler, 1984; Crutzen et al., 1985). This range is tentative mainly because of the large uncertainties in the estimation of burning biomass. However, a new study presently under way seems to confirm this range with a likely mean around 50 Tg (Hao, personal communication).

Methane is further released in substantial amounts from fossil fuel exploitation and transport. Natural gas, the major constituent of which is methane, is a source of atmospheric CH_4 since losses through leaks in the distribution systems are common. It is estimated that in Europe 2-4 per cent of natural gas is lost during distribution. If this range is extrapolated to the global scale, some 20-30 Tg are released to the atmosphere (Seiler, 1984; Crutzen, 1986). During coal exploitation gas is liberated which consists primarily of methane. According to Koyama (1963) 18-19 m^3 of CH_4 are liberated to the atmosphere per ton of coal produced; this estimate has been used to derive a present methane release rate of 30 Tg (Seiler, 1984) and 35 Tg (Crutzen, 1986). However, mines vary considerably in methane emissions; in the USA emissions range from insignificant to a mean rate of 76 m^3 for the ten mines with highest emissions, with a national mean rate of only 4.5 m^3 in 1980 (Grau III and LaScola, 1980). This is much less than the emission rate given by Koyama (1963). From the presently known data we may thus only derive an upper annual emission of 50-70 Tg from coal mining and natural gas leaks.

However, this estimate of the fossil methane source is in conflict with radiocarbon (^{14}C) data of atmospheric methane, which suggest a larger contribution from fossil sources. Fossil methane contains no ^{14}C in contrast to recent methane, so that fossil sources dilute the naturally occurring isotope composition of atmospheric methane. After correction for fractionation processes of biogenetic methane and for atmospheric contamination from nuclear bomb tests and nuclear power plants, Lowe *et al.* (1988) estimate that 32 per cent (with a possible range between 22 per cent and 41 per cent) of the yearly methane input to the atmosphere is of fossil origin. The data in Table 5.1 yield, at best, a fraction of 24 per cent when the minimum estimates of non-fossil sources are combined with the upper value of 70 Tg from fossil sources. In this case the total source of 300 Tg is considerably smaller than the lower limit of the methane sinks in Table 5.1.

It is hypothezied that decay of ancient organic matter in tundra regions may lead to the liberation of ^{14}C free methane, which would add to the total fossil source (Pearman and Fraser, 1988). It is further suggested that methane hydrates at high latitude regions may contribute substantial amounts to the fossil fuel source (Ehhalt, 1988). If we take the non-fossil sources in Table 5.1 to be equivalent to 68 per cent of a total source (100 per cent minus 32 per cent), fossil sources should account for 100-190-290 Tg (Table 5.1). This adds up to a total range of 320-900 Tg as the yearly global methane input to the atmosphere. As the upper value is much higher than the estimated upper value for methane destruction of 640 Tg, a contribution of fossil sources in the range of 100-200 Tg would be in better agreement with the sinks. Refinements of present estimates of methane from coal mining and gas leakage as well as studies of the possible fossil source from tundra and methane hydrates are certainly warranted. It is also necessary to narrow down the large range of possible radiocarbon fractions to fill the present gaps in our knowledge of the methane budget.

Although considerable uncertainties exist, the data in Table 5.1 suggest that present day emissions are mainly caused by man. This is in agreement with the observed drastic increase of atmospheric methane over the past 300 years. Methane emissions must have been primarily of natural origin in the 18th and 19th centuries when the global human population was much smaller.

5.8. Pre-industrial Methane Emissions

Of course, only general remarks on past emissions can be made and the values presented in Table 5.4 have to be taken as rough estimates.

Table 5.4. Methane sources in 1825. (In Tg year^{-1}.)

Natural:	
Natural wetlands	30 - 75 - 170
Wild herbivores	
and small fauna	$\leq$36
Anthropogenic:	
Rice paddies	17 - 26 - 40
Domestic animals	14
Biomass burning	5 - 11 - 35
Total sources:	87 -147 - 295

Fuel exploitation was insignificant in pre-industrial times and methane emissions from this source are assumed to be zero for the 1820s. Landfill is also a relatively recent source, starting with the rapid growth of urban population in industrial nations and particularly with rising wealth and consumption after World War II. It is reasonable to assume that there were negligible wastes in pre-industrial societies, as in the case of present day rural populations in the developing world, and that dumping played no significant role globally. We may therefore ignore wastes as a source of methane for pre-industrial times. The role of swamps at the beginning of the last century is highly uncertain mainly due to the following points. Conversion of wetlands has taken place during the last 150 years. This conversion was not confined to Europe (documented by Gore, 1983) but also occurred in North America, Australia and New Zealand, commonly for the purpose of agriculture and peat exploitation. In Asia part of the expansion of rice paddies was at the expense of natural wetlands, particularly in valley bottoms and along the large river deltas. The overall reduction in mire area during the past 150 years is not known, not only because precise data on human conversion are lacking, but also because climate may possibly have changed the global distribution of wetlands. The period since the Industrial Revolution has shown an increase in Northern Hemisphere mean temperature of some 0.6°C (Lamb, 1984; Jones *et al.*, 1986) with possible impacts on wetlands, including extension and emission characteristics. Tentatively, we may take the same range of emissions from wetlands for the 1820s as calculated for present day conditions by assuming that anthropogenic reduction has been compensated by natural expansion and/or increase in emissions mainly in circumpolar regions of tundra and taiga. For the contribution of wild herbivores and other fauna, the upper limits as calculated for present day conditions have been assumed. This takes account of a somewhat larger wildlife population in the last century (Crutzen *et al.*, 1986). In the case of anthropogenic sources methane emissions have been scaled down according to a world population of 1×10^9 people in 1825. The area of rice harvested in 1890 was 0.61×10^6 km^2 (Darmstadter *et al.*, 1987) and the population at that time accounted for 1.6×10^9 (Gates, 1981). This yields an area of 0.38×10^6 km^2 for harvested rice when scaled down proportionally to the population in 1825. Assuming the same emission characteristics in 1825 as in the 1980s yields a range of 17-40 Tg for the pre-industrial rice paddy source. Emissions from livestock can likewise be inferred from the 1890 emission estimated by Crutzen *et al.* (1986), which calculates them as 14 Tg in 1825 (Figure 5.6). The pre-industrial emission from burning biomass is based on the data in Seiler and Crutzen (1980). Their estimates of burning biomass due to man's activities in the mid-1970s with a world population of 4×10^9 have been divided by four to calculate the possible value for 1825. Yet, in the case of wild fires, the value for 1975 has been adopted unaltered. Based on these data one may infer that the total methane emission from burning biomass accounted for some 5-35 Tg in 1825.

According to the above calculation, methane emission has quadrupled over the last 150 years or so. What can be expected for the future?

5.9. **Future Trends**

Taking the period between 1980 to 1985 as a reference for future projections, there is no sign of a changing trend in the major anthropogenic methane sources. The world population has increased by 8.8 per cent, cattle, sheep and buffalo by some 3 per cent, 3 per cent and 5.7 per cent respectively; and rice paddies have increased by 6.4 per cent in this time period (FAO, 1986). Largest increases are reported for the developing nations while the values for mainly OECD member countries show decreases or stagnation. For example, the total

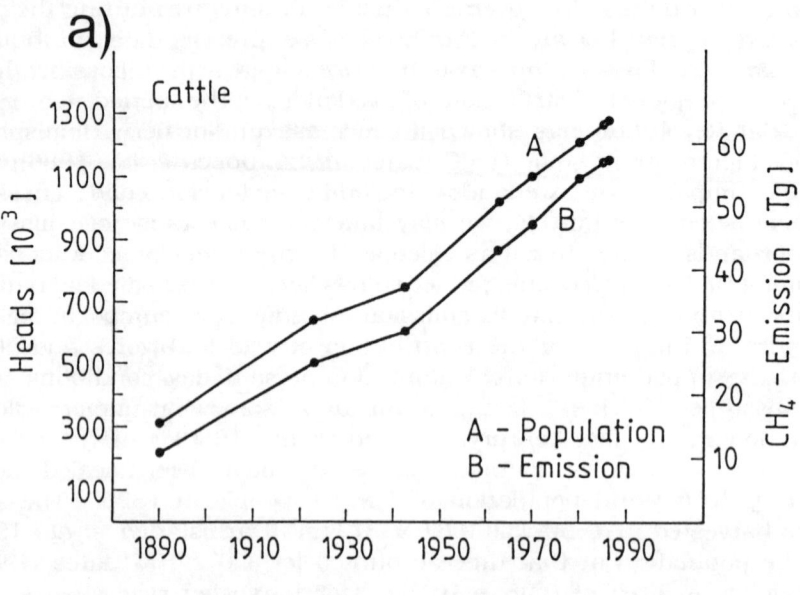

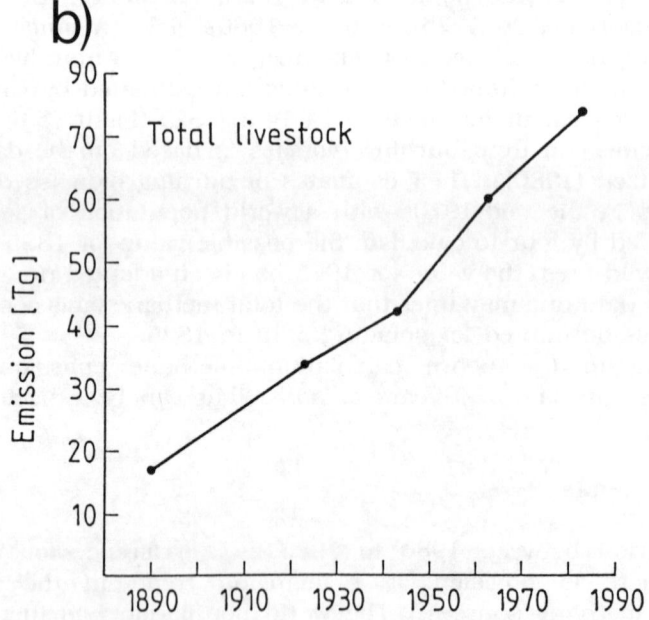

Figure 5.6. a) Trend of global cattle population and associated methane emissions between 1890 and 1985. During this period emissions have increased by some 600%. b) Increases in CH_4 emissions from all domestic livestock. (Based on data in Crutzen *et al.*, 1986.)

number of cattle in South America has grown by 7.4 per cent, but has decreased in Canada, the USA and Australia, and stagnated in Europe. Although rice paddies are increasing worldwide with the potential of larger methane emissions, these could partly be compensated for if present efforts in rice breeding for less water requirement are successful, thereby reducing the area of wet rice cultivation (Professor J. Bardon, personal communication). It is certain that a growing population concentrated in urban areas will generate more wastes. Projections for the year 2000 show that municipal solid wastes will have doubled in developing countries. Landfills may therefore become one of the most important methane sources in the next century (Bingemer and Crutzen, 1987). Also the growing population in rural areas in the developing world will intensify the use of natural resources. Consequently, biomass burning will increase and remain an important source of methane in the future. Projections for the energy demand and actual use are numerous. Whatever the scenario and projections of changing technologies, it is certain that fossil fuel consumption will be a growing source of methane in the foreseeable future.

In brief, anthropogenically induced methane emissions will continue to rise in the next decades. This is not as clear with natural sources. Natural wetlands may be affected by changing climate and land reclamation by man. Large areas of tropical wetlands will vanish or be modified if plans for large hydrological power schemes, for example, in the Amazon basin, are realized (Junk, 1989). Furthermore, increasing cattle farming in flooded savannas and conversion of other wetland types for land reclamation all over South America is a major threat to wetlands in this region. The same threat exists in other tropical regions (Howard-Williams and Thompson, 1985). Thus, tropical wetlands will certainly decrease in extent as have temperate wetlands. However, the effect on methane emissions is not clear since the impact on wetlands due to changing patterns of precipitation and river discharge in the course of a predicted change in climate is not known. Furthermore, as assumed for the past, rising temperatures may enhance decay rates and the release of methane, particularly in the Arctic and boreal zone. This may compensate for part of the wetland conversion.

5.10. **Concluding Remarks**

Summarizing the foregoing discussion it can be concluded that the major methane sources have been identified, but the overall range of current emissions of 320-900 Tg shown in Table 5.1 is larger than the range given for sinks. The quantification involves many uncertainties so that the individual source strengths remain tentative in most cases. This is particularly true for the figures for pre-industrial emissions. Yet, the calculated fourfold increase over the past 150 years due to man's activities is in agreement with the observed increase in atmospheric methane. The present geographic distribution of all sources is about equally divided between temperate/boreal and tropical/subtropical regions. Future changes in the source distribution will depend on the response of emissions from northern wetlands and tundra to global warming as well as on anthropogenic releases in the developing world, namely from waste production, biomass burning and fuel consumption. The environmental implications were discussed in the first Section. Rising levels of methane will enhance greenhouse forcing in the future and increase the importance of methane relative to CO_2. Through its chemical impact methane is a major candidate for triggering changes in the chemistry of the atmosphere, namely a reduction of the cleansing ability of the troposphere through a reduction of OH radicals, and, at least on regional levels, an increase in ozone levels.

ACKNOWLEDGEMENTS

I am indebted to Dr. Ch. Brühl for careful review of the first part of this Chapter and for valuable suggestions. I also thank Karen Valentin very much for time-consuming and repeated proof-reading of the manuscript. This work was partly supported by the Ministry for Research and Technology of the Federal Republic of Germany through grants BMFT/KF 20128.

REFERENCES

Aselmann, I., 1989, Global scale extrapolation: A critical assessment. In M.O. Andreae and D.S. Schimel (eds.) *Exchange of Trace Gases Between Terrestrial Ecosystems and the Atmosphere*. Dahlem Conference, Berlin, February 19-24. Wiley & Sons, Chichester. 119-133.

Aselmann, I. and Crutzen, P.J., 1989, Freshwater wetlands: global distribution of natural wetlands, rice paddies, their net primary productivity, seasonality and possible methane emissions. *J. Atmos. Chem.* **8**(4): 309-358.

Bingemer, H.G. and Crutzen, P.J., 1987, The production of methane from solid wastes. *J. Geophys. Res.*, **92**(D2), 2181-2187.

Blake, D.R. and Rowland, F.S., 1988, Continuing worldwide increase in tropospheric methane, 1978-1987. *Science*, **239**, 1129-1131.

Bolin, B., 1981, *Carbon Cycle Modelling*. SCOPE 16. John Wiley and Sons, Chichester/New York.

Bolin, B., Döös, B.R., Jäger, J. and Warrick, R.A., 1986, *The Greenhouse Effect, Climatic Change and Ecosystems*. SCOPE 29. John Wiley and Sons, Chichester/New York.

Brasseur, G. and Solomon, S., 1984, *Aeronomy of the Middle Atmosphere*. Reidel, Dordrecht, The Netherlands.

Brühl, Ch. and Crutzen, P.J., 1988, Scenarios of possible changes in atmospheric temperatures and ozone concentrations due to man's activities, estimated with a one-dimensional coupled photochemical climate model. *Climate Dynamics*, **2**, 173-203.

Chamberlain, J.W., Foley, H.M., MacDonald, G.J. and Ruderman, M.A., 1982, Climate effects of minor constitutents. In W.C. Clark (ed.), *Carbon Dioxide Review 1982.* Clarendon Press, Oxford. 255-277.

Clark, W.C., 1982, *Carbon Dioxide Review 1982.* Clarendon Press, Oxford.

Crutzen, P.J., 1986, Role of the tropics in atmospheric chemistry. In R.E. Dickinson (ed.), *The Geophysiology of Amazonia.* John Wiley and Sons, New York. 107-132.

Crutzen, P.J., 1988, Tropospheric ozone: an overview. In I.S.A. Isaksen (ed.), *Tropospheric Ozone.* D. Reidel, Dordrecht, The Netherlands. 3-32.

Crutzen, P.J., Heidt, L.E., Krasnec, J.P., Pollock, W.H. and Seiler, W., 1979, Biomass burning as a source of atmospheric gases. *Nature*, **282** (5763), 253-256.

Crutzen, P.J. and Gidel, L.T., 1983, A two-dimensional photochemical model of the atmosphere 2: The tropospheric budget of anthropogenic chlorocarbons CO, CH_4, CH_3Cl and the effect of various NO_x sources on tropospheric ozone. *J. Geophys. Res.*, **88**(C11), 6,641-6,661.

Crutzen, P.J. and Schmailzl, U., 1983, Chemical budgets of the stratosphere. *Planet. Space Sci.*, **31**, 1,009-1,032.

Crutzen, P.J., Delany, A.C., Greenberg, J.P., Haagensen, P., Heidt, L.E., Lueb, R., Pollock, W.H., Seiler, W., Wartburg, A. and Zimmerman, P.R., 1985, Tropospheric chemical composition measurements in Brazil during the dry season. *J. Atmos. Chem.*, **2**, 233-256.

Crutzen, P.J., Aselmann, I. and Seiler, W., 1986, Methane production by domestic animals, other herbivorous fauna, wild ruminants and humans. *Tellus*, **38B**, 271-284.

Darmstadter, J., Ayres, L.W., Ayres, R.U., Clark, W.C., Crosson, R.P., Crutzen, P.J., Graedel, T.E., McGill, R., Richards, J.F. and Torr, J.A., 1987, *Impacts of World Development on Selected Characteristics of the Atmosphere: An Integrated Approach.* Oak Ridge National Laboratory, 2 volumes, ORNL/Sub/ 86-22033/1/V2, Oak Ridge, Tennessee 37831, USA.

Dickinson, R.E., 1986, How will climate change? In B. Bolin, B.R. Döös and R.A. Warwick (eds.), *The Greenhouse Effect, Climatic Change and Ecosystems.* SCOPE 29. John Wiley and Sons, Chichester/New York. 207-270.

Dickinson, R.E. and Cicerone, R.J., 1986, Future global warming from atmospheric trace gases. *Nature*, **319**, 109-115.

Ehhalt, D.H., 1988, How has the atmospheric concentration of CH$_4$ changed? In F.S. Rowland and I.S.A. Isaksen (eds.), *The Changing Atmosphere (Dahlem Konferenzen).* John Wiley and Sons, Chichester. 25-32.

Ehhalt, D.H. and Schmidt, U., 1978, Sources and sinks of atmospheric methane. *Pure Appl. Geophys.*, **116**, 452-464.

FAO (Food and Agriculture Organization), 1986, *Production Yearbook.* Rome.

Fraser, P.J., Rasmussen, R.A., Creffield, J.W., French, J.R. and Khalil, M.A.K., 1986, Termites and global methane - another assessment. *J. Atmos. Chem.*, **4**, 295-310.

Gates, D.M., 1981, *Energy and Ecology.* Sinauer Associates, Sunder - land.

Gore, A.J.P., 1983, Introduction (Chapter 1). In A.J.P. Gore (ed.), *Ecosystems of the World (4A). Mires: Swamp, Bog, Fen and Moor.* Volume 1, Elsevier, Amsterdam. 1-34.

Grau III, R.H. and LaScola, J.C., 1980, Methane emissions from US coal mines in 1980. *Information Circular, United States Department of the Interior, Bureau of Mines*, **8987**.

Hansen, J. and Lebedeff, S., 1987, Global trends of measured surface air temperature. *J. Geophys. Res.*, **92**(D11), 13,345-13,372.

Harriss, R.C., Sebacher, D.J. and Day, F.P., 1982, Methane flux in the Great Dismal Swamp. *Nature*, **297**, 673-674.

Houghton, R.A., Hobbie, J.E., Melillo, J.M., Moore, B., Petersen, B.J., Shaver, G.R. and Woodwell, G.M. (1983). Changes in the carbon content of terrestrial biota and soils between 1860 and 1980. A net release of CO$_2$ to the atmosphere. *Ecological Monographs*, **53**(3), 235-262.

Howard-Williams, C. and Thompson, K., 1985, The conservation and management of African wetlands. In P. Denny (ed.), *The Ecology and Management of African Wetland Vegetation.* Dr. W. Junk Publ., Dordrecht, The Netherlands. 203-210.

Jones, P.D., Wigley, T.M.L., and Wright, P.B., 1986, Global temperature variations between 1861 and 1984. *Nature*, **322**, 430-434.

Junk, W.J., 1989, Wetlands of northern South America. In D. Whigham (ed.), *Wetlands of the World.* (In preparation.)

Keller, M., Goreau, T.J., Wofsy, S.C., Kaplan, W.A. and McElroy, M.B., 1983, Production of nitrous oxide and consumption of methane by forest soils. *Geophys. Res. Lett.*, **10**, 1156-1159.

Khalil, M.A.K. and Rasmussen, R.A., 1983, Sources, sinks and seasonal cycles of atmospheric methane. *J. Geophys. Res.*, **88**(C9), 5131-5144.

Khalil, M.A.K. and Rasmussen, R.A., 1987, Atmospheric methane: Trends over the last 10,000 years. *Atmospheric Environment*, **21**(11), 2445-2452.

Koyama, T., 1963, Gaseous metabolism in lake sediments and paddy soils and the production of hydrogen and methane. *J. Geophys. Res.*, **68**, 3971-3973.

Lamb, H.H., 1984, Climate in the last thousand years: Natural climatic fluctuations and change. In H. Flohn and R. Fantechi (eds.), *The Climate of Europe: Past, Present and Future*. Reidel, Dordrecht, The Netherlands. 25-64.

Levy, H. II., 1971, Normal atmosphere: Large radical and formaldehyde concentrations predicted. *Science*, **173**, 141-143.

Logan, J.A., 1985, Tropospheric ozone: seasonal behavior, trends and anthropogenic influence. *J. Geophys. Res.*, **90**, 10,463-10,482.

Logan, J.A., Prather, M.J., Wofsy, S.C. and McElroy, M.B., 1981, Tropospheric chemistry: A global perspective. *J. Geophys. Res.*, **86**(C8), 7210-7254.

Lowe, D.C., Brenninkmeijer, C.A.M., Manning, M.R., Sparks, R. and Wallace, G., 1988, Radiocarbon determination of atmospheric methane at Baring Head, New Zealand. *Nature*, **332**, 522-525.

Matthews, E. and Fung, I., 1987, Methane emission from natural wetlands: global distribution, area and environmental characteristics of sources. *Global Biogeochemical Cycles*, **1**(1), 61-86.

McConnell, J.C., McElroy, M.B. and Wofsy, S.C., 1971, Natural sources of atmospheric CO. *Nature*, **233**, 187-188.

McDowell, R.E., 1976, Importance of ruminants of the world for non-food uses. *Cornell International Agricultural Mimeograph 52*.

Molina, M.J. and Rowland, F.S., 1974, Stratospheric sink for chloro-fluoromethane: chlorine atom catalysed destruction of ozone. *Nature*, **249**, 810-812.

Pearman, G.I., 1988, Greenhouse gases: Evidence for atmospheric changes and anthropogenic causes. In G.I. Pearman (ed.), *Greenhouse, Planning for Climate Change*. E.J. Brill, Leiden. 3-21.

Pearman, G.I. and Fraser, P.J., 1988, Sources of increased methane. *Nature*, **332**, 489-490.

Perner, D., Platt, U., Trainer, M., Hübler, G., Drummond, J., Junker-mann, W., Rudolph, J., Schubert, B., Volz, A., Ehhalt, D.H., Rumpel, K.J. and Helas, G., 1987, Measurements of tropospheric OH concentrations: A comparison of field data with model predictions. *J. Atmos. Chem.*, **5**, 185-216.

Ramanathan, V., Cicerone, R.J., Singh, H.B. and Kiehl, J.T., 1985, Trace gas trends and their potential role in climate change. *J. Geophys. Res.*, **90**(D3), 5,547-5,566.

Ramanathan, V., Callis, L., Cess, R., Hansen, J., Isaksen, I., Kuhn, W., Lacis, A., Luther, F., Mahlman, J., Reck, R. and Schlesinger, M., 1987, Climate-chemical interactions and effects of changing atmospheric trace gases. *Rev. Geophys.*, **25**, 1141-1482.

Rasmussen, R.A. and Khalil, M.A.K., 1984, Atmospheric methane in the recent and ancient atmospheres: Concentrations, trends and interhemispheric gradients. *J. Geophys. Res.*, **89**(D7), 11,599-11,605.

Seiler, W., 1984, Contribution of biological processes to the global budget of CH_4 in the atmosphere. In M.J. Klug and C.A. Reddy (eds.), *Current Perspectives in Microbial Ecology.* American Society for Microbiology, Washington D.C. 468-477.

Seiler, W. and Crutzen, P.J., 1980, Estimates of gross and net fluxes of carbon between the biosphere and the atmosphere from biomass burning. *Climatic Change*, **2**, 207-247.

Seiler, W., Conrad, R. and Scharffe, D., 1984, Field studies of methane emissions from termite nests into the atmosphere and measurements of methane uptake by tropical soils. *J. Atmos. Chem.*, **1**, 171-186.

Sheppard, J.C., Westberg, H., Hopper, J.F. and Ganesan, K., 1982, Inventory of global methane sources and their production rates. *J. Geophys. Res.*, **87**(C2), 1,305-1312.

Tucker, G.B., 1988, Climate modelling: How does it work? In G.I. Pearman (ed.), *Greenhouse, Planning for Climate Change.* E.J. Brill, Leiden, The Netherlands. 22-34.

WMO (World Meteorological Organization), 1985, *Atmospheric Ozone 1985.* Global Ozone Research and Monitoring Project No. 16, Vol. 1-3. WMO, Case Postale No. 5, CH-1211 Geneva.

Wolin, M.J. and Miller, T.L., 1987, Bioconversion of organic carbon to CH_4 and CO_2. *Geomicrobiology J.*, **5**(3/4), 239-259.

Woodwell, R.A., Whittaker, R.H. and Reiners, W.A., 1978, The biota and the world carbon budget. *Science*, **199**.

Zimmerman, P.R., Greenberg, J.P., Wandiga, S.O. and Crutzen, P.J., 1982, Termites: A potential large source of atmospheric methane, carbon dioxide and molecular hydrogen. *Science*, **213**, 563-565.

Chapter 6

PERSPECTIVES ON A CHANGING HYDROCLIMATE: LAND USE IMPLICATIONS

M. Falkenmark

6.1. Introduction

In many considerations of sustainable development, hydrological phenomena have often been treated as peripheral, seen only as indirectly involved in ecosystem functioning. However, the global ubiquity and circulatory nature of water present strong arguments for not only the inclusion of it and related phenomena in such studies but even investigation with a hydrological focus.

The objective of this Chapter is to address the dynamics of land use with respect to ecosystems of all scales using water as the pivotal factor. Water is basic to all life; it is the linchpin of climatic models, the medium for the transport of nutrients and pollutants, a habitat and water-bodies are a major sink for the products of anthropogenic activity.

Methods using factors, which represent crucial phenomena to simplify complex relationships, are traditionally used to analyse the effects of major perturbations on a system. Biophysical factors mainly determine the supply of land or the effective extent of the resource base whereas socio-economic factors mainly determine demand which, in turn, influences land utilization.

This Chapter discusses the water dependency of land use in terms of:

1)   hydrological characteristics directly related to land attributes; and

2)   the impacts of land use practices on hydrological phenomena.

Sustainable development becomes a question of striking a balance between acceptable impacts caused by particular land uses and the water dependency of other land uses. A conceptual framework is presented to help address the problems involved in determining such an equilibrium. Matrices are employed to clarify the complex phenomena involved where disturbances of the geosphere and atmosphere are

F. M. Brouwer et al. (eds.), Land Use Changes in Europe, 127–151.
© 1991 *Kluwer Academic Publishers. Printed in the Netherlands.*

translated into environmental impacts.

This study will not limit its focus to Europe; such an approach is transferable to other regions. The role of water in the generation of negative feedbacks should be explicitly brought into a general framework of human interaction with land and water.

6.2. The Hydrological Cycle

6.2.1. Hydrological pathways

Water circulates in the global system between the oceans, the atmosphere and the continents (Figure 6.1). The cycle has three main aspects:

i)   wind driven aerial water vapour flux causing continuous exchange of water via processes of precipitation and evapotranspiration;

ii)   precipitation provides the essential medium for plant production in terrestrial ecosystems;

iii)   excess flow recharges terrestrial water systems; underground storage aquifers and surface water courses, such as rivers and lakes, finally returning water to the sea (Falkenmark, 1987a).

All aspects of the hydrological cycle are closely related, particularly ii) and iii). The incoming precipitation is partitioned into two subflows:

i)   a short or vertical path or branch, returning water to the atmosphere through plants and soil surfaces;

ii)   a long or horizontal branch involving the recharge of terrestrial freshwater underground and surface systems where water eventually reaches the sea to be evaporated once again.

The transpiration stream, that is, the passage of water into roots and out through stomata and other surfaces of plants, represents the short branch of the water cycle. Plant growth is impeded if evaporation demand at plant surfaces exceeds water supply in the root zone. As transpiration is reduced the short branch pathway may become insignificant. The long branch of the water cycle includes water surplus to plant requirements that percolates through the soil profile to groundwater storage bodies plus surface runoff to surface water courses.

The land surface can be viewed as the station where incoming precipitation is partitioned (Figure 6.2) between the short and long branches (Falkenmark, 1986). All vegetation changes of any magnitude, brought about by alteration in climate, may be reflected not only in

a)

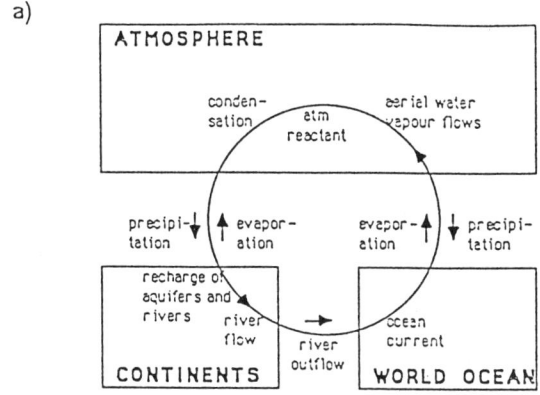

b)

c)

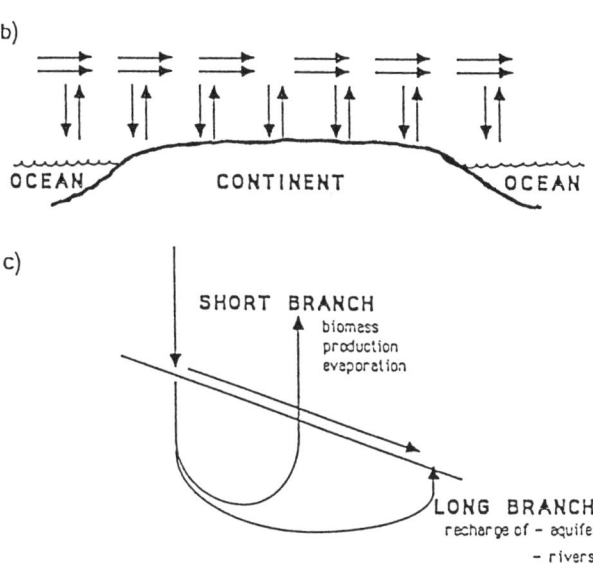

Figure 6.1. The global water circulation system (part a) brings water to the continents (part b), moisturizing them and feeding their ecosystems (part c). The surplus that does not return to the atmosphere (the so-called short branch of the water cycle) will recharge terrestrial water systems in aquifers and rivers (the so-called long branch of the water cycle).

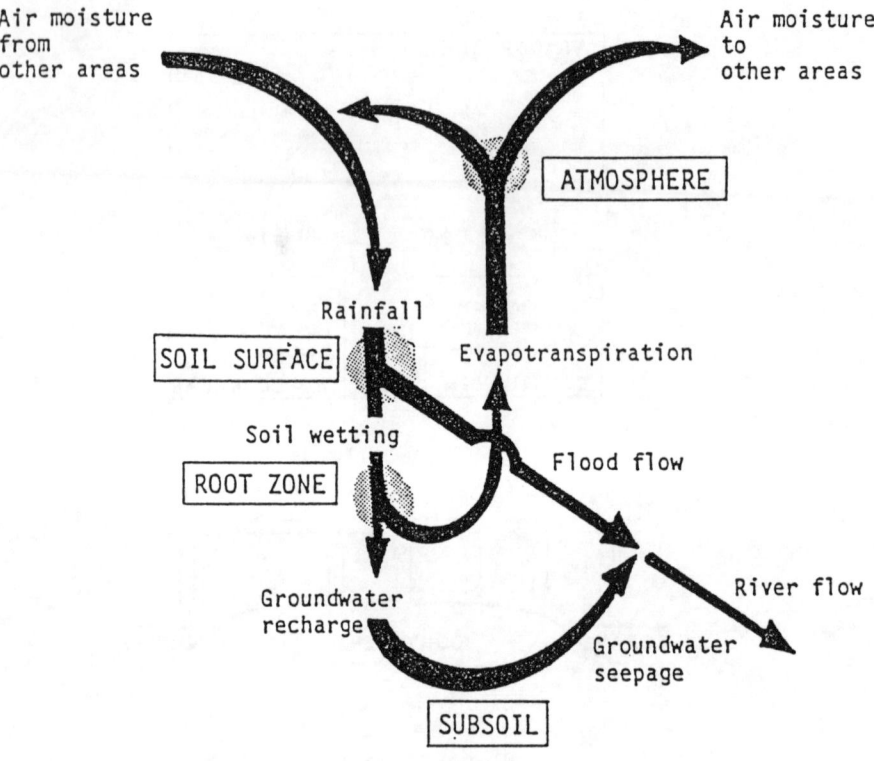

Figure 6.2.   Continuity relations in the terrestrial part of the water cycle.

altered water yields but also in the seasonality of water courses. Seasonal changes are particularly important on the hydrological margin between different climates, especially between humid and arid climates where small absolute changes to the water regime can cause significant biome shifts.   Perturbations of the water cycle, primarily generated by land use, can be oriented around three main partitioning zones (Figure 6.2):

SOIL SURFACE
Infiltration versus overland flow (flood flow).   Perturbations that may affect the balance and the infiltration capacity of the soil surface include: deforestation/devegetation - producing more runoff; land mismanage- ment - erosion and compaction = more runoff due to crust formation and sealed structures.

ROOT ZONE
Evapotranspiration versus subsurface runoff (short branch versus long branch pathway).   Partitioning is affected by perturbations that alter root

zone conditions especially vegetation changes and changes influencing the soil water storage capacity and permeability. Vegetation removal may reduce the short branch routing and thus more precipitation is likely to follow the long branch pathway. Changes in tillage practices, crop type, tree cover all potentially affect the proportions.

ATMOSPHERE
Recirculated precipitation versus evapotranspiration. Perturbations are in the form of any changes in atmospheric conditions.

The continuous and pivotal nature of the water cycle ensures that any alteration of soils or vegetation is propagated and reflected through a whole series of water cycle phenomena. Such changes are manifested by something of a "domino effect" where water flow and quality characteristics are mutually adjusted.

6.2.2. Catchment hydrology

The geographical layout and pattern of hydrological features represent the catchment characteristics of a region. The catchment network is vital to the potential for human activity and is consequently of major importance in decisions regarding the land use of an area.

The catchment water cycle links land users upstream with those living downstream since the water resource of the latter largely comprises "long branch" water originating from rainfall higher up in the catchment area. Upstream activities that increase evaporation or water consumption will therefore reduce water availability downstream. Similarly, upstream pollution contaminates the downstream resource base.

A catchment may be conveniently divided into recharge areas, receiving incoming precipitation and allowing percolation to the water table, and discharge or seepage areas where underground water returns to the surface (Figure 6.3).

The hydrological structure of the landscape is of fundamental importance in determining ecosystem characteristics since water availability is a major factor determining the type of natural vegetation and land capability.

6.2.3. Chemical composition generated in the root zone

The path followed by water determines its chemical content. Underground water is chemically quite different in the recharge areas, where the content of dissolved solids is low, and in the discharge areas, where it is high as a result of the chemical interaction with the geological surroundings all along the water pathways through the underground landscape.

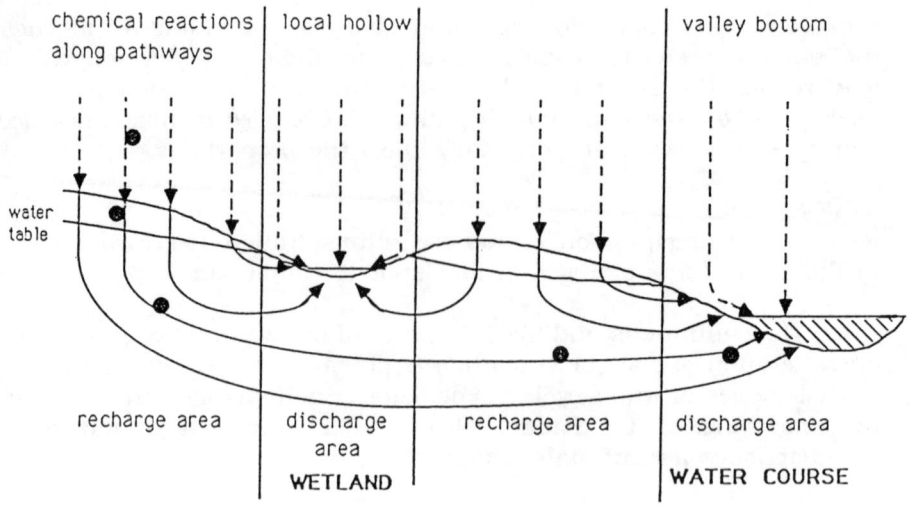

chemical reactions | local hollow | valley bottom
along pathways

water
table

recharge area | discharge area | recharge area | discharge area

WETLAND | WATER COURSE

Figure 6.3. Water flow through a landscape under humid conditions. All along its pathways, water reacts chemically with its local environment.

The water continuum and its biogeochemical environment form a complex multicomponent system illustrated in Figure 6.4 (Falkenmark and Allard, 1988).

A major determinant of the system is the product of the interaction in the root zone of the infiltrating water with carbon dioxide and humates produced by the roots and from decomposing organic matter respectively. Another determinant is the "output" component comprising weathering products from the mineral matrix through which the water is passing. This output system is composed of:

1)   a slow weathering silicate mineral system;

2)   a fast ion exchange system controlled by the clay
         minerals.

Such interactions largely determine the acidity or alkalinity of the water as it reaches the aquifer as well as its redox potential, that is, its ability to act as an oxidizing or reducing agent. These fundamental chemical characteristics control the behavior of metals, and organic

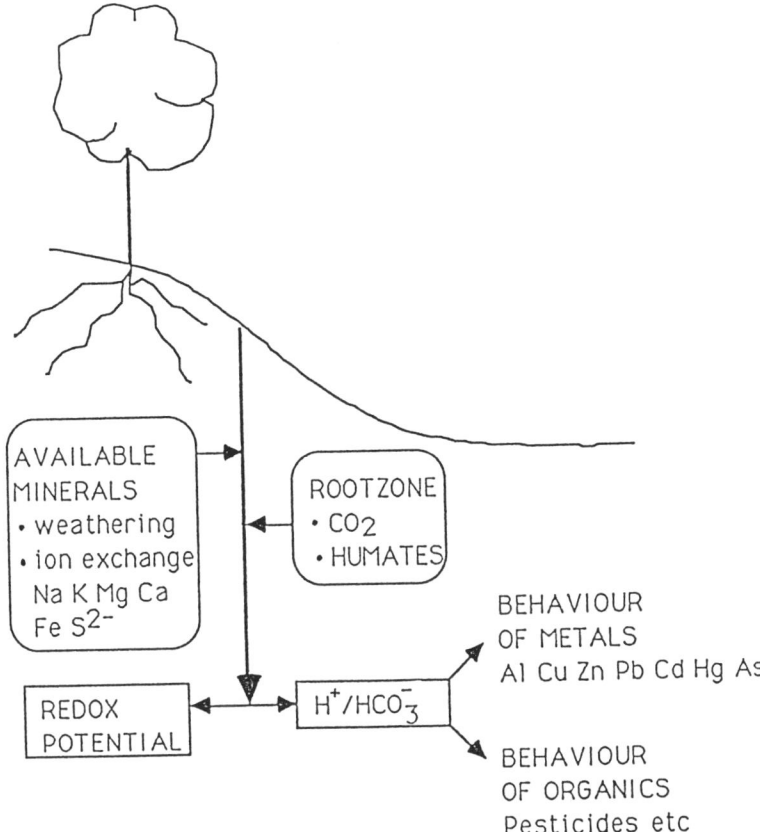

Figure 6.4. A simplified sketch of the water quality genesis that takes place in the soil.   Main groups of reactions of the complex multi-component system are indicated, as well as crucial chemical components, influencing the mobility of metals and organics.

compounds such as pesticides, as the groundwater moves towards the discharge area.   Disturbances of water chemistry can produce higher order effects on flora, fauna and human health.

6.2.4. Interregional differences in groundwater recharge

Before discussing the interactions between the land user and the hydrology of an area, it is necessary to be aware of regional differences in recharge characteristics (Falkenmark and Chapman, 1989).   In humid areas groundwater is recharged from all topographical fractions of a catchment, excluding valleys and local hollows where groundwater returns to the surface (Figure 6.5).   The situation is quite different in

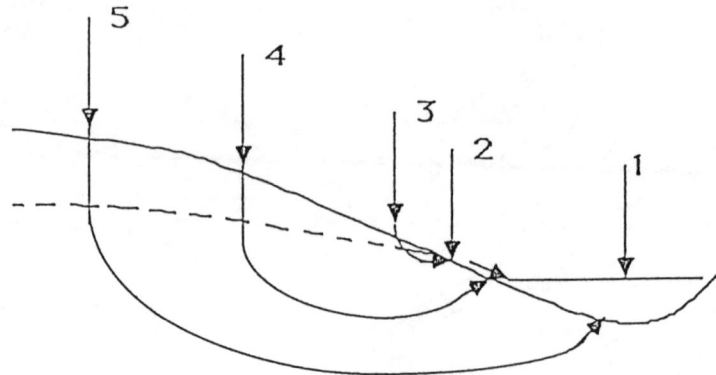

1.   Precipitation over the river surface.
2.   Precipitation over the valley bottom.
3.   Precipitation over the foothill region.
4.   Precipitation over the midslope region.
5.   Precipitation over the hilltop region.

Figure 6.5.   River water is a composite mixture of water fractions appearing in continuously changing proportions, reaching the river via different pathways and therefore having different chemical histories.

arid regions since precipitation often occurs on the higher more humid parts of the terrain; thus, subsequent recharge is from more remote sources. However, the main recharge is from flash floods along the river bottoms of small water courses, a common phenomenon of the precipitation regime in arid regions. A second form of groundwater is formed along floodplains while inundated. Rainfall, and therefore recharge, tend to be more intermittent and less predictable than in humid zones.

6.3.   Interdependency of Catchment Hydrology and Land Use

6.3.1. Land and water interactions

The complexity of the interactions between land and water implies that any manipulation by Man will be reflected not only through intended benefits but also in negative feedbacks (Figure 6.6).

The increasingly more frequent and widespread introduction of exogenous substances into the environment inevitably influences water quality, particularly in view of the excellent solvent properties, chemical activity and circulatory nature of water (Falkenmark and Allard, 1988).

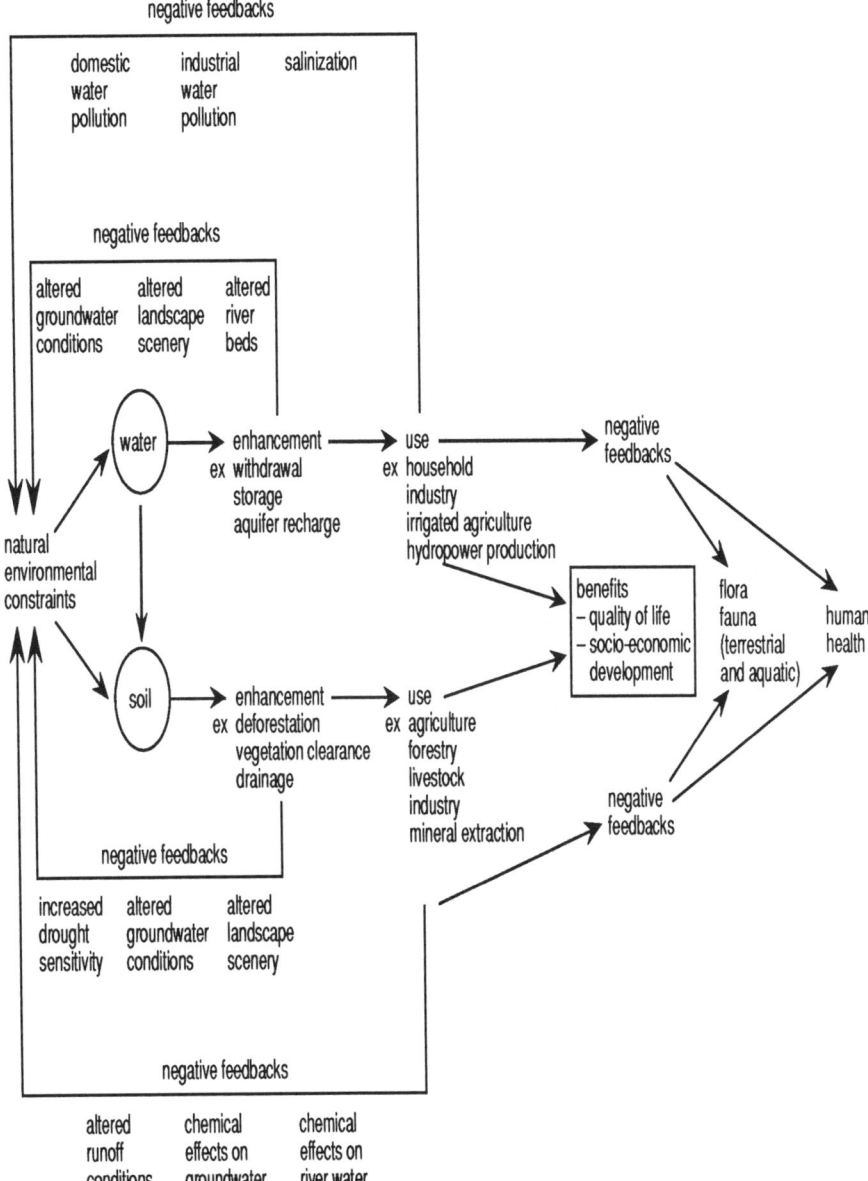

Figure 6.6. Natural constraints to resource use have to be overcome by enhancement measures, increasing accessibility and productivity and reducing obstacles. When such measures interfere with complex natural systems, negative feedbacks are produced besides the benefits intended.

Once suspended or dissolved, pollutants will be transported unless chemical transformations or physical barriers influence their mobility.

Figure 6.7 illustrates the consequence for the integrity of the water cycle when pollutant producing land use activities are undertaken. On a catchment or local scale, water soluble products of agricultural or industrial practices may accumulate in groundwater aquifers and remain undetected until "tapped", for example, by a well. Similarly, such compounds in the discharge area may be leached or disposed of into surface water and carried through the river system. Effects of the pollutants are manifested downstream in the river where the water may be abstracted for domestic use or irrigation.

On a wider scale, pollution of inland streams and rivers tend to accumulate and cause deleterious effects at the eventual destination of the water in coastal regions. The North Sea environment illustrates these connections. Concentrations of nutrients in agricultural runoff reach pollutant proportions causing eutrophication of the estuaries and coastal habitats receiving the load; here, massive algal blooms have been observed with associated fish kills due to the anoxic conditions. This is inevitably a late stage of pollution. It could have been avoided if the polluting activities had been stopped on land. Wiser land use policies critically depend on both the general public and the politicians really understanding the integrity of the water cycle.

6.3.2. Water quality

Combining the knowledge of the components responsible for water quality with the knowledge of the catchment characteristics allows predictions regarding the likely water quality at different points in the system. Figure 6.8 illustrates the two major questions to be asked.

1)   What is the destination of water infiltrating at a certain point? In other words, where in the system would it be expected that the land use-generated pollution of the point in question will be mirrored?

2)   Where did the water passing a certain river section originally infiltrate? In other words, what and from where are the constituent sources?

Water in a river system comprises water from different points in the catchment or river basin representing different flow pathways and chemical histories since infiltration. The river water mix will include:

1)   net rainfall directly on to the water surface;

2)   net rainfall on to nearby surfaces which then reaches the river by overland flow; and

a Mesoscale perspective

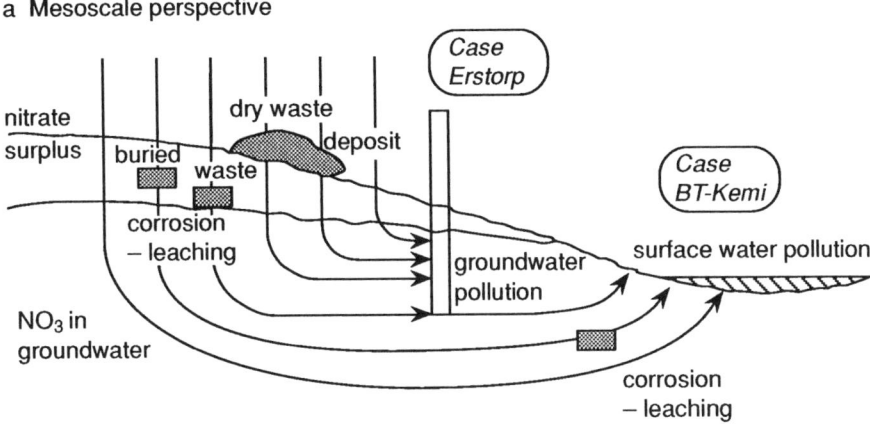

b Macroscale perspective

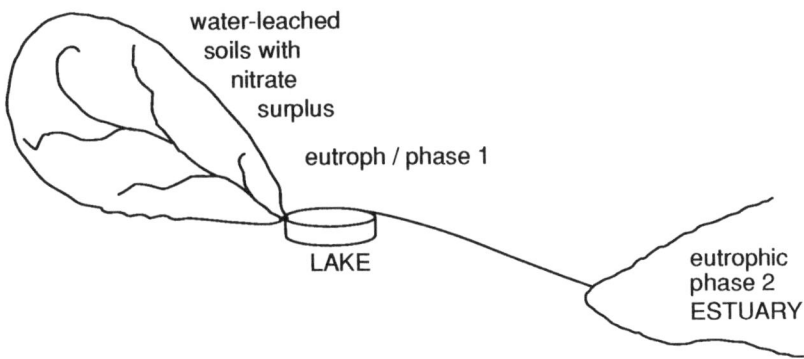

Figure 6.7.   Consequences of water cycle integrity in terms of the impacts of polluting land based activities.   a) Mesoscale perspective, indicating different types of pollution.   Case Erstorp refers to a Swedish case where buried industrial refuse in a recharge area polluted a village water supply; case BT Kemi refers to a case where industrial refuse had been tacitly buried in a discharge area close to river Braån.   b) Macroscale perspective, indicating upstream-downstream progression of water pollution.

to where?

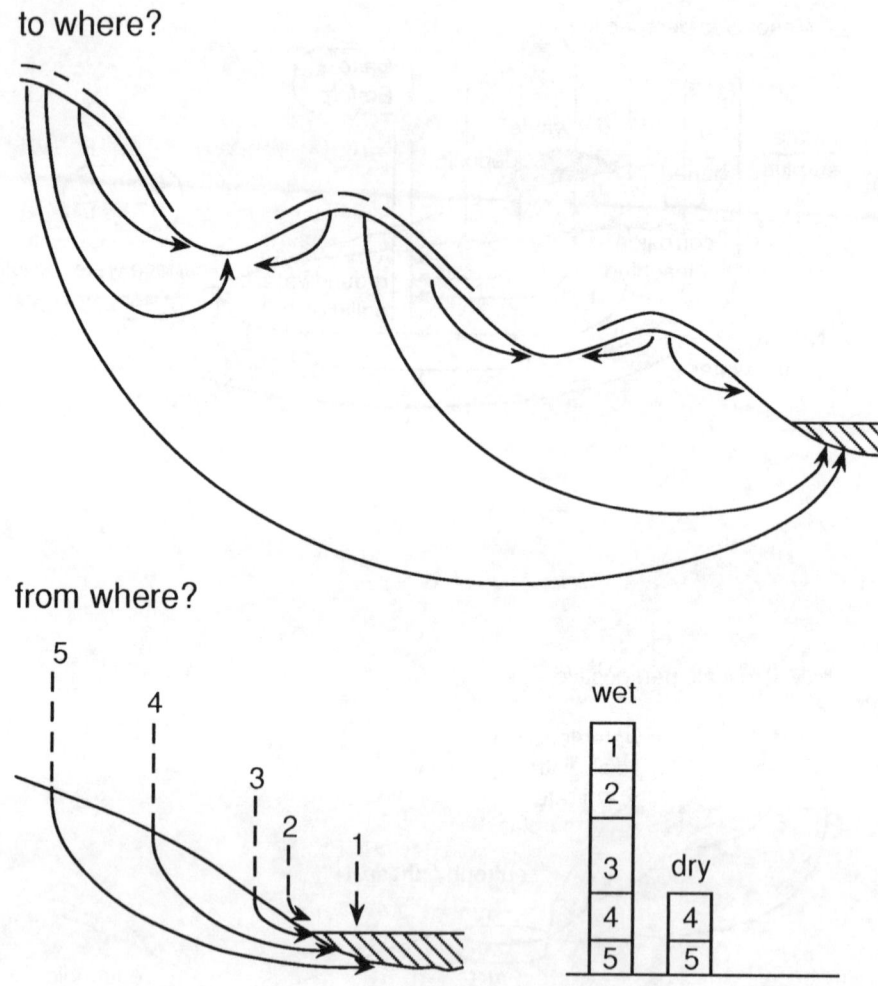

from where?

Figure 6.8. Complimentary aspects of the hydrological cycle: to where does the water go and from where does it originate?

3)   various types of groundwater, eventually entering the river system via subsurface flow.

During the flood season rivers mainly comprise water from direct and overland flow categories. During the dry season, when the groundwater table is low, (4) and (5), in Figure 6.8, assume importance explaining the large seasonal fluctuations often observed in surface water quality.

6.4. Environmental Manipulation

Man manipulates the environment with the aim of enhancing the quality of human life (Falkenmark *et al.*, 1987): water is made more easily accessible by sinking wells and establishing pipelines and canals. Water is also made more continuously available by building dams, reservoirs and by using flow control devices. Water is used as a transport medium for waste disposal. Man introduces chemicals into the environment for a variety of reasons including to improve agricultural productivity or by burning of fossil fuels, having the inadvertent effect of disturbing natural equilibria.

Deterioration of the natural environment is generally induced by three main categories of manipulations or perturbations (Falkenmark, 1989):

1)   introduction of chemical substances into the biosphere;

2)   manipulation of the soil-vegetation system;

3)   utilization of both renewable and finite natural resources.

Table 6.1 clarifies some disturbances with higher order effects on flora, fauna and human health. Such effects are the consequences of the life supporting functions of water and reflect its versatility. Table 6.1 highlights the fact that human intervention is water-impacting.

6.5. Influence of Climatic Change

6.5.1. Impact of climatic change on the hydrological cycle

The projected large-scale climatic changes as a result of the accumulation of greenhouse gases in the atmosphere are likely to be manifested as shifts in the global distribution of precipitation and an overall increase in potential evapotranspiration. The expected magnitude and direction of the changes are, however, subject to debate.

If annual precipitation for a region increases, more water is likely to follow the long branch pathway of the water cycle except where precipitation was previously a limiting factor to plant growth. In the latter situation the maximum plant water requirement will be met before any excess percolates to groundwater storage. If regions experience an increase in potential evaporation, the transpiration stream - neglecting for the time being the effect of increased CO_2 in the atmosphere - through vegetation and evaporation from the soil surface will enhance the attraction of the short branch pathway for incoming precipitation. This may, in turn, result in depleted water availability downstream.

In general, two groups of higher order consequences will follow:

Table 6.1. Categories of environmental manipulations, some phenomena disturbed and higher-order effects.

Disturbing activity	Disturbed phenomena	Higher-order effects	Mitigation strategy
(1) introduction of chemical substances	chemical composition of water	flora, fauna and human health	avoid
(2) intervention with soil and vegetation	partitioning of water (short and long branches) soil permeability	"domino" effects on water-related phenomena	balance benefit against damage
(3) utilization of natural resources (water, hydropower)	water pathways river profile	higher-order effects on flora, fauna, human health	

1)   changes in potential plant productivity - resulting in adjustments of the land use pattern;

2)   changes in terrestrial water systems - affecting groundwater table, quality and flow characteristics.

6.5.2. Consequential societal changes

The versatility of water will expose every sector of human society to water-related impacts of climate change. Table 6.2 exemplifies some of the major implications, listing expected problems and the required engineering measures to mitigate or counteract the effects. Table 6.2 lists water-related disturbances in two main groups: water on land and water in rivers. Society's use of water ranges from domestic use to hydro-electric power production, sanitation to irrigation and as an

Table 6.2. Implications of water-related disturbances.

	Land			Rivers and lakes
	soil moisture	ground-water	overland flow	flow-related uses
Societal activities concerned	agriculture forests buildings roads	rural water supply local irrigation	urban activities	water supply hydropower irrigation
Problems encountered under wetter conditions	reduced fertility	water logging building damages	urban storm runoff erosion	failing flow control floodings
under drier conditions	droughts crop failures	drying wells pumping costs foundation problems subsidence		water deficiencies reduced hydropower production
Engineering measures to mitigate under wetter conditions	drainage	drainage	urban drainage	flow control
under drier conditions	irrigation	deeper wells		water transfer schemes

Table 6.2. (continued)

		Rivers and lakes		
Societal activities concerned		carrier of effluents	use of water bodies	use related to water depth
Societal activities concerned		sanitation industrial waste water disposal	fishing recreation	navigation
Problems encountered	under wetter conditions		erosion/sedimentation increasing currents flushing of lake phosphorus	floodings inundations
	under drier conditions	reduced dilution aeration problems	changing fish population reduced water removal unsafe for bathing	collapsing navigation reduced traffic
Engineering measures to mitigate	under wetter conditions		river training	increased levee height reservoirs
	under drier conditions	flow control waste treatment aeration	dredging	flow control barrages sluices dredging

industrial cooling source to recreation. Man's water use will have to adjust to climatic changes and the consequential new hydrological circumstances. If changes are not recognized and mitigating actions are not taken, increased environmental degradation is a certain outcome, leading to a permanent depletion of the resource base. For example, if rainfall decreases and river flows are reduced, industrial waste pumped into the system will not be diluted to a satisfactory extent. Concentration of toxic or undesirable compounds is likely which will, in turn, promote secondary impacts on natural components. Similarly, if evapotranspiration increases dramatically, the water holding capacity of soils will be affected. Agriculturalists will need to shorten irrigation intervals and maintain levels of soil moisture to avoid drought stress.

6.6. Sustainable Interaction between Society and the Water Cycle

6.6.1. Criteria for sustainability

A basic criterion for sustainable development is the satisfaction of present needs without compromising future needs (WCED, 1987). Applied to the water domain this requires that a number of fundamental water-related needs are met in the present environment and in that projected for the future:

i)   the soil resource must be maintained to protect fertility, permeability and water holding capacity; at the same time the water supply must be adequate for biomass production at a non resource-depleting level;

ii)   groundwater has to be drinkable;

iii)   there has to be enough water accessible to allow general hygiene;

iv)   the fish in rivers and lakes has to be edible.

Many factors of the modern world threaten sustainability. Economic pressures demanding increased agricultural productivity over the last twenty years are now manifested as serious deterioration of the soil resource and therefore the inherent biological potential of the land resource. Soil erosion and decreasing soil fertility highlight the imbalances created by modern agriculture. Some replacement activities, such as fertilizing and the use of pesticides and herbicides, now show secondary impacts in the form of the eutrophication and toxification of water courses that results from polluted agricultural leachate (Hyams, 1976).

6.6.2. **Water quality and quantity**

Water quality and quantity are indicators of stability. The quality and quantity of three types of water are equally important for human activity: i) soil water in the root zone; ii) groundwater; and iii) surface water.

Local scarcity may be of different type; scarcity for humans by too large a population in relation to the available amount of renewable water; and scarcity for plants by destruction of the water storage in the root zone. Table 6.3 shows the extent to which these parameters act as determinants of sustainable development.

Table 6.3.   Water-related determinants of and threats to sustainable development.

Type of water	Determinant of	Threatened by
Soil water	crop production forest production fuelwood supply	crust formation, by overgrazing overuse deforestation

Terrestrial systems:
  Flow
	water supply
	hydropower
	irrigation potential,
	  i.e. dry region crop
	  production
		vegetation changes,
		e.g. upsetting
		partitioning of
		rainwater

Quality	water supply irrigation potential aquatic ecosystems fish production, i.e. food supply	
- rivers		waste water return flow from irrigation
- groundwater		leaching: - waste deposition - fertilizers - pesticides leaking pipes vegetation changes (in arid regions)

6.6.3. **Land use activities**

In general terms, three types of human activity may threaten sustainability:

i)   interaction with soil-vegetation systems by, for example, the removal of vegetation for agricultural, industrial or domestic purposes; such actions interfere with the local water balance; impermeable crusts may form at the soil surface, runoff may be increased with the removal of the protective, intercepting layer encouraging erosion of the top soil;

ii)   the introduction and handling of polluting substances threatens the equilibrium of the ecological environment and therefore threatens overall stability;

iii)   activities involving the direct abstraction of water for various uses and its return to natural water systems after use.

Removal of groundwater at a rate exceeding that of natural recharge will eventually lead to overexploitation and groundwater depletion. The use of water for agriculture and industry, and the subsequent return of this water to the river system, introduce salts and organic substances, leached from soils in the case of agriculture, and introduce by-products and waste in the case of industrial use. The irrigation of land with insufficient drainage causes waterlogging and anaerobic conditions, reducing soil fertility and land capability.

6.6.4. **Environmental limitations**

The concept of sustainable development also requires the recognition of scale limitations imposed by the environment (WCED, 1987). One such environmental constraint is the amount of water available for human use (Falkenmark, 1988). This amount is a function of the geographical location of a region which dictates the precipitation regime, the river basin hydrography and the effect of the aerial water vapour flux. Such constraints limit the freedom of action for development, often setting a maximum value to the population size that a region can successfully support with food and water without depleting essential resources. Past experience has shown that even a western semi-arid society is not able to manage above a certain population pressure on water of about 2,000 people per flow unit of 1 million m^3 of water per year.

Another critical environmental limitation is related to the robustness or sensitivity of the soil resource. Identifying safety margins for maintaining soil structure, texture and fertility is essential. In principle, human activities must only work within such limits and not exceed thresholds which will initiate deterioration and therefore instability. In general, extraction must not exceed replacement in the longer term.

6.7. Matrices to Clarify the Impacts of Changes

6.7.1. Matrices as a communication tool

A basic understanding of the limits and interactions of environmental components is necessary in the design of effective social responses to practical problems. The propagation of a disturbance throughout the water cycle can be illustrated by using matrices.

Following an example presented by Clark (1985), where effects on critical environmental indicators of perturbations of atmospheric chemistry were visualized, a similar series of matrices has been developed for land-water disturbances (Falkenmark, 1989). The framework suggested by Clark (1985) is a further development of an environmental assessment matrix derived by Crutzen and Graedel (1985). Their matrix aims to focus on the atmospheric environment and to illustrate how chemical processes influence its nature, composition and response to disturbances. The objective of the land-water matrices is to demonstrate how a disturbance at one point in the system is propagated and translated into impacts on remote or apparently unrelated phenomena.

6.7.2. Conceptual base

The overall framework was centered around a set of basic concepts (Clark, 1985):

1)   *valued environmental components*:   properties most worthy of attention or protection in a given assessment context; the chosen components reflect the judgement of broader political and social communities as well as that of scientific experts;

2)   *potential activities and fluctuations*:   perturbations leading to environmental disturbances;

3)   *mutual interactions between environmental constituents*;

4)   a final *overall framework* tying together the implications of environmental disturbances with resultant changes in selected environmental components.

In a primary matrix A (Figure 6.9), potential sources of environmental perturbations (for example, coal combustion), were related

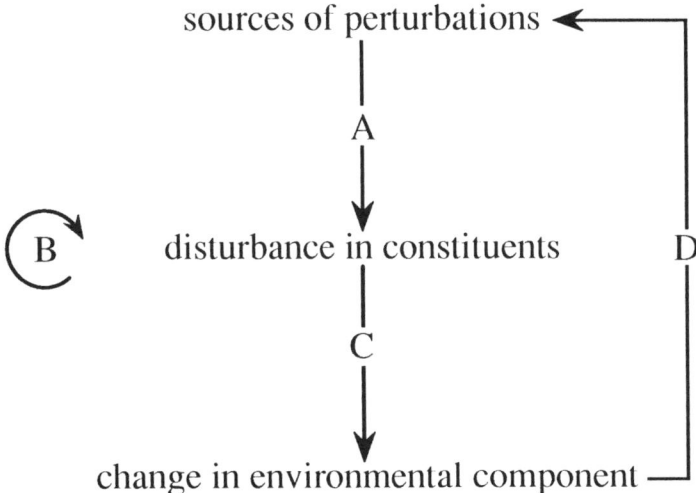

Figure 6.9. Diagram to represent interlinkages between perturbations and environmental components.

to the environmental constituents affected (such as CO_2 or O_3 concentration). In the matrix B, the mutual interactions between various constituents were linked (such as a change in halocarbons having an impact on ozone). A third matrix C shows disturbances in selected constituents linked to changes produced in environmental components (as in the impact of a change in ozone concentration on the thermal radiation budget and photochemical oxidant formation).   The final overall matrix D links changes in environmental components back to potential sources of perturbations.

6.7.3. Land-water disturbances

For the land and water matrix system the starting point is the recognition of the soil as the key zone in the terrestrial phase of the water cycle. Indeed, the processes in the upper soil determine not only the amount of water available for groundwater recharge but also its quality, the pathway followed through the catchment to the river and the proportion of incoming precipitation that reaches surface water bodies by runoff.

At the same time, soil productivity is directly dependent on soil moisture conditions. There is increasing awareness of the importance of the soil-water mutual interplay and consequently of the need to treat water and soil in an integrated fashion. This requires that an environmental water matrix incorporates soil characteristics, especially those relating to fertility and land use attributes. The conceptual base is summarized in Figure 6.10.

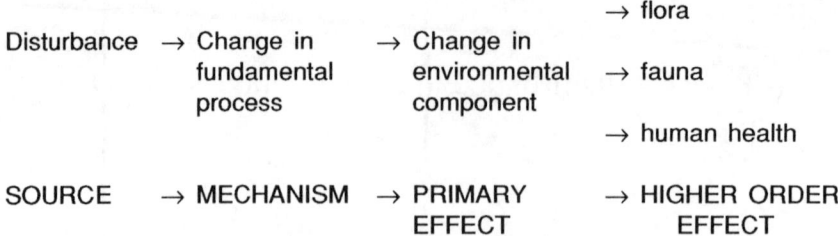

Figure 6.10.   Conceptual framework for matrix analysis.

The cyclical nature of water implies that a change in a primary hydrological process will be propagated through a sequence of water cycle steps to produce changes in secondary hydrological processes. The outcome of such models has been discussed in general terms by Kovacs (1989), by Salati and Vose (1984) for the Amazon Basin, by Lvovich (1979) for the USSR Steppe and by Heathcote and Mabbutt (1988) for the Australian savanna.

6.7.4. Environmental processes and disturbance

The criteria employed for the selection of the processes and environmental components to be used as indicators in a land-water matrix are discussed in this Section.

The most relevant water-related environmental components are those relating to land fertility and land use attributes on the one hand, and to water use and aquatic ecosystems on the other. Thus, processes most appropriate for the matrices are grouped under three headings:

i)   soil in terms of water and nutrient supply;

ii)   groundwater in terms of recharge and quality;

iii)   river flow in terms of formation and quality.

The pertinent activities would then include a wide range of disturbances which would influence processes such as precipitation, potential evaporation, infiltration, percolation through soil profile, underground pathways and water quality genesis.

Table 6.4. Perturbations-disturbances and the changes in various environmental properties generated by those disturbances.

Environmental property	Atmospheric perturbations changing		Terrestrial perturbations changing						Aquatic perturbations changing		
	a	b	c	d	e	f	g	h	i	j	k
precipitation	0										
soil fertility	*	0	0	0	0	0	0				
groundwater recharge	*		*	0	0						
groundwater quality		*			*	0					
river flow	*		*	*	*	*	0	0			
river extremes			*	*	*	*	*				
river quality		*	*	*	*	*	*	0	0		
water levels	*		*				*	0	0	0	0
sediment yield						0					
water temperature	0										
estuary quality	0	*				*	*			*	

Key:

0 = 1st order effect according to A in Figure 9
* = higher order effect according to B in Figure 9

a = air temperature
b = air quality
c = impermeable surface
d = soil structure
e = vegetation
f = fertilizer

g = chemical dry waste
h = groundwater withdrawal
i = river water withdrawal
j = waste water
k = river works

6.7.5. Final land-water disturbance matrices

Table 6.4 illustrates the final matrix linking perturbations associated with human activities with environmental components. The matrix shows the changes generated as a consequence of the domino effect set in motion by original disturbances.

6.8. Conclusion

Rajiv Gandhi said, "Development can hardly be sustained when the natural resources of soil, water and vegetation, the basic economic capital of a country, are depleted". It must be regarded as unwise to assess land use changes expected in response to a climatic change without, simultaneously, analyzing the outcome from a sustainability perspective.

Projected land use changes over the next fifty years will result from two origins. Changes in land use patterns will be the result of Man-induced deterioration of soil fertility and water quality reflected in higher order effects. The second origin of land use change will be climatic change forcing an adjustment in land use patterns. The second might well exacerbate the inevitable effects of the first. The interdependency of water and land suggests that water-related changes will influence land use changes to a much greater extent than the traditional factor-approach shows. In conclusion, land use and water cannot be comprehensively studied separately; an integrated analysis is essential. Failure to do this induces a dry approach to land use studies, promoting "water blindness" and leading to omissions of important impacts. One major example is the ignorance of the massive water penury now threatening socio-economic development in sub-Saharan Africa (WCED, 1987). Water cannot be taken for granted as an ubiquitous abundant resource in land use planning - it should be regarded as the factor determining the manipulations necessary to allow sustainable development strategies to be implemented.

REFERENCES

Clark, W.C., 1985, Sustainable development of the biosphere: managing interactions of the global economy and the world environment - a program plan. International Institute for Applied Systems Analysis, Laxenburg, Austria (mimeographed).

Crutzen, P.J. and Graedel, T.E., 1986, The role of atmospheric chemistry in environment-development interactions. In W.C. Clark and R.E. Munn (eds.), *Sustainable Development of the Biosphere*. Cambridge University Press, Cambridge, 213-251.

Falkenmark, M., 1986, Fresh water - time for a modified approach. *Ambio*, **15** (4), 192-200.

Falkenmark, M., 1987a, Hydrological Phenomena in Geosphere-Biosphere Interactions - Outlooks to Past, Present and Future. IUGG Pre-meeting Workshop. Monographs and Reports No. 1, International Association of Hydrological Sciences, Wallingford 1989.

Falkenmark, M., 1988, Sustainable development as seen from a water perspective. In *Perspective of Sustainable Development. Some Critical Issues*. Stockholm Group for Studies on Natural Resources Management, 1, 71-84.

Falkenmark, M., 1989, Global-change induced disturbances of water-related phenomena - The European perspective. CP-89. International Institute for Applied Systems Analysis, Laxenburg, Austria.

Falkenmark, M. and Allard, B., 1988, Water quality disturbances of natural freshwaters. In O. Hutzinger (ed.), *The Handbook of Environmental Chemistry. 5. Water Pollution*. Springer Verlag, Berlin (in press).

Falkenmark, M. and Chapman, T. (eds.), 1989, *Comparative Hydrology. An Ecological Approach to Land and Water Resources*. UNESCO, Paris.

Falkenmark, M., da Cunha, L. and David, L., 1987, New water management strategies needed for the 21st Century. *Water International*, September 1987.

Heathcote, R.L. and Mabbutt, J.A. (eds.), 1988, *Land, Water and People*. Surveys for the Academy of the Social Sciences in Australia.

Hyams, E., 1976, *Soil and Civilization*. Harper and Row Inc., New York.

Kovacs, G., 1989, In M. Falkenmark and T. Chapman (eds.), *Comparative Hydrology. An Ecological Approach to Land and Water Resources*. UNESCO, Paris.

Lvovich, M.I., 1979, *World Water Resources and their Future*. Translation by the American Geophysical Union. LithoCrafters Inc., Chelsea, Michigan: 250.

Salati, E. and Vose, P.B., 1984, Amazon Basin: A system in equilibrium. *Science*, **225**(4,658).

WCED, 1987, *Our Common Future*. Oxford University Press, Oxford.

Chapter 7

DYNAMICS IN LAND USE PATTERNS:
SOCIO-ECONOMIC AND ENVIRONMENTAL ASPECTS
OF THE SECOND AGRICULTURAL LAND USE REVOLUTION

P. Nijkamp and F. Soeteman

7.1. Introduction

Our economic and environmental system seems to be governed by
antagonistic forces, so that current demographic, industrial, social and
technological processes exhibit incompatible economic and environmental
implications. Environmental, resource and land use policies are fraught
with many conflicts that threaten the ideal of an *ecologically sustainable
economic development*, advocated *inter alia* in the Brundtland Report
(WCED, 1987).

The intertwined nature of all processes in an economic-
environmental system calls for due attention to be given to economic and
ecological paradigms from a steady-state and/or long-term perspective
(Nijkamp and Soeteman, 1988; Repetto, 1986). Conventional economic
accounting schemes (such as marginal cost or shadow cost principles)
often neglect the intriguing problem of environmental externalities and
of qualitative shifts in dynamic economic-ecological systems. Long-term
strategic considerations (multigenerational effects, irreversibilities) are
thus usually left out in environmental policy analysis (or are, at best,
incorporated in the social rate of discount in cost-benefit calculations
(Gijsbers and Nijkamp, 1988)).

Despite the global nature of environmental problems, it is the
environmental externalities on a local scale that provide foundations and
propagate cause and effect relationships on a broader scale. For
example, global problems such as acidification, sedimentation,
desertification, stratospheric ozone depletion, eutrophication, ocean
pollution and resource extraction are often the result of small-scale, local
activities (without control by an environment watchful constituency); also
far-reaching environmental impacts can be observed most clearly at a
local or regional scale. Consequently, the problems of land use
(interpreted in a broad sense, including landscape, cityscape, soil quality,
marine environments) are of central importance in environmental
management (Bartelmus, 1986).

F. M. Brouwer et al. (eds.), Land Use Changes in Europe, 153–175.
© 1991 *Kluwer Academic Publishers. Printed in the Netherlands.*

Unfortunately, the pluriform nature of dynamic ecological and economic processes cannot generally be described in a monodisciplinary framework due to differences in precision of measurement, spatial scale, time horizon and the adjustment speed of different variables (Braat and van Lierop, 1987; Brouwer, 1987). The relationship between economic development and ecological sustainability is often regarded as one of conflict, thus it is a daunting task to derive a methodology (and a related environmental policy analysis) which emphasizes compatibility instead of antagonism between development and sustainability.

In this context a *formal welfare concept* as a joint frame of reference for economic development and environmental sustainability will be introduced (Section 7.2). Based on this paradigm, attention is focused on land use problems. Section 7.3 will outline the nature of various land use transformations in the light of recent economic, agricultural, regional and urban development patterns. In Section 7.4 the position of agricultural land use will be dealt with in more detail from the viewpoint of agricultural growth and related policies in the EEC countries. Various serious future constraints and threats to environmental development will be identified in Section 7.5. The Chapter concludes with some options for strategic land use management.

7.2. Development and Sustainability: A Methodological Framework

7.2.1. Formal welfare concept

Economic development and environmental transformations are key aspects of industrialized countries. The performance of these countries is usually measured by means of gross national product (GNP) per capita. However, average GNP does not include social costs outside the market realm, so that environmental externalities are not regarded as components of GNP. Needless to say this may lead to a biased measurement and perception of actual welfare patterns in a country. Especially in a long-term perspective, characterized by dramatic (quantitative and qualitative) environmental consequences, the uni-dimensional measuring rod of GNP does not provide meaningful information for strategic policy-making. This shortcoming of conventional welfare indicators has in the past decade led to the popularity of multi-criteria decision methods in environmental policy analysis (Nijkamp, 1981).

To ensure a full account of environmental externalities (including ecological sustainability) in a welfare context, a *formal* welfare concept is required. Such a concept takes for granted that all elements which are utility constituents (including toxic material loads, ionizing radiation, deforestation, species diversity, beauty of landscape) are to be included as arguments of a social welfare function, regardless of whether such elements can be measured in monetary terms or not (provided these elements lead at least to a satisfaction of needs for scarce goods or services). For instance, in a more limited context of agricultural activities the welfare gains from agriculture should not only be measured by income generated in agriculture, but should also be corrected for negative impacts on landscape, species diversity or eco-stability (Dahlberg, 1986). Clearly various changes in land use patterns may be due to factors outside the agricultural system itself, for example, changes in climatic conditions (such as a projected rise in temperature or a change in precipitation pattern).

It is evident that a formal welfare concept does not *a priori* imply a conflict between conventional goods (for example, a house or a car) and environmental goods (such as a forest, or clear water); both types may contribute to human welfare.

Admittedly, since there are mutual interactions between the use of conventional and environmental goods, their effects are not by definition mutually supportive (at least not in the short run), so that it is ultimately the trade-off between these types which determines the final welfare change. From a long-term perspective, however, the question arises of whether a trajectory of economic development can be found that is in harmony with ecological sustainability, so that a mutually supportive evolution of both the economic and the environmental systems may arise (see also the concept of "co-evolutionary development", introduced by Norgaard, 1984). The question of whether structural changes (including morphogenetic transformations) in economic and/or environmental systems will enhance "quantity" without affecting "quality" (or enhance quality without affecting quantity) is not easy to answer. Especially in the case of morphogenetic (nonlinear dynamic) transformations in environmental or economic systems a welfare trade-off is difficult to make.

In this context *economic development* refers to a situation marked by qualitative shifts in the economy which lead to a positive contribution to welfare. The same applies to *ecological sustainability*: this refers to a situation which involves long-term maintenance or improvement of the quality of an ecosystem which have a positive welfare impact (Clark and Munn, 1986). Clearly, development and sustainability are concepts which do not automatically take for granted a stable evolution: morphogenetic transformations may imply turbulent systems' behavior in a transition period, caused by cyclical dynamics and complicated feedback relationships. Consequently, economic or environmental policies aiming at a permanent steady-state of a dynamic system may threaten the ultimate long-term stability, because its resilience potential may then decline.

It is important to note that irreversible processes, such as the extinction of rare species, subjected to a formal welfare approach, should also incorporate the interest of future generations. Such an equity consideration implies that the next generation should not be deprived of the potential of enjoying certain valuable environmental commodities, the so-called bequest value in option theory, (Nijkamp, 1988). This idea of maintaining at least a minimum bequest value in strategic environmental policies was also advocated by Ciriacy-Wantrup (1952), in particular in relation to establishing safe minimum standards of conservation by avoiding critical areas brought on by human activities which make it uneconomical to halt and reverse depletion.

It is evident that with the notions of economic development and ecological sustainability latent variables are essentially being dealt with, which can only be measured more precisely by using observable indicators. For instance, in economics such indicators might include the evolution of income, the changing pattern in income distribution, the composition of the labor force and the evolution of labor force participation. In the context of environmental considerations various other indicators may be used, such as sustainable yield, carrying capacity, multifunctionality and resilience (Brooks, 1986; Cozijn, 1986). All measures serve to provide quantitative indicators for judging whether the long-term quality of a dynamic system is affected or not. Clearly, the notions of development and sustainability are not mechanical measures, but refer to the value system (including risk behaviour) of society (reflected in a formal welfare approach) (Kleindorfer and Kunreuther, 1987; Wynne, 1987).

7.2.2. **Preventive environmental research**

In light of the formal welfare approach outlined above, strategic and preventive environmental research (conducted from a social science perspective) should concentrate on the following methodological focal points:

1) an investigation of internal and external *key forces* which act as major driving forces for the long-term evolution (including perturbations) of both the economic and the environmental system;

2) an exploration of the conditions under which *unanticipated surprises* in the dynamics of both economic and environmental systems may be brought about (both endogenous and exogenous surprise phenomena);

3) an identification of long-term feasible *boundaries* (technical, economic, demographic, social, ecological) within which economic and environmental evolution (including shocks) may take place.

Clark (1986, p.11), stated:

"... we have learned just enough about the planet and its
workings to see how far we are from having either the
blueprints or the operator's manual that would let us turn
that diffuse and stumbling management into the confident
captaincy implied by the "spaceship" school of thought."

Clearly, many attempts have been made in the past decade to
model or to replicate the complexity of dynamic economic-environmental
systems, but the strategic components (the key forces, surprises and
boundaries mentioned above) were not adequately included, so that these
models failed to provide effective and preventive environmental policies
(Braat and van Lierop, 1987). Consequently, mega-trend analysis at a
meso level, focusing on the qualitative changes and major directions of
influence is, from a strategic policy viewpoint, more important than
seemingly precise model predictions which are usually bound to fail.
This implies that joint expert views (based on strategic scientific forum
analysis) and long-term cross-national comparative studies may often
provide more appropriate information than conventional analytical tools.
These ideas will be elaborated in subsequent Sections focusing mainly
on land use problems.

7.3. Land Use and Economics: An Historical Orientation

It is interesting to see the shifts in perception of importance of land use
in economic history. For instance, in the early stages of economic
theory (in particular by the physiocrats) the production capacity of the
natural environment (notably land) was regarded as the main, if not
exclusive, source of welfare. Later on classical economists extended the
scope of economic theory by introducing, in addition to land, capital and
labor as complementary production factors for generating commodities
(and hence income and welfare). In the latter view, the government
plays only a minor role: it serves to maintain the institutional and
structural conditions within which market decisions can be taken.
Classical economists also mention the possibility of a stagnant economic
development caused by limits on available natural resources, in
particular agricultural land.

In later phases of economic theory building, especially in the
postwar neoclassical thought, it was asserted that the final source of
welfare does not rest with nature as such, but with the productive
capacity which is mainly determined by the quality and quantity of labor
and capital. This does not imply that in the neoclassical view nature
has become irrelevant. Randall and Castle (1985) clarify this as follows:

"... there seemed no reason to accord land any special
treatment that would suggest its role is quite distinct from
that of the other factors.   Land could safely be subsumed
under the broader aggregate of capital, since i) its
productivity was clearly responsive to investment and the
application of technology, and ii) the increasing economic
importance of non-food-and-fiber commodities together with
the increased use of capital inputs in the food and fiber
industries suggested very substantial possibilities for
substitution between land and capital."

In contrast to neoclassical thinking, Keynesian economics, with the
emphasis on macro-economic equilibrium phenomena, neglected mainly
supply limiting (such as environmental) factors.   In past decades,
however, especially as a result of the "limits to growth" discussion in the
1970s, the role of the natural environment in the process of economic
development has again become a focal point in economic research: in the
first place considering non-renewable resources (oil, materials), and later
focusing on renewable resources such as fisheries and forestry. It was
increasingly realized that the natural environment is not only a utility
component in a formal welfare approach, but also a production factor in
a normal economic-technological system (for example, a supplier of raw
materials as well as a recipient of waste materials).   However, since the
market does not provide appropriate signals for a proper allocation of
scarce environmental resources, overexploitation seems to be a logical
consequence (which reinforces the emergence of social costs in resource
exploitation).
        In conventional welfare economics, such market failures are
usually denoted as negative externalities.   However, it is not an easy
task for a government to cope with such externalities in the practice of
policy-making, because i) operational insight into the long-term
structural relationships between the economic and the environmental
system if often lacking, and ii) the nature and type of public or
institutional stimuli (charges, subsidies, regulations, quota systems,
environmental standards) do not often prompt congruent responses from
the public.   A clear exposition on such issues can be found in a
classical article by Hardin on the "Tragedy of the Commons", where it is
claimed that a free entry to a common agricultural market (an unpriced
or underpriced use of scarce common resources) will inevitably lead to
overexploitation, unless certain rules are established such as quota
systems or property rights; although such a view has recently been
challenged (Berkes, Feeny, McCay and Acheson, 1989).
        In this context, it is interesting to observe that the so-called
"enclosure movement" at the end of the medieval period was the
manifestation of the first major revolution in agricultural land use in
Europe with a major impact on environmental quality.   It was a logical

response to strong competition among farmers who were induced to act as "free riders" in agricultural resource use. By introducing a system of user and property rights, more care for economic continuity (economic development) and soil quality (environmental sustainability) could be ensured.

It is interesting to observe that in past centuries agricultural land use has not shown revolutionary or even significant changes in terms of land use institutions, despite the large-scale introduction of mechanization, automation and modern biotechnology, and despite changes in settlement and urbanization patterns. However, in the past decade, various qualitative (structural) changes which will be described in the next Section have led to an agricultural land use which is a direct and large-scale threat to ecological sustainability all over Europe. The opinion is put forward here that society now faces the beginnings of a second agricultural revolution (the first one being the enclosure movement), which will be induced by the unacceptable social costs (in the form of environmental externalities) of modern farming activities.

7.4. Agricultural Land Use and the Environment

7.4.1. Environmental potential and utilization form

This Section will attempt to provide a more coherent framework for connecting the two key concepts of "economic development" and "ecological sustainability" by introducing two intermediate auxiliary terms, *viz.* "*environmental potential*" and "*utilization form*". The environmental potential refers to the capacity of the natural environment to offer a structural contribution to socio-economic development without affecting environmental components that would reduce ecological sustainability. In addition to this Pareto-optimality concept, the utilization form refers to the extent to which production or consumption in an economic system does absorb components of the environmental system (the degree to which production and/or consumption exert a claim on the environmental potential). Thus environmental potential and utilization form are not independent of each other, as a high level of environmental potential is often accompanied by a low level of socio-economic functions, and vice versa (Figure 7.1).

For instance, the transition from a nomadic culture towards an industrialized society has meant a transformation of landscapes from natural to man-made (Wilkinson, 1973). Agricultural key factors acting as driving forces in this context include modernization, economies of scale (notably concentration), and specialization in monoculture. Clearly, the upper limits of agricultural production (or productivity) may still be raised (de Wit *et al.*, 1987), but such a rise would no doubt affect the environmental potential (or the sustainability) in the long run. A further increase in land productivity tends to lead to a lack of resilience caused

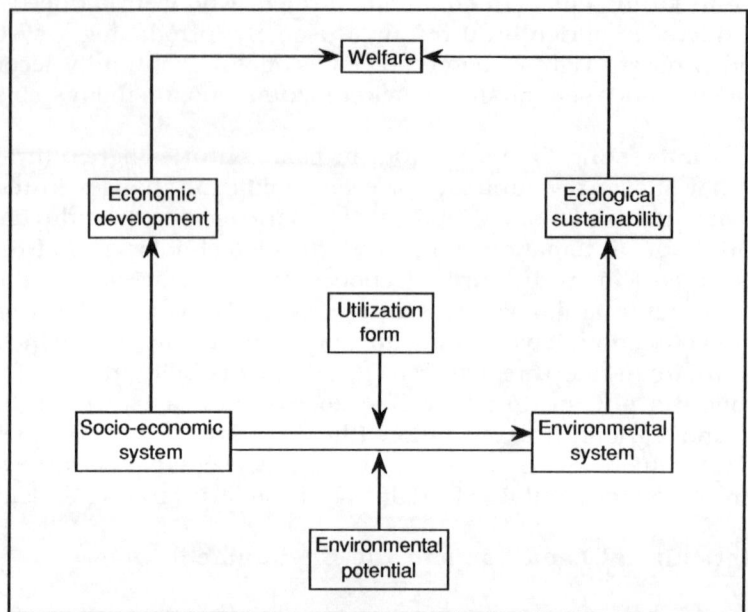

Figure 7.1. A formal welfare-theoretic framework of economic development and ecological sustainability.

by soil degradation. Seen from this perspective, the environmental potential may also, beyond a critical limit, become a key force for further socio-economic development.

7.4.2. Productivity limitations

When land productivity reaches a limit the concept of marginal land is introduced (Brouwer, 1989). These limits may stem from various constraints:

1) *physical*: caused by climatic, physiographic or soil conditions in a certain area (soil erosion or desertification after deforestation);

2) *social*: caused by lack of necessary skills, traditional family patterns, demographic processes and the like in an agricultural society; the latter type of limit may be removed in the long run, as is shown by the history of agriculture in Europe;

3) *technological*: caused by the lack of appropriate tools in agriculture, for example, environment-friendly or biologically-oriented approaches to pests;

4)   *economic*: caused by efficiency motives, taking for granted the necessity of marginal costs of production not exceeding marginal benefits (especially in cases of a fully operating market mechanism).

The recent history of agricultural land use has demonstrated that considerable parts of European agriculture have reached (or are about to reach) one or more of the above limits, which means that environmental potential and utilization forms tend to come into conflict.

7.4.3. Transfer of externalities

The question of feasibility of further land use claims in the light of ecological sustainability and economic development is an intricate one, which has not been studied satisfactorily in European countries. In fact, the situation is even more complicated, as spatial substitution effects, in the form of a geographical transfer of negative externalities, may occur. For instance, part of the environmental potential of region A may be used for an expansion of utilization forms in region B.

To some extent this spatial substitution effect for land use is comparable to the bubble concept in industrial environmental policies, although in the case of transfer of externalities connected with the environmental potential and utilization form no explicit policy has been adopted so far among European countries. But the previous observations also show that such transfer processes are so far-reaching that the environmental issues which often emerge at local scales really become visible on a world-wide scale. This inclusion can also be found in the Brundtland Report (WCED, 1987), in which it is convincingly demonstrated that the geographically interwoven pattern of environmental potentials and utilization forms leads to global resource problems manifesting desertification, deforestation, soil erosion, acidification and so forth. But in terms of strategic and preventive policies, the notion of "think globally while acting locally" has not yet reached any stage of maturity!

7.4.4. Environmental standards for agriculture

Clearly, in a way analogous to environmental policies for the industrial sector, governments might be willing to impose maximum limits on agricultural production which comply with environmental standards, but even such a seemingly simple policy choice would include various disadvantages from an environmental viewpoint:

-    even strict norms seldom lead to no-effect levels of environmental degradation;

- agricultural production standards are more oriented towards market interests than to preventive environmental protection measures;

- production limits adopted in only one country do not reconcile transboundary environmental effects;

- the spatial distribution of environmental externalities and related social costs may be quite uneven in case of a system of uniform production limits.

7.4.5. Integrated land management

The previous observations demonstrate clearly that operational and policy-oriented research is badly needed in the area of integrated agricultural land use planning and environmental management. Also from an economic viewpoint, Europe is facing an unfavorable situation of a lack of insight into social costs of various forms of land use and into the social benefits of alternative environment-friendly land use policies. For example, the estimated social costs of acidification in The Netherlands range from 150 to 3,000 million Dutch guilders per year. On the other hand, the total management costs for public policy actions by the Dutch Government in this area amounted to 557 million Dutch guilders in 1988.

There is, of course, one main problem in the field of land use policy, namely the interference of agricultural policy with environmental policy. Agriculture has direct and an indirect impacts on the quality of the environment: there is no other sector so dependent for its inputs on the environmental potential. Unfortunately, agricultural utilization forms are often out of step with environmental potential.

According to Odum (1969) one may regard agricultural development as an ecosystem transformation process in which the number of species diminishes, the efficiency of food recycling gradually declines, the production increases but the vulnerability of the production also increases and the biomass production is reduced. The post-war developments in the agricultural sector have shown a clear transformation from natural equilibrium mechanisms towards man-induced equilibrium mechanisms, which have affected the diversity and stability of ecosystems. An agriculturally advanced country, such as The Netherlands, presents a clear illustration of the points mentioned above, as agriculture in this country is increasingly turning into a high-tech sector.

7.4.6. The example of The Netherlands

Taking the Dutch case as a representative example, the following factors which have acted as the main driving forces for the recent evolution in the agricultural sector may be listed:

i) *concentration tendencies caused by economies of scale*: for instance, the number of Dutch farms specializing in milk production declined between 1973-1985 from 99,000 to 61,000 (approximately a 40 per cent decrease), while the average number of cows per farm increased during that period from 22.8 to 39.8 and the total production of milk increased from 9 million to 12.5 million tons; similar observations can be made regarding related industrial sectors such as food processing; thus, the agricultural sector has followed the pathway of the industrial sector toward large-scale activities and for the moment there is no reason to suggest that this development will soon come to an end;

ii) *modernization and intensification*: the capital intensity, as well as the share of intermediate deliveries for agricultural production has increased significantly in the recent past, a situation strongly induced by the emergence of the high-tech sector (notably biotechnology); in various subsectors of agriculture the land productivity has almost been doubled in the past 15 to 20 years; without a clear environmental perspective, this may lead to serious degradation of soils, not only because of depletion of nutrients by the exhaustion of fertile soil but also because of compaction by the use of heavy machinery;

iii) *lack of diversification*: the diversity of spatial and environmental structures is increasingly affected by new cultivation methods and far-reaching physical planning in agricultural areas; the rise of big agribusiness-complexes contributes to a uniformity of the agricultural landscape (Post *et al.*, 1987); this trend towards a leveling out of traditional environmental variety in rural areas is closely connected with the above mentioned specialization (induced by automation and mechanization);

(iv) *socialization*: social backgrounds, notably drastic changes in the socio-economic position and the image of farmers' families, have exerted a deep impact on the life style and attitude of farmers, which in turn has impacted environmental conditions; in The Netherlands around 1930, approximately 20 per cent of the labor force was agriculturally oriented, whereas today this figure is approximately 4 per cent; at the same time the gap between rural and urban life styles has diminished, the agricultural community has, from a social-cultural viewpoint, developed towards an urban-oriented community; this emancipation of the agricultural community has caused an abandonment of traditional family patterns, and has stimulated a modern attitude towards risk taking in business; consequently, farmers have become innovative entrepreneurs of the Schumpeterian type; the strong competition on a national and international market has led to a rationalization process, in which environmental concerns are subordinate to survival strategies; the resulting soil degradation implies a loss of environmental potential which, in the long run, may become a serious threat for both the future economic perspectives (the development option) and the future quality of the environment (the sustainability option).

In conclusion, the environmental potential of the soil and the utilization forms are interconnected phenomena which may be in conflict in the short-term. In the long-term, however, it should be stressed that a meaningful compromise between these elements has to be reached in order to support both economic development and ecological sustainability in the agricultural sector. Whether or not this is a feasible strategy in a European setting, will not only depend on the entrepreneurial decisions in the agricultural sector, but also on public policy decisions taken at the level of the European Community. The question of whether the international political arena in Brussels will act as a key force (or constraint) for a sound and balanced agricultural development will be discussed in the next Section.

7.5. Overproduction in the Agricultural Sector: An International Perspective

7.5.1. Case study

In the previous Sections no explicit attention has been given to a major determinant of changes in agricultural land use and production, the supra-national policies pursued at the level of the European Economic Community (EEC). In this context, the paradoxical problem of unacceptable environmental degradation on the one hand and overproduction in the agricultural sector on the other hand deserves closer attention, with a particular view on shifts in land use patterns. In this Section we will again take agriculture in The Netherlands as a frame of reference, as this sector most clearly reflects the problems caused by the strictly regulated Common Agricultural Policy (CAP) of the EEC.

Despite the decline of employment in agriculture in The Netherlands, this sector is still providing a significant contribution to national income. This is due mainly to the sharp rise in land productivity over the past few decades. Figure 7.2 shows that the average agricultural land productivity in The Netherlands is approximately four times as high as the EEC average, and the Dutch agricultural labor productivity is approximately three times as high as the EEC average.

It should be added, however, that these figures provide a biased picture, as they neglect the fact that The Netherlands is heavily dependent for its agricultural sector on imports of intermediate products from abroad (see Section 7.4). But as far as domestic land use development in The Netherlands is concerned, a continuing trend towards a decrease in the agricultural labor force, a decline in agricultural land use (and farm units) accompanied by a significant rise in production (reflected in an average annual growth of land productivity in the period 1961-1981 of 4.3 per cent in The Netherlands) is invariably found.

For the EEC as a whole, the average growth rate of production for the period 1961-1981 was 2.3 per cent. Since the hectarage of agricultural land use is gradually declining, the actual growth rate per hectare is even higher. The Netherlands has, as mentioned before, the highest growth rate per hectare mainly because of its advanced organization of research, education, information and technology in the agricultural sector, as well as its well established marketing, logistical and distributional strategies, a situation which is mainly due to the strong and efficient societal support for this sector. For the next 20 years, there have been estimates for The Netherlands suggesting that the annual growth of production will be about 2 per cent, the decrease of labor force will be about 2.5 per cent, while the growth of the labor productivity is estimated to be about 4 per cent. The surplus of cultivated land may rise to 25 per cent of the present hectarage.

In recent years, the position of the agricultural sector has increasingly been questioned for two reasons:

i)   the enormous subsidies (domestic and EEC) given to this sector;

ii)   the environmental degradation caused by this sector (in a period where pollution by the industry is being increasingly controlled).

Thus both the environmental potential and the utilization forms of agriculture are becoming a source of controversial debate in The Netherlands. In this framework, the wisdom of the current EEC agricultural policy is being increasingly questioned.

The Common Agricultural Policy of the EEC has various objectives:

-    increase in agricultural productivity;
-    maintenance of agricultural income at an acceptable level;

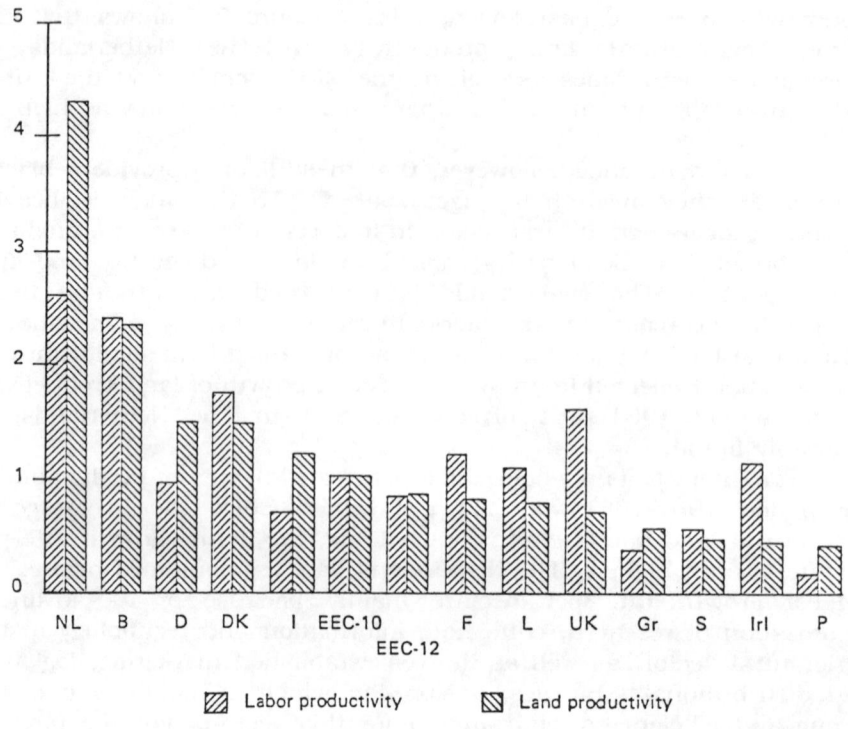

Figure 7.2. Labor productivity and land productivity for the member countries of the EEC in 1985 (EEC-10 = 1). (Source: Commission of the European Community, 1987.)

-   stabilization of agricultural markets;
-   safeguarding the provision of agricultural products;
-   a reasonable price level for consumers.

Such a diversity of objectives would require a broad spectrum of instruments. However, in practice the EEC has only one major instrument, a price policy. It is also noteworthy that some of these objectives (for example the rise in productivity) have been realized even without the use of specific EEC instruments. Clearly, certain regions in the Community may be lagging behind in terms of productivity, and the self-provision rate is not evenly spaced over the member countries. However, the requirement for agricultural products can be almost totally covered by internal production. A stable price level of agricultural products has never been a serious problem in the Community. Nevertheless, there are some severe tensions in the Common Agricultural Policy (CAP), which have serious implications for land use in the Community.

7.5.2. Implications of the CAP for land use in the community

The EEC has the *dual aim* of using price policies for achieving a situation of both stable markets for agricultural products and of acceptable income levels for farmers (comparable to non-agricultural income). However, since economic development in the EEC member states does not run parallel, and in fact, shows significant discrepancies, a complicated system of compensating monetary transfers was designed in order to meet the income target. But the latter policy measure implied that a uniform price policy became illusory, a situation which was more recently coped with by introducing an indirect system of income subsidies via a reduction in value added tax in several countries. Clearly, this situation of artificially low prices may stimulate agricultural overproduction with negative implications for environmental quality in rural areas. Thus instead of incorporating social costs of environmental externalities, the agricultural market is even further destroyed by indirect price subsidies.

A second problem concerns the aim of *stable markets* for agricultural products. Equilibrium of such markets can in principle only be reached if prices become flexible so as to meet a balance between demand and supply (which would most probably affect the income objective) or if supply was strictly regulated. But the latter policy is problematic, as it introduces a planned submarket in an otherwise free market system, while it does not ensure a maintenance of acceptable income levels for the agricultural sector.

The current situation of a market disequilibrium is mainly caused by the strong rise in agricultural productivity. Despite the very moderate increase in the demand for agricultural products in the EEC (approximately 0.5 per cent per year) and despite the gradual decline in real prices of agricultural products (approximately 2.0-2.5 per cent per year), the supply of agricultural products in the Community has risen by approximately 2.0-2.5 per cent per year. As a consequence, there is hardly any shortage of agricultural products in the EEC, as indicated in Figure 7.3. The markets for agricultural products are apparently saturated, the internal demand does not increase and the supply to the world market of agricultural products from outside the EEC is increasing mainly due to lack of purchasing power in Third World countries. Consequently, stock control and supply control are challenging but extremely difficult tasks for the Community.

A third problem emerges from the significant *differences in agricultural income* between the member states of the EEC. The average net agricultural income per capita shows a large variation and ranges from 22,000 ECU in The Netherlands to 3,400 in Portugal. The latter situation indicates that agriculture in some countries is a (sub) marginal activity, which requires complementary income earned in other sectors.

Finally, in recent years the *financing of agricultural subsidies* has become a source of many tensions. The growth in these expenditures has caused severe political friction. The gross EEC agricultural expenditures have increased from 13,056 million ECU in 1982 to 23,916 million ECU in 1987. Although EEC revenues (import taxes, value added taxes) on products have increased from 21,240 million ECU in 1982 to 35,672 million ECU in 1987, the agricultural part of the revenues increased only from 2,228 million ECU to 3,297 million ECU in this period.

7.5.3. Future agricultural policy

In the future it is plausible that the following issues in agricultural policy will gain importance:

-       export restitutions may tend to increase due to the weak position of the dollar and the high level of autarky of the Community;

-       a decline in the world market price of agricultural products will increase the net deficit between import taxes and export subsidies

-       a tendency towards a more uniform Community policy for the agricultural sector (including Spain and Portugal) will necessitate the development of a more structurally-oriented and strategic agricultural policy;

-       price compensation, quota systems and related policy measures (including land set aside programs) may become standard practice for most agricultural products, if productivity increases are not arrested;

-       in a stagnating economy, the growth in value added tax will not run parallel to agricultural expenditures in the EEC, so that international, intersectoral and intrasectoral conflicts for the agricultural sector may become sharper;

-       the increasing practice of fraud in agricultural subsidies may lead to structural shifts in policy.

In view of the saturation levels for almost all agricultural products, it is evident that a continuation of current trends would be a major

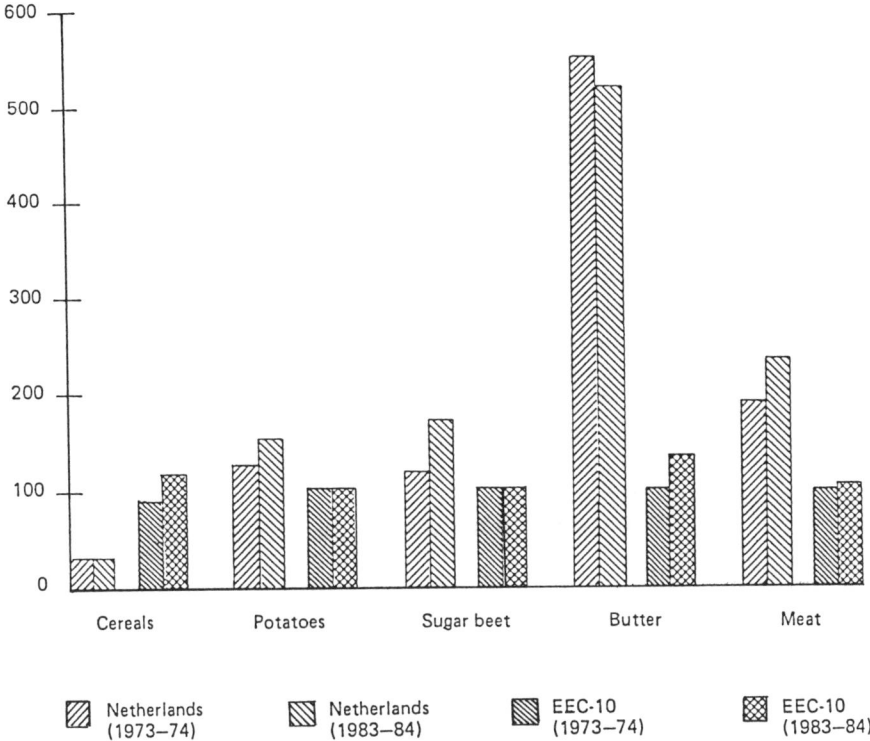

Figure 7.3. Internal supply of some important products (percentage). (Source: Commission of the European Community, 1987.)

failure from an economic viewpoint, while the negative externalities of a further rise of the agricultural sector would also become excessively high. Consequently, the European agricultural sector badly needs a structural reorientation. Some selected issues related to the latter point will be discussed in the next Section.

7.6. Strategic and Scientific Options for Co-evolutionary Development

In view of the serious socio-economic and environmental frictions inherent in agricultural overproduction, various strategic options may be considered. Three different options will be outlined here:

i) modern (high-tech oriented) agriculture; this option takes for granted the necessity for the use of modern technology in order to remain competitive by reducing production costs (and eventually also by coping with environmental degradation, transforming chemical farming to a more biologically oriented farming practice);

ii)   traditional agriculture; this strategy would imply a gradual
        development of the sector, but would require price compensating
        measures in order to comply with the income targets;

iii)   "green" agriculture; in this way the environmental repercussions of
        agricultural activities would be minimized by establishing a more
        soil-extensive cultivation mode.

In the third option a non-transferable and land-related quota
system, or a forced extensification of agricultural land use, would be
plausible to cope with overproduction in this sector, especially because
the alternative, a transition towards a market mechanism, might lead to
price reductions and hence would stimulate more competitive behavior
(including large-scale concentration, mechanization and intensification).
However, it is also often claimed that the quota system option does not
necessarily lead to an extensification of agricultural land use, but may
cause further intensification (and hence even more environmental
threats) in case of strong competition. Furthermore, both options may
lead to serious price and market distortions, so that an equilibrium on
a European scale would be even more difficult to attain. Compensating
policy measures (price measures, individual income subsidies) may be
necessary; these also have many socio-economic and financial
disadvantages. One thing is clear: the EEC budget cannot bear the
burden of huge transfers to the agricultural sector for much longer, so
that in the near future a forced solution to agricultural overproduction,
based on a closer market orientation, seems inevitable. This option may
benefit the modern agricultural production sectors. No doubt this will
lead to a new problem: excess supply of rural land. On the basis of
ongoing technological development and a more market oriented policy,
the surplus of cultivated land may reach 25 per cent within the next 20
years in the EEC. This seems to be an option which is in agreement
with environmental interests, but it involves a great many social and
financial adjustments. Moreover, the modern agricultural sector (if it
survives) will use the most economically practical and best technological
means to cope with lower prices, and generally this will not be in
agreement with environmental objectives. Traditional agriculture may
then be another meaningful option but here serious income problems
(again inducing unlimited competition) may emerge and preclude a
balanced solution.

Hence, it seems that a greater emphasis on the market
mechanism accompanied by satisfactory policy measures regarding
ecological sustainability is the only viable option. This implies that the
modern technological evolution (biotechnology, energy technology,
information technology, genetic manipulation, robotization) in the
agricultural sector will continue, provided it becomes more oriented
towards safeguarding long-term environmental interests. Thus high-tech

agricultural technology would have to find a compromise with environmental technology.

Finally, the question is, "What is to be done with vacant agricultural land?". This problem of so-called *area management* has received increasing attention in past years. Of course, part of the available land can be used for new urbanization, industrialization and infrastructure plans, part for recreational and leisure purposes and another part for extensification of agricultural uses induced by lower land prices. But even then a large stock of vacant land may remain. One obvious potential use would be a reconversion into "environmental capital" (via afforestation), but the financial and management implications of such far-reaching policy decisions may be excessive. Even on the present moderate scale this already leads to major problems in the EEC.

Clearly, for environmentally sound land management, much information and research is needed in order to gain more precise insights into the potential reconciliation of agricultural economic development and ecological sustainability. Whether or not utilization forms will then be more in harmony with environmental potential is still an open question. Hence this new research field of a co-evolutionary development of a modern and sound agricultural sector and of an ecologically sustainable environment deserves a multidisciplinary and cross-comparative analysis in various European countries. Some relevant items on such an ambitious research agenda would be:

-      identification of environmental components which in the long run are critical for balanced land use development;

-      analysis of economic consequences of changes in environmental potential (caused by changes in multifunctionality);

-      assessment of long-term land use implications of shifts in environmental potential;

-      analysis of the changing role of agriculture with respect to the changing quantitative and qualitative needs of the public regarding both the type and the mode of agricultural production;

-      compatibility of changes in land use with other societal objectives (the use of vacant agricultural land for bioenergy purposes);

-      socio-economic analysis of both efficiency and equity questions emerging from policy options on environmental sustainability in the framework of agricultural policy;

-      study of the role of modern technology (biotechnology) in enhancing the environmental potential of agricultural land use;

-       investigation of possible strategic EEC options regarding
        agricultural utilization forms, which, given economic objectives for
        this sector, would ensure a long-term sustainable land use at a
        European scale.

        In conclusion, it seems that the second agricultural revolution has
not only induced a great many environmental issues (such as
sustainability) but it also seems to offer an option for a redirection of
socio-economic and environmental goals.   Sustainable agricultural
development may even be possible within the limits of the environmental
potentials.   But this requires a comprehensive view on the utilization
form of land.   Dynamics always offer opportunities.   It is a challenge for
scientists and policy makers to select the sustainable opportunities.

ACKNOWLEDGEMENT

The authors acknowledge the support for preparing this Chapter given
by the Dutch Council for Environment and Nature Research (RMNO).

REFERENCES

Bartelmus, P., 1986, *Environment and Development.* Allen and Unwin,
        London.
Berkes, F., Feeny, D., McCay, B.J. and Acheson, J.M., 1989,   The
        benefits of the commons. *Nature,* **340,** 91-93.
Braat, L.C. and van Lierop, W.F.J. (eds.), 1987,   *Economic Ecological
        Modelling.* North-Holland, Amsterdam.
Brooks, H., 1986, The typology of surprises in technology, institutions
        and development.   In W.C. Clark and R.E. Munn (eds.),
        Sustainable Development of the Biosphere. Cambridge University
        Press, Cambridge. 325-348.
Brouwer, F.M., 1987, *Integrated Environmental Modelling: Design and
        Tools.* Kluwer, Dordrecht, The Netherlands.
Brouwer, F.M., 1989, Determination of broad scale land use changes by
        climate and soils. *Journal of Environmental Management,* **29,** 1-
        15.
Ciriacy-Wantrup, S.V., 1952, *Resource Conservation: Economics and
        Policies.* University of California Press, Berkeley.
Clark, W.C., 1986, Sustainable development of the biosphere: themes for
        a research program.   In W.C. Clark and R.E. Munn (eds.),
        Sustainable Development of the Biosphere. Cambridge University
        Press, Cambridge. 5-48.
Clark, W.C. and Munn, R.E. (eds.), 1986, *Sustainable Development of the
        Biosphere.* Cambridge University Press, Cambridge.

Commission of the European Community, 1987, *De Toestand van de Landbouw in de Gemeenschap, Verslag 1986.* Bureau voor officiële publikaties der Europese Gemeenschappen, Brussels.

Cozijn, P., 1986, An analysis of key terms of the World Conservation Strategy. Institute of Taxonomic Zoology, University of Amsterdam, Amsterdam, The Netherlands (mimeographed).

Dahlberg, K.A. (ed.), 1986, *New Directions for Agricultural Research: Neglected Dimensions and Emerging Issues.* Rowman and Allanheld, New York.

Gijsbers, D. and Nijkamp, P., 1988, Non-uniform social rates of discount in natural resource models: an overview of arguments and consequences. *Environmental Systems*, **17**(3), 221-235.

Hardin, J., 1968, The tragedy of the commons. *Science*, 1243-1248.

Kleindorfer, P. and Kunreuther, H. (eds.), 1987, *Insuring and Managing Hazardous Risks.* Springer-Verlag, Berlin.

Nijkamp, P., 1981, *Environmental Policy Analysis.* Wiley, New York.

Nijkamp, P., 1981, Culture and region: a multidimensional evaluation of monuments. *Environment and Planning B: Planning and Design*, **15**, 5-14.

Nijkamp, P. and Soeteman, F., 1988, Ecologically sustainable economic development: key issues for strategic environmental management. *International Journal of Social Economics*, **15**, (3/4), 88-102.

Norgaard, R.B., 1984, Co-evolutionary development potential. *Land Economics*, **60**(2), 160-173.

Odum, E.P., 1969, The strategy of ecosystem development potential. *Science*, **164**, 262-270.

Post, J.H., Breeveld, J., van der Ploeg, B., Strijker, D. and de Vlieger, J.J., 1987, *Agribusinesscomplexen in Nederland*, Research Report 32, Agricultural Economics Research Institute, The Hague, The Netherlands (in Dutch).

Randall, A. and Castle, E.N., 1985, Land resources and land markets. In A.V. Kneese and J.L. Sweeny (eds.), *Handbook of Natural Resource and Energy Economics*, Volume II. North-Holland, Amsterdam. 571-620.

Repetto, R., 1986, *World Enough and Time.* Yale University Press, New Haven.

WCED, 1987, *Our Common Future.* Oxford University Press, New York.

Wilkinson, R.G., 1973, *Poverty and Progress: An Ecological Model of Economic Development.* Methuen, London.

de Wit, C.T. Huisman, J. and Rabbinge, R., 1987, Agriculture and its environment: are there other ways? *Agricultural Systems*, **23**, 211-236.

Wynne, B. (ed.), 1987, *Risk Management and Hazardous Waste.* Springer-Verlag, Berlin.

APPENDIX

1.   Development of employment in the agricultural sector in The Netherlands, EEC-12, EEC-10, USA and Japan (per cent of total labor force).

Year	Netherlands	EEC-12	EEC-10	USA	Japan
1960	9.8	21.1	18.4	8.5	30.2
1970	6.3	13.8	11.4	4.5	17.4
1973	6.0	11.9	10.0	4.2	13.4
1980	4.9	9.6	8.0	3.6	10.4
1982	5.0	9.1	7.6	3.6	9.7
1983	5.0	9.1	7.6	3.5	9.3
1984	5.0	8.9	7.4	3.3	8.9
1985	4.9	8.6	7.2	3.1	8.8

Source: Eurostat, several years.

2.   Number of farms in The Netherlands and EEC-6 (in thousands) (in brackets the relative index compared to 1966-67).

Year	Netherlands		EEC-6	
1966-67	341.7	(100.0)	10,110.3	(100.0)
1975	253.7	(74.2)	6,415.5	(63.5)
1983	243.4	(71.2)	5,124.7	(50.7)

Source: Eurostat, several years.

3.   Area of cultivated land in The Netherlands (1900-1983) (in 1,000 ha) (the percentage area of cultivated land from the total land area is given in brackets).

	1900	1920	1940	1960	1977	1983
total area	3,255	3,265	3,335	3,613	3,719	3,729
area cultivated	2,085	2,184	2,324	2,314	2,060	2,008
	(64.1)	(66.9)	(69.7)	(64.0)	(55.4)	(53.8)

Source: Agricultural Economics Research Institute of The Netherlands, several years.

4.   Gross investment in real assets in the agricultural sector (1973-1985) (1980 = 100).

	1973	1981	1983	1985
Netherlands	54.7	82.5	102.0	95.0
Belgium	76.7	80.3	109.4	115.9
Denmark	54.5	68.4	74.1	97.5
FRG	67.3	92.3	117.2	104.8
France	60.2	104.2	138.0	129.4

Source: Eurostat.

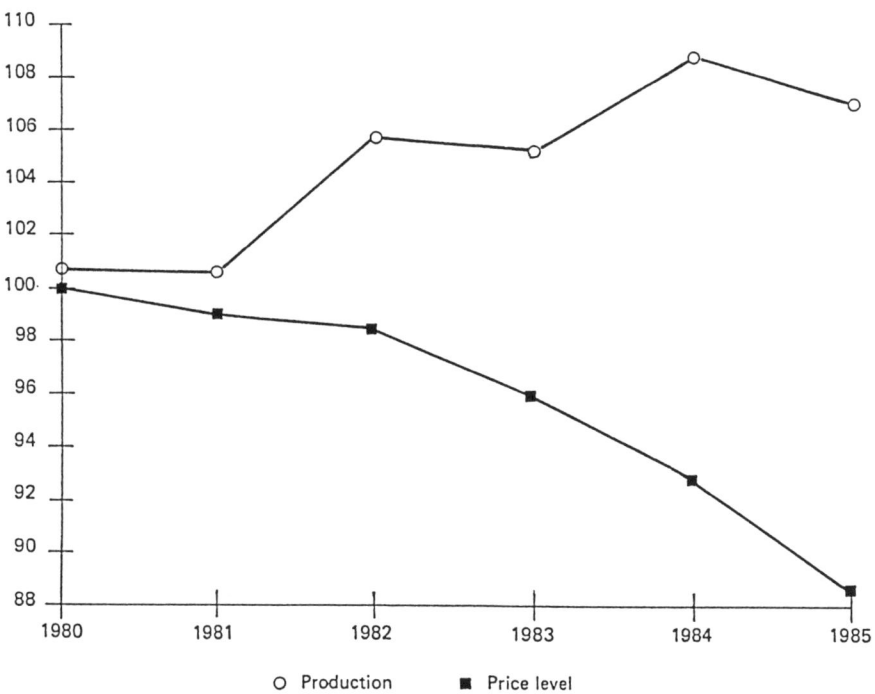

O Production   ■ Price level

5.   Development of production (constant prices) and product prices (after correction of inflation) in the EEC-10. The average of the years 1979, 1980 and 1981 has been set equal to 100. (Source: Eurostat.)

Chapter 8

CLIMATIC CHANGE AND LAND USE IMPACT IN EUROPE

G.P. Hekstra

8.1. Introduction

Scenarios of climatic change (based on general circulation models) have
projected a 4-5°C warming over the Mediterranean area and up to 8-
10°C over the northern tundra, due to a doubling of atmospheric CO_2 or
equivalent greenhouse gases over the next 5-7 decades (Bach, 1988).
However, recent research shows that such predictions may be an
exaggeration. It is considered that a rise of 1.5 - 4.5°C on a global basis
up to 2050 is more likely with a further 1 - 2°C rise by the end of the
twenty-first century. The models differ in detail, more so for the
hydrological components such as evapotranspiration, clouds, precipitation
and soil moisture than for insolation, radiation balance and surface
temperatures.

The global effects of current trends in climatic change sound
potentially alarming (Hekstra, 1988a and 1988b):

- higher surface temperatures will cause the
  oceans to warm and thermal expansion of the
  water will cause a rise in sea level; densely
  populated coastal lowlands (comprising about 20
  per cent of the present world population) are
  threatened;

- changes in the hydrological cycle may profoundly
  affect natural ecosystems, agriculture, domestic
  and industrial water supply and the control of
  salinization and desertification;

- extreme weather events are likely to become
  more severe and more frequent (droughts, floods,
  storms, cold and heat waves); average conditions
  will become less important than the extremes;

- major disasters could occur; a sudden
  disintegration of parts of Antarctica might cause
  a tidal wave with devastating impact on coasts

F. M. Brouwer et al. (eds.), Land Use Changes in Europe, 177–207.

or a change of direction of an oceanic or
atmospheric current might lead to major regional
climatic shifts;

-       significant changes in geopolitical and socio-
economic relationships could take place as
secondary impacts.

What confuses the issue of socio-political and economic impacts
of climatic change is a false perception of "winners" and "losers".
Climatic changes will upset all present societies and economies but
some of the required responses will be more costly than others.
Individuals within a society may become "winners" to the detriment of
others. Globally, those who contribute least to the causes of climatic
change may suffer most from the consequences. A decisive factor in
responding to changing conditions will be the financial-economic ability
to adapt.

Climate change is just a part of global changes in the
environment. Climatic changes act on a pre-stressed world. The land
degradation features most often considered to have a direct connection
with climate are all environmental problems in their own right deserving
political attention.

Some of the degradation processes such as desertification are
recognized by environmental ministries and agencies and are well
established topics on policy agendas. Desertification is now regarded to
be a result of land mismanagement (Stebbing, 1953; Acocks, 1953). Up
to 6 million ha of arable or grazing land is lost to desert annually.
Also, 11 million ha of land is deforested annually and, after a transition
period of use for agriculture, may end up as wasteland because of
erosion, salinization and loss of soil fertility (WCED, 1987).

Land use management agencies and agricultural ministries monitor
such changes but fail to control them. In addition, more insidious and
less dramatic processes such as acidification, eutrophication, chemical
changes and the loss of biotic diversity can alter the climate-vegetation-
soil balance towards land degradation. These processes need to be
better understood throughout Europe and elsewhere (Stigliani, 1988).

8.2. **European Climate-Vegetation-Soil Relationships**

The world biomes and natural vegetation types are primarily determined
by *climate* (radiation, temperature, precipitation, wind and waves) and
substrate (geology, soil-type, mineral content and water table condition).
The components of climate and substrate are ranked from stable to
highly variable. The natural biota are directly dependent on the locally
predominant components of climate and substrate. They constitute the
living link between climate and soil unless there is interference by

human action.   *Culture/cultivation* is now the third driving force in evolution.

All types of world biomes, as determined by temperature and precipitation (Figure 8.1), can be observed in Europe with the exception of the tropical biomes. North Africa and the northern part of the Middle East belong bioclimatically to Europe (Figure 8.2).

Figure 8.3   shows the zone classification first proposed by Holdridge   (1967).   This   takes   into   account   the   potential evapotranspiration ratio (PET/P) and includes latitudinal and altitudinal

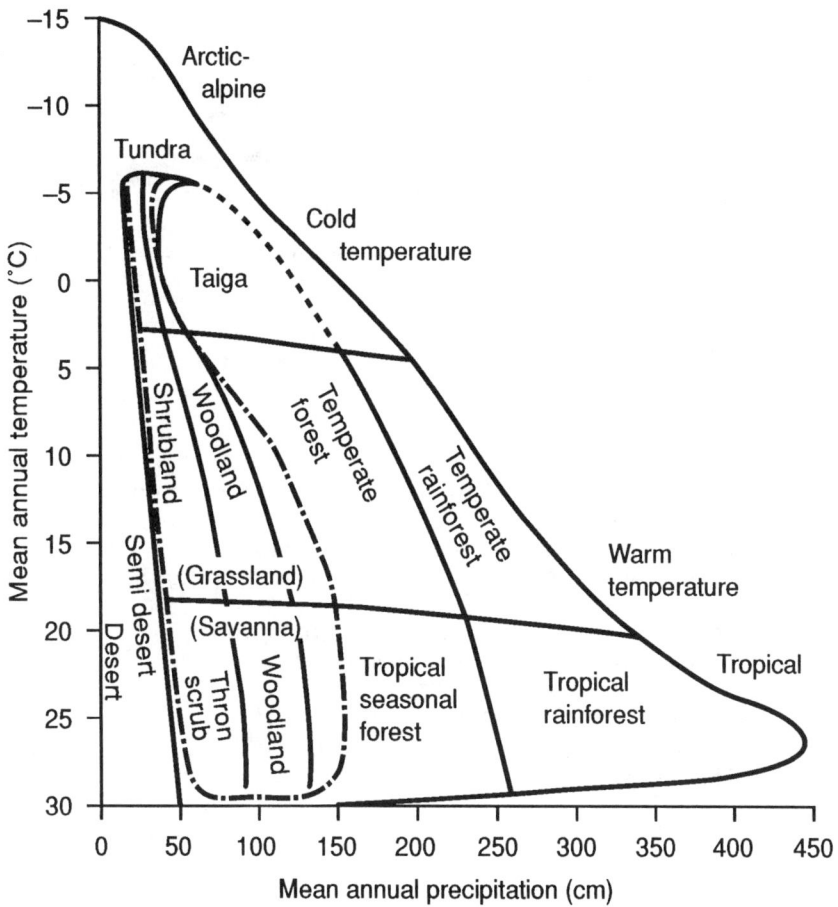

Figure 8.1. World biome types as determined by temperature and precipitation (adapted from Whittaker, 1975).

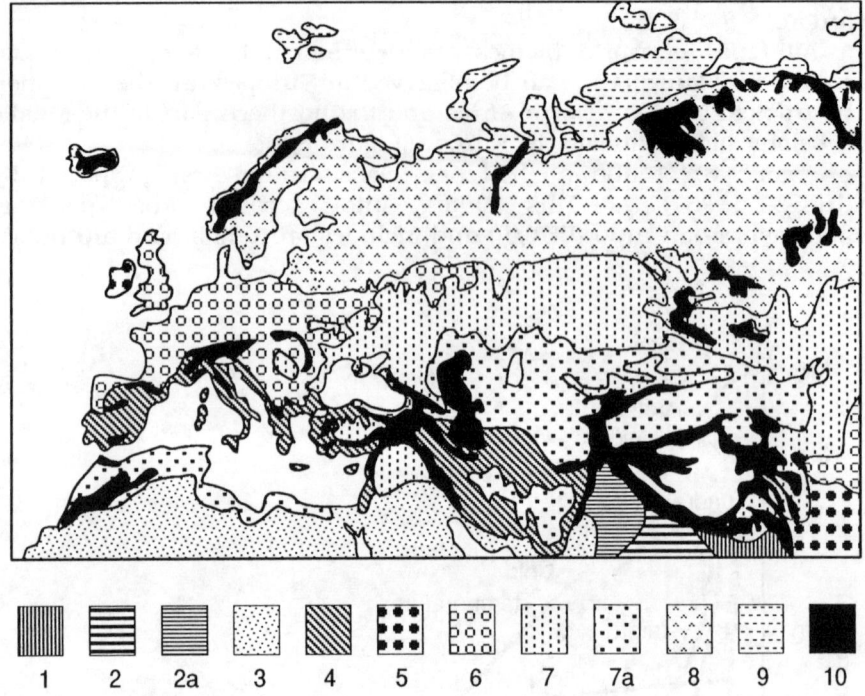

1.   Evergreen forests of the lowlands and mountain
     sides.
2.   Semi-evergreen and deciduous forests.
2a. Dry woodlands, natural savannas or grassland .
3.   Hot semi-deserts and desert.
4.   Scleryphyllous woodland with winter rain.
5.   Moist warm temperature woodlands.
6.   Deciduous (nemoral) forest.
7.   Steppes of the temperate zone.
7a. semi-desert and desert with cold winter.
8.   Boreal coniferous zone.
9.   Tundra.
10. Mountains.

Figure 8.2. Major vegetational zones. (Source: Walter, 1964.)

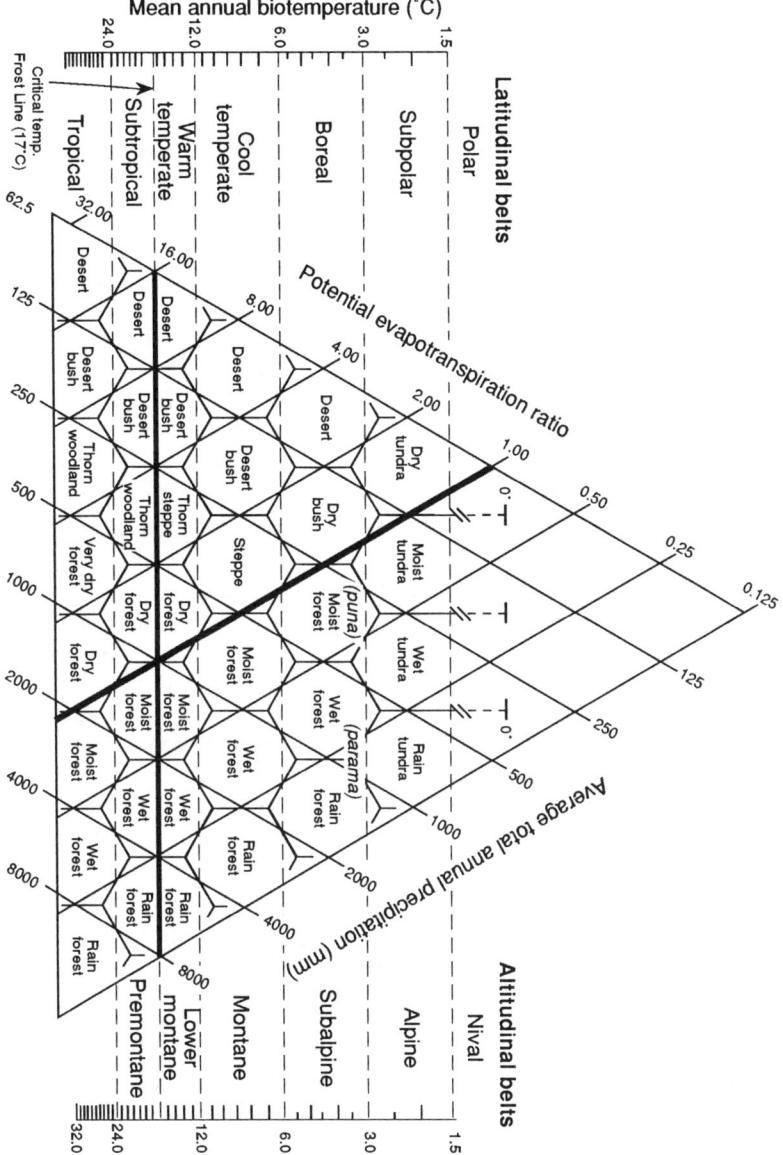

Figure 8.3.   Holdridge (1967) life-zone classification (adapted from Emanual *et al.*, 1985). The horizontal heavy line separates the tropical from the temperate biomes. The oblique heavy line (where potential evapotranspiration equals total annual precipitation) separates the dry and humid biomes. Notice exponential scales on all axes.

zonation. The heavy horizontal line in Figure 8.3 marks the limit between temperature and subtropical vegetation which coincides with the southern limit of frosts; the other heavy line marks the limit between dry and humid types of vegetation (PET/P > 1 or < 1 respectively). The scales for temperature differ; *biotemperature* of Holdridge (1967) excludes the period of no growth during the winter, whereas the mean annual temperature values used by others span the whole year. Therefore, the Holdridge classification is ecologically more informative and is more appropriate for correlation with biotic processes of humification and retention of soil carbon. This correlation assessment was made by Post *et al.* (1982) and is shown in Figure 8.4.

A potential evapotranspiration ratio of about 1 corresponds to approximately 10 kg m^{-2} carbon in mineral soil. The carbon level in mineral soils is an indicator of its water- and mineral-holding capacity. 7 kg m^{-2} is the minimum for ecologically acceptable dryland-crop farming. Below that limit the soils should be kept permanently under grass and herb vegetation and only extensive and controlled grazing allowed. This limit excludes virtually all tropical dryland from crop production, except for oases and dry forest regions where sufficient organic materials can be obtained to replenish the soil carbon content (stubble and green manure). But also in temperate dry regions (the triangle enclosed by PET/P = 1 and biotemperature line of 17°C in Figure 8.3), carbon depletion of farmed soils might occur so rapidly that constant replenishment with organic material would be required. As this can only be achieved on a limited scale as in domestic gardens, the conditions enclosed by this triangle (Figure 8.3) should also be regarded as unsuitable for arable farming and left to extensive grazing. Yet, such areas are often used for arable crops and devastating results can ensue over large areas.

Figure 8.2 shows a marked difference in the north-south zonation of biomes in Europe, east (continental) and west (oceanic) respectively. The north to south transect drawn by Schennikow for eastern Europe from the Arctic to Turkestan (Walter, 1970), as given in Figure 8.5, shows a very gradual transition from the tundras in the north to the deserts in the south. The zonation applies in principle to the western part of Europe, although it is interrupted by mountains (Pyrennees, Alps and Carpathians). These mountains are the major reason for extended broad-leaved forests in that part of the continent.

Almost completely lacking over most of western Europe, except in parts of central Spain and North Africa, are the temperate cool steppes and semi-deserts. In contrast to this, south-western Europe is particularly rich in sclerophyllous evergreen forest and scrub ("macchia") and, more to the north, rich in humid broad-leaved deciduous forests; both biomes are being replaced by temperate agro-ecosystems and tree plantations.

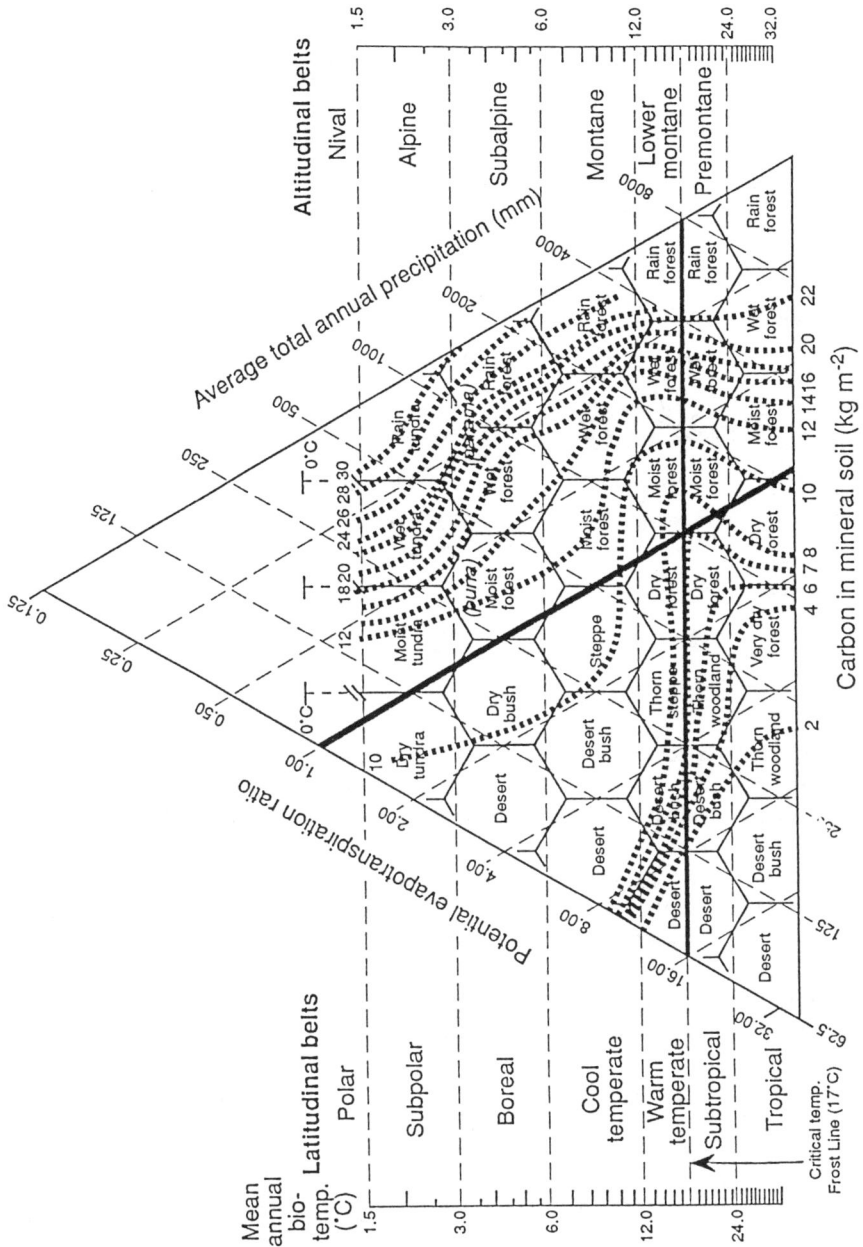

Figure 8.4. Carbon in mineral soil (in kg m^{-2}) overlaid with the Holdridge life-zone classification (adapted from Post *et al.*, 1982). Notice the position of the line indicating 7 kg m^{-2}, beyond which only extensive grazing and no tillage should be allowed.

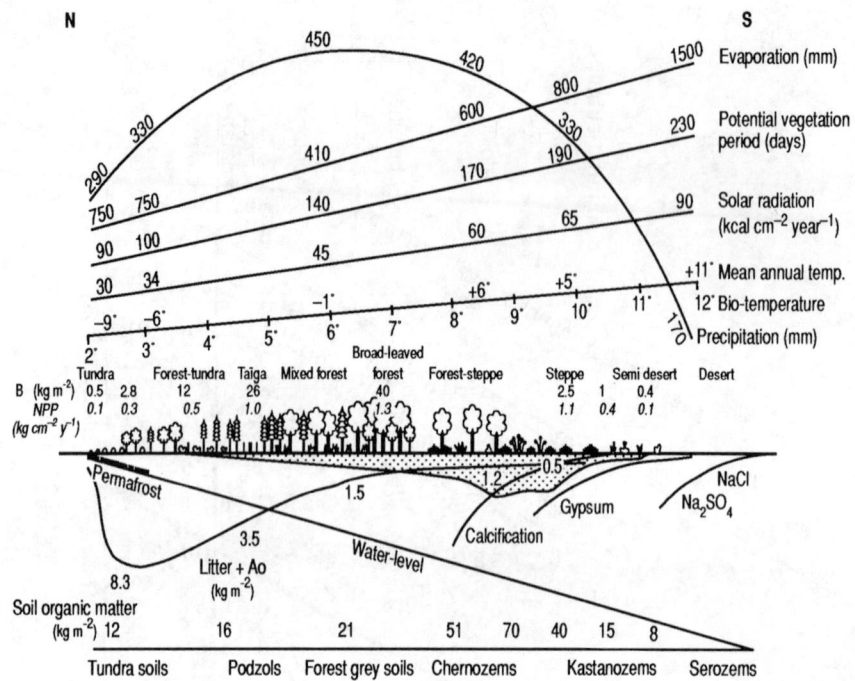

Figure 8.5. North-to-south succession of biomes in eastern Europe from tundra to desert. Indicated are i) climate parameters, including precipitation, evaporation, potential vegetation period, solar radiation and mean annual temperature, ii) vegetation types, with B = biomass and NPP = net primary productivity, and iii) soil parameters (water level, permafrost, calcification, gypsification, salinization/sodification, litter-accumulation in upper soil horizon, humus layer (in gray), soil organic matter and soil types). Added to this are mean annual biotemperatures to fit the diagram with the Holdridge life-zone classification in Figures 8.3 and 8.4. (After Schinnikow in Walter (1970), modified.)

Another important east-west difference in Europe is geological in nature, namely the differential displacement since the Early Tertiary period of the northern boundary of the arid zone (Figure 8.6). This difference is important ecologically and the soil formation processes as summarized in Figure 8.7 have to be considered; in particular, the process of calcification and podzolization.

Calcified soils are formed over millions of years under cool prairies to hot deserts at low moisture and deep water tables, so that soil minerals do not leach out (so-called closed soils). Peats and podzols are soils formed underneath boreal and temperate forests. Minerals are leached out easily from such soils, rendering them slightly acidic.

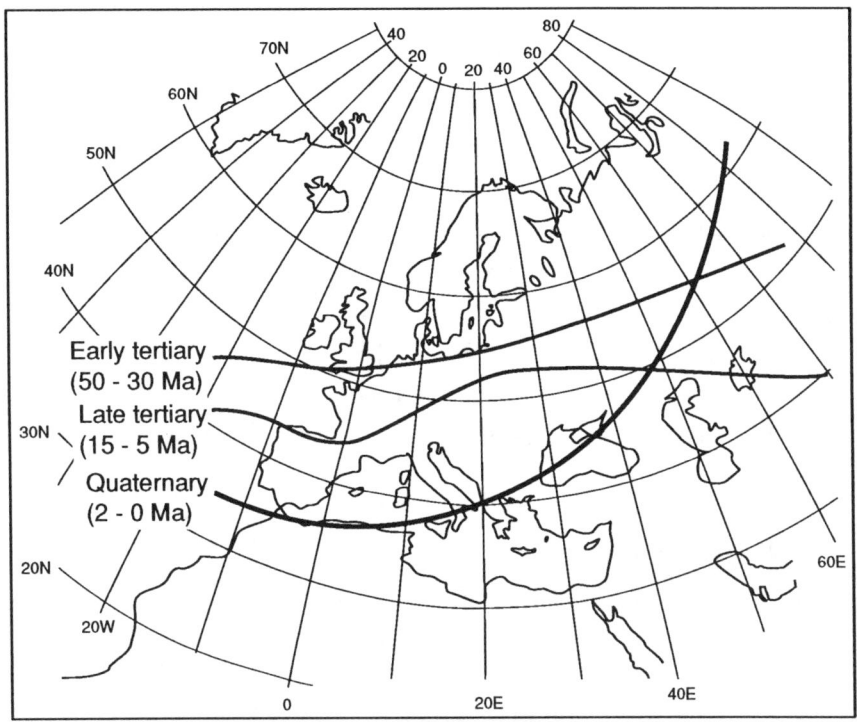

Figure 8.6. Displacement of the northern boundary of the Northern Hemisphere arid zone, based on the distribution of evaporites such as gypsum, salt, etc. (Modified after Lotze, 1963).

Podzolization occurs by accumulation of humus in the upper soil horizon. The B horizon of a podzol is leached of base nutrients and there is often a zone of deposition farther down the profile characterized by the accumulation of clays, iron compounds and organic material. In the Early Tertiary period, many very fertile calcified soils were formed as far north as the line from southern England to southern Urals (Figure 8.6). In the Late Tertiary period, this type of pedogenesis was largely restricted to south of the line following the Pyrennees-Alps-Carpathians-Caucasus-Aral to the sea. During the Pleistocene glaciations about 2 million years ago, boreal and temperate forests were pushed south on the mineral-rich calcified soils of western Europe and North Africa but not so far into eastern Europe. Here, the cool prairies and steppes shifted north, even beyond the line of the Early Tertiary.

The temperate forests on the west European calcified soils are relatively young and the process of podzolization was short compared with that of calcification. South of the Pyrennees-Alps-Carpathians-Caucasus-line, podzolization has always been less intensive than north

Climate			Pedogenic process	Characteistic plant formation class	Zonal soil group	Main soil class
Moisture	Temperature	Climate type				
High	Cold	ET	GLEIZATION	Tundra	Tundra	PEDALFER
	Cool	D	PODZOLIZATION	Needle-leaf forest	Podzol	
	Warm	C		Deciduous forest	Podzolic	
	Hot	A	LATERIZATION	Rain forest	Lateritic	
Low	Cool-warm	BS	CALCIFICATION	Prairie	Chernozem	PEDOCAL
				Steppe	Chestnut and brown	
	Hot	BW		Desert	Desert and red desert	

Figure 8.7. Soil formation processes (pedogenesis) in relation to climate and vegetation. (Adapted after Cox and Atkins, 1979).

of it and calcification was more prolonged and more intensive in nature. This is still reflected in the fact that southern soils (except where they are locally eroded to granitous bedrock) do not easily acidify.

Peat and humus accumulation is greatest under the forest- tundra (see the line for Litter+Ao in Figure 8.5) and decreases towards the steppes. This is because decomposition by soil bacteria is very slow under tundra temperatures, about twice the tundra rate under forest-steppe, and again increases by a factor of two under desert conditions (doubling of activity at each 10°C increase). The distribution of peat areas in Europe is shown in Figure 8.8. The decreased deposition in mountainous areas is particularly noteworthy. The climate would favor peat formation in the mountain regions; however, erosion counteracts accumulation.

Bacterial decomposition of plant material in the soil leads to accumulation of fine soil carbon, reaching a maximum level under grasslands and steppes with little or no woody material to be decomposed. Under deserts the oxidation is so rapid that not only raw litter but also fine soil carbon is almost completely lacking (see the

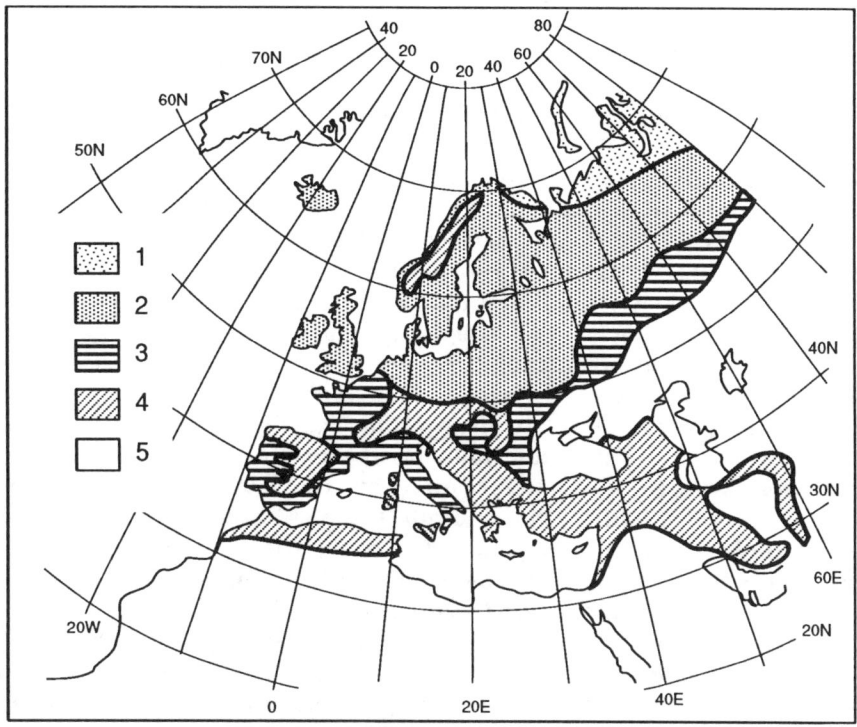

1.   Polar areas and tundra with low rate of peat accumulation.
2.   Generally low lying areas with intensive peat accumulation
      under forest-tundra, taiga and mixed forests.
3.   Regions with medium to minor peat accumulation under
      temperate deciduous forest and open woodland.
4.   Mountainous regions with little peat accumulation due to
      erosion.
5.   No peat accumulation under steppe and desert vegetation.

Figure 8.8.   Major peat areas in Europe.   (Adapted from Heikurainen,
1973).

shaded layer in Figure 8.5, which is inversely related to the Litter+Ao
line as far as the steppes, and then reduces to the desert region).
       As a consequence of what has been shown in Figures 8.6, 8.7
and 8.8, a marked east-west difference will also occur as a result of the
man-induced climatic change over the next centuries. If south-western
Europe loses its sclerophyllous forest cover, largely acquired during the
Pleistocene glacial periods, steppes and deserts will invade the relatively
thin, mineral-poor and young podzols that are rapidly oxidized and
eroded to expose the thick Tertiary calcareous rock ("Karst"). But if the
eastern steppes penetrate deep into the southern Russia black earth
(thick podzols) with only little or no underlying Tertiary calcification, the

soil carbon loss will be huge and much of the minerals would be leached.   Eutrophication as well as salinization and acidificiation of groundwater and surface water will become serious problems.

8.3.   Major Future Shifts of Biomes and Land Use

The climate scenario presented in Section 8.1 will now be considered for the biomes represented schematically in Figure 8.5. If there were to be an increase of 4-5°C, then broad-leaved forest would become a steppe; similarly, tundra would turn into a mixed or broad-leaved forest. The former is more likely as not only temperature but also the light intensity during the growing season is a limiting factor in the latter case.

The incoming solar radiative energy at the Arctic Circle is only one-third of that at the Tropic of Cancer and it is effective only during the short summer. The gain in growing season will mainly be at low light intensity and hence low photosynthesis. Although total primary production will increase, respiration by the whole community will increase more rapidly as community respiration is temperature dependent but not light dependent. Thus, the present tundras and taigas will become major sources of CO_2 rather than sinks as has been their traditional role since the Pleistocene.

Major uncertainties still exist with respect to the shifts in the hydrological balance, or more precisely precipitation, as evaporation and runoff are mathematically related to temperature and wind (Olejnik, 1988 compared present and future conditions of evapotranspiration and runoff for a changing climate) but rainfall is not. Here the problem is faced that the global circulation models are too crude to deal with regional hydrology. In the following Sections some of the changes in biomes and land use patterns responding to a climate change will be discussed for several regions of Europe.

8.3.1. Coastal lowlands

The regions that are likely to be most affected by sea level rise are estuaries with important industrial harbors and rich agricultural hinterlands (Figure 8.9). Losses of present natural wetlands will probably not be compensated for as in most places human activities will not permit the formation of new wetlands (RIVM, 1987). Losses of agricultural production, however, can partially be compensated by the promotion of mariculture in the new lagoons. This is a worst case scenario with storm peaks five meters higher than at present.

Some workers suggest that average sea level rise will be between 0.5 and 1 cm per year over the next century, but that the major damage will be from floods as extreme events become more frequent (Hekstra, 1986). Further urban and industrial development in the vulnerable areas should be discouraged, but for most countries such

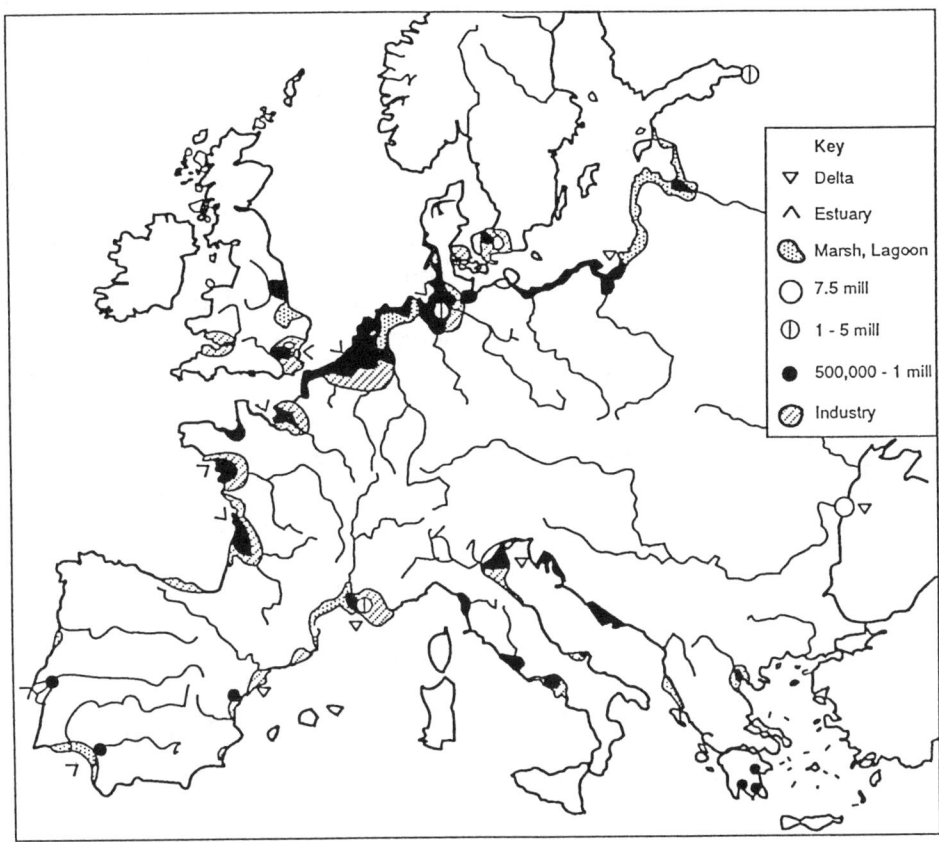

Figure 8.9.   European areas vulnerable to sea level rise.   (Source: Jelgersma, 1987.)

development will be difficult to avoid (RIVM, 1987).

    It is not only the coastline that is at risk and will need costly protection.   Salt penetration into groundwater will affect lowland agriculture and drinking water production (Chapter 13).   Technically and economically Europe is capable of coping with these problems, but developing countries such as Indonesia, Thailand, Bangladesh and the Maldives do not have the necessary resources (Hekstra, 1988b).

8.3.2. Mediterranean region

Increasing aridity is the major problem in the Mediterranean region. The implications are diverse and include lower rates of organic matter production in the soil, deterioration of soil structure, changes in the salt and water balances and particularly shifts in the partitioning of rainfall between infiltration and runoff. Higher amounts of overland flow will increase the risk of erosion and flooding, sedimentation in river channels and dike breaks. Agriculture will require more irrigation, but the water will become increasingly more saline (Chapter 13).

Episodic forest fires will become more frequent and more devastating, helped by man-induced acidification, because of a build-up of burnable litter due to lower rates of decomposition. The shift from subhumid forest (pines and sclerophytes) to desert in the Mediterranean will cause a large reduction in tree biomass; however, species diversity will not be reduced as the woodland and shrub deserts are more species rich than the forests they replaced (Figure 8.10). Further aridization will then lead to biotically more impoverished true desert.

In the Mediterranean rivers, cold water species assemblages will move upstream or disappear. Lower river courses will become more sluggish or dry out. Most vulnerable are the Plecoptera insects and

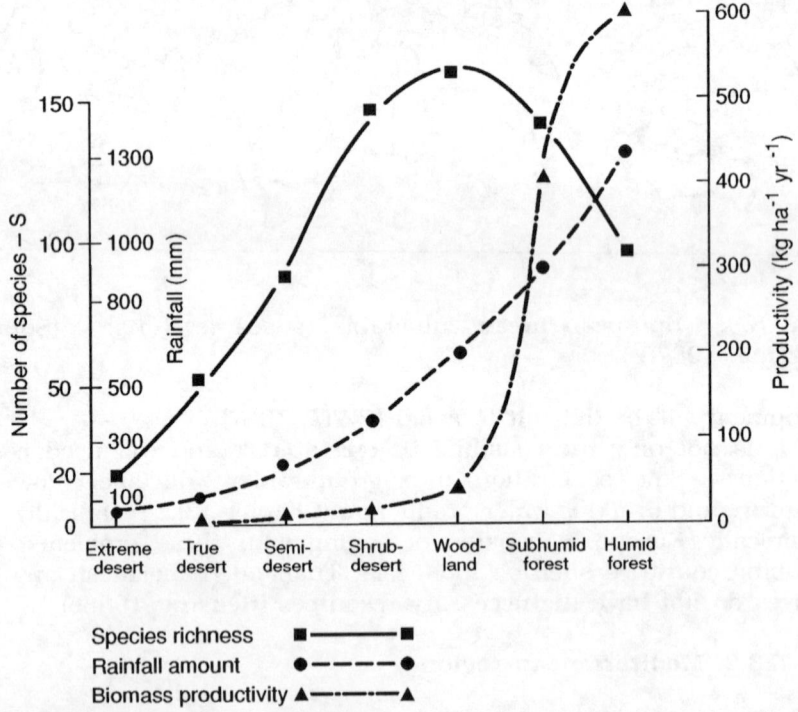

Figure 8.10. General relationship between vegetation types (number of dominant plant species), biomass production and annual rainfall. (Adapted from: Imeson and Dumont, 1987.)

Salmonid fish. Shallow lakes and reservoirs will experience algal blooms, periods of anoxia and massive fish kills more frequently. Estuarine swamps and lagoons will become increasingly saline and, consequently, will lose biotic diversity. Even in deep lakes, the hypolimnion (lower, colder layer) will warm up so much that salmonids and coregonids will disappear (Imeson and Dumont, 1987).

Some general circulation models (such as the GISS) forecast an increase in precipitation over the Mediterranean region. Generally, this would have more beneficial effects on the already heavily eroded parts of North Africa and eastern Mediterranean and more negative effects (increased erosion) on the northern and western parts. Such effects would occur if the southerly limit of the polar front rains were to penetrate deeper into the Sahara. But, on the other hand, the trajectories of the Sudano-Sahelian depressions could also extend farther north and bring more Saharan dust over the Mediterranean (Figure 8.11).

A special consequence of sea level rise and an increase in sea water temperature will be the colonization of many tropical plankton and fish species and even of corals from the Red Sea through the Suez Canal into the Mediterranean. However, the present extent of sea water pollution would probably prevent settlement of any coral species.

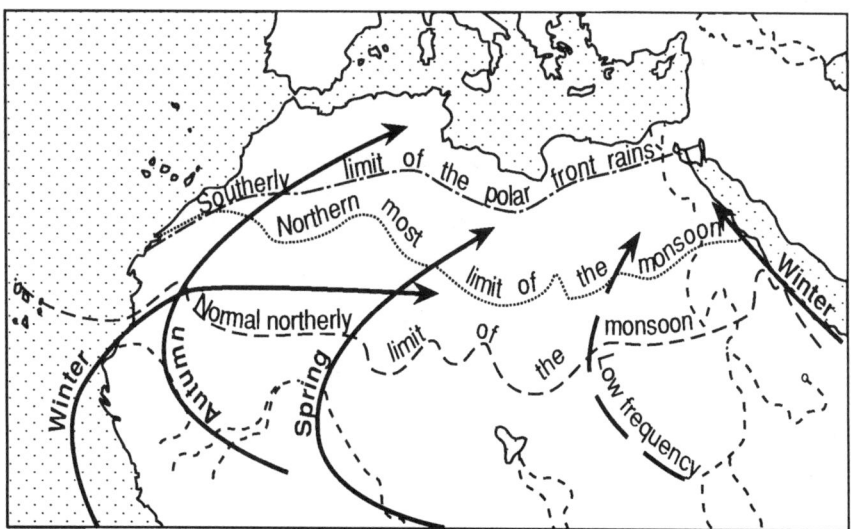

Figure 8.11. Trajectories of Sudano-Sahelian depressions and limits of the polar front and monsoon rains. (Adapted from: Dubief, 1979.)

Agriculture and horticulture throughout the Mediterranean will become less reliable; however, it is presumed that new technologies (hydroponics, biotechnological and genetic manipulation) will provide the necessary adaptations to cope with greater climatic instability and increasing physiological stress due to salinization. The outstanding feature for agriculture, however, will be that the area will become virtually frost-free, removing one of the major risks involved with fruit tree cultivation.

Different climatic models give completely different results for biomass potential (that is, the potential for plant productivity) in particular for the Mediterranean region. The GISS-model shows that countries such as Italy and Greece would gain 19 per cent and 11 per cent respectively, whereas the BMO model predicts that they would lose 5 per cent and 36 per cent respectively (Briggs, 1983). The differences between the model results are less pronounced north of the Alps.

van Huis and Ketner (1987) have shown that areas of natural vegetation would be most severely affected (Figure 8.12). But at the desert fringe, such as the Negev in Israel, the acuteness of aridity is not necessarily directly correlated with annual precipitation. The sensitivity of an arid ecosystem to climatic change may be greater for regions where loess soil cover is predominant and where infiltration and water absorbtion capacity are higher than on rocky soils. Hence, traditional agriculture favored the rocky areas where the low levels of precipitation is more readily available in the root zone of plants (Yair and Berkowicz, 1988).

8.3.3. Alpine regions

A mean annual warming of about 5°C over the Pyrennees, Alps, Carpathians and Caucasus will theoretically shift all vegetation zones higher up the mountain by approximately 500 m on north facing slopes and by 800 m on south facing slopes, as the potential growing seasons at all elevations will start earlier and last longer (Figure 8.13). The entire area below 1,500 m (northside) to 1,800 m (southside) would become frost-free. The total amount of water stored in the snow pack will decrease considerably as a result of the increase in winter temperature. Even if more intensive episodic snowfall occurs, melting will start earlier, causing higher spring peak runoff. Ozenda (1987) assumes a 300 to 500 km northward shift of circulation patterns which should decrease the contrast of high summer to low winter precipitation that is presently typical for the majority of the Alps.

At the same time the locally fixed center of cyclogenesis over the Gulf of Genoa will continue to steer summer precipitation towards the Alps, thus saving the central alpine valleys from desertification. Increased aridity will be noticeable on south facing slopes but to a lesser extent in the Pyrennees due to oceanic influences. Precipitation

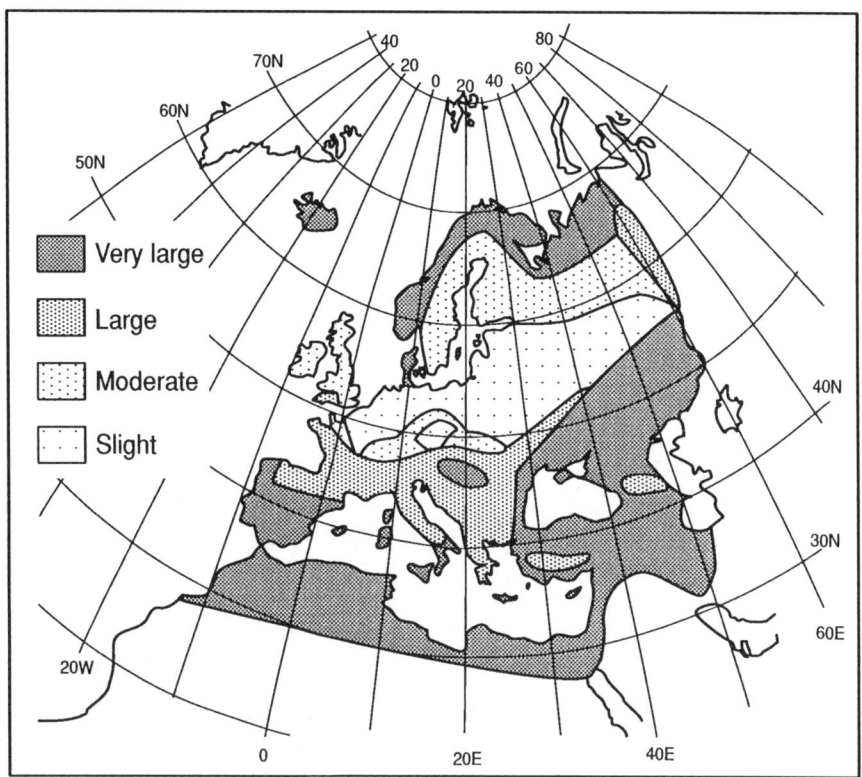

Figure 8.12. Extrapolation of the potential impact of climatic change on Europe natural vegetation (GISS scenario, 2 x CO$_2$). (Adapted from: van Huis and Ketner, 1987.)

would tend to become more episodic: less frequent and more intense, with more snow (and avalanches) at high altitudes and much less snow cover at medium and lower altitudes (the winter sport resorts).

A reduction of the area under permafrost would lead to a stabilization of scree slopes and a cessation of boulder creep. The expected drastic diminution of the surface area of glaciers would expose vast area of unstable morainic material, which is an active source of sediment into local water courses, lakes and reservoirs. Given more episodic peak flows, more erosion in the upper water course would be expected.

So far, it has been assumed that vegetation would react to climatic change as if independent of anthropogenic activity. However, the dieback of forests from pollution and other human influences will accelerate rates of soil erosion on deforested slopes and increase the occurrence of mudflows and the consequent sediment deposition on alluvial cones (present areas of human settlements). Only in the

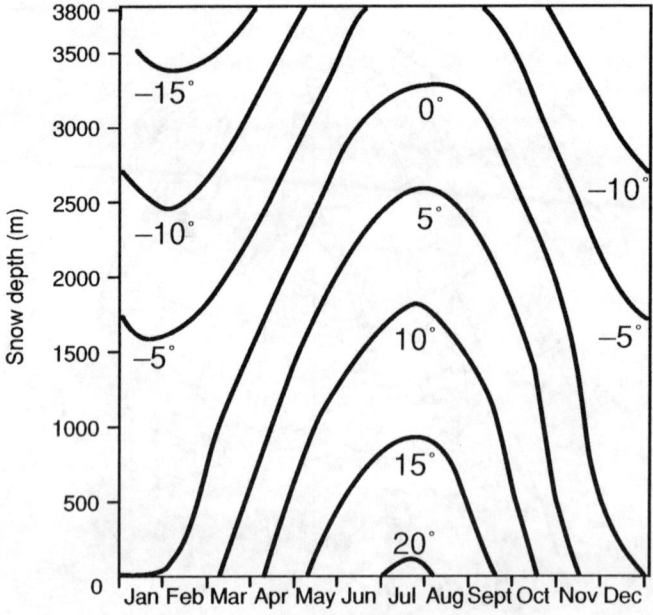

Figure 8.13.   Isoplethes of air temperature in the eastern Alps from average values of snow depth up to 3,800 m over the period 1851-1950. (Adapted from: Ozenda, 1987.)

situation where natural vegetation is allowed to adapt undisturbed to the new climate, particularly in the northern and central Alps, would a stabilization of the potential sources of morainic material and a decrease in the frequencies of mudflows be more likely (Kerschner, 1987).

Alpine agriculture will tend to shift from pasture land to higher temperature high valued vegetables, fruits, wine and tobacco and some areas may require irrigation (eastern Alps, Carpathians).   This would, however, affect the traditional pastoral economy and also result in an increase in salinization.

Alpine forestry will be most detrimentally affected; the upward shift of the biomes will reduce the area of land available for forestry and the trees will be affected by less reliable episodic precipitation and by an increase in acidification of groundwater (as a result of aridization). Rising temperatures alone do not cause significant physiological stress to the trees as long as water availability is unlimited.   Forest renewal would be threatened only if the timing of flower development, meiosis, anthesis, fruit set and seed development were disturbed (Cannell, 1987). However, genetic engineering and biomanipulation might help to overcome some of the Alpine forestry problems.

Of particular significance in the central European mountainous areas is the intensification and acceleration of soil erosion from the combined effects of industrial acidification and agricultural practice, as discussed by Pǎces in Chapter 14. Thus, forest damage is thought to be caused by direct aerial acid deposition and more indirectly by the acidification of soils through agriculture. In addition, climatic warming will enhance the loss of the acid buffering capacity of soils.

8.3.4. Oceanic temperate zone (western Europe)

The temperate zones east (continental) and west (oceanic) are so contrasting that they warrant separate discussion. The western temperate zone was almost completely covered with broad-leaved forests until the ninth century. Agriculture gradually replaced the forests and now the remnant forests are disappearing due to air and water pollution, aridization by man-induced lowering of water table level, and recreation. Higher temperatures, more episodic rainfall and more frequent storms will help to complete the full transition to a "culture-steppe" divorced from any indication of natural biome shifts. Almost the entire landscape will be influenced by man, such that the real factor determining future land use patterns is no longer climate but technology. The patches of relic forests will lose much of their species diversity and shift towards northern Mediterranean types of dry woodlands (van Huis and Ketner, 1987).

The south-western temperate zone (France, northern Spain) is likely to become increasingly drier, leading to higher but shorter duration peak flows in rivers and longer periods of extremely low flows, which will seriously affect fisheries and lead to salinization under irrigation. At the same time, chemical pollutants, compounds causing eutrophication and acidic deposition will be increasingly concentrated in the environment.

The most important changes in the north-western river basins (Meuse, Rhine, Weser and Elbe) will be the incidence of shorter and higher peak flows in spring after the high-alpine snow-melt, and the increased frequency of extremely low discharges in late summer to winter, affecting the water supply for the densely populated areas of north-western Europe (Jongman, 1987).

The change in peak flows will strongly alter erosion and sedimentation patterns which will particularly affect the more inland harbors. Sediment will not be carried all the way to the estuaries and sea, and the consequential lack of sediment at the coast will increase erosion rates in these sensitive ecosystems. This, coupled with the likely sea level rise, makes coastal harbors all the more vulnerable (Kwadijk, 1987; Hekstra, 1988b).

Areas potentially susceptible to increased soil erosion under the predicted climatic change mirror those areas which experience problems today, such as the hilly loess landscapes and the Fore-Alps. It is quite possible that the gross overproduction problem in Europe may actually come to the rescue since both surplusses and soil erosion may be tackled and controlled by the single action of taking marginal land out of production (as discussed by Lee in Chapter 1).

There is ample evidence (Klijn *et al.*, 1987) suggesting that increased crop productivity will result as higher ambient CO_2 concentrations stimulate both photosynthesis and water use efficiency, but the potentially enhanced productivity may be strongly hampered by restricted availability of soil moisture. Therefore, larger water storage capacity (surface water and groundwater) will be needed in low-lying areas; not only for crop production but also to counteract greater penetration of salt water due to sea level rise.

8.3.5. Continental temperate zone (eastern Europe)

Temperate grasslands dominated the Hungarian plains, the northern Black Sea coast to the southern Urals and far into southern Siberia. This vegetation type also extended south from the highlands of Anatolia through Iran into Afghanistan. The whole area is now under agriculture and through bad management practices much is becoming semi-desert as seen in most of Kazachstan and Uzbekistan (including the present interior of Iran-Afghanistan).

Mesopotamia, the cradle of western civilization, obtains its water supply from the sources of the Euphrates and Tigris in eastern Turkey, which, according to some climate models, will be affected by a decrease in precipitation for most of the year (exceptional peak flows excepted), so much so that soil moisture would be reduced by between 50-70 per cent. Desertification would then spread from Mesopotamia westwards to the presently well forested north-western part of Turkey and Bulgaria and along the Lower Danube, establishing a connection to the Great Plains of the Danube-Tisza (Figure 8.14).

The whole area will show a large increase in potential evapotranspiration, but the models do not agree on shifts in precipitation. The groundwater in the semi-arid Danube-Tisza interfluve is very sensitive to short-term fluctuations in climatic conditions. The groundwater may disappear during prolonged arid periods, at least as far as the first impermeable layer at several meters depth. Salt would accumulate at that depth but the top soil would be more or less free from salt. In spite of this doubtful advantage, longer-term shortage of water would become an obstacle to soil fertility. Irrigation would cease to be a good option because of increased water shortage and salinization. A high groundwater table is only of direct benefit to the vegetation in areas where the salt content of the groundwater does not threaten to contaminate the soil (Kerényi, 1987).

If the climate in the Great Plains shifts towards greater aridity, the

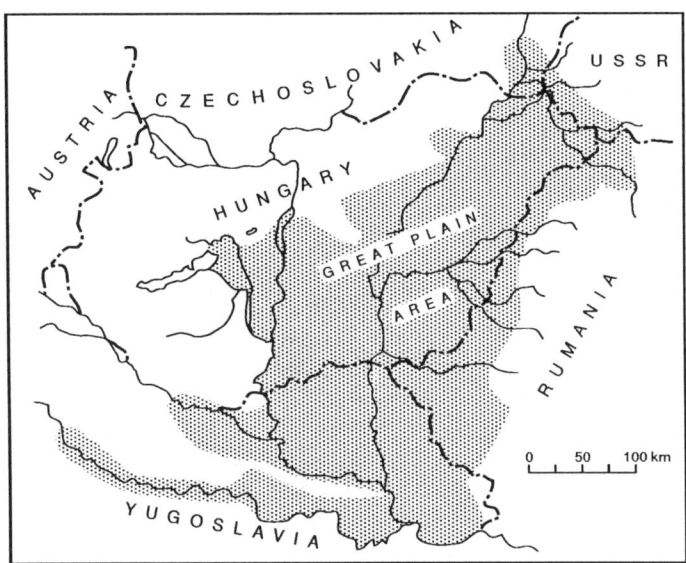

Figure 8.14. The central European Great Plain area (Hungary, Yugoslavia, Romania). (Adapted from: Pinczés, 1987.)

effect on most of the soils would be a reduction in fertility. But if, according to other models, there is more rainfall, then higher groundwater levels with a reduction in its salt content might be the beneficial consequence (Kerényi, 1987).

It is interesting to speculate on the sensitivity of lakes in these semi-arid zones to climatic conditions. If Lake Balaton experiences a rise of 0.5°C in annual temperature and a 5 per cent reduction in annual precipitation, evaporation from the lake surface would become greater than the surface inflow, and the outflow would soon cease, rendering Balaton a closed inland lake (Kindler, 1987).

Farther to the east are the fertile southern Ukrainian and Lower Volga loesses of mixed peri-glacial and Saharan origin. If this area were to experience winter temperatures typical for northern Italy, as indicated by some models, then an accordant increase in evapotranspiration would be expected strongly affecting water availability and soil moisture in summer (Kindler, 1987). Salinization in this zone will become a major feature (Szabolcs, Chapter 13). This does not hold, however, for the entire Volga river basin, which has shown an increasing trend in precipitation over the past decade. The higher precipitation will only fall over the upper Volga basin which is influenced by oceanic air circulation over the Baltic Sea.

The sedimentation history of the rivers that flow into the Black Sea has been extensively researched (Figure 8.15). Sedimentation rates

a)

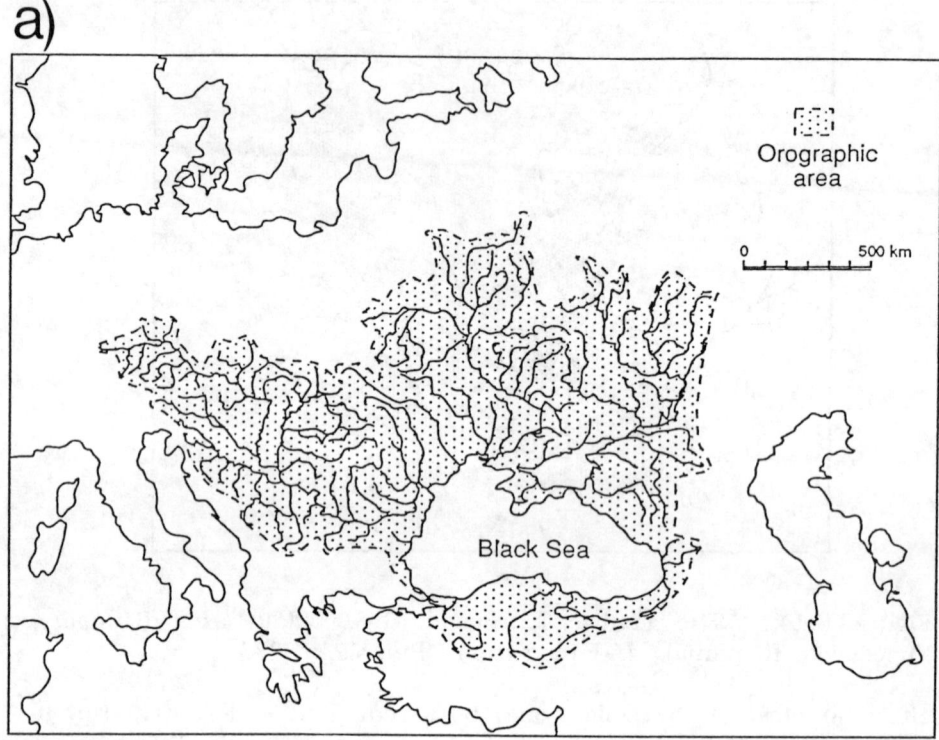

b)

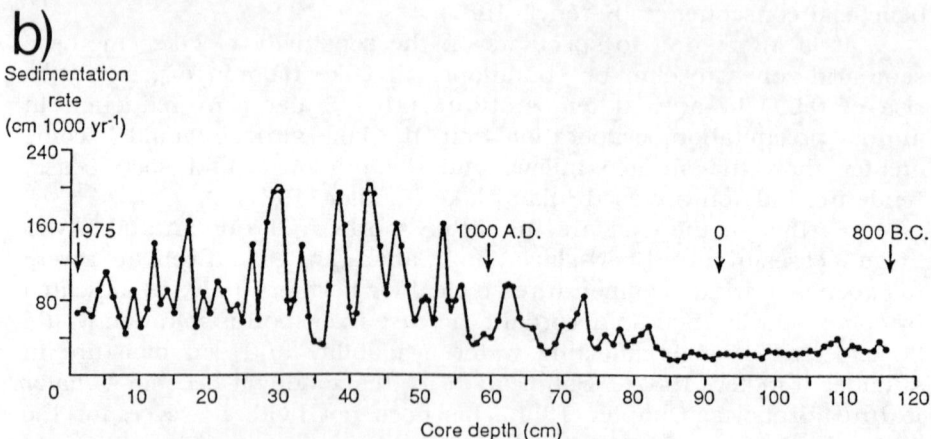

Figure 8.15. a) Orographic map of the Black Sea drainage area; and b) sedimentation rates since 800 B.C. as derived from varve model ages in a core taken from the Black Sea. (Source: Degens *et al.*, 1976.)

were uniformly low until about 200 A.D., after which the successive colonization waves, manifested by increased sediment load from deforested areas, show the transition to the present "culture-steppe". Denudation rates are excessively high for the easternmost short steep rivers Coruh and Rioni, which have a very low salt content; they are very low for the broad flat Dneiper and Don which have comparatively high salt contents. Farther north the grassy plains of the Ukraine gradually become more wooded. This is a transition to the temperate broad-leaved forests on black earth (Chernozems) rich in loess and in soil carbon. This loess (several decimeters to more than a meter thick) is of peri-glacial origin (Figure 8.16) and supports broad-leaved forests exhibiting some features common to the forests in the oceanic temperate zone; however, the continental zone is characterized by species more tolerant of low winter temperatures. In addition, the eastern temperate forests are comparatively more intact than oceanic counterparts as they have been under less pressure from rural land use changes.

Contiguous to the boreal zone is a mixed coniferous-deciduous type forest bordering the southern Baltic Sea and extending through northern Russia (Upper Volga) to the Urals. The podzolic soil is very peaty with little or no loess, as the area was mostly ice-covered during the Würm glaciation. Most rivers from this area, except the Volga, flow into the Baltic Sea. Their precipitation/runoff ratios (Odra 23 per cent, Vistula 24 per cent) are generally much lower than west European rivers (Rhine 44 per cent).

The south-eastern Baltic and Upper Volga area will benefit greatly from the projected increases in temperature and precipitation (resulting from a deeper penetration of oceanic air). But, as a consequence, the risk of erosion will increase markedly. Other climatic scenarios, however, do not project a major increase in precipitation, at least not in summer, so agriculture may become more vulnerable to water shortage (Kindler, 1987). Pitovranov *et al.* (1988) made an assessment of the agricultural outlook for the areas around Leningrad and Cherdyn. In the Leningrad region, Winter rye yields would be much reduced; Spring wheat yields would only decrease slightly in the Cherdyn region west of the Ural Mountains. Both assessments are based on present day technologies and crop varieties.

Karavayeva *et al.* (Chapter 16) examines present land use pressures on the Russian Plain and observes a dramatic decrease in humus content over the past century in the order of between 23-60 per cent in the forest-steppe chernozems with highest figures in the Volga Region and lowest around Moscow. In the real steppes of the south-east, losses of humus content over the past century range from 17-40 per cent, with highest rates in Moldavia. These losses are mainly attributed to mismanagement of land but a slow climatic warming could well have contributed.

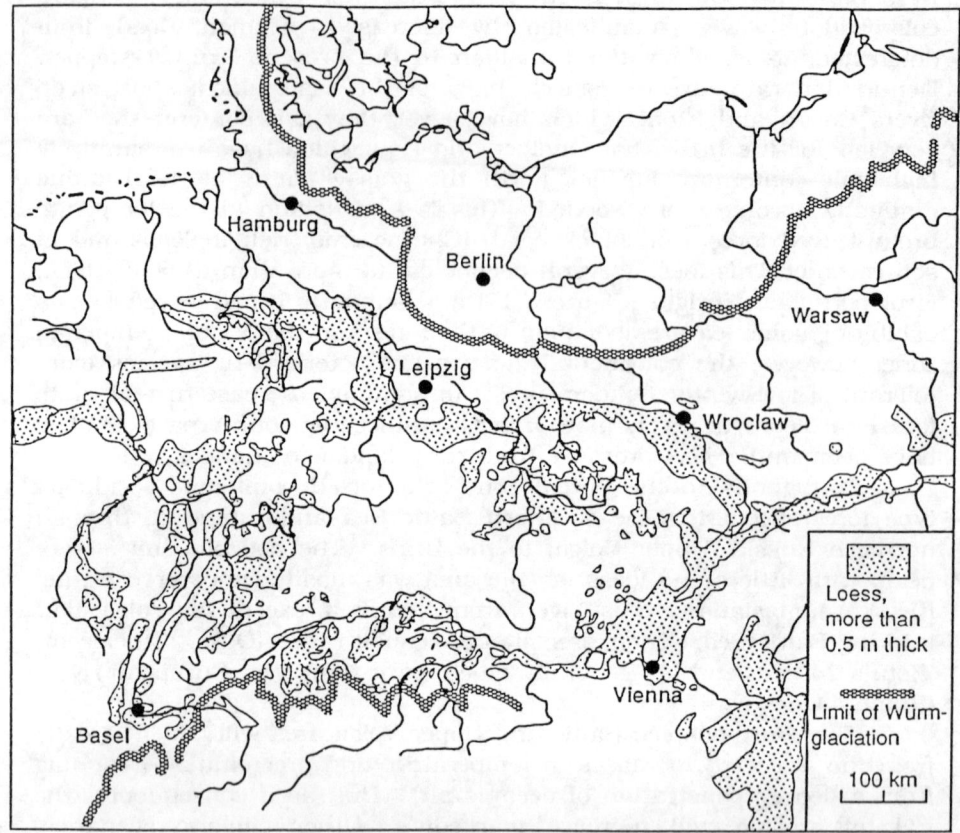

1.   Shaded area = loess, thickness more than 0.5 m.
2.   Limit of Würm glaciation.

Figure 8.16. Distribution of loess in Central Europe. (Source: Junge, 1979.)

8.3.6. **Boreal zone (Taiga)**

In the Boreal zone the natural vegetation is cold-tolerant pine forest on very peaty and humic soils with little or no buffering capacity (slight to strongly acidic). Cold-tolerance increases from the west (oceanic) to the east (continental). Climate scenarios for this part of the continent project an increase in summer temperature of 5°C and an increase in winter temperature of 8-10°C. The projected increase in annual precipitation is 20-40 per cent. The temperature-limited growing season will be prolonged by up to 70-100 days (Heino, 1987). Crops and weeds can easily adapt to the new climatic conditions but the rate of migration of trees is slow. Active planting and seed selection may ameliorate this situation (Dahl, 1987).

The potential growth of forest production is considerable: up to 25 per cent in the south and 50 per cent in the north (Lundmark, 1987). Forests will become more diverse with many species migrating from the south, but indigenous species that depend on the protective snowcover will become more vulnerable. The natural vegetation will change dramatically (Ingelög, 1987).

The potential for agriculture will increase markedly and the area could become a net food exporter (Chapter 1). Many crops from the temperate zone can be introduced, but pests and weeds will rapidly take advantage of the new situation. Detrimental factors for farming could be the decreasing permeability and trafficability of clay soils and a rainy harvest season (Heinonen and Torssell, 1987).

As runoff is expected to increase (Norway 5-10 per cent; Sweden 15 per cent; Finland 30-40 per cent) there is potential for greater use of hydropower. The floodpeak in spring will be diminished because of thinner snow cover, which means that water power could be used more efficiently. Other effects include an increased dilution of pollutants and larger groundwater recharge (Eriksson, 1987).

Fire has always played an important role in the taiga ecosystem because of the highly organic terrain (peat bogs, mires) and the flammable nature of the terpenes in pine trees. Forest dieback from air pollution and groundwater acidification augment the destructive role of fire. An *Abies* dominated forest has a relatively low fire hazard relative to a *Picea* or *Pinus* dominated forest. As *Abies* is likely to be more advantaged by climatic change and as soils and vegetation become wetter, forest fires may become less frequent (Wein and MacLean, 1983). Zachrisson (1977) has made it clear that fires have been instrumental in increasing the area covered by pines during the last 900 years in northern Sweden, as their seeds germinate more quickly after the fire than those of the mountain tundra species. However, an obscured climatic warming signal can be postulated as having contributed to the expansion of the pine forests.

8.3.7. Arctic zone (Tundra)

This biome zone will probably change most dramatically. The disappearance of permafrost will make the establishment of taiga tree species possible, although this would be hampered by low light intensity in spring and autumn. There is much speculation about the increased biomass potential. The fact that much of the European Arctic tundra is near sea level makes it necessary to consider what the impact of sea level rise might be in combination with a disappearance of permafrost.

Great losses will occur in the typical tundra wildlife as the land area available for northward expansion is limited except on Iceland, Svalbard and Novaya Zemlya, where the cold semi-desert may be transformed into tundra. In the mountains of Scandinavia the tree line will migrate altitudinally and most of the mountain tundra may disappear (van Huis and Ketner, 1987). Lifestyles of the indigenous northern people will change.

8.4. Climate Related Acidification, Eutrophication and Aridification

Will a change in predominant weather patterns influence long-range transport of air pollutants and hence acid deposition? This was studied within the IIASA RAINS model (Regional Acidification Information and Simulation model for Europe) with the input of climate data from the GISS 2 x CO_2 model (Pitovranov, 1988). In particular, the decrease of zonal circulation in winter could result in decreasing transport of air pollutants from western to eastern Europe, but computations showed insignificant deviations (in the 10 per cent range) in total deposition in comparison with basic scenario depositions. But what about the soils?

Alcamo *et al.* (1987) concluded that there will be continued acidification of soils in spite of reduced acid emissions. As much as 60 per cent of the soils in central Europe could have pH values of less than 4 by the year 2040 even with a 30 per cent reduction of the sulfur emissions from 1980. The problem is that although the reduction policies will certainly slow down acidification, the depletion of buffering capacity (base concentrations) will still occur at rates more rapid than the rate at which the soil can replenish them. In addition, there is a marked difference between the way wetlands and dry lands accommodate the nitrate and sulfate deposited from the air. Figure 8.17 illustrates how wetlands function as a sink and dry lands as a transporter of nitrate and sulfate. Thus, in areas of future aridification one can expect more rapid runoff of nitrate and sulfate into surface water and groundwater. In areas where soils become wetter the water will tend to improve, not only due to more rainwater runoff but also as microbial transformation rates of nitrate and sulfate increase.

Nitrogen fertilizers also influence the redox potential of soils and sediments and at low redox potential can stimulate remobilization of heavy metals from sediments into water in a form that is more bio-available than it was in the sediment. Also, aridification of soils (either climate-induced or by artificial lowering of the groundwater table) can strongly increase the bio-availability of toxic heavy metals by oxidation effects in the soil. Therefore, it is worth examining linkages of climate change with the widespread problems of acidification, eutrophication and the dispersal of toxic chemicals (Stigliani, 1988).

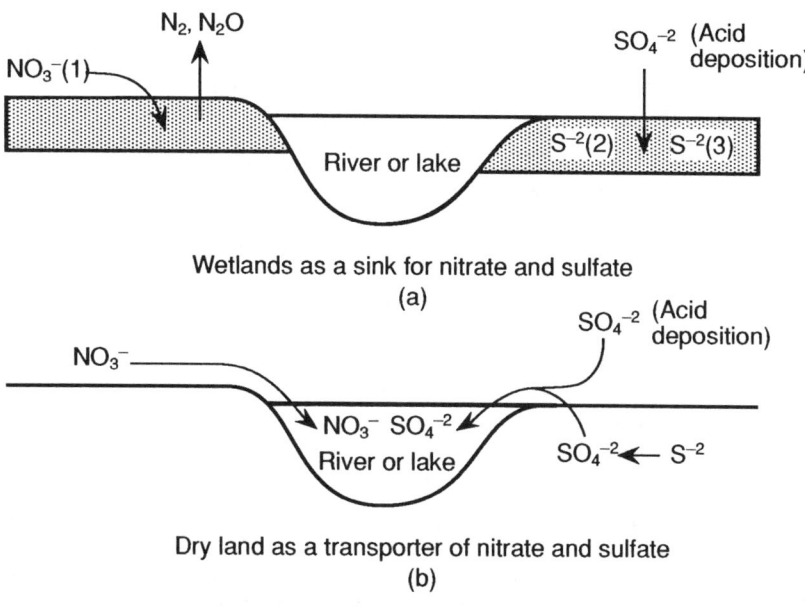

(1) Nitrate from runoff of nitrogenous fertilizer
(2) Sulfide minerals from former marine sediments
(3) Sulfide reduced from sulfate inputs of acid deposition

Figure 8.17. a) Ability of wetlands to buffer against nitrate and sulfate inputs to water bodies. Nitrate is reduced to molecular nitrogen (N_2) or nitrous oxide (N_2O), both of which are vented to the atmosphere. Sulfate (SO_4^{2-}) is reduced to sulfide (S^{2-}), which forms an insoluble precipitate in wetland soils. b) Under conditions where wetlands become dry, none of the protective reducing reactions occurs. In addition, sulfides that were stored in wetland soils may become oxidized and leach into water bodies. Shaded area indicates region of wetland. (Adapted from: Stigliani, 1988.)

ACKNOWLEDGEMENT

This Chapter is largely based on the findings of the European Workshop on *Inter-related Bioclimatic and Land Use Changes* (Kwadijk and de Boois, 1989).

REFERENCES

Acocks, P.J., 1953, Veld types of South Africa. *Union of South Africa Botanical Survey Memoirs*, **28**, 1-192.

Alcamo, J., Amann, M., Hettelingh, J.-P., Holmberg, M., Hordijk, L., Kämäri, J., Kauppi, L., Kauppi, P., Kornai, G. and Mäkelä, A., 1987, Acidification in Europe: a simulation model for evaluating control strategies. *Ambio*, **16**(5), 232-245.

Bach, W., 1988, Development of climatic scenarios. A: From general circulation models. In M.L. Parry, T.R. Carter and N.T. Konijn (eds.), *The Impact of Climatic Variations on Agriculture: Assessments in Cool Temperate and Cold Regions*. Kluwer, Dordrecht, The Netherlands. 125-157.

Briggs, D.J., 1983, Biomass potential of the European Community. *Report for the Environment and Consumers Protection Services of the European Community*. Sheffield, U.K.

Cannell, M.G.R., 1987, Possible effects of CO_2-induced climatic change on the temperate forests of Europe. Paper presented at the European Workshop on Inter-related Bioclimatic and Land Use Changes, Noordwijkerhout, The Netherlands. 17-21 October.

Cox, G.W. and Atkins, M.D., 1979, *Agricultural Ecology*. W.H. Freeman, San Francisco.

Dahl, E., 1987, Impact analysis of climatic change in the Fennoscandian part of the boreal and sub-arctic zone: evolution and climatic change. Paper presented at the European Workshop on Inter-related Bioclimatic and Land Use Changes, Noordwijkerhout, The Netherlands. 17-21 October.

Degens, E.T., Paluska, A. and Eriksson, E., 1976, Rates of soil erosion. In B.H. Svensson and R. Söderland (eds.), *Nitrogen, Phosphorus and Sulphur: Global Cycles*. SCOPE Report 7, Ecological Bulletin, **22**, 185-191.

Dubief, J., 1979, Review of the North African climate with particular emphasis on the production of aeolian dust in the Sahel zone and in the Sahara. In C. Morales (ed.), *Saharan Dust: Mobilization, Transport, Deposition*. SCOPE 14, John Wiley and Sons, Chichester/New York, 27-48.

Emanuel, W.R., Shugart, H.H. and Stevenson, M.P., 1985, Climatic change and the broad-scale distribution of terrestrial ecosystem complexes. *Climatic Change*, **7**, 29-43.

Eriksson, E., 1987, Impact analysis of climatic change in the Fennoscandian part of the boreal and sub-arctic zone: impact on water resources management. Paper presented at the European Workshop on Inter-related Bioclimatic and Land Use Changes, Noordwijkerhout, The Netherlands. 17-21 October.

Grigg, D., 1987, The industrial revolution and land transformation. In M.G. Wolman and F.G.A. Fournier (eds.), *Land Transformation in Agriculture*. SCOPE 32, John Wiley and Sons, Chichester/New York. 79-109.

Heikurainen, L., 1973, *Skogsdikning*. P.A. Norstedt and Söners Fölag, Stockholm, Sweden.

Heino, R., 1987, Impact analysis of climatic change in the Fennoscandian part of the boreal and sub-arctic zone: climate scenarios. Paper presented at the European Workshop on Inter-related Bioclimatic and Land Use Changes, Noordwijkerhout, The Netherlands. 17-21 October.

Heinonen, R. and Torssell, B.W.R., 1987, Impact analysis of climatic change in the Fennoscandian part of the boreal and sub-arctic zone: impact on agriculture. Paper presented at the European Workshop on Inter-related Bioclimatic and Land Use Changes, Noordwijkerhout, The Netherlands. 17-21 October.

Hekstra, G.P., 1984, Productivity and land use in semi-arid regions; lessons from the Sahel. Ecoscript 27, Mondiaal Alternatief, Zandvoort, The Netherlands.

Hekstra, G.P., 1986, Will climatic changes flood The Netherlands? Effects on agriculture, land use and well-being. Ambio, 15(6), 316-326.

Hekstra, G.P., 1988a, Effects of future climatic changes. Paper presented at the Deutscher Bundestag Enquete-Kommission "Vorsorge zum Schutz der Erdatmosphäre", Bonn, Germany. 6-7 June.

Hekstra, G.P., 1988b, Prospects of sea level rise and its policy consequences. Paper presented at the Workshop on Controlling and Adapting to Greenhouse Warming, Resources for the Future, Washington, D.C. 14-15 June.

Holdridge, L.R., 1967, Life Zone Ecology. Tropical Science Center, Costa Rica.

van Huis, J.C. and Ketner, P., 1987, Sensitivity of natural ecosystems in Europe. Department of Vegetation Science, Plant Ecology and Weed Science, Agricultural University, Wageningen.

Imeson, A. and Dumont, H., 1987, The sensitivity and vulnerability of the Mediterranean environment to bioclimatic change. Paper presented at the European Workshop on Inter-related Bioclimatic and Land Use changes, Noordwijkerhout, The Netherlands. 17-21 October.

Ingelög, T., 1987, Impact analysis of climatic change in the Fennoscandian part of the boreal and sub-arctic zone: impact on nature conservation. Paper presented at the European Workshop on Inter-related Bioclimatic and Land Use Changes, Noordwijkerhout, The Netherlands. 17-21 October.

Jelgersma, S., 1987, The impact of a future rise in sea-level on the European coastal lowlands. Paper presented at the European Workshop on Inter-related Bioclimatic and Land Use Changes, Noordwijkerhout, The Netherlands. 17-21 October.

Jongman, R.H.G., 1987, The Rhine-Maas megalopolis and its water resources: consequences of climate-induced hydrological changes. Paper presented at the European Workshop on Inter-related Bioclimatic and Land Use Changes, Noordwijkerhout, The Netherlands. 17-21 October.

Junge, C., 1979, The importance of mineral dust as an atmosphere constituent. In C. Morales (ed.), *Saharan Dust: Mobilization, Transport, Deposition.* SCOPE 14. John Wiley, Chichester/New York.

Kerényi, A., 1987, Expectable pedological effects of climatic changes in the Hungarian Great Plain. Paper presented at the European Workshop on Inter-related Bioclimatic and Land Use Changes, Noordwijkerhout, The Netherlands. 17-21 October.

Kerschner, H., 1987, Impact analysis of climatic change in the Central European mountain ranges: the Alps 4°C warmer. Paper presented at the European Workshop on Inter-related Bioclimatic and Land Use Changes, Noordwijkerhout, The Netherlands. 17-21 October.

Kindler, J., 1987, Impacts of CO_2-induced climatic change on water resources in the central European lowlands. Paper presented at the European Workshop on Inter-related Bioclimatic and Land Use changes, Noordwijkerhout, The Netherlands. 17-21 October.

Klijn, J.A., Bannink, M.H. and van Lanen, H.A.J., 1987, CO_2-induced climatic changes: impact on soil moisture deficit and grass yield in a sample area in The Netherlands; perspectives for extrapolation to larger areas. Paper presented at the European Workshop on Inter-related Bioclimatic and Land Use Changes, Noordwijkerhout, The Netherlands. 17-21 October.

Kwadijk, J., 1987, Identification of climate sensitive processes and landforms. Paper presented at the European Workshop on Inter-related Bioclimatic and Land Use Changes, Noordwijkerhout, The Netherlands. 17-21 October.

Kwadijk, J. and de Boois, H., 1989, *European Workshop on Inter-related Bioclimatic and Land Use Changes.* RIVM, Bilthoven.

Lotze, F., 1963, The distribution of evaporites in space and time. In A.E.M. Nairn (ed.), *Problems in Palaeoclimatology.* Interscience Publishers, London. 491-507.

Lundmark, J.E., 1987, Impact analysis of climatic change in the Fennoscandian part of the boreal and sub-arctic zone: impact on forest production in Sweden. Paper presented at the European Workshop on Inter-related Bioclimatic and Land Use Changes, Noordwijkerhout, The Netherlands. 17-21 October.

Olejnik, J., 1988, Present and future estimates of evaporation and run-off for Europe. WP-88-037. International Institute for Applied Systems Analysis, Laxenburg, Austria.

Ozenda, P., 1987, Impact analysis of climatic change in the Central European mountain ranges: altitudinal effects. Paper presented at the European Workshop on Inter-related Bioclimatic and Land Use Changes, Noordwijkerhout, The Netherlands. 17-21 October.

Pinczés, Z., 1987, The expectable effects of climatic changes on soil erosion on the Hungarian Great Plain. Paper presented at the European Workshop on Inter-related Bioclimatic and Land Use Changes, Noordwijkerhout, The Netherlands. 17-21 October.

Pitovranov, S.E., 1988, The assessment of impacts of possible climatic changes on the results of the IIASA RAINS sulfur deposition model in Europe. *Water, Air and Soil Pollution,* **40**(1-2), 95-119.

Pitovranov, S.E., Iakimets, V., Kiselev, V.I. and Sirotenko, O.D., 1988, Effects of climatic variations on agriculture in the sub-arctic zone of the USSR. In M.L. Parry, T.R. Carter and N.T. Konijn (eds.), *The Impact of Climatic Variations on Agriculture. Volume 1. Assessments in Cool Temperate and Cold Regions.* Reidel, Dordrecht, The Netherlands. 617-722.

Post, W.M., Emanuel, W.R., Zinke, P.J. and Stangenberger, A.G., 1982, Soil carbon pools and world life zones. *Nature,* **298**, 156-159.

RIVM, 1987, European Workshop on Inter-related Bioclimatic and Land Use Changes: Statement of Findings and Recommendations. National Institute for Public Health and Environmental Protection, Bilthoven, The Netherlands.

Stebbing, E.P., 1953, *The Creeping Desert in the Sudan and Elsewhere in Africa.* McCorquodale, Khartoum.

Stigliani, W.M., 1988, Changes in valued "capacities" of soils and sediments as indicators of non-linear and time-delayed environmental effects. *Journal of Environmental Monitoring and Assessment,* **10**, 245-307.

Walter, H., 1964, *Die Vegetation der Erde.* Volume 1 (second edition). Gustav Fischer, Jena.

Walter, H., 1970, *Vegetationszonen und Klima.* Ulmer Verlag, Stuttgart.

WCED (World Commission on Environment and Development), 1987, *Our Common Future.* Oxford University Press, Oxford/New York.

Wein, R.A. and Maclean, D.A. (eds.), 1983, *The Role of Fire in Northern Circumpolar Ecosystems.* SCOPE 18. John Wiley and Sons, Chichester/New York.

Whittaker, R.H., 1975, *Communities and Ecosystems.* MacMillan, New York.

Yair, A. and Berkowicz, M., 1988, Climatic change and resulting environmental effect: the case of the northern Negev of Israel. Paper presented at the IIASA Workshop on Land Use Changes in Europe, Poland. 5-9 September.

Zachrisson, O., 1977, Influence of forest fires on the North Swedish boreal forest. *Oikos,* **29**, 22-32.

Chapter 9

CLIMATIC CHANGES AND LAND USE POTENTIAL IN EUROPE

M.L. Parry and T.R. Carter

9.1. Introduction

The purpose of this Chapter is to consider the effects of the possible changes of climate discussed in previous Chapters on land use potential in Europe. The focus is on agriculture and the emphasis is on identifying approaches that are useful in the development of appropriate policies for managing the so-called greenhouse effect or for responding to it.

In the following Sections some methods that have been employed in previous studies to evaluate the effect of climatic changes on agricultural potential are presented. There are advantages in investigating the effects of climatic variations using a range of models, from simple descriptive indices regarding the character of the climatic resources for agricultural production, through statistical models that facilitate the prediction of plant responses to climatic variations (though with little explanation of the mechanisms) to more complex physiological models that consider the processes of plant growth and development. It is helpful to consider the effects of a range of climatic scenarios providing information both on the magnitude and on the rate of climatic change. Finally, in order to make the results more relevant for policy-makers, they should be interpreted in the light of a range of adaptations that could be implemented to limit, to anticipate or to react to climatic change.

9.2. Specific Approaches

In this Section previous studies which have related climate and plant response, both for the present day climate and under conditions of a changed climate, are reviewed. The scope of the review extends beyond agricultural crops to include other plants, as a number of studies have also considered the potential pattern and productivity of natural vegetation. Three broad approaches can be identified: i) descriptive, ii) correlational and iii) physiological.

F. M. Brouwer et al. (eds.), Land Use Changes in Europe, 209–231.

9.2.1. **The descriptive approach**

In essence, the descriptive approach involves describing and comparing the spatial patterns of a plant attribute and the spatial patterns of those environmental phenomena (including climate) thought to influence this attribute. The approach comprises five elements:

1)   the spatial pattern of the plant attribute (for example, plant growth rate, productivity or spatial distribution) is mapped based on measurements and observations from ground survey, aerial photographs or satellite imagery;

2)   the spatial patterns of important environmental phenomena are mapped (for example, soils on the basis of survey information and long-term average climate);

3)   the mapped distributions in 1) and 2) are compared visually to identify any qualitative similarities;

4)   those environmental factors displaying a spatial pattern that corresponds to a particular plant attribute are selected as potential proxy indicators of that attribute;

5)   where the proxy indicators are climatic, the effects of a climatic change can be simulated by adjusting the observed or derived climatic variables according to the "scenarios" of climatic change; the resulting pattern of climate can then be used qualitatively to infer changes in the spatial distribution of the plant attribute under study.

Individually, however, such direct measures of climate are seldom appropriate for describing the important climatic effects on plants. It will be more useful to combine these variables into agroclimatic indices that bear a closer relation to the plant response observed. For example, effective temperature sum (the sum of air temperatures throughout the year above a threshold temperature) can be related to the growth requirements of agricultural crops and forests and has been mapped for Finland by Kettunen *et al.* (1988) under both the present day (1931-1960) and GISS 2 x CO_2 scenario climates (Figure 9.1). Similarly, the length of the potential growing season under present day and 2 x CO_2

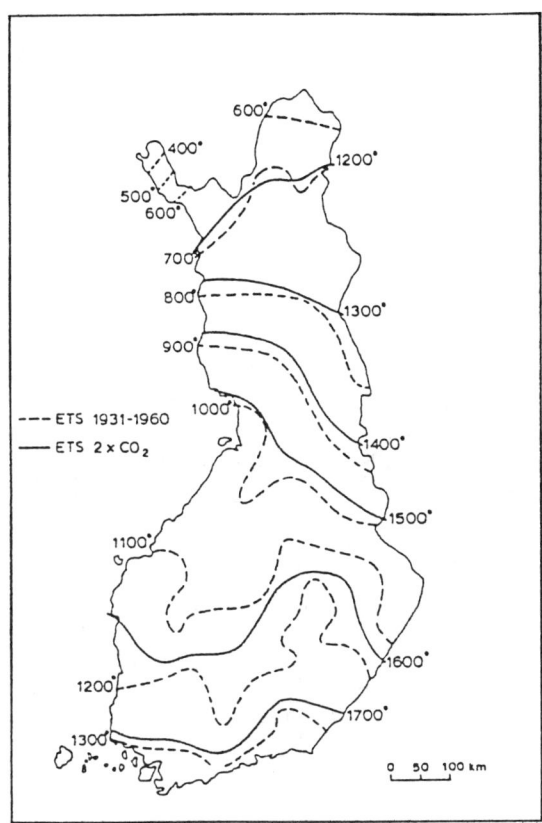

Figure 9.1. Effective temperature sums (degree-days) in Finland for the period 1931-1960 (after Kolkii, 1969) and estimates for the GISS 2 x CO_2 climatic scenario. (Source: Ketunen *et al.*, 1988.)

climatic conditions has been computed for the whole of Europe (Brouwer, 1989). A third example is the hydrothermal coefficient (a measure combining information on growing season temperatures and precipitation) which has been used by Nikonov *et al.* (1988) to characterize agroclimatic regions according to the degree of aridity in the Stavropol territory of the European USSR (Figure 9.2).

9.2.2. The correlational approach

The descriptive approach can be used to identify those climatic variables that appear, qualitatively, to be related to certain plant attributes. In order to quantify these relationships, the conventional method employed is to use statistical techniques of correlation and regression. In traditional ecology this type of analysis involves relating certain climatic

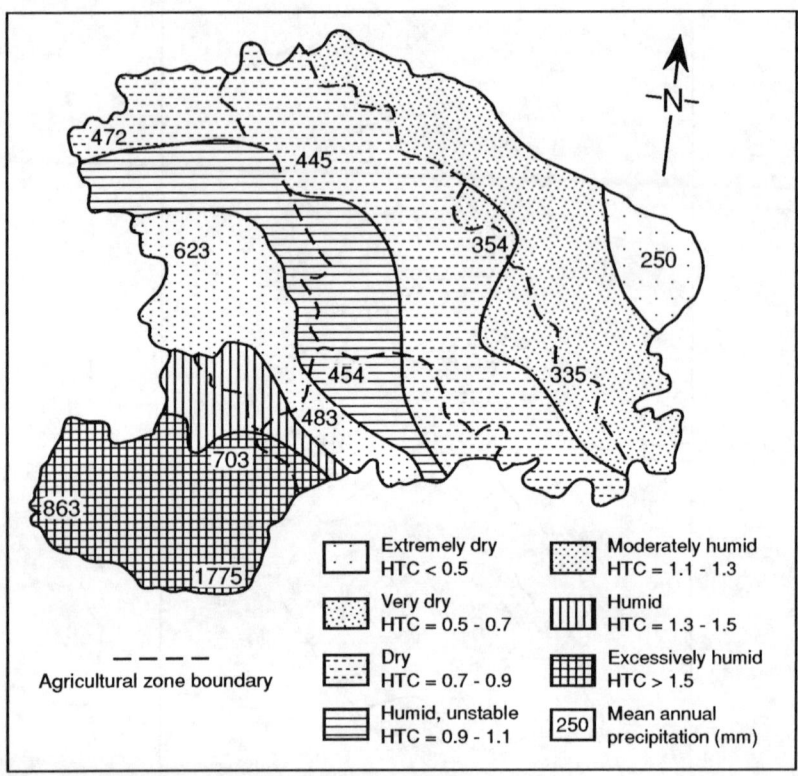

Figure 9.2. Agroclimatic regions of the Stavropol Territory, southern European USSR, defined according to the hydrothermal coefficient (HTC). (Source: Nikonov *et al.*, 1988.)

elements (for example, air temperature, effective temperature sum, precipitation, potential evapotranspiration) to observed plant attributes (for example, presence/absence, productivity, growth rate, yield) across a global or regional set of locations to produce a single prognostic model. Validation of the predictive performance of the model is conducted by using climatic data from an additional site (not used in the model construction) to "predict" the plant attributes at that site which are then compared with actual (observed) plant data. Following validation, global sets of long-term mean climate can be input to the model to predict plant attributes at as many sites as there are meteorological stations supplying data, hence providing a basis for mapping. Similarly, the mean climate can be adjusted according to specified climatic scenarios and these altered values input to the model.

There are many examples of this approach including relationships that have been identified between vegetation zones and climate, for example, the Holdridge Life Zone system (Holdridge, 1947, 1964), between net primary productivity (NPP) of natural vegetation and climate (for example, the "Miami Model" of NPP derived by Lieth, 1975) and between agricultural potential and climate (for example, the Climatic Index of Agricultural Potential used by Turc and Lecerf, 1972). The correlational approach has also been employed in estimating the effects of climatic change. For example, a semi-empirical model of biomass potential (Briggs, 1983), combining elements of the methods to compute potential evapotranspiration (Turc and Lecerf, 1972), and to relate evapotranspiration to NPP (Lieth and Box, 1972), was used to estimate the percentage changes in theoretically achievable biomass production of a mixed grass sward (relative to the present day) under two $2 \times CO_2$ scenario climates in the European Community (Meinl *et al.*, 1984; Santer, 1985) (Figure 9.3). A second example is the distribution of vegetation zones or "zonobiomes" as defined by Walter (1985) which has been mapped for Europe (de Groot, 1988) under both the present day (1931-1960) climate and a hypothetical $2 \times CO_2$ climate (Figure 9.4).

There are a number of significant drawbacks in using the correlational approach, the first being that it does not attempt to explain the mechanisms and processes whereby climate affects plant response. The relationships are statistically derived, presenting a black box with climate as an input and plant attributes as outputs. Secondly, results are strongly dependent upon the implicit assumption that it is valid to employ a statistical relationship developed on the basis of *spatial* variations in climate to estimate plant responses to *temporal* variations in climate at individual sites. Thirdly, the large changes in climate implied in some future projections are well outside the observed range of climatic variations at individual sites, and may well either lie outside the range or not resemble the combination of conditions to be found anywhere on the Earth at the present day. Thus, the predictions of future climatically determined plant attributes rely strongly on uncertain model extrapolations.

However, partial solutions can be found to the above problems. They involve the use of temporally- (rather than spatially-) based correlation models. These models are commonly employed to estimate regional crop yields and are constructed by relating time series of observed crop yields to time series of one or more observed climatic variables over an equivalent period, using statistical techniques such as regression or principal components analysis. Such models should be tested in regions other than that for which they were constructed to illustrate general applicability, but such global relationships are rarely obtainable for single crops. Alternatively, separate models can be constructed for different regions and operated in parallel to provide a large area assessment of crop responses to climatic variations. For example, 294 regression equations were developed by Hanus (1978) to relate climatic variables to wheat yields at 42 sites in the European

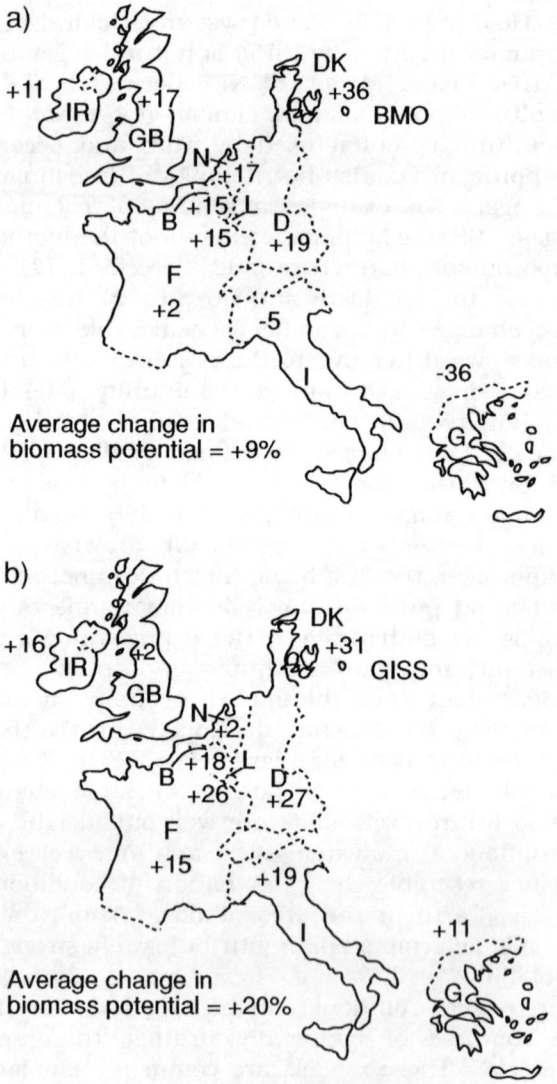

Figure 9.3.   Modeled changes in average biomass potential of ten European Community countries relative to the present day under: a) the BMO 2 x CO$_2$ climatic scenario and b) the GISS 2 x CO$_2$ climatic scenario (after Briggs, 1983). (Source: Santer, 1985.)

Community, and subsequently used to estimate the possible effect of two 2 x CO$_2$ scenario climates on yields (Meinl *et al.*, 1984; Santer, 1985).   The results were averaged across individual countries and are presented as percentage changes in yield for the period January to July

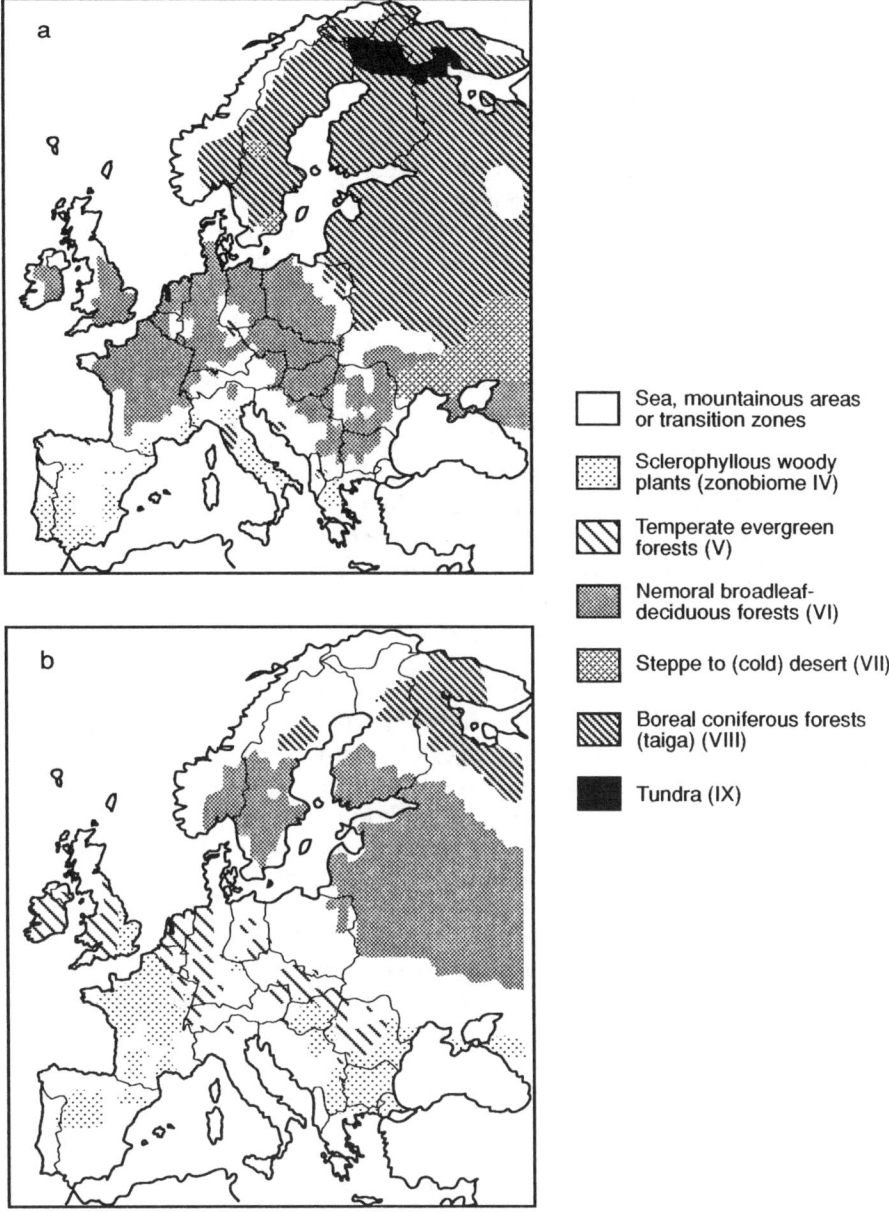

Figure 9.4. The potential distribution of natural vegetation ("zonobiomes" - Walter, 1985) for: a) the 1931-1960 period and (b) hypothetical increases in mean annual temperature and annual precipitation of 5°C and 10% respectively. (Source: de Groot, 1988.)

and as monthly averaged changes during the same period, both relative to recorded national average yields (Table 9.1).

A second technique, which has some potential for testing the validity of extrapolating temporally derived statistical relationships outside the range of observed regional conditions, is the use of regional analogues (Parry and Carter, 1988). This involves seeking those regions that have a climatic regime at the present day which resembles the future climate projected for a region of interest. Examination of present-day crop responses to climatic conditions in the "analogue region" may provide useful information about the robustness of model relationships in the study area.

To illustrate this, mean monthly temperatures in Iceland estimated for the GISS 2 x CO_2 scenario were found to be markedly warmer than the warmest months observed in the instrumental record in Iceland, resembling instead those of northern Britain at the present day (Bergthórsson et al., 1988). In studies of the likely effects of the 2 x CO_2 climate on grass yields in Iceland, information on temperatures and grass growth in northern Britain were thus used to justify the extrapolation of relationships developed between temperature and grass growth under present day conditions in Iceland.

9.2.3. The physiological approach

In contrast to the correlational approach, the physiological approach attempts to incorporate our understanding of plant physiological processes within a model framework. Some of these processes are represented as deterministic functions that are well enough understood to be regarded as accepted biophysical laws. Others that are either poorly understood or of secondary interest are frequently represented by empirical functions. Thus, no physiological approach can be described as truly deterministic; all incorporate at least some empirical (black box) elements. The different plant processes are brought together into a model framework that is usually *dynamic* and frequently *non-linear*. For example, this approach is often capable of representing critical levels of plant tolerance to climate. If these levels are exceeded only slightly, this may have a disproportionate effect on plant response. Such non-linearities and discontinuities of response are rarely incorporated in correlation-based schemes.

There have been few simulation studies in Europe which employ physiological models to evaluate the potential effects of climatic change. Models have been used, however, to evaluate the climatic control on plant attributes in Europe under the present day climate, and several are potentially applicable to climatic change experiments. For example, a semi-empirical grass yield model was developed for Europe by Hume et al. (1986). This requires daily climatic data as input and has been used to evaluate and map grass yields for much of western Europe on

Table 9.1. Mean monthly and cumulative changes in yields of wheat and spelt (in percentages) estimates for eight European Community countries in response to two scenarios based on global climate model projections of the future climate under doubled concentrations of atmospheric carbon dioxide (GISS and BMO). Values are expressed as percentages of 1975-1979 average yields[a] (after Santer, 1985).

Country	Average yield of wheat and spelt	Mean monthly changes (%)		Cumulative changes (%)	
		GISS 2 x CO_2	BMO 2 x CO_2	GISS 2 x CO_2	BMO 2 x CO_2
Belgium	4.72	-1.5	-1.1	-9.5	-6.8
Denmark	5.25	+2.7	+0.2	+18.7	+1.1
France	4.38	-1.4	-1.8	-9.6	-12.3
FRG	4.66	-0.2	-1.1	-1.1	-8.6
Ireland	4.79	-1.3	-2.5	-8.8	-17.5
Italy	2.58	0.0	-0.4	-0.8	-1.2
Luxembourg	3.09	+1.0	+1.0	+7.8	+6.1
Netherlands	5.82	+0.2	0.0	+1.2	+0.3

[a]All average yields are for Winter wheat and spelt, except for Ireland, for which only wheat and spelt data were available.

the basis of data from the 1951-1980 period.

Elsewhere in the World there are a number of illustrations of physiological models being used in climatic change experiments. For example, the so-called "Chikugo model" has been used to estimate net primary productivity of natural vegetation, both at a global scale (Uchijima and Seino, 1988) for the present day (1931-1960) climate, and at a national scale (Yoshino *et al.*, 1988) for scenarios of changed climate in Japan. In form, the Chikugo model closely resembles correlational models of NPP, but it differs from these in that the simple model equation is derived from physiological theory. Use of this model in Europe would provide a useful comparison with the results of the Miami model (Leith, 1975).

In the United States an intensive study by the Environmental Protection Agency (EPA) is employing generic crop-climate models to assess the possible effects of future long-term climatic change on crop yields in four regions of the United States (EPA, 1989). These include the CERES models for wheat and maize (Jones and Kiniry, 1986; Ritchie and Otter, 1985), which have been tested across a wide range of environmental conditions and appear to offer promising possibilities for adoption in climatic change experiments in regions such as Europe, providing that appropriate input data are available to run the models.

9.3. A Summary of Possible Effects of Climatic Change on Agricultural Potential in Europe

In this Section the results of some recent studies of effects of climatic change on agricultural potential in Europe are discussed briefly. It should be emphasized that a high degree of uncertainty is attached to all such results. They are included here for illustrative purposes only. To facilitate comparison of the results only those that relate to the same scenario of climatic change are discussed here. This is the GISS 2 x CO_2 scenario (see above) which has been employed in two studies of European agriculture: an EC-funded study of the European Community countries completed in 1983 (Meinl *et al.*, 1984) and an IIASA/UNEP-funded study, which included four regional case studies in Europe (among eleven studies worldwide) completed in 1987 (Parry *et al.*, 1988a, 1988b).

The performance of the GISS model in simulating present day climate in Europe has been tested by comparing the GISS 1 x CO_2 model estimates with observed 30-year averaged data (Figure 9.5). It may be noted here that while the GISS model reproduces the observed values of mean monthly air temperature reasonably well, the fit for mean monthly precipitation is poor. The deviations between estimated and observed precipitation in certain areas may exceed 100 per cent in some months. It follows that our confidence in the GISS 2 x CO_2 estimates is also subject to large uncertainties. It should, therefore, be

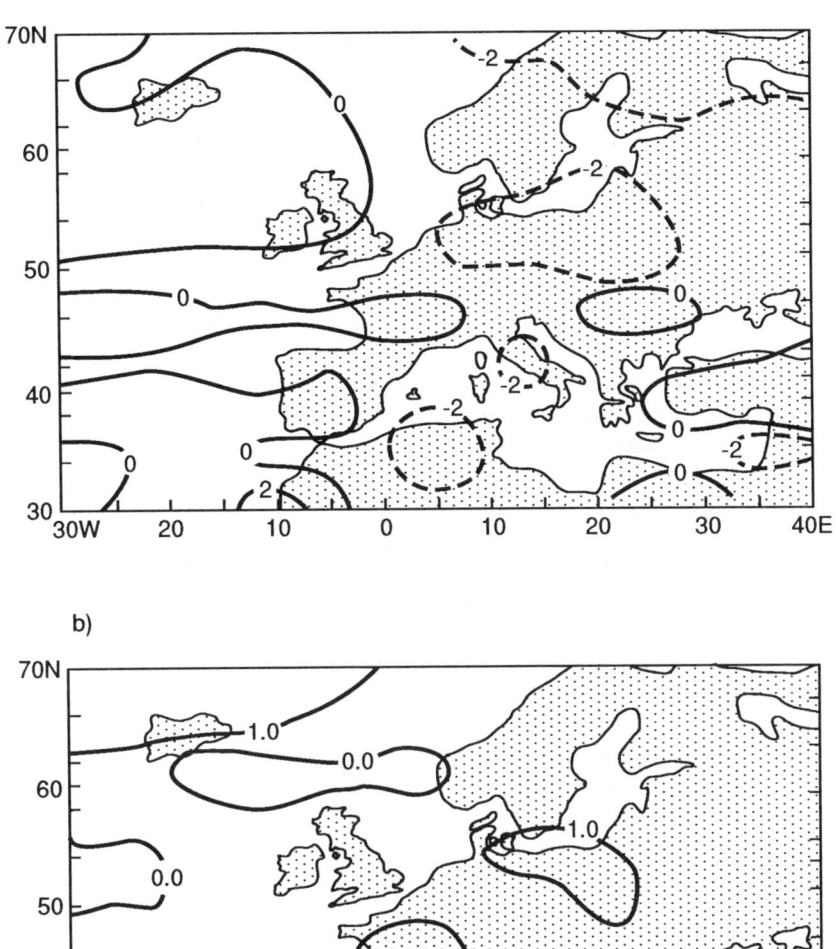

Figure 9.5. Differences between the GISS GCM control climate and the observed climate (1931-1960) over Europe for: a) mean annual temperature (°C) and b) mean annual precipitation rate (mm day^{-1}). (Source: Bach, 1988.)

emphasized that they are regarded in all the studies described below not as predictions but as scenarios of future climate that are useful in extending our understanding of the methods of estimating biophysical and economic effects of climatic change.

9.3.1. Estimated effects in the European USSR

These results are summarized from two IIASA/UNEP case studies in the European USSR. Most results are from the first study of three regions in the northern half of the area (the Leningrad, Cherdyn and Central (Moscow) regions (Pitovranov et al., 1988a)), while the remainder are for a second study in the Saratov region in the south-east of the area (Pitovranov et al., 1988b).

The GISS model-derived data were utilized in two ways in the USSR study: i) as a discrete, step-like perturbation between the present climate and the doubled CO_2 climate, and ii) as the basis of a scenario of transient changes of climate, when combined with an empirical approach developed by Vinnikov and Groisman (1979). Assuming a doubling of CO_2 to occur in the year 2050, smooth linear changes in climate were assumed between the present day and the year 1995 and between 1995 and 2005, according to the Vinnikov and Groisman estimates, followed by a third linear trend from 2005 to 2050, to reach the levels of change estimated by the GISS model. The period 1951-1980 was adopted as a reference against which to compare the scenario results.

Several arbitrary scenarios of climatic change (+1.0°C and +1.5°C relative to the baseline) were also considered in the northern USSR study. These are summarized here but not discussed in detail.

Estimates of effects are summarized in Figure 9.6. In the Leningrad Region, Winter rye yields (estimated using the semi-empirical VNIISI-Obukhov environmental model) are reduced by 13 per cent as a result of higher temperatures and increased rates of evapotranspiration, assuming present day technology (Pitovranov et al., 1988a). The response of Winter rye yield and some environmental parameters were also considered in relation to the transient climatic scenario over 1980-2035. Winter rye yields (relative to the upward trend estimated for improved technology and management) increase up to 2010, but are 23 per cent below trend by 2035. An explanation for this is that relatively small increases in temperature appear to be beneficial, but these are subsequently countervailed by large increases in precipitation, leading to increased soil degradation (due to increased waterlogging and erosion) and increased nitrate leaching with concomitant increases in surface water pollution and reduced soil fertility (Table 9.2).

In the Cherdyn region, Spring wheat yields (estimated using a process-based crop-growth simulation model) decreased slightly relative to yields simulated for the 1951-1980 baseline period, owing to water

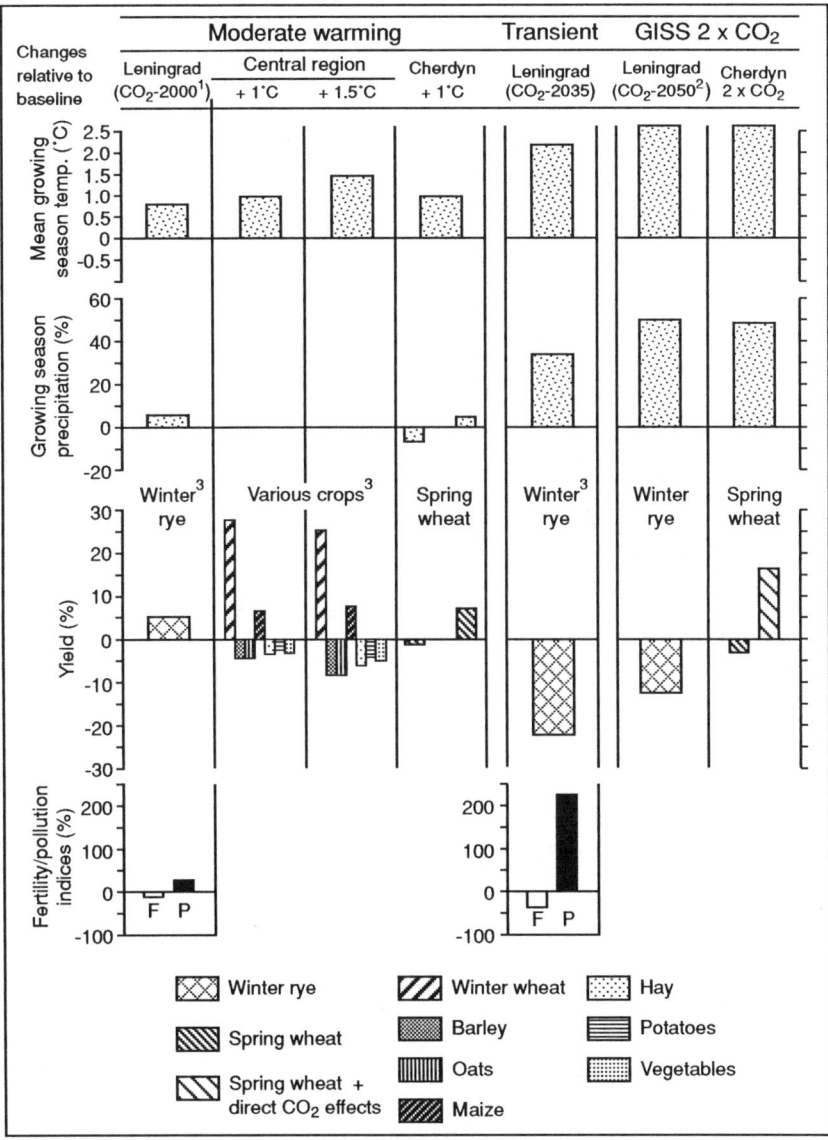

Figure 9.6. Estimated effects of climatic variations on agricultural production in the northern European USSR. Baseline climate is 1951-1980 for the Leningrad and Cherdyn regions and 1931-1960 for the Central Region. 1 = climatic change estimated using an empirical method (Vinnikov and Groisman, 1979); 2 = assumed data of CO_2-doubling; 3 = relative to technology trend. Data from Pitovranov *et al.* (1988a). (Source: Parry and Carter, 1988.)

stress and to the premature development of the crop under the increased temperatures. When the "direct" effects of doubled CO_2 concentrations on crop photosynthesis were combined with the climatic effects, however, a 17 per cent increase in yields was estimated (Pitovranov et al., 1988a).

Similar experiments using the same model, but in the warmer and drier Saratov region (some 1,000 km to the south of Cherdyn), showed a yield increase of about 14 per cent relative to the baseline (80 per cent if direct CO_2 effects are included), mainly due to the 22 per cent increase in growing season precipitation implied by the GISS 2 x CO_2 scenario for this region (Pitovranov et al., 1988b).

Table 9.2. Changes relative to trend of Winter rye yield (Y), an index of soil fertility (SI), groundwater level (GW) and surface water pollution by nitrogen from agricultural watersheds (POL) in response to a scenario of "transient" changes in growing season air temperature (T) and precipitation (P) relative to the 1951-1980 baseline period (adapted from Pitovranov et al., 1988a).

			Changes relative to trend			
Year	T (°C)	P (mm)	Y (%)	SI (%)	GW (%)	POL (%)
1990	+0.4	+8	+4	-1	+2	+18
2000	+0.8	+16	+5	-4	+4	+30
2010	+1.2	+38	+5	-11	+9	+75
2020	+1.6	+74	-4	-22	+14	+213
2035	+2.2	+127	-23	-42	+38	+325

Further estimates for arbitrary changes in temperature are summarized for the Central Region (Moscow area) in Figure 9.6. In general, temperature increases of 1.0°C and 1.5°C lead to increases in yields (estimated using a process-based crop production model) of Winter wheat and maize for silage but decreases in yields of barley, oats, potatoes, hay and vegetables (Pitovranov et al., 1988a).

9.3.2. Estimated effects in western Europe

The majority of these results are summarized from the 1983 EC study (Meinl et al., 1984). Two crop-climate models (introduced in Section 9.3) were used to assess the effects of changes in temperature and precipitation estimated under both the GISS and the BMO 2 x CO_2 scenarios: an empirical/statistical (correlational) model for estimating

Winter wheat yields (Hanus, 1978) and a semi-empirical model of biomass production (Briggs, 1983). Changes in average annual yields of Winter wheat and spelt by country are presented in Table 9.1. Results for biomass potential are given in Figure 9.3. Comparison between the results is difficult for such aggregated (national level) data and no clear conclusions can be drawn. There are indications that the effect of the GISS scenario on wheat yields is more negative than that of the BMO scenario, while the effects on biomass potential appear to be more conservative for the GISS than for the BMO scenario. However, perhaps the most important lesson to be drawn from the discrepancies between results is that they emphasize the uncertainties surrounding GCM projections of future climate.

In a further study a crop-climate simulation model developed by Hough (1980) was employed to estimate the response of Winter wheat yields to the GISS 2 x CO_2 scenario temperature changes in northern England (Carter, 1988). Assuming no change in crop varieties, the mean annual temperature increase of about 3.7°C is estimated to shorten the growing season of crops in lowland areas by about two months and to reduce mean yields of well watered crops by up to one-third. In contrast, the temperatures at upland sites above about 500 meters (where no crops are grown today) would be close to optimum for Winter wheat growth. Furthermore, the effects of water stress on crops (not modeled here) could be expected to intensify this implied upland-lowland reversal in relative crop potential. Similar results have been obtained using another Winter wheat productivity model for a range of hypothetical temperature increases at sites in the United Kingdom (Squire and Unsworth, 1988).

9.3.3. Estimated effects in northern Europe

These results are summarized from two of the IIASA/UNEP case studies: Iceland (Bergthórsson et al., 1988) and Finland (Kettunen et al., 1988).

In both countries the estimated increase in both mean annual and growing season temperatures under the GISS 2 x CO_2 scenario is greater than further south in Europe (exceeding 4°C in nearly all months). Precipitation is also estimated to increase throughout the year in both countries.

In Iceland, effects under the GISS 2 x CO_2 scenario exceed those estimated for anomalously warm years in the instrumental record (Figure 9.7). This is due both to the large degree of warming and to the relatively low present day variability inherent in Iceland's highly maritime climate.

Under the GISS 2 x CO_2 scenario hay yields are estimated (using an empirical statistical model) to increase by 66 per cent relative to the present day (compared with a range of -13 per cent to +18 per cent for the 10 coldest years and the 10 warmest years in the period 1931-1984

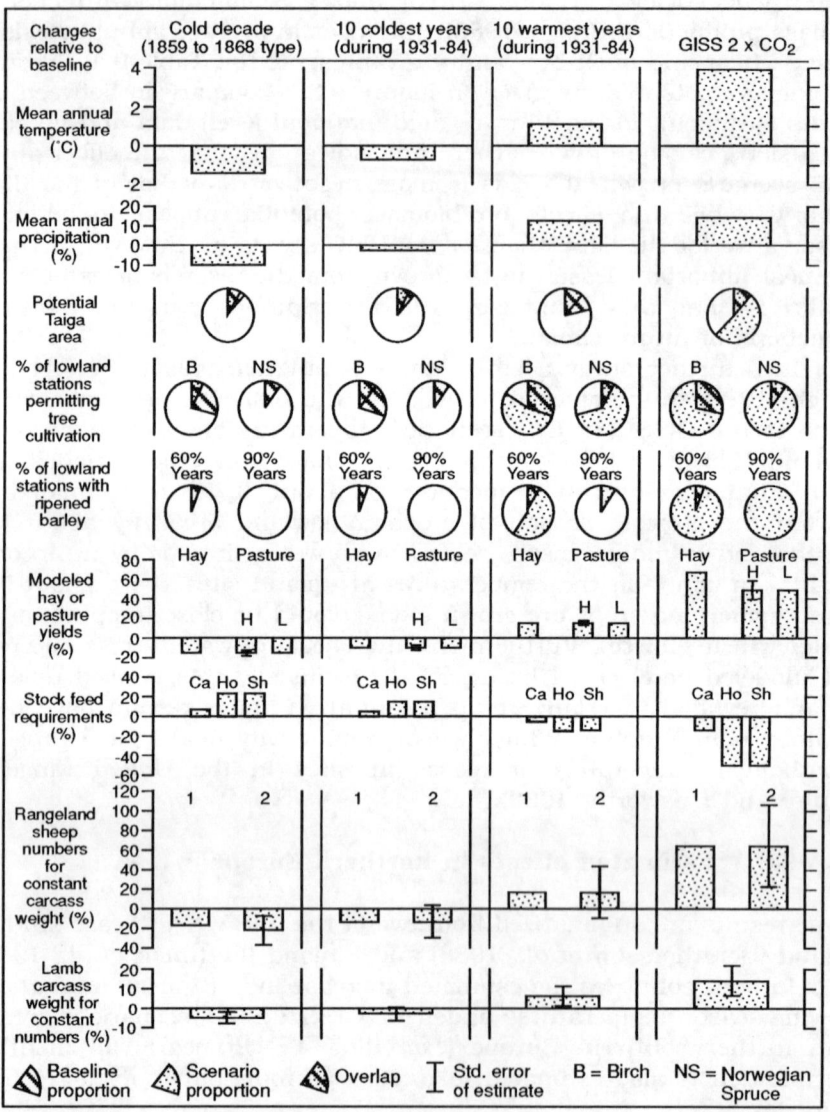

Figure 9.7.   Estimated effects of climatic variations on agricultural production in Iceland.   Baseline climate is 1951-1980.   H = high input (120 kg N ha⁻¹); L = low input (80 kg N ha⁻¹); Ca = cattle; Ho = horses; Sh = sheep; 1 = from national model; 2 = from refined national model. Data from Bergthórsson *et al.* (1988).   (Source: Parry and Carter, 1988.)

respectively).   The respective estimates for rangeland carrying capacity and sheep carcass weight are increases of 64 per cent and 12 per cent (compared with ranges of -13 per cent to +18 per cent and -2 per cent to +3 per cent). Under this scenario barley would ripen 6 years in 10 at all lowland locations (compared with only 4 per cent of lowland locations at present), and the theoretical forested area would increase sixfold (Bergthórsson et al., 1988).

In Finland, effects under the GISS 2 x CO_2 scenario exceed those for anomalously warm periods under the present day climate (Figure 9.8), although the difference is less marked than in Iceland. The latter is due partly to a weaker temperature/yield relationship for grain crops in Finland than for grass crops in Iceland, partly to the sensitivity of Finnish grain yield to precipitation changes, and partly to the greater inter-annual variability of temperatures in Finland than in Iceland.

Yields of grain crops were estimated using empirical-statistical models in this study. Under the GISS 2 x CO_2 scenario, barley yields are estimated to increase by between 9 per cent and 21 per cent (depending on location and on the model employed). This compares with an equivalent range of +1 per cent to +12 per cent in an anomalously warm period such as the 1930s. Spring wheat yields (assuming that longer-season varieties are introduced under the warmer climate) are estimated to increase by about 10 per cent in southern Finland and by about 20 per cent in central Finland (compared with +5 per cent and +15 per cent respectively for present varieties during a recent anomalously warm period (Kettunen et al., 1988)).

9.4. **Potential Technological and Management Responses**

Changes in management practices or technology, which are assumed to be fixed in the majority of impact experiments, are the natural farm-level responses to cope with climatic changes. These can be simulated by some models in so-called "adjustment experiments", a number of which were employed in the IIASA/UNEP studies (Parry and Carter, 1988).

Changes in crop variety -

a) changes from spring-sown to winter-sown cereal varieties to take advantage of warmer winters and withstand better the increased frequency of moisture stress in some regions resulting from higher temperatures;

b) changes to varieties with higher thermal requirements, to exploit longer and warmer growing seasons;

c) changes to varieties giving less variable yields.

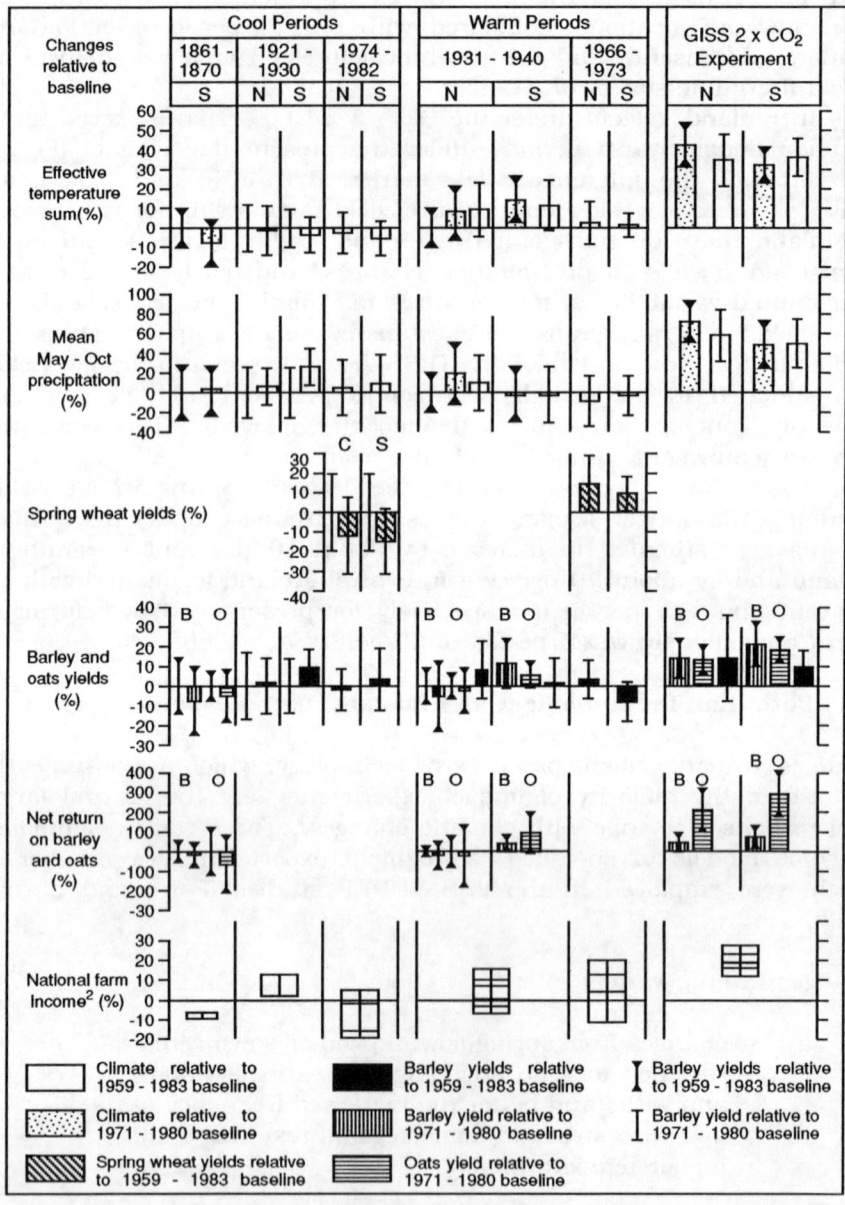

Figure 9.8.   Estimated effects of climatic variations on agricultural production in northern (N), central (C) and southern (S) Finland. Baseline climate is 1959-1983 or 1971-1980, as shown.   1 = estimated farm income based on yield estimates from all regions.   Data from Kettunen *et al.* (1988).   (Source: Parry and Carter, 1988.)

Changes in fertilizing and drainage -

a)   altered fertilizer applications, to optimize yields
        under the changed climate;
b)   improved field drainage, to reduce the risk of
        water erosion, offset the risks of waterlogging
        from increased precipitation and to dispose more
        efficiently of nitrate pollutants.

Changes in land allocation -

a)   changes of land use to optimize production;

b)   changes of land use to stabilize production.

9.5. Conclusions

9.5.1. Extrapolating assessments of effects

A necessary conclusion from the above findings is that at present it is
inappropriate to attempt any "predictions" of the state of agriculture in
Europe under a future altered climate.   Rather, the results so far
obtained represent a useful sensitivity analysis of present day
agriculture, enabling the identification of those aspects and areas that
may be especially vulnerable to climatic variations.   However, when
viewed as a whole, the findings do indicate the following:

1)   the varying degrees of absolute change in climate at different
        locations are not necessarily reflected in the spatial pattern of their
        impacts; instead, they tend to be a function of the change in
        climate relative to the existing (baseline) conditions;

2)   yield-climate relationships for different crops are almost always
        non-linear, differ between locations, and are closely tied to levels
        of technology and management (for example, fertilizer application
        levels, planting dates), which also vary spatially;

3)   future changes in climate will lead to a spatial shift of crop
        productivity potential requiring a relocation of cropping patterns
        from the present situation;

4)   changes in mean climate (even assuming no changes in inter-
        annual variability) can lead to changes in crop yield variability; for
        example, while mean grain yields are estimated to increase in all
        parts of Finland under the GISS 2 x CO_2 scenario, inter-annual
        yield variability decreases in the south but increases in the north

of the country, probably due to the large precipitation increases projected for the north (Kettunen *et al.*, 1988);

5)   most of the assessments summarized above assume that factors such as technology and management remain fixed at the present-day levels; this assumption is unrealistic in the context of the long-term nature of the projected climatic changes.

9.5.2. **Specific research priorities**

Given the uncertainties described above, the following items (which are not necessarily specific to Europe) merit further research:

i)   the provision of improved scenarios of future climatic change, incorporating regional (subgrid-scale) detail and information on anomalous climatic events that can have a major effect on agricultural output;

ii)   the refinement and validation of impact models, with further experimentation for a wider range of agricultural activities, reflecting not only biophysical effects but also "downstream" economic and social effects of climatic variations;

iii)) consideration of the effects of climatic changes in combination with other related or unrelated effects occurring simultaneously (for example, the direct effects of CO_2 on crop yields, the effects on crop yields of climatically-related changes in the incidence of pests and diseases);

iv)   further experiments with a range of potential farm-level adjustments in agriculture, to explore their efficacy in mitigating impacts or exploiting options;

v)   more exploration of the potential economic and political responses to climatic change (for example, restructuring of price support systems or insurance mechanisms).

REFERENCES

Bach, W., 1988, Development of climatic scenarios: A. From general circulation models. In M.L. Parry, T.R. Carter and N.T. Konijn (eds.), *The Impact of Climatic Variations on Agriculture. Volume 1. Assessments in Cool Temperate and Cold Regions.* Kluwer, Dordrecht, The Netherlands. 125-157.

Bergthórsson, P., Björnsson, H., Dyrmundsson, O., Gudmundsson, B., Helgadóttir, A. and Jónmundsson, J.V., 1988, The effect of climatic variations on agriculture in Iceland. In M.L. Parry, T.R. Carter and N.T. Konijn (eds.), *The Impact of Climatic Variations on Agriculture. Volume 1. Assessments in Cool Temperate and Cold Regions.* Kluwer, Dordrecht, The Netherlands. 383-509.

Briggs, D.J., 1983, Biomass potential of the European Community. *Report for the Environment and Consumer Protections Service of the European Community.* Sheffield, U.K.

Brouwer, F.M., 1989, Determination of broad-scale land use changes by climate and soils. *Journal of Environmental Management*, **29**, 1-15.

Carter, T.R., 1988, Climatic change and cropping margins in upland Britain. *Unpublished Ph.D. Thesis.* University of Birmingham, U.K.

EPA, 1989, *The Potential Effects of Global Climate Changes on the United States.* Report to Congress. Office of Policy, Planning and Evaluation. USEPA, Washington, D.C.

de Groot, R.S., 1988, Assessment of potential shifts in Europe's natural vegetation due to climatic change and some implications for nature conservation. WP-88-105. International Institute for Applied Systems Analysis, Laxenburg, Austria.

Hanus, H., 1978, Forecasting of crop yields from meteorological data in the EC countries. *Statistical Office of the European Communities, Agricultural Statistical Studies, No. 21.* Brussels.

Holdridge, L.R., 1947, Determination of world plant formations from simple climatic data. *Science*, **105**, 367-368.

Holdridge, L.R., 1964, *Life Zone Ecology.* Tropical Science Center, San Jose, Costa Rica.

Hough, M.L., 1980, A proposed cereal yield simulation model. *Agricultural Memorandum No. 892*, Meteorological Office, Bracknell, U.K. (mimeographed).

Hume, C.J., 1986, A quantitative study of the effects of climate variability and climatic change on herbage production from intensively managed grassland in western European countries. *Commission of the European Communities, Contract No. CLI-058081-UK (H).*

Jones, C.A. and Kiniry, J.R. (eds.), 1986, *CERES - Maize: A Simulation Model of Maize Growth and Development.* Texas A and M University Press, College Station.

Kettunen, L., Mukula, J., Pohjonen, V., Rantanen, O. and Varjo, U., 1988, The effects of climatic variations on agriculture in Finland. In M.L. Parry, T.R. Carter and N.T. Konijn (eds.), *The Impact of Climatic Variations on Agriculture. Volume 1. Assessments in Cool Temperate and Cold Regions*, Kluwer, Dordrecht, The Netherlands. 513-614.

Kolkki, O., 1969, Katsaus Suomen ilmastoon. *Ilmatieteen Laitoksen Tiedonnantoja*, **18**, 1-18 (in Finnish).

Lieth, H., 1975, Modeling the primary productivity of the World. In H. Lieth and R.H. Whittaker (eds.), *Primary Productivity of the Biosphere.* Springer-Verlag, New York. 237-263.

Lieth, H. and Box, E.O., 1972, Evapotranspiration and primary productivity. C.W. Thornthwaite Memorial Model. *Publications in Climatology*, **25**(2), C.W. Thornthwaite Associates, Centerton/Elmer, New Jersey. 37-46.

Meinl, H., Bach, W., Jäger, J., Jung, H-J., Knottenberg, H., Marr, G., Santer, B. and Schwieren, G., 1984, *Socio-economic Impacts of Climatic Changes due to a Doubling of Atmospheric Co₂ Content.* Commission of the European Communities, Contract No. CLI-063-D.

Nikonov, A.A., Petrova, L.N., Stolyarova, H.M., Levedev, V.Yu, Siptits, S.O., Milyutin, N.N. and Konijn, N.T., 1988, The effect of climatic variations on agriculture in the semi-arid zone of the European USSR. A. The Stavropol Territory. In M.L. Parry, T.R. Carter and N.T. Konijn (eds.), *The Impact of Climatic Variations on Agriculture. Volume 2. Assessments in Semi-arid Regions.* Kluwer, Dordrecht, The Netherlands. 579-627.

Parry, M.L. and Carter, T.R., 1988, The assessment of effects of climatic variations on agriculture: aims, methods and summary of results. In M.L. Parry, T.R. Carter and N.T. Konijn (eds.), *The Impact of Climatic Variations on Agriculture. Volume 1. Assessments in Cool Temperate and Cold Regions.* Kluwer, Dordrecht, The Netherlands. 11-95.

Parry, M.L., Carter, T.R. and Konijn, N.T. (eds.), 1988a, *The Impact of Climatic Variations on Agriculture. Volume 1. Assessments in Cool Temperate and Cold Regions.* Kluwer, Dordrecht, The Netherlands.

Parry, M.L., Carter, T.R. and Konijn, N.T. (eds.), 1988b, *The Impact of Climatic Variations on Agriculture. Volume 2. Assessments in Semi-arid Regions.* Kluwer, Dordrecht, The Netherlands.

Pitovranov, S.E., Iakimets, V., Kisilev, V.I. and Sirotenko, O.D., 1988a, The effects of climatic variations on agriculture in the sub-Arctic zone of the USSR. In M.L. Parry, T.R. Carter and N.T. Konijn (eds.), *The Impact of Climatic Variations on Agriculture. Volume 1. Assessments in Cool Temperate and Cold Regions.* Kluwer, Dordrecht, The Netherlands. 615-722.

Pitovranov, S.E., Maximov, A.D., Sirotenko, S.E., Aashina, E.V., Pavlova, V.N. and Carter, T.R., 1988b, The effect of climatic variations on agriculture in the semi-arid zone of the European USSR. B. The Saratov Region. In M.L. Parry, T.R. Carter and N.T. Konijn (eds.), *The Impact of Climatic Variations on Agriculture. Volume 2. Assessments in Semi-Arid Regions.* Kluwer, Dordrecht, The Netherlands. 629-664.

Ritchie, J.T. and Otter, S., 1985, Description and performance of CER ES-wheat: a user-orientated wheat yield model. In W.O. Willis (ed.), *ARS Wheat Yield Project*. USDA-ARS. ARS-38. 159-175.

Santer, B., 1985, The use of general circulation models in climate impact analysis - A preliminary study of the impact of a CO_2-induced climatic change on west European agriculture. *Climatic Change*, **7**, 71-93.

Squire, G.R. and Unsworth, M.H., 1988, Effects of CO_2 and climatic change on agriculture. *Contract Report to the Department of the Environment*. University of Nottingham, Sutton Bonington, U.K.

Turc, L. and Lecerf, H., 1972, Indice climatique de potentialite agricole. *Science du Sol*, **2**, 81-102.

Uchijima, Z. and Seino, H., 1988, Probable effects of CO_2-induced climatic change on agroclimatic resources and net primary productivity in Japan. *Bulletin National Institute Agro-Environmental Science*, **4**, 67-88.

Vinnikov, K. and Groisman, P., 1979, An empirical model of present-day climatic changes. *Meteorologia i Gidrologia*, **3**, 25-28 (in Russian). English translation available in *Soviet Meteorol. Hydrol.*, **3**.

Walter, H., 1985, *Vegetation of the Earth and Ecological Systems of the Geobiosphere* (3rd ed). Springer-Verlag, New York.

Yoshino, M., Horie, T., Seino, H., Tsujii, H., Uchijima, T. and Uchijima, Z., 1988, The effect of climatic variations on agriculture in Japan. In M.L. Parry, T.R. Carter and N.T. Konijn (eds.), *The Impact of Climatic Variations on Agriculture. Volume 1. Assessments in Cool Temperate and Cold Regions*. Kluwer, Dordrecht, The Netherlands. 725-868.

Chapter 10

ENVIRONMENTAL CONSTRAINTS ON AGRICULTURAL PRODUCTION

N. Brink

10.1. Introduction

For many years there have been discussions on leaching losses of nutrients to inland waters.   Until about 1970 the discussions mainly concerned phosphorus and eutrophication of rivers and lakes because of the heavy discharge from sewage plants.   Following large investments in phosphorus precipitation in such plants and increasing use of fertilizers in agriculture, the main interest turned to nitrogen and to groundwater quality, and later to water quality in bays and marine areas, such as those on the Swedish west coast.   The pollution of groundwater with nitrate is an increasing problem, as well as the associated aquatic faunal and floral changes.   Nitrogen seems to be the main cause of this in marine areas (Rosenberg and Loo, 1986; Fleischer *et al.*, 1987; Graneli *et al.*, 1988).

The use of pesticides in agriculture has also become a major problem since residues are transferred to rivers and wells used as a source of drinking water.   In addition, pesticides pose a threat to terrestrial flora and fauna as well as to aquatic organisms.   Some pesticides have been shown to occur at such high concentrations in stream water that its use for irrigating sensitive greenhouse crops can be harmful (Brink, 1986; Solyom, 1986).

In addition to the present concern about the deleterious effects of pollution from agriculture, there is also a growing awareness of grain production surpluses.   The question now is whether there should be a decrease in either the intensity of land use (extensification), or in the amount of arable land.   The objective of this Chapter is to discuss the available options for agricultural development in the future in order to diminish leaching losses of nutrients and pesticides.

10.2. Materials and Methods Applied

Experiments on the fluxes of nutrients, pesticides and heavy metals in agriculture are being carried out in whole-field and plot experiments. The objective is to explore methods to prevent or at least diminish

F. M. Brouwer et al. (eds.), Land Use Changes in Europe, 233–252.
© 1991 *Kluwer Academic Publishers. Printed in the Netherlands.*

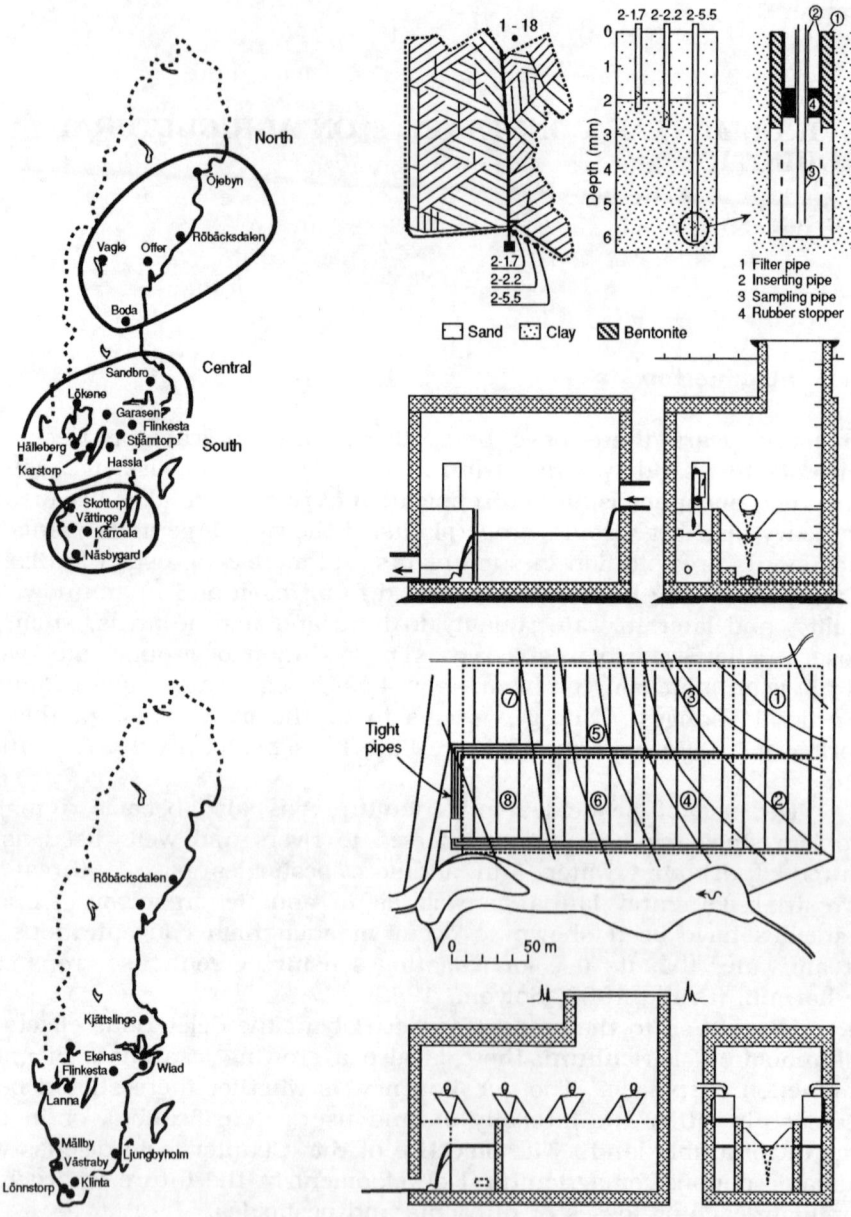

Figure 10.1. *Top.* Sites of whole-field experiments. Tile-drained field, soil profile, sampling pipe and an underground measuring station with a triangular weir. *Bottom.* Sites of experimental fields with separately drained plots. Measuring station with tilting-vessels and a triangular weir.

losses. In the whole-field experiments both surface runoff and drainage discharge from tile systems are collected in underground concrete cisterns and are continuously registered with triangular weirs and water level recorders (Figure 10.1, *top*). At one or more sites, piezometric and sampling wells are placed at different depths. The size of the fields varies between 4.5 and 36 hectares. Many soil types are represented in the analysis. In the plot experiments, either surface runoff or drainage water or both are collected. Separate watertight furrows along the lowest edges of the plots and separate drainage tiles run from each plot to an underground flow meter for continual registration of the discharge. The groundwater is checked (Figure 10.1, *bottom*).

10.3. **Results of the Experiments**

10.3.1. **Leaching losses of nitrogen and phosphorus**

The *nitrogen flux* in agriculture is very important, both from an ecological and an economic point of view.

Figure 10.2 shows the nitrogen flux through an average hectare of arable land in Sweden. The total annual input is 125 kg ha^{-1}. Of that, 20 kg is used by man (export 5, sludge 3, sewage water 12 kg). The total loss from soil and stables is 119 kg; 77 kg to the air and 42 kg to water. The efficiency is very low.

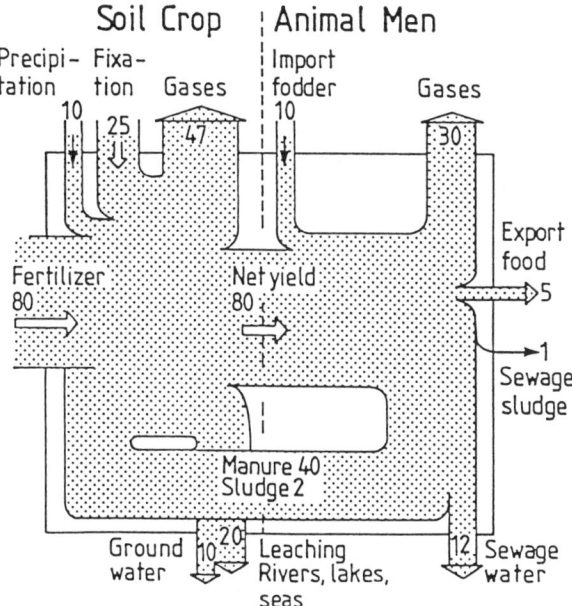

Figure 10.2. Nitrogen balance sheet of an average hectare of arable land in Sweden (kg ha^{-1} yr^{-1}).

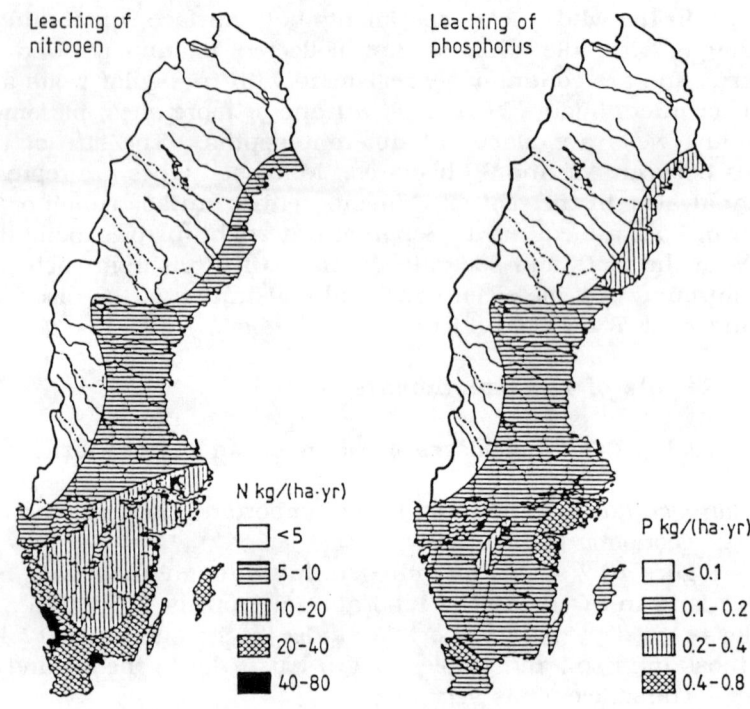

Figure 10.3. Leaching losses of nitrogen and phosphorus from arable land in Sweden.

Only the leaching loss of nitrogen will be considered despite the fact that serious consequences also result from evaporation of ammonia and nitrous gases (Nihlgård, 1985).

Climatic factors are crucial for the leaching of nitrogen but not for phosphorus, as indicated by Figure 10.3. The long winter in northern Sweden effectively limits the nitrogen loss since most of the water, poor in nitrogen, runs off on the surface in spring. The nitrogen losses are greatest in southern Sweden, especially on the sandy soils on the west coast and in a relatively small area on the east coast. In the western region the substrate comprises shallow sandy soils on clay with low permeability. Therefore, nearly all water is collected in the drain tiles. The same situation is found in the eastern part which also acts as a recharge area. On the other hand, only a small proportion of the water can be collected on deep sandy soils in discharge areas. Heavy contamination of the groundwater would be expected in such areas.

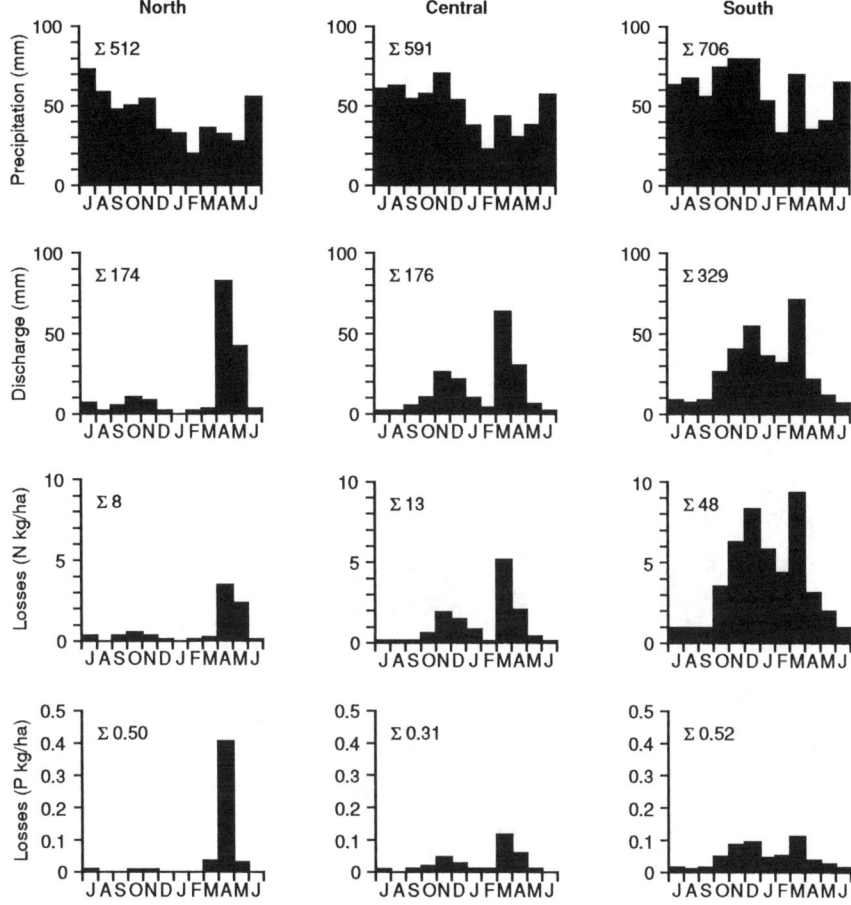

Figure 10.4. Precipitation, water discharge and nutrient losses from arable land in different parts of Sweden.

The total annual phosphorus loss does not differ greatly between north and south (Figure 10.3). In northern Sweden, erosion is one of the essential factors for determining phosphorus losses. More than 90 per cent of the phosphorus runs off on the surface in spring. In the south, leaching through the soil profile predominates, more than 85 per cent of the total being leached. The transport mode and speed of nutrients and pollutants through the soil depend on the structure, texture and permeability of the substrate as well as on its temperature or frozen status.

The erosion from clay is much greater than from sand. In particular, sloping fields supplied with animal manure in late winter can lose fairly large amounts of phosphorus. The highest value measured

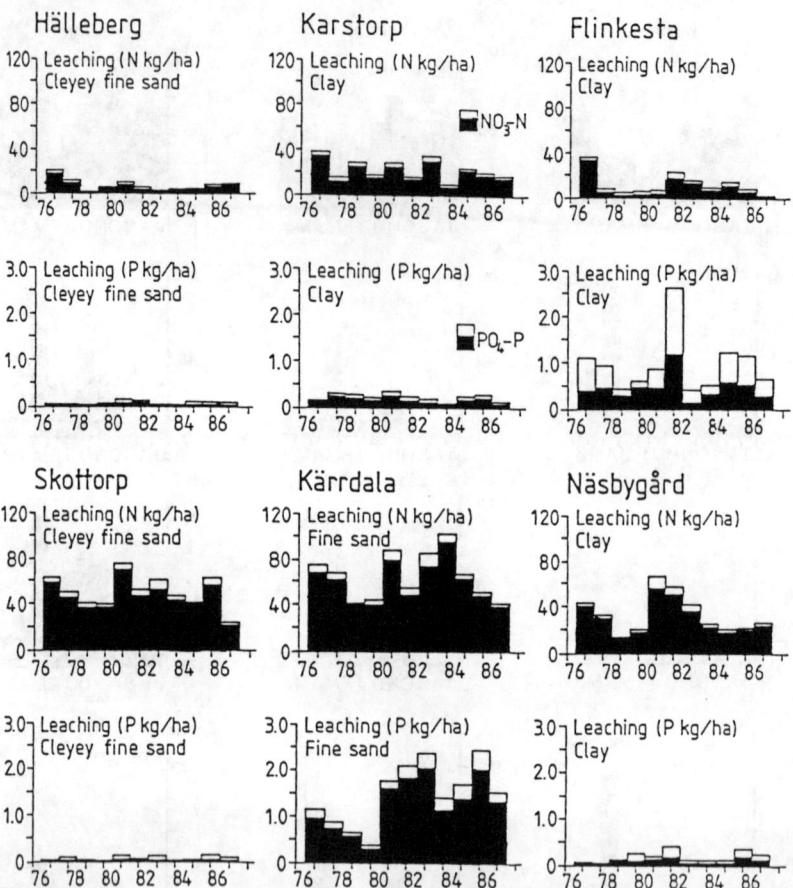

Figure 10.5. Losses of nitrogen and phosphorus from experiments on some whole-fields.

was about 7 kg P ha^{-1} yr^{-1} (Brink *et al.*, 1983). Another important factor is the freezing out of phosphorus from the plant residues on the ground (Uhlen, 1979). The erosion of phosphorus may be as high from such areas as from bare soil (Ulén, 1985).

The annual distribution of nitrogen and phosphorus losses are quite different in the north and south of Sweden. The nutrient flux is dependent largely on water discharge and, of course, on precipitation (Figure 10.4). In Sweden there are two peaks, one in autumn and one in spring. The autumn peaks become increasingly similar to and even larger than the spring peaks when moving from north to south. The differences depend primarily on the climatic conditions. Different stations gave a wide variation in results, as can be seen in Figure 10.5.

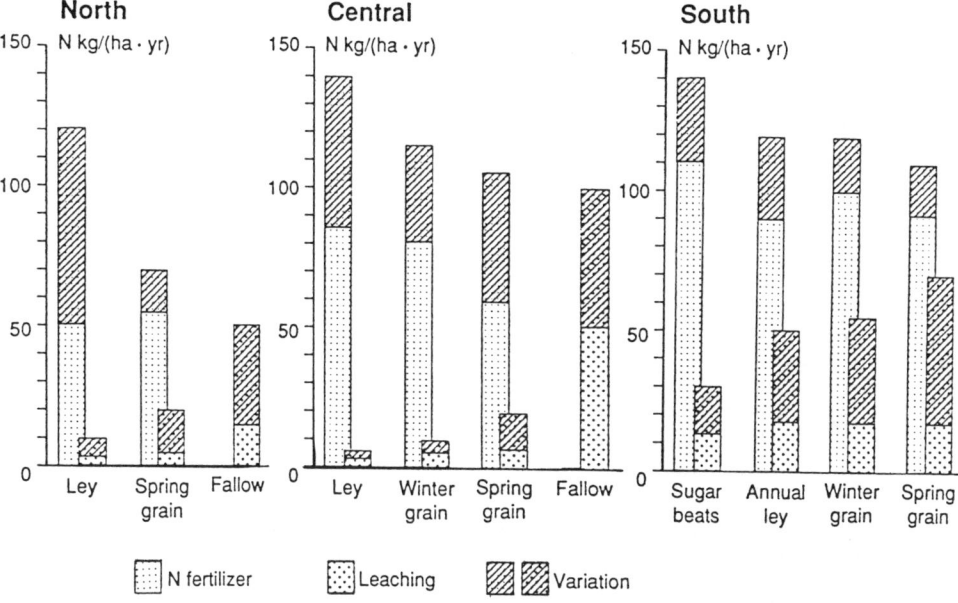

Figure 10.6. Importance of crops on the leaching losses of nitrogen in different parts of Sweden.

The soil type is important. Sandy soils deliver large amounts of nitrogen to the hydrological system throughout the year, as at Skottorp and Kärrdala (Figure 10.5). The difference in losses between fields depends primarily on the amount of nitrogen applied. On the other hand, the kind of fertilizer greatly influences the leaching of phosphorus. At Skottorp only commercial fertilizers have been used, whereas at Kärrdala manure was also used. Organically bound P is more mobile than the PO_4^- ion.

Crop type is important in determining how much N is leached (Figure 10.6). Perennial plants use nitrogen most efficiently and annual plants, with short vegetative periods, are often least efficient. Crops can be ranked from good to bad as follows: Perennial ley, Sugar beet, Annual ley ≈ Winter grain, Spring grain, and Fallow.

The leaching of nitrogen depends on the *yield and the amount of fertilizer applied*. Figure 10.7 shows how the yield (q(N)) and the protein content (c(N)) increased considerably in a plot experiment at low N doses and leveled out at high doses (Brink, 1987). At doses where the yields increased strongly, the leaching 1(N) was low but rose considerably where there was a smaller yield increment or no change. The Figure also shows increased accumulation of mineral-N (min-N) to a depth of

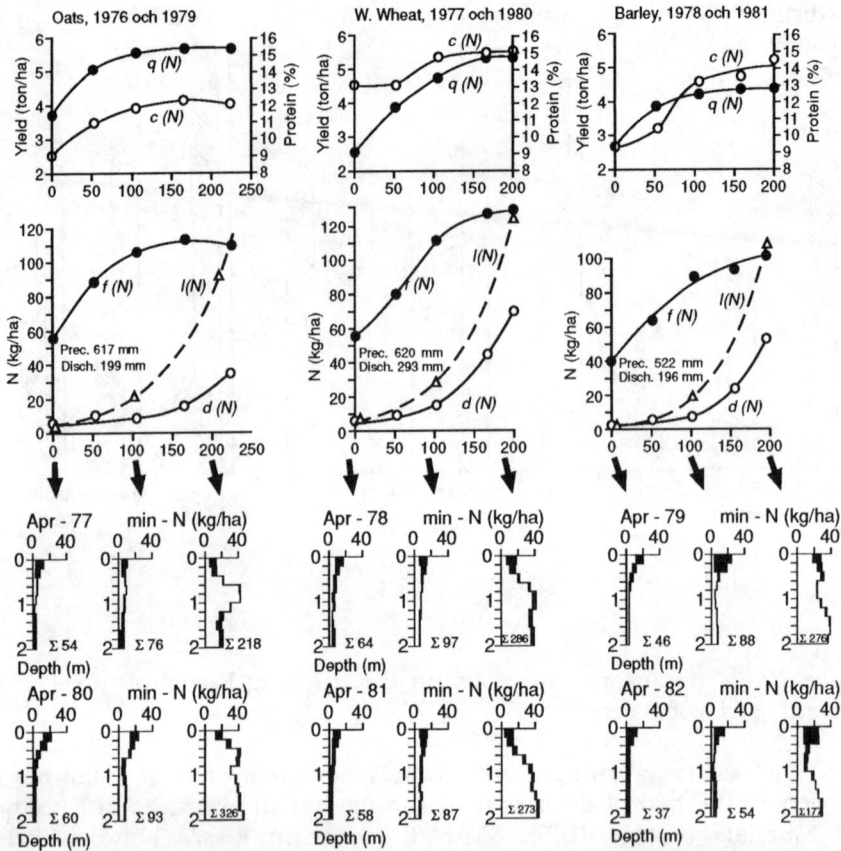

Figure 10.7. Yield q(N), protein content in grain c(N), harvested protein f(N), leaching with drainage water d(N), total leaching to surface water and groundwater l(N) as functions of applied nitrogen (N). Annual values for oats, Winter wheat and barley on a clay soil.

2 m. The upper dashed curve has been calculated on the basis of leached mineral-N to a depth of 1.2-3 m (Bergström and Brink, 1986).

The leaching curves shown in Figure 10.7, with increasing loss from zero, are not always valid. Sometimes there is a minimum loss at a higher input of nitrogen, such as in Figure 10.8, owing to poor crop growth when no fertilizers are applied. A small initial dose would probably give a much better yield and a lower leaching loss.

As regards phosphorus, there are no positive relationships between the applied and leached amounts. In fact, the opposite is often the case: decreasing loss of phosphorus with increasing nitrogen loss, and vice versa (Figure 10.8). Both pH and yield are important causal factors. Lower pH gives more adsorption in the soil and a higher yield means a greater requirement for phosphorus.

There are other, often more harmful phosphorus sources on farms. Large cow stables, dungyards and silage bunkers are frequently found to be large polluters of streams (Brink and van der Meulen, 1987). At least three times more phosphorus was measured from catchments with stables than from arable farmland. Another example concerns two experimental stations with cows, pigs, hens or mink where the water pollution from dung heaps and silage bunkers was extremely high.

As far as long-term changes are concerned some early measurements (Brink, 1965; Wiklander and Hallgren, 1971) provide an opportunity to make comparisons with more recent ones (Brink, 1983). The increase in the application of annual fertilizers between 1965 and 1975 was followed by increasing leaching losses (Table 10.1).

Table 10.1. Changes in fertilizer addition and loss from 1960-65 to 1975-81.

Years	1960-65	1975-81
Runoff (mm) yr^{-1}	206	199
N-fertilizing (kg N ha^{-1})	65	117
N-loss (kg NO_3-N ha^{-1})	11	18
P-fertilizing (kg P ha^{-1})	26	30
P-loss (kg Tot-P ha^{-1})	0.14	0.20

10.3.2. Leaching losses of major constituents

The flux of major constituents in drainage water from arable land has been monitored for five years. Mean values of those elements from eleven mineral soils and one organic soil are shown in Table 10.2.

The similarity in results between mineral and organic soils, apart from S and Ca, is striking. Obviously there are large amounts of sulfur and calcium bound in the organic soil (a fen peat) owing to rich surrounding land and the low mobility of the two elements.

10.3.3. Leaching of heavy metals

Heavy metals have been measured at twelve of the whole-field stations, all with mineral soils (Andersson and Gustafson, 1982). The samples gave a few extremely high values, thus the local conditions could be described in terms of median values rather than by averages. The median values are given in Table 10.3.

Table 10.2.   Mean values (kg ha⁻¹ yr⁻¹) of seven elements in drainage
water from Swedish soils.

Soil	N	S	Cl	Na	K	Mg	Ca	n[a]
Mineral	21	29	41	22	14	15	120	55
Organic	20	300	46	23	11	25	460	5

[a]Number of samples.

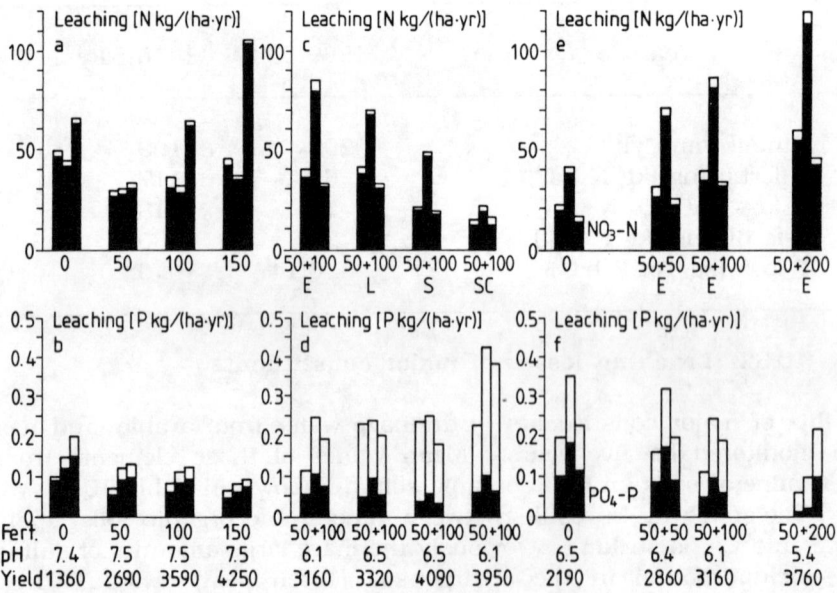

Figure 10.8.   Leaching losses of nitrogen and phosphorus from sandy
soils at different inputs of commercial fertilizers and manure; a) and b):
nitrogen fertilizers in kg ha⁻¹. (c-f): fertilizers 50 kg N ha⁻¹ + nitrogen in
manure. E: early spreading in autumn; L: late spreading in autumn; S:
spreading in spring; SC: spreading in spring after a catch crop (Winter
wheat). Mean annual yield in kg ha⁻¹.

Table 10.3. Median values of heavy metals (μg l^{-1}) in drainage water from twelve fields in Sweden (11 = neutral, 1 = acid).

	Cr	Mn	Fe	Ni	Cu	Zn	Cd	Pb	pH
neutral	1.6	7.1	71	1.8	1.6	1.3	0.014	0.25	6.5-7.6
acid	2.2	4000	405	57	21	200	0.51	0.14	4.3

Field pH was usually between 6.5 and 7.6. Extremely high values were found for almost all metals at the station with a mean value of 4.3 pH (Table 10.3). Low values were found at several sites due to the fact that the field had recently been tile drained, coupled with a low water table and the oxidation of sulfides in the soil.

A rough estimation of annual loss in g ha^{-1} of heavy metals is found by multiplying values in Table 10.3 by a factor of 2 (200 mm discharge). The concentrations observed are usually fairly low, but the question remains whether these concentrations are harmful or not.

10.3.4. Leaching of pesticides

There is a widespread theory that, in general, pesticides do not leach through soils to surface and groundwater (Hallberg, 1987). A fast breakdown or adsorption is thought to preempt leaching. Nonetheless, contamination of wells and streams has been observed in the past (Erne, 1970). At the time these events were explained as the result of incorrect handling of application equipment and pesticide residues and of spray drift. Such explanations might, of course, be valid but the question of leaching and the possibility of surface runoff from sprayed ground must be addressed.

Numerous plot experiments have been conducted in order to discover to what extent the applied pesticides leach into surface and groundwater (Brink, 1986; Kreuger and Brink, 1987). After spraying in autumn or spring, surface runoff and drainage water were analysed as well as soil samples. The results are summarized in Table 10.4.

The observed concentrations of pesticides in the waters were usually below 5 μg l^{-1} and the total loss generally amounted to less than 1 per cent of the amount sprayed. However, it should be noted that the loss of cyanazine in surface runoff was 4 per cent and that of TBA in drainage water was 10 per cent.

Thus, although cyanazine is a very immobile substance in soil, as shown in Figure 10.9 (top), the loss to the environment can be substantial. It is also interesting to note that not only passive transport

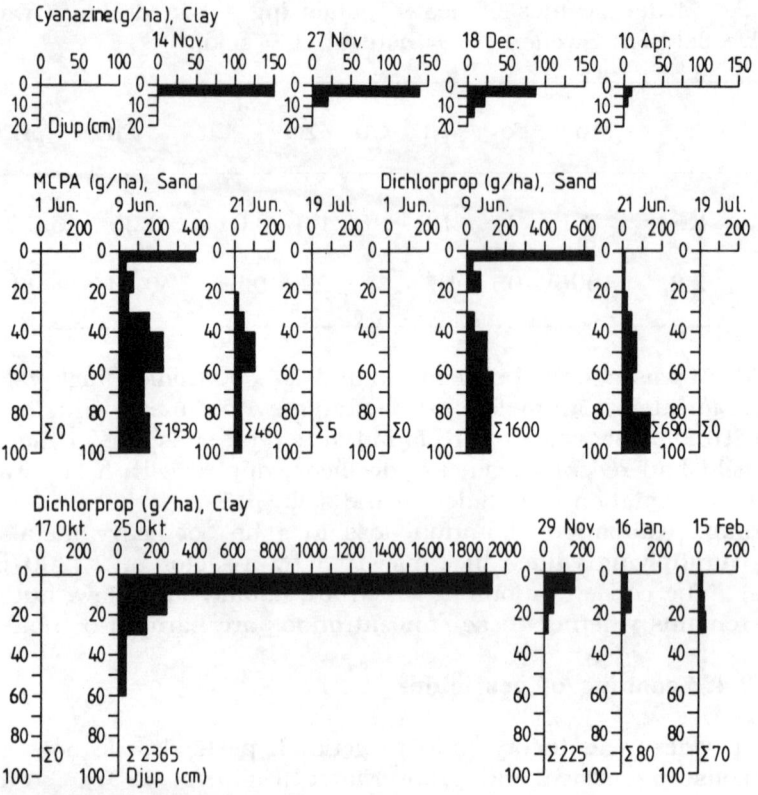

Figure 10.9. Pesticide movement in soils.

via percolating water occurs, but also that active transport through plants may be considered as highly probable.

The rapid transport, of MCPA and dichlorprop in a sandy soil as indicated by soil analyses (Figure 10.9, middle), could be explained in the following way. One day after spraying on June 8 the substances could be detected to a depth of 1.0 m despite a very low water discharge. After just over one month there were no detectable amounts left in the soil. On a clay soil the process was considerably more prolonged (Figure 10.9, bottom). In this case, the dichlorprop was sprayed on October 24, 1983, and on the following day the substance had reached 0.6 m depth. No water discharge occurred until the end of December, when the highest concentration, 2.5 µg l⁻¹ was detected. In one month with 66 mm discharge the soil content had dropped to below the detection limit.

Table 10.4. Leaching losses of pesticides from arable land. Measurements in plot experiments. (Dose in kg ha^{-1}, concentration in µg l^{-1} and loss in %.)

Pesticide[a]	Spraying	Dose	Concentration	Loss
Surface runoff, clay soil				
Cyanazine	November	0.16	3-34	4
Drainage water, clay soil				
MCPA	October	2	<0.3	0
Dichlorprop	October	2	0.2-2.5	0.06
Fenvalerate	June	0.1	0	0
TCA	April	21	60	0.01
Drainage water, sandy soil				
TBA	October	0.2-0.6	2-78	10
MCPA	November	2	1	0.4
Dichlorprop	November	2	1	0.9
TCA	April	20	100-900	1.1
MCPA	June	1.5	0.3	-
Dichlorprop	June	1.5	0.3	-

[a]Fenvalerate is an insecticide; all the others are herbicides.

A program was started in 1985 to monitor the leaching of pesticides into rivers. Samples were taken once a month from April to October from 30 water courses (Kreuger and Brink, 1987). Eighteen compounds were identified (Table 10.5). The phenoxy acids MCPA and dichlorprop were most frequently observed around the period of spraying (May and June). The reason for their occurrence in stream waters is usually explained as a result of bad handling or of spray drift.

Nonetheless, there can be no doubt that leaching from arable land is important. Durations of more than two months cannot be blamed solely on something so brief as the flushing-out of sprayers. It would also be remarkable if all the pesticide residue in the rivers had originated from leakages in storage facilities.

The occurrence of pesticides in stream water may have major impacts on aquatic organisms and on the quality of irrigation water and drinking water.

Insecticides such as endosulfan, lindane, and permethrin are toxic to fish and other organisms. Thus, the level set by International Joint Commission (IJC) for endosulfan to protect aquatic life is 0.003 µg l^{-1} (Frank *et al.*, 1982) which is far below the detection limit of 0.1 µg l^{-1} used in this investigation. The concentrations found for lindane exceeded the US Environmental Protection Agency (EPA) criteria of 0.01 µg l^{-1} for freshwater aquatic life for several months. Further, the

Table 10.5. Maximum pesticide concentrations (in µg l^{-1}), and number of positive samples between April and October in Swedish stream waters. Measurements taken between 1985 and 1987.

Pesticide	Concentrations	April	May	June	July	August	September	October
Herbicides:								
Atrazine	6.0	0	5	9	14	14	13	2
Bentazone	0.5	NM	NM	NM	NM	18	18	NM
Cyanazine	0.7	0	0	10	3	1	1	0
2,4-D	0.9	0	1	3	0	0	1	0
Dichlorprop	16.0	0	14	44	24	7	5	0
Dimethachlor	0.3	NM	NM	NM	NM	NM	3	NM
MCPA	8.0	0	12	45	32	7	3	0
Mecoprop	6.0	0	13	16	19	7	3	0
Metazachlor	7.0	0	0	3	2	0	8	0
Simazine	1.1	0	0	0	1	1	1	0
Terbuthylazine	0.7	0	0	1	1	0	0	0
Fungicides:								
Metalazyl	1.3	0	0	1	2	1	0	0
Propiconazole	1.2	0	0	2	1	0	0	0
Insecticides:								
Endosulfan	0.1	1	0	0	0	0	0	0
Fenitrothion	0.1	0	0	1	1	0	0	0
Lindane	0.6	0	2	2	2	2	0	0
Permethrin	0.6	0	0	0	0	1	0	0
Pirimicarb	3.7	1	0	1	2	0	1	0
Number of samples		4	42	58	56	52	45	2
Maximum concentration (µg l^{-1})		4	14	25	5	4	10	6

herbicide atrazine is an inhibitor of photosynthesis and, at concentrations of 1-5 µg l^{-1}, may affect phytoplankton (Denoyelles et al., 1982).

Herbicides which contaminate stream water may have injurious effects if used to irrigate greenhouse cultures. Studies have shown that phenoxy acids can disturb the growth of tomatoes, onions and lentils at concentrations of 2-10 µg l^{-1} (Solyom, 1986). Also atrazine and simazine have a similar effect at 0.4-0.6 µg l^{-1} (GEFO, 1987). These concentrations are often exceeded in Swedish streams.

Great attention has been focused on pesticides in drinking water since water is abstracted from some streams for human consumption. A few private wells have also been shown to contain phenoxy acids. There are no guidelines in Sweden for drinking water regarding pesticide content. However, the European Community (EC) has proposed guidelines based on the detection limits, for example 0.1 µg l^{-1} for a single pesticide and a maximum total content of 0.5 µg l^{-1}. EPA has proposed maximum contamination levels for about twenty organic chemicals. Two of these chemicals have been observed in the streams studied in our investigation, namely, 2,4-D (EPA norm 70 µg l^{-1}) and lindane (EPA norm 0.2 µg l^{-1}).

10.4. Measures to Prevent Leaching

The most urgent environmental task is to minimize the pesticide and phosphorus fluxes to rivers and lakes, and the nitrogen flux to bays and the sea. At the same time, the existing overproduction of cereal crops should be taken into account. In principle, there are three options available in agricultural practice:

1) decreasing the intensity of agriculture;

2) changing the use of land (in terms of the cropping system);

3) instituting special restrictions on fertilizer and pesticide use.

The Swedish Government intends to use all three options. The instruments include increasing taxes on nitrogen fertilizers and pesticides, as well as payments for changing land use from the production of cereals, oil plants, potatoes, sugar beet, leguminous plants, and certain other crops to grassland farming and afforestation. Special restrictions are placed on the handling of manure in the catchments of Lake Ringsjön in Skåne and Laholm Bay on the west coast, two very sensitive waters. In those areas, the stocking rate is to be restricted to 1.6 dairy cows per hectare of farmland after 1990, and

similar rates for other animals.   From January 1, 1993, manure
spreading will be banned between December 1 and February 28.   From
the same date impermeable manure pits and liquid manure tanks
should be made available with storage capacities of at least 8 months,
an increase from the presently required period of 6 months.

The objectives for the future include a reduction in the use of
commercial fertilizers by 30 per cent and placing restrictions on animal
farming in Sweden similar to those mentioned above.   In addition, the
use of plant regulators such as chlomequat chloride for cereals will not
be permitted (rye is exempt from this restriction for a period of five
years).   These restrictions are being introduced because residues have
been observed in food, and because this use is an incentive for excessive
fertilizing.   In 1985 the Swedish Government proclaimed the aim to
reduce the use of pesticides to 50 per cent of present levels within a five
year period.

Taxes, called environmental charges, today amount to 0.60
Swedish crowns (SEK) kg^{-1} for nitrogen (as N) and 1.20 SEK kg^{-1} for
phosphorus (as P).   This tax adds about 10 per cent to the prices, and
an increase of 5 per cent in 1988. These charges average 66 SEK ha^{-1}.
Taxes on pesticides are 4 SEK kg^{-1} active ingredient, for instance, 100
SEK ha^{-1} for TCA and 0.02 SEK ha^{-1} for Glean (chlorsulfuron), a highly
effective preparation.

There are also other charges on both fertilizers and pesticides
aimed at subsidizing the export of grain surplus on the world market.
These charges vary heavily for fertilizers but have been constant for
pesticides at 29 SEK ha^{-1} since 1986 when the charges were introduced.

Annual payments for changing land use are in the range of 700
to 2,900 SEK ha^{-1}, depending on the agricultural district in which the
farm is situated.   The payment will run for three years if the change is
to organic farming, that is, without commercial fertilizers and pesticides.
From the environmental point of view, payment for fallowing is
misdirected because of heavy nitrogen leaching (Figure 10.6).   It should
be mentioned that despite the taxes on pesticides, their use has
increased, when measured in treated area.   So the goal to reduce their
use remains elusive.

Is a reduced intensity or a changed land use the most efficient
way to reduce the grain surpluses and minimize leaching?   This
fundamental question can be answered by making use of results shown
in Figure 10.7.

The curves in the figure have been described mathematically
(Brink, 1987), and the formulae have been used for further calculations.
Some results are shown in Figure 10.10, which shows that optimum
fertilizer levels and leaching of nitrogen would decrease with an
increasing nitrogen price and a decreasing wheat price.   A doubling of
the nitrogen price from 7.5 to 15 SEK kg^{-1} would, for instance, reduce
the losses by half, provided that the fertilizing rates were optimal.

If environmental costs are also taken into account, the lower dashed curves (Figure 10.10) are obtained. That is when every kilogram of nitrogen causes as much damage to the environment as benefit to farming. But the taxes and charges on fertilizers are not enough to reach this goal.

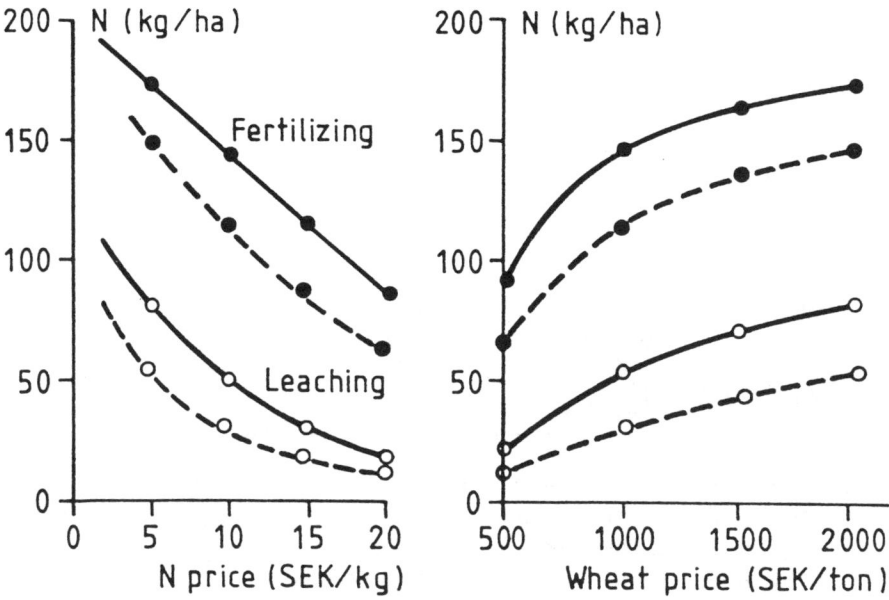

Figure 10.10. Optimum fertilizer levels and leaching losses of nitrogen as functions of N price and wheat price. The lower dashed curves are if environmental costs are included.

Furthermore, as there is a parabolic relationship between agricultural intensity and leaching of nitrogen, and a linear relationship between cultivated area and leaching of nitrogen, a reduction of intensity would be more effective than a reduction of the area. This is demonstrated in Table 10.6, based on Figure 10.7. The yield of Winter wheat is 5,300 kg ha^{-1} at an optimum fertilizer level of 160 kg N ha^{-1} and a leaching loss of 70 kg N ha^{-1}.

The leaching of nitrogen from an annual ley (Ley 1) is also 70 kg N ha^{-1}; from a long ley (Ley 1-4) 15 kg N ha^{-1}; from a deciduous wood 5 kg N ha^{-1} yr^{-1}; and from a coniferous forest 1 N kg ha^{-1} yr^{-1}.

Among the cropping systems mentioned in Table 10.6, conifers appear to be the most effective from an environmental point of view, followed by deciduous trees and long lying leys. An annual ley would not improve leaching losses. Even if this analysis concerns a specific

Table 10.6. Leaching losses of nitrogen (N) from 100 ha of arable land
at different yield levels of Winter wheat, different intensity and different
cropping systems.

Yield (tonnes)	530	477	424	371
Intensity				
Fertilizer (kg N ha⁻¹)	160	112	72	48
Leaching (kg N 100 ha⁻¹)	7,000	3,100	1,680	1,190
Cropping system				
Wheat area (ha)	100	90	80	70
Changed area (ha)	0	10	20	30
Leaching (kg 100 ha⁻¹)				
Wheat + Conifers	7,000	6,310	5,620	4,930
Wheat + Deciduous trees	7,000	6,350	5,700	5,150
Wheat + Ley 1-4	7,000	6,450	5,900	5,350
Wheat + Ley 1	7,000	7,000	7,000	7,000

case, the principle of the outcome is general. Thus, there is no question
that a cut in the intensity would be the best way to reduce both the
grain surpluses and the nitrogen leaching.

Cultivation of *catch crops* is recommended as a special measure
against nitrogen leaching. Such crops should either be sown in spring
with cereals or sown as early as possible after the harvest of the
preceding crop, and plowed under in late autumn or early spring.

A series of plot experiments has been conducted to assess the
nitrogen leaching losses with and without catch crops (Gustafson and
Torstensson 1984, 1987; Kreuger and Brink, 1984). Results are given
in Table 10.7. The improvements show a wide variation, from about 0
to 50 per cent, (with an average improvement of 20 per cent). An
improvement of a few kilograms is, under such conditions, much easier
and probably cheaper to achieve by decreasing intensity.

Finally, what should be done in relation to pesticides? To stop
spraying would be the simplest way to eliminate all harmful problems
associated with their use. However, even though pesticides are artifacts
in nature the most realistic way is to successively decrease their use in
real terms (not by weight). This is important because one substance
may be needed in large amounts whereas another only in small
amounts. For example, 20 kg ha⁻¹ of TCA is required compared with
only 4 µg ha⁻¹ of the active ingredient chlorosulfuron.

Table 10.7.   Nitrogen leaching losses with and without catch crops, plowed or harrowed in late autumn or early spring. Annual fertilizer 90-150 kg N ha^{-1}.

| Catch crop | Head crop | Nitrogen leaching (kg ha^{-1} yr^{-1}) | | | |
		Without	With	Gain	Years
Winter rye	Potato[a]	47	37	10	9
Winter rape	Barley	46	44	2	1
Rye grass	Oats[b]	21	16	5	2

[a]Potato, barley or Spring rape.
[b]Oats and Spring rape.

10.5. Conclusion

It is evident that environmental protection from the adverse effect of agricultural chemicals (fertilizers and pesticides) will require financial disincentives to modify particular agricultural practices.   The careful monitoring of fertilizer, pesticide and other losses to drainage water will help to establish these on a realistic basis.

REFERENCES

Andersson, A. and Gustafsson, A., 1982, Metal contents in drainage water from cultivated soils. *Ekohydrologi*, **11**, 13-18.   (In Swedish).

Bergström, L. and Brink, N., 1986, Effects of differentiated applications of fertilizer N on leaching losses and distribution of inorganic N in the soil. *Plant and Soil*, **93**, 333-345.

Brink, N., 1965, The role of agriculture for the eutrophication of waters. *Nord. Jordbrugsforskning*, **47**, 197-207.

Brink, N., 1983, Nutrients and organic matters from farmland and woodland. *Ekohydrologi*, **14**, 21-30. (Abstract in English).

Brink, N., 1986, Factors affecting mass transport from farmland.   In *Saavalainen and Vakkilainen (eds.). Proceedings of an International Seminar on Land Drainage* .   Helsinki University of Technology, Finland.

Brink, N., 1987, Economic measures against leaching losses of nitrogen. In W. van Duijvenboden and H.G. Waegeningh (eds.), *Vulnerability of Soil and Groundwater to Pollutants*.   Proceeding and Information, Volume 38.   TNO Committee on Hydrological Research, The Hague, The Netherlands,   555-562.

Brink, N. and van der Meulen, K., 1987, Losses of phosphorus and nitrogen to Lake Ringsjön. *Ekohydrologi*, **25**, 5-15.

Brink, N., Gustavsson, A. and Ulén, B., 1983, Surface transport of plant nutrients from a field spread with manure. *Ekohydrologi*, **13**, 3-14. (Abstract in English).

Denoyelles, F., Kettle, W.D. and Stinn, D.E., 1982, The responses of plankton communities in experimental ponds to atrazine. *Ecol.*, **63**, 1285-1293.

Erne, K., 1970, Herbicides in the Swedish environment. *Statens Veterinärmedicinska Anstalt*, Stockholm. (In Swedish).

Fleischer, S., Hamrin, S. and Kindt, T., 1987, Coastal eutrophication in Sweden: reducing nitrogen in land runoff. *Ambio*, **5**, 146-251.

Frank, R., Braun, H.E., van Hove Holdrinet, M., Sirons, G.J. and Ripley, B.D., 1982, Agriculture and water quality in the Canadian Great Lakes Basin. *J. Environm. Qual.*, **11**(3), 497-505.

GEFO, 1987, Pesticides in surface water and groundwater. *Inst. Georesurs- og Forurensningsforskning*, Norway (In Norwegian).

Graneli, E., Schulz, S., Schiewer, U., Gedziorowska, D., Kaiser, W. and Plinski, M., 1988, Is there the same nutrient limiting potential phytoplankton biomass formation in different coastal areas of the southern Baltic? *Kieler Meereforschungen*. (In press).

Gustafson, A. and Torstensson, G., 1984, Catch crop after barley. *Ekohydrologi*, **15**, 13-20. (Abstract in English).

Gustafson, A. and Torstensson, G., 1987, Catch crop after harvest. *Ekohydrologi*, **24**, 4-14. (Abstract in English).

Hallberg, G.R., 1987, Agricultural chemicals in groundwater: extent and implications. *Alternative Agriculture*, **2**, 3-14.

Kreuger, J.K. and Brink, N., 1984, Catch crop and divided N-fertilizing when growing potatoes. *Ekohydrologi*, **17**, 3-14. (Abstract in English).

Kreuger, J.K. and Brink, N., 1987, Losses of pesticides from agriculture. International Symposium on *Changing Perspectives in Agrochemicals*, Nuremberg, Germany, F.R.

Nihlgård, B., 1985, The ammonium hypothesis - an additional explanation of the forest dieback in Europe. *Ambio*, **14**(1), 2-8.

Rosenberg, R. and Loo, L.O., 1986, Effects on the bottom fauna in Laholm Bay, Kattegat and Skagerrak. In R. Rosenberg (ed.), *Eutrofieringsläget i Kattegatt*. SNV Rapport **3272**, Stockholm. (In Swedish).

Solyom, P., 1986, Aspects on irrigation water quality demands. Kons. Rapp. Allm. 84. SLU Uppsala. (In Swedish).

Uhlen, G., 1979, Nutrient leaching and surface runoff in field lysimeters on a cultivated soil. *Meld. Norges Landbrugshögskole 57 o 58*. Norway.

Ulén, B., 1985, Erosion of phosphorus from arable land. *Ekohydrologi*, **20**, 26-35. (Abstract in English).

Wiklander, L. and Hallgren, G.,1971, Leaching of nutrients. *Grundförbättring*, **24**, 95-111. (In Swedish).

Chapter 11

POTENTIAL EFFECTS OF CLIMATE AND LAND USE
CHANGES ON THE WATER BALANCE STRUCTURE IN
POLAND

L. Ryszkowski, A. Kedziora and
J. Olejnik

11.1. Introduction

The global increase of energy consumption is causing an increase in the
concentration of CO_2 in the atmosphere. This concentration may double
before the end of the next century (Zimen, 1979; Rotty, 1979; Blasing,
1985). Carbon dioxide is a major factor determining the energy balance
of the Earth and increases in atmospheric CO_2 will therefore alter global
climate. Changes in global temperature have received much attention but
seasonal changes in precipitation are also likely to occur.

Changes affecting hydroclimate phenomena may occur in response
to increasing concentrations of CO_2. The order of magnitude of these
changes is difficult to assess but some effects can be anticipated. Vapor
pressure and water vapor pressure deficit could increase resulting in an
increase of potential evapotranspiration. An increase in air temperature
will affect the amount of energy used for evapotranspiration in
ecosystems, since the part of net radiation that is used for this process
would also be likely to increase (Monteith, 1975; Lockwood, 1979).

Increased carbon dioxide content in the atmosphere causes a
reduction in stomatal opening of plants and therefore increases stomatal
resistance to water loss, diminishing plant transpiration levels (Callaway
and Currie, 1985). But an increase in potential evaporation and soil
moisture may also cause a substantial increase in evaporation which
would outweigh this transpiration decrease. Thus an overall decrease in
evapotranspiration is not foreseen. Increases in evapotranspiration of
between 15 per cent and 70 per cent may result under these new
conditions.

Projections on the behavior of plant communities in response to a
rise in atmospheric CO_2 and the accompanying climatic changes are
complicated by the fact that increasing rates of global deforestation will
result in a change in the amount of CO_2 that is released to the
atmosphere (Hampicke, 1979). It may also result in structural changes
of the heat balance components in landscapes. Evapotranspiration is

F. M. Brouwer et al. (eds.), Land Use Changes in Europe, 253–274.

most intensive in forest ecosystems (Calder, 1986). Deforestation of an area may raise the surface albedo and at the same time decrease the radiation balance. One of the difficulties that remains is the projection of climatic change on a regional/national scale, since the scenarios of climatic change differ strongly in spatial detail.

The objective of this Chapter is to project changes in heat and water balance components in relation to changes in climatic conditions and land use patterns. The study will focus on projected changes in Poland. Any model that is based on a number of assumptions will include an element of uncertainty. Major uncertainties still exist concerning the *absolute* changes in heat and water balance components. Therefore attention is focused, mainly on *relative* changes in the water and heat balance components, as compared to present conditions.

11.2. Methods of Assessing the Heat and Water Balance Structure

An assessment of the heat and water balance structure has been prepared for 28 regions in Poland (Figure 11.1 and Table 11.1). The regionalization is based on climatic characteristics (Romer, 1949). The regions in the southern and northern part of the country are analogous to the climatic regions as they have been developed by Romer (1949). A subdivision was made for the central part of the country because of significant differences in land use patterns.

Information on the water balance structure, in terms of precipitation, evaporation and runoff has been collected and assessed for two periods of the year: i) April-September (growing season); and ii) October-March (winter period). Annual values are also presented. Averages were calculated for each of the 28 regions and for the whole country. The values for the whole country are weighted averages of the regional values determined by using the following equation:

$$\bar{x} = \frac{\sum\limits_{i=1}^{i=28} (x_i \, w_i)}{\sum\limits_{i=1}^{i=28} w_i}, \tag{1}$$

where x_i is the value of the hydrological component for region i, and w_i is a weighting factor representing the ratio of the size of region i to the national total (Table 11.1).

Figure 11.1. Climatic regions of Poland.

The heat and water balance equations are as follows:

$$R_n + LE + C + G = 0 \qquad \text{heat balance}$$

$$P - E - R = 0 \qquad \text{water balance}$$

where R_n is net radiation, LE is latent heat flux, C is sensible heat flux, G is soil heat flux, P is precipitation, E is evapotranspiration and R is runoff.

All quantities in the heat balance equation are expressed as watts per square metre (W m^{-2}). The incoming fluxes to the active surface are taken as positive, while the outgoing fluxes are negative. The components of the water balance equation are expressed in millimeters.

Calculations of heat balance components for the growing season were carried out using the method described by Olejnik (1988) and Ryszkowski and Kedziora (1987).   This   method   is   based   on   the

Table 11.1.   Size of the areas (km²) and projected level of percentage
deforestation for the year 2050.

Region	Size (km²)	Deforestation (%)
1	7,066.9	10
2	4,827.5	25
3	4,170.7	50
4	22,011.8	10
5	17,957.0	0
6	7,646.2	0
7	14,867.6	25
8	12,859.5	25
9	30,893.7	50
10	16,026.1	10
11	18,922.4	0
12	3,977.6	75
13	14,790.4	50
14	25,796.3	10
15	19,347.2	25
16	14,983.5	25
17	8,959.2	100
18	4,672.7	100
19	21,432.5	75
20	5,483.6	50
21	3,861.7	100
22	11,276.2	50
23	4,093.4	50
24	6,835.2	25
25	4,093.4	25
26	1,853.6	25
27	3,282.5	25
28	695.1	25
Poland	312,663.5	32

interdependence between the Bowen ratio (defined as the ratio between C and LE) and a complex agrometeorological index:

$$\beta = \frac{12.75}{W + 3.9} - 0.19 \tag{2}$$

where β is the Bowen ratio and W is an agrometeorological index given by the equation:

$$W = \frac{100 \ (d \ \sqrt{v})^{\arctan\left(\frac{\pi}{2}\,f\right)}}{T \ (u + 0.4)}, \tag{3}$$

where d is the saturation vapor pressure deficit (in millibars), v is wind speed (m s^{-1}), T is air temperature (°C), u is relative sunshine (dimensionless) and f is a parameter describing the plant development stage, this parameter ranges from 0.01 (bare soil) to 1.0 (fully developed plant cover). The plant development stage depends on the land use category and is calculated on a regional scale on a monthly basis (Chapter 4). The index W shows the relationship between plant growth and meteorological conditions and the proportion of energy that is used for evapotranspiration and air heating in the radiation balance. The higher the index value, the greater the part of net radiation used for evapotranspiration.

All components of the heat balance were calculated independently. Evapotranspiration was calculated by using the following equation:

$$E = \frac{LE}{L}, \tag{4}$$

where L is latent heat of vaporization, 2.5 MJ kg^{-1}.

For the winter period, the radiation balance was calculated by using the Black and Brunt equations and the latent heat flux by calculating evaporation and multiplying it by:

$$L \ (LE = L \times E)$$

Sensible heat flux was calculated from the following equation:

$$C = -(R_n + LE + G) \tag{5}$$

Soil heat flux was obtained by using the empirical equation (Olejnik, 1986). A major level of accuracy to soil heat flux, either for a half year period or on an annual basis, is of no vital importance since it is close to zero for a long period.

Changes of the albedo are also taken into consideration. During the winter period, these changes will be influenced by a shorter period of snow cover because of the projected increase in temperature.

All information on present climatic conditions originates from the Climatic Atlas of Poland (1973), while the plant development stage has been assessed for different land use patterns. The assessments are based on different land use scenarios and climatic conditions for the present day (1985 used as the reference year), the year 2050 and the year 2100.

The year 2100 reflects the period of strong climate warming, and is based on a scenario of a general circulation model with an equilibrium response of a doubling of atmospheric CO_2.   The year 2050 is a transient period of climate change.

Changes in heat and water balance components are assessed in four scenarios; two are for the year 2050 and two are for the year 2100. The scenarios take changes in land use patterns and climatic conditions into account. Five indicators are incorporated (in Table 11.2 the scenarios for land use and climate change are summarized).

1)   Ratio of dryland to wetland.

2)   Percentage of forest area. Projected levels of deforestation for the year 2050 are given in Table 11.1. Forest dieback in Poland due to nitrogen and sulfur oxides has been assessed by the National Statistical Office (1986). Coniferous forests for example, are presently under major stress due to air pollution. The largest amount of deforestation is found in three highly industrialized areas of southern Poland. No deforestation was considered to occur in three regions of north-east Poland because of the low level of air pollution.

3)   Ratio of relative sunshine compared to the 1985 data. Relative sunshine is either considered to be unchanged or to be increased by 5 per cent.

4)   An increase in air temperature in response to an increase of CO_2 and other greenhouse gases. The scenario with an annual temperature increase of between 2 and 6°C is based on WMO-UNEP (1988).

5)   Changes in precipitation for the year 2100 (also based on WMO-UNEP, 1988). It is considered that an increase of 0.6 mm day^{-1} for March, April and May, and of 0.4 mm day^{-1} for the remaining months is a plausible climatic scenario for this area in response to a doubling of CO_2. Such a scenario for temperature and precipitation changes in Poland fits the results from the GISS

Table 11.2. Scenarios of land use (ratio of dryland to wetland, and level of afforestation in percentage of total area) and changes in climatic conditions (increase in temperature in °C, increase in precipitation in mm day⁻¹).

Scenario	Year	Ratio of dryland to wetland	Forestation in % of total area	Ratio of actual relative sunshine to relative sunshine in 1985	Increase in Temperature in °C	Increase in Precipitation mm day⁻¹
Present	1985	75:25	28	1.00	-	-
1	2050	75:25	according to Table 11.1	1.05	1-3	0.2
2	2050	75:25	0	1.05	1-3	0.2
3	2100	25:75	30	1.05	2-6	0.4
4	2100	0:100	0	1.00	2-6	0.4

general circulation model. For the year 2050 half the projected changes for the year 2100 have been assumed.

A distinction has been made between seven major land use categories, including forests, water, urban land, degraded plant areas, meadows and pastures, root crops and cereal crops. The present patterns of these land use categories, as shown in Table 11.3, are based on the Agricultural Atlas of Poland (1978). The length of the growing season is based on phenological information (National Atlas of Poland, 1964). Figure 11.2 gives details for 28 regions.

A projected increase in temperature will result in an increase in the concentration of water vapor in the atmosphere, as well as in an increase in the saturation vapor pressure deficit. The equation below has been used in order to assess these, it is based on average monthly data (Figure 11.3).

$$e = 5.3 \exp(0.0615 \ T), \qquad\qquad (6)$$

where e is water vapor pressure, T is air temperature (both for present and projected conditions).

The method of evaluating heat and water balance components of the areas presented above was tested against information in the Wrzesnica river catchment (in central Poland). The method proved to be very effective as the results for this catchment area are similar to those found in other assessments (Kirchner, 1984), as well as to hydrological calculations carried out by Stachy (1965). The annual runoff difference for example, (as calculated by the method presented above and the results obtained by Stachy) is approximately 4 mm (Kedziora *et al.*, 1987).

11.3. Scenarios of Heat and Water Balance Components

Heat and water balance components and their ratios, calculated for four scenarios of climatic change and land use patterns, are described in Tables 11.4, 11.5 and 11.6 for different parts of the year. Table 11.4 shows the components during the growing season and Table 11.5 shows the components during the winter period. Annual values are given in Table 11.6. The ratios of latent and sensible heat to net radiation, as well as the ratios of evapotranspiration and runoff to precipitation, are also shown to express the relative contribution of these components to the structure of the heat and water balance.

The order and magnitude of changes in cloudiness in response to changes in atmospheric CO_2 are still uncertain. On the one hand, an increase in precipitation may stimulate cloudiness, which may cause a decrease in the amount of solar energy that is reaching the Earth's surface. A decreased radiation balance might, therefore, result from changes in cloudiness. Recently (Mitchell, Senior and Ingram, 1989), the

Table 11.3. Land use patterns in Poland for each of the 28 regions (in percentages of the total area).

Region	Water	Forest	Urban area	Degraded plant area	Meadows and pasture	Root crops	Cereal crops
1	5.2	35.5	7.0	3.0	16.8	8.7	23.8
2	6.0	25.4	7.2	3.1	27.6	10.5	21.1
3	7.4	30.2	7.1	3.7	17.0	9.6	25.0
4	3.9	33.7	6.5	2.4	12.5	9.7	26.6
5	4.8	28.6	5.6	2.7	27.8	6.3	24.2
6	7.5	31.6	4.6	3.8	25.9	3.8	22.8
7	2.9	45.7	7.0	2.0	10.2	6.9	25.3
8	2.2	27.8	6.5	1.3	16.1	13.2	32.9
9	2.8	20.7	6.6	2.1	17.7	14.7	35.3
10	1.4	24.0	5.1	2.2	36.6	8.5	22.2
11	1.1	26.6	5.4	1.7	15.8	12.5	36.9
12	1.9	38.4	9.0	1.5	20.8	6.5	21.7
13	2.3	25.1	8.5	0.6	22.8	14.9	35.9
14	1.6	20.2	7.1	1.1	12.5	16.4	41.1
15	1.8	21.4	7.5	1.0	11.4	17.2	39.7
16	1.6	21.7	5.6	1.3	12.0	17.4	40.5
17	1.6	33.0	8.6	1.2	29.9	6.6	18.9
18	2.3	28.6	12.9	1.0	16.8	6.8	31.3
19	1.4	29.1	7.8	0.5	10.7	15.1	34.8
20	1.1	21.8	5.5	0.6	11.7	13.2	46.2
21	1.3	22.0	9.3	0.7	13.2	12.1	41.5
22	1.5	26.3	6.3	0.3	27.3	10.8	27.2
23	3.3	37.9	7.5	0.6	26.9	6.9	17.2
24	1.4	24.5	7.2	0.5	18.8	10.9	35.6
25	1.4	37.2	6.0	1.0	22.3	7.0	25.5
26	1.7	37.2	5.0	0.5	26.6	6.1	17.2
27	1.6	48.7	4.6	0.4	26.0	3.7	15.0
28	1.7	42.4	5.0	1.0	27.8	4.7	17.4

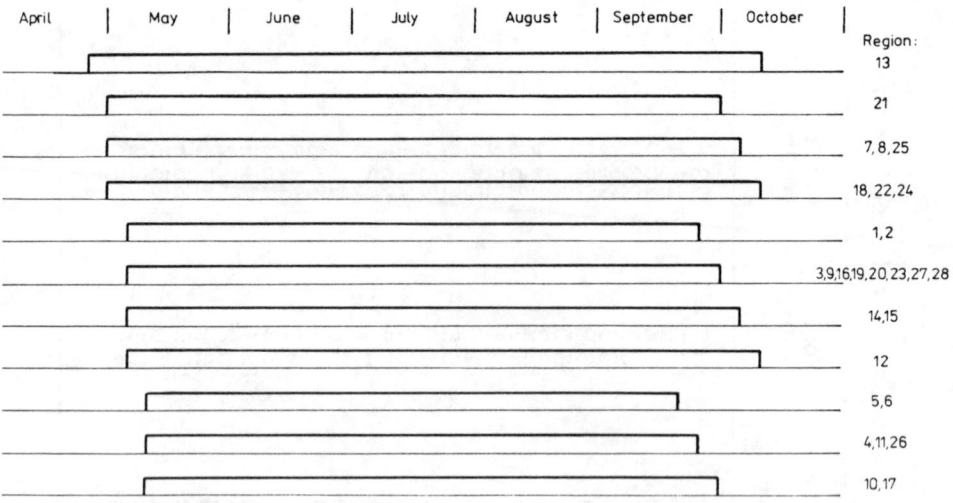

Figure 11.2. Growing season in 28 regions of Poland.

effect of changes in cloud cover and cloud properties as possible sources of climatic feedback have been discussed. Difficulties in reconciling the two divergent trends of projected temperature can be resolved if no changes in the quality (not quantity) of cloudiness are assumed. An increase in temperature may cause an intensive turbulent exchange in the atmosphere, to be followed by an increasing rate of vertically developed clouds (Cumulus and Cumulonimbus) and a decreasing rate of horizontal cloud formation (Stratus and Stratocumulus). Relative sunshine and precipitation might both increase due to such changes in cloudiness.

The maximum change of net radiation, R_n, during the growing season (Table 11.4) was 5 W m^{-2} (comparing present conditions with Scenario 3). This is equivalent to the energy required for the evaporation of 32 mm of water during this period. The division of net radiation into latent heat flux (LE) and sensible heat flux (C) is of major importance from an ecological point of view. Presently, about 66 per cent of net radiation is used in the evapotranspiration process. Changes in environmental conditions (changes in land use, increase of habitat moisture, variation in temperature and precipitation) may result in an increase in the LE/R_n ratio from 62 per cent to 75 per cent (Scenario 3, Table 11.4). This Scenario is based on the assumption that degraded land is re-afforested and an additional 2 per cent of wetland will also be afforested. The results from Scenario 2 seem surprising: for example, evapotranspiration by the year 2050 under land use conditions where total deforestation is assumed (Table 11.2). This result can be explained by the increase in potential evaporation caused by increased temperature and water vapor deficit. Although transpiration will decrease, evaporation

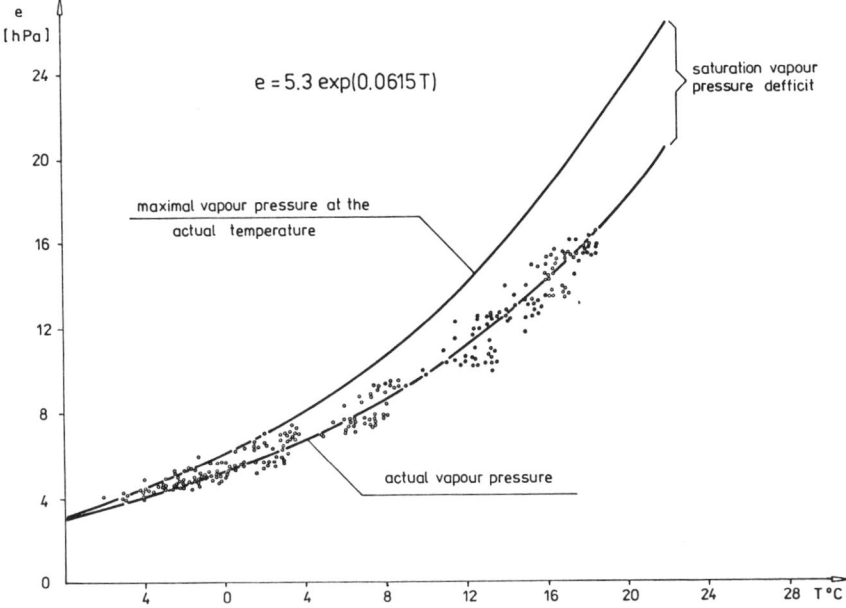

Figure 11.3. Relationship between vapor pressure and air temperature.

will increase because of changed atmospheric conditions and habitat moisture. This conclusion is also confirmed by a decrease of the Bowen ratio under such conditions (Monteith, 1975; Lockwood, 1979).

Table 11.6 is based on heat and water balance components on an annual basis and the maximum increase of net radiation is 3 W m^{-2} (comparing present conditions with Scenario 3), but the increase in the LE/R_n ratio from 78 per cent to 94 per cent results in an increase in evapotranspiration. However, the increase of precipitation causes the initial decrease of the E/P ratio (evapotranspiration to precipitation) because of plant cover degradation through the first half of the twenty-first century. During the second half of the next century the E/P ratio will increase to 75 per cent.

The distribution of the E/P ratio over the whole country during the growing season for present conditions, the year 2050 and the year 2100 are presented in Figures 11.4, 11.5 and 11.6 respectively.

The ratio of evapotranspiration to precipitation is not considered to show major changes because both components are increasing. The hydrological inputs barely match the outputs during the growing season in central Poland. Most rainfall is consumed in evapotranspiration during

Table 11.4. Present and projected heat and water balance components during the growing season. (Heat balance components in W m^{-2}, and water balance components in mm.)

Scenarios	Heat balance components					Water balance components				
	R_n	LE	C	$\dfrac{LE}{R_n}$	$\dfrac{C}{LE}$	P	E	P-E	$\dfrac{E}{P}$	$\dfrac{P-E}{P}$
Present	88	-56	-31	-0.64	0.55	397	364	33	0.92	0.08
1	91	-61	-29	-0.67	0.48	439	369	44	0.90	0.10
2	90	-59	-31	-0.66	0.53	439	380	59	0.87	0.13
3	93	-70	-23	-0.75	0.33	481	452	29	0.94	0.06
4	87	-54	-33	-0.62	0.61	481	347	134	0.72	0.28

Table 11.5. Present and projected heat and water balance components during the winter period. (Heat balance components in W m^{-2}, and water balance components in mm.)

Scenarios	Heat balance components					Water balance components				
	R_n	LE	C	$\frac{LE}{R_n}$	$\frac{C}{LE}$	P	E	P-E	$\frac{E}{P}$	$\frac{P-E}{P}$
Present	4	-16	9	-4.00	-0.56	214	103	111	0.48	0.52
1	5	-21	14	-4.20	-0.67	270	136	134	0.50	0.50
2	5	-21	14	-4.20	-0.67	270	136	134	0.50	0.50
3	6	-23	17	-3.86	0.74	310	149	161	0.48	0.52
4	7	-22	15	-3.14	-0.68	310	142	168	0.46	0.54

Table 11.6. Present and projected heat and water balance components on an annual basis. (Heat balance components in W m^{-2}, and water balance components are in mm.)

Scenarios	Heat balance components					Water balance components				
	R_n	LE	C	$\frac{Le}{R_n}$	$\frac{C}{LE}$	P	E	P-E	$\frac{E}{P}$	$\frac{P-E}{P}$
Present	46	-36	-11	-0.78	0.31	610	461	149	0.76	0.24
1	48	-41	-8	-0.85	0.20	711	527	183	0.74	0.26
2	48	-40	-8	-0.84	0.20	711	510	200	0.72	0.28
3	49	-46	-3	-0.94	0.07	789	596	193	0.75	0.25
4	47	-38	-9	-0.82	0.24	789	489	300	0.62	0.38

the growing season. A lowering of the water table is observed under such conditions. A projected increase in runoff especially during spring and a projected increase in water deficit during the growing season require proper planning of artificial reservoir systems to store excess water. An increase in runoff could exacerbate the erosion hazard on arable land, especially in the highlands of south-east Poland. Figures 11.7, 11.8 and 11.9 show the spatial distribution of annual runoff for present conditions and the projected levels for the two Scenarios. Runoff is projected to increase all over Poland under these Scenarios: for example, in central Poland annual runoff is presently less than 100 mm over large areas; this is projected to increase to between 100 and 200 mm over the next 100 years.

A clear relationship between the LE/R_n (the energetic ratio of evapotranspiration or ERE) and the E/P (moisture ratio of evapotranspiration or MRE) ratios is observed both for the growing season and for the whole year (with the exception of the mountainous regions). This relationship is presented in Figures 11.10 and 11.11. Figure 11.10 shows that a 10 per cent increase in the MRE during the growing season would presently cause a 1.2 per cent increase of the

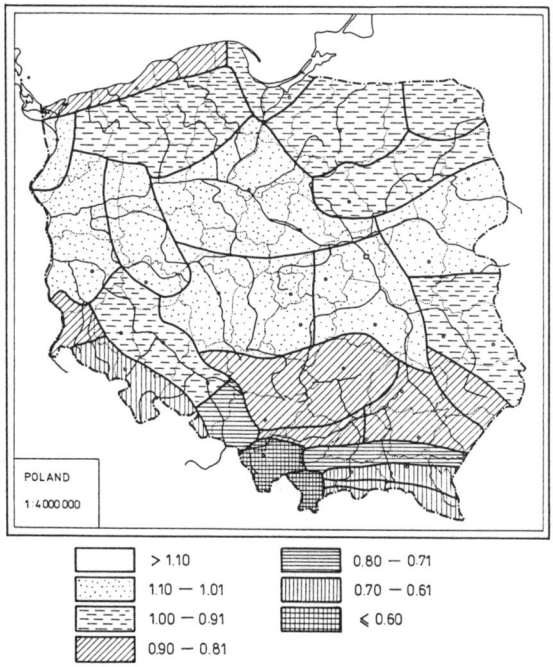

Figure 11.4.   E/P ratio spatial distribution in Poland for the growing season (present conditions).

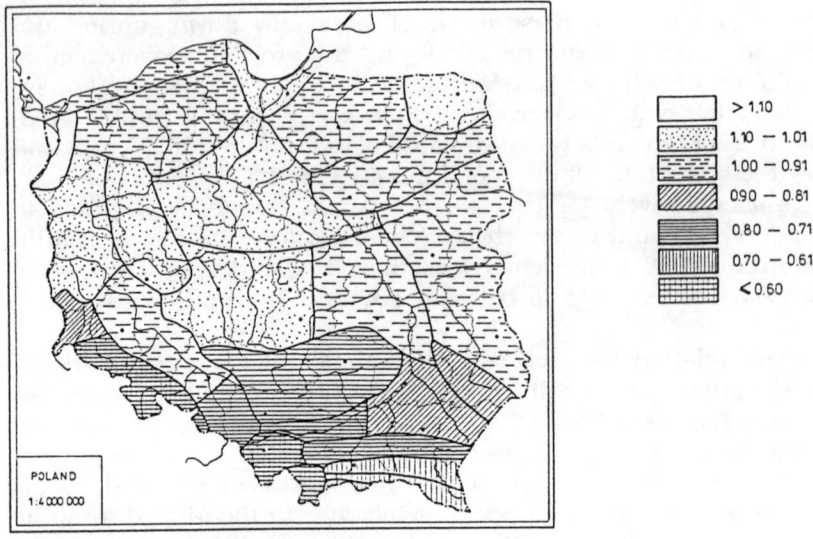

Figure 11.5. E/P ratio spatial distribution in Poland for the growing season (the year 2050).

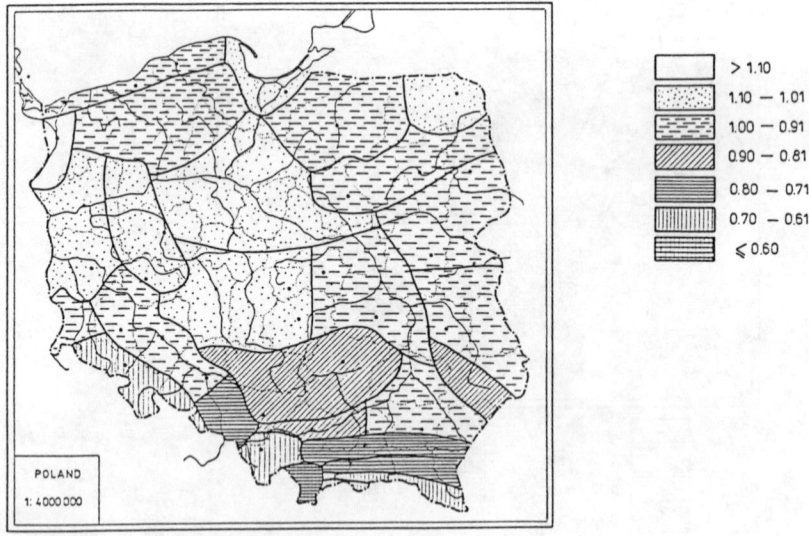

Figure 11.6. E/P ratio spatial distribution in Poland for the growing season (the year 2100).

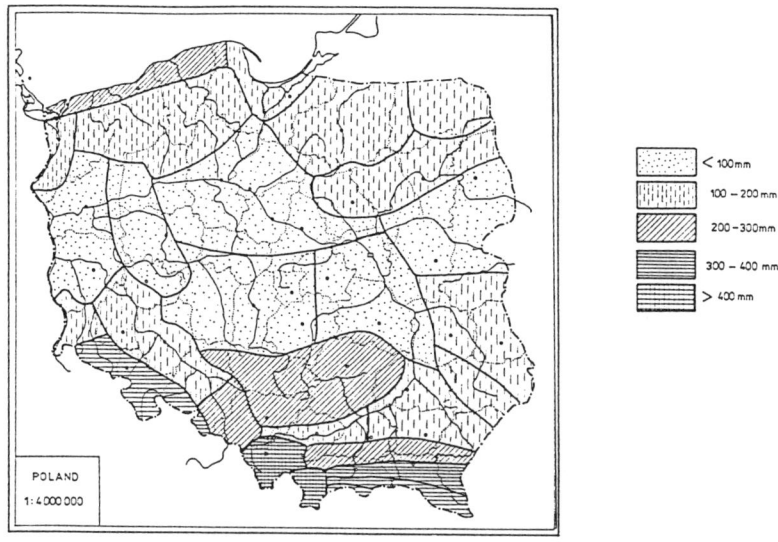

Figure 11.7.   Annual runoff in Poland (present conditions).

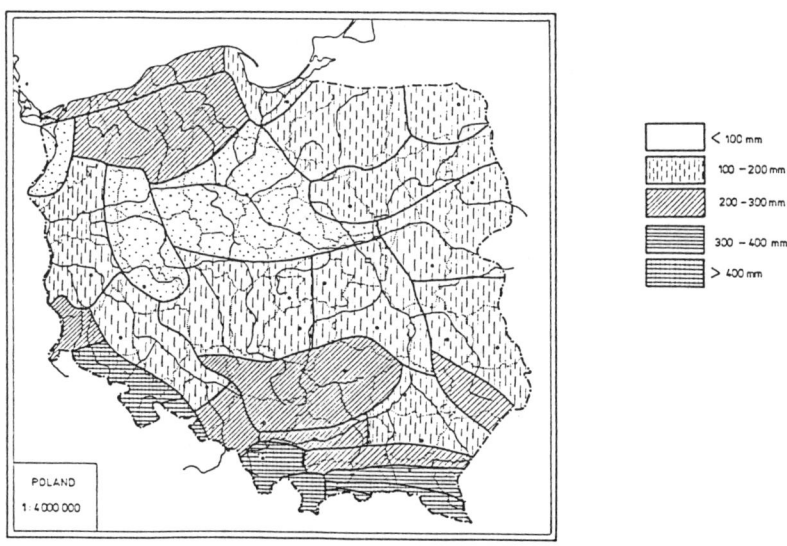

Figure 11.8.   Annual runoff in Poland (the year 2050).

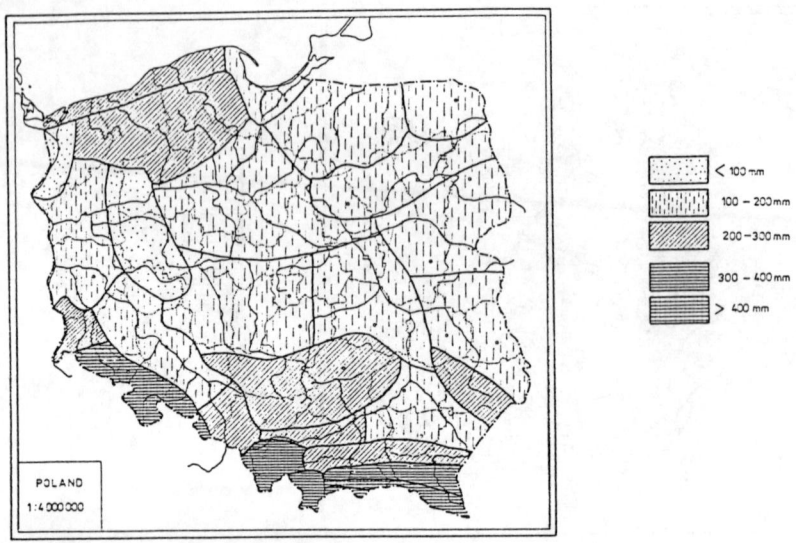

Figure 11.9.   Annual runoff in Poland (the year 2100).

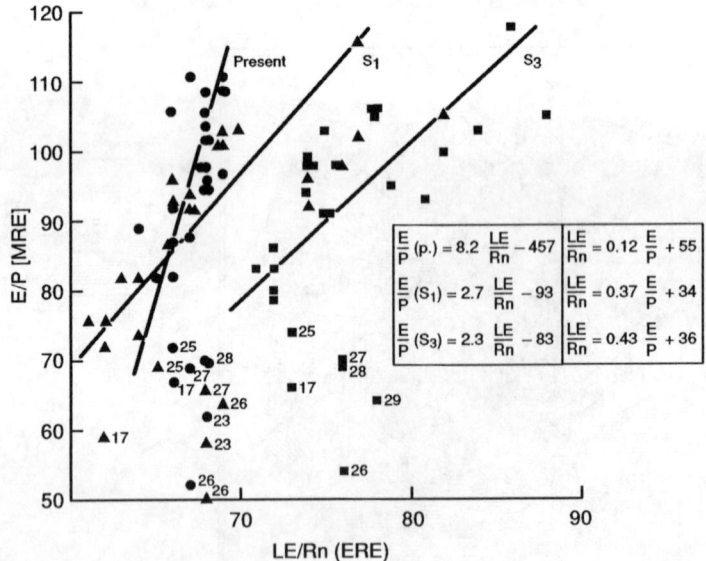

Figure 11.10.   Correlation between ratios E/P and LE/R$_n$ during the growing season for present conditions, Scenario 1 (S$_1$) and Scenario 3 (S$_3$).   Major outliers are mountain areas with number in the figures.

ERE. Under Scenario 1 (2050), the increase will be equal to 3.7 per cent and under Scenario 3 (2100) the increment will be 4.3 per cent. In relation to the whole year these values will be 2.8 per cent, 8.0 per cent and 8.3 per cent respectively (Figure 11.11).

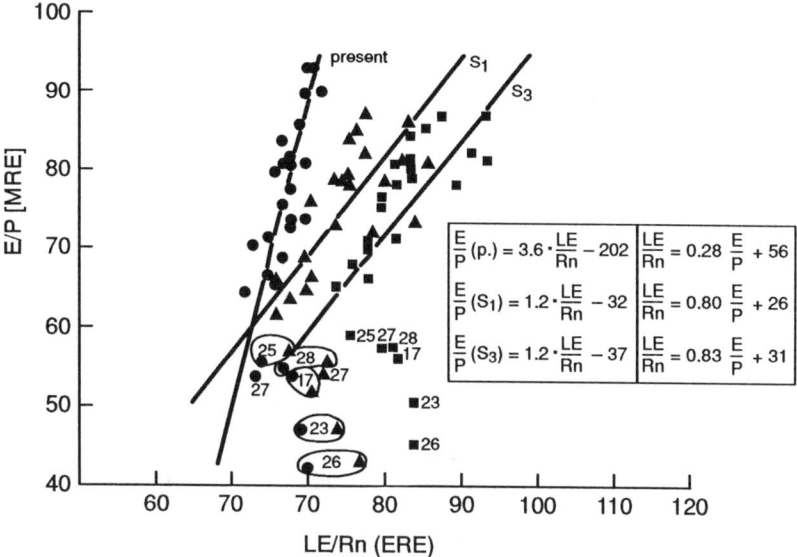

Figure 11.11. Annual values of correlations between E/P and LE/R_n for the present and Scenario 1 (S_1) and Scenario 3 (S_3).

This analysis shows that the relative importance of precipitation to the heat balance structure is rising but will level off with increasing rates of precipitation. When the precipitation is about 500 mm during the growing season or 800 mm annually, the consequent increase in the ERE, as a result of a 10 per cent increase of MRE, will reach 4.5 per cent for the growing season and 8.5 per cent for the whole year (Figure 11.12).

11.4. Conclusions

The following conclusions can be drawn regarding the major hydrological shifts in response to climate change and changes in land use patterns:

i)   a projected increase in temperature and precipitation will result in an increase in evapotranspiration despite plant cover degradation; re-afforestation, with a further increase in precipitation, will result in an increase in annual evapotranspiration by as much as 135 mm by the year 2100 in comparison with the present level of 461 mm;

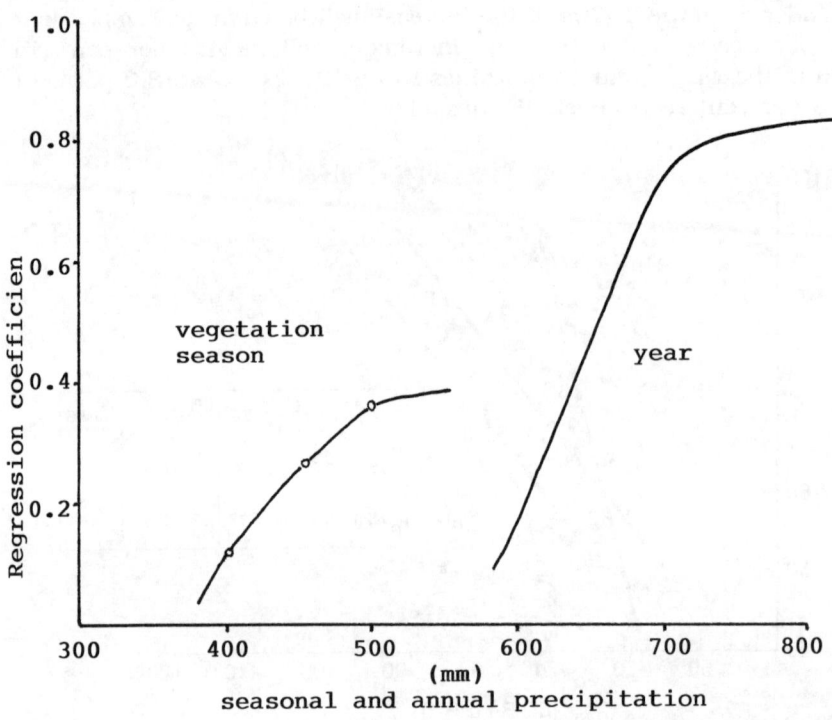

Figure 11.12. Influence of precipitation on changes of ERE as a function of MRE.

ii)   the runoff component of the water balance may change significantly and the intensity of erosion could expand rapidly;

iii)   an increase in precipitation will result in a stronger relationship between ERE and MRE; despite deforestation the increase in evapotranspiration by the year 2050 will result in increased evaporation as an effect of higher evaporative atmospheric demand.

REFERENCES

Agricultural Atlas of Poland, 1978, Wydawnictwo Geologiczne, Warsaw (in Polish).

Blasing, T.J., 1985, Background: Carbon cycle, climate and vegetation response.   In M.R. White (ed.), *Characterization of Information Requirements for Studies of CO$_2$ effects: Water Resources, Agriculture, Fisheries, Forests and Human Health.* US Department of Energy, Washington. 9-23.

Calder, J.R., 1986, What are the limits of forest evaporation? *Journal of Hydrology,* **89**, 33-36.

Callaway, J.M. and Currie, J.W., 1985, Water resource systems and changes in climate and vegetation. In M.R. White (ed.), *Characterization of Information Requirements for Studies of CO₂ Effects: Water Resources, Agriculture, Fisheries, Forests and Human Health.* US Department of Energy, Washington. 23-67.

Climatic Atlas of Poland, 1973, Panstwowe Przedsiebiorstwo Wydawnict Kartograficznych, Warsaw (in Polish).

Hampicke, U., 1979, Man's impact on earth's vegetation cover and its effects on carbon cycle and climate. In W. Bach (ed.), *Man's Impact on Climate.* Series Developments in Atmospheric Science, vol. 10, Elsevier Scientific Publishing Company, Amsterdam, The Netherlands. 139-159.

Kedziora, A., Kapuscinski, J., Olejnik, J. and Moczko, J., 1987, Estimation of the micro- and macro-scale structure of different ecosystems. Department of Agrobiology and Forestry, Polish Academy of Sciences, Poznan, Poland (mimeographed).

Kirchner, M., 1984, Influence of different land use on some parameters of the energy and water balance. *Progress in Biometerology,* **3**, 65-74.

Lockwood, J.G., 1979, *Causes of Climate.* Edward Arnold Ltd., London.

Mitchell, J.F.B., Senior, C.A. and Ingram, W.J., 1989, CO₂ and climate: a missing feedback? *Nature,* **341**, 132-134.

Monteith, J.L., 1975, *Vegetation and the Atmosphere. Vol. 1: Principles.* Academic Press, London.

National Atlas of Poland, 1964, Polish Academy of Sciences, Institute of Geography, Warsaw (in Polish).

National Statistical Office, 1986, *Environmental Protection and Water Management* (in Polish).

Olejnik, J., 1986, Plant cover influence on turbulent transport of heat and vapor in the boundary layer. Ph.D. thesis, Agricultural University, Poznan, Poland.

Olejnik, J., 1988, The empirical method of estimating mean daily and mean ten-day values of latent and sensible-heat fluxes near the ground. *Journal of Climate and Applied Meteorology,* **10**.

Romer, S., 1949, Climatic regions of Poland. Prace Wroclawskiego Towarzystwa Naukowego, Seria B, nr. 16 (in Polish).

Rotty, R., 1979, Energy demand and global climate change. In W. Bach (ed.), *Man's Impact on Climate.* Series Developments in Atmospheric Science, vol. 10. Elsevier Scientific Publishing Company, Amsterdam, The Netherlands. 269-283.

Ryszkowski, L. and Kedziora, A., 1987, Impact of agricultural landscape structure on energy flow and water cycling. *Landscape Ecology,* **1**(2), 85-94.

Stachy, J., 1965, The distribution of the mean runoff in Poland. Prace PIHM, Warsaw, Z. 88. 3-41 (in Polish).

WMO-UNEP, 1988, *Developing Policies for Responding to Climatic Change.*
    WCIP-1 and WMO/TD - No. 225, World Meteorological Organization
    and United Nations Environment Programme.

Zimen, K.E., 1979, The carbon cycle, the missing sink, and future CO_2
    levels in the atmosphere. In W. Bach (ed.), *Man's Impact on
    Climate.* Series Developments in Atmospheric Science, vol. 10,
    Elsevier Scientific Publishing Company, Amsterdam, The
    Netherlands. 129-137.

Chapter 12

SOIL EROSION, SOIL DEGRADATION AND CLIMATIC CHANGE

J. de Ploey, A. Imeson and L.R. Oldeman

12.1. Introduction

Erosion and soil degradation form an increasing threat to land throughout
the world. The severity of this threat is difficult to quantify as basic data
have not been collected or analysed. In this Chapter the present condition
and future perspectives for soil erosion in western Europe and elsewhere
are addressed by presenting what is probably the first survey of the regional
pattern of soil erosion in western Europe based on information supplied by
local experts. The potential impact of climatic change on soil erosion in two
regions of Europe where soils are potentially sensitive to degradation is then
considered and a program whereby data on land degradation are being
collected to enable the world scale of the problem to be identified is then
described. These three topics are complementary and they have been
integrated in an attempt to make a combined statement about the impact
of land use changes on erosion. The Chapter is divided into a number of
Sections. Section 12.2 presents a consideration of the distribution of
erosion and land degradation and discusses the underlying explanation for
the observed patterns, and Section 12.3 discusses the available information
base. Section 12.4 considers the problem of mapping erosion and land
degradation on a European and world scale. Section 12.5 considers the
potential impact of climatic change on erosion in two regions of Europe
where erosion is currently a hazard.

12.2. Soil Erosion in Western Europe

12.2.1. The distribution of soil erosion in western and central Europe

From a preliminary map of the extent of soil erosion in Europe, it is
estimated that 25 million hectares of land are threatened. Annual soil loss
is in the order of tens of tons per hectare for some millions of hectares in
Europe. Erosion is concentrated in the following belts:

   i)   a belt of wind erosion on sandy soils of the Baltic

F. M. Brouwer et al. (eds.), Land Use Changes in Europe, 275–292.
© 1991 *Kluwer Academic Publishers. Printed in the Netherlands.*

Plain, extending westward from Denmark into the Coversand area of the eastern Netherlands and including areas in central Europe;

ii)   an area of sheet and rill erosion in the loess belt of northern Europe, extending from northern France eastward into Germany and Poland;

iii)   an area of slope instability centered in the Alps, where mass movements are encouraged in mountainous areas by forest degradation and deforestation;

iv)   areas of badlands and severe gully and rill erosion at certain locations in the Mediterranean region (for example, in the Apennines and south-east Spain). Argillic rocks and marls are particularly prone to the combined effects of gullying and mass-wasting;

v)   Mediterranean soils located on degraded stony slopes and in which organic matter and clay contents are low and/or the exchangeable sodium content of the soil is high;

vi)   steep slopes in wine-growing areas (for example, the Moselle Valley).

12.2.2. **Soil erosion and climate in Europe**

On a European or world scale, it has long been known that the sediment yields of large rivers are related to climate (Fournier, 1960; Langbein and Schumm, 1958). Maximum yields were considered to occur with mean annual precipitation of 250 mm and were expected to decrease to a minimum in the temperate zone where annual rainfall amounts to 1,000 mm. Consequently, most erosion might be expected in the drier, Mediterranean areas which have a pronounced climatic seasonality. However, recent soil erosion research concerned with seasonal variation and extreme events has shown that conditions are more complex. For example, in temperate areas, conditions that characteristically cause high erosion rates in semi-arid areas can also exist at certain seasons. Thus, the summer erodibility of loamy arable soils in north-west Europe may be increased by one or two orders of magnitude when heavy rainfall is preceded by a long dry period. Erodibility has a memory and is related to antecedent weather and soil moisture conditions (Govers *et al.*, 1987; Gerard, 1987). This concerns the conclusion of Fournier (1960) that regions with maximum erosion rates are those where monthly precipitation varies greatly. Hence, erodibility with respect to rainwash erosion is at a maximum in the summer when periods of desiccation alternate with periods of aggressive rainfall. The awareness of this has led to recent research emphasizing the time-dependence of erodibility and soil loss. Future soil conservation strategies will have the advantage of increased understanding of the nature of the complex effects of extreme events. Improved indices of erosivity could usefully focus on the idea of Ahnert (1986) whereby the frequency of daily rainfall amounts are considered with respect to the corresponding recurrence interval to identify different morphoclimatic zones.

12.2.3. Soil erosion and land use

The present and future pattern of erosion in Europe is closely related to land use. In temperate areas, except on steep slopes, the erosion is considerably less under forest and grassland than it is on cultivated land. However, this is not necessarily so on steep slopes, underlain by plastic, pelitic rocks, where mass movement on forested land may form a potential hazard.

Considering mass movement, the high infiltration rates of forest soils may promote percolation and during wet periods give rise to low shear strengths in the bedrock and regolith (de Ploey, 1979, 1981). Conversely, in an ambivalent way during a large part of the year forests may have an overall positive influence by increasing drainage and shear strength. Whether forests will have a positive or negative effect is difficult to predict since many factors play a role, for example, the type of root system and the root matrix. The effect of land use on mass-movements is confined to shallow slides and surface creep. Bonnard (1985) showed that in the Valais region of Switzerland all alpine slopes were moving along deep seated plains regardless of land use. In the Carpathians or Apennines in southern Italy, tectonic and neotectonic movements also seem to trigger landslides independently of land use (Balteanu, 1983; Carrara et al., 1982).

Considering cultivated land there are also many anomalous relationships which give rise to complex relationships between land use practices and erosion by sheetwash. For example, plowing to increase percolation can lead to subsurface pipe flow and to pipe and rill erosion at locations downslope. Also, many crops whilst protecting the soil from raindrop impact may also generate stemflow which may activate turbulence in overland flow and enhance sediment entrainment or rill initiation. Govers (1986) describes the processes involved and the effect of different crops and cover. In general, in north-west Europe there is a critical period between April and mid-June when a high rainfall erosivity and low plant cover combine to create a potentially hazardous situation. This situation is considered later with respect to climatic change. Also, in many parts of Europe (for example, southern England and north-west Spain), the autumn period can be critical when heavy rains fall on tilled soils. Wet autumn conditions and delayed harvesting can lead to the development of a poor soil structure and to increased susceptibility of the soil to erosion during the following year.

Vegetation is also of critical importance in relation to wind erosion. Much research has been reported on this subject from the Baltic Plain, central England (Morgan, 1985; Morgan *et al.*, 1987), Denmark (Hensen, 1983) and Hungary (Loczy, 1987).

12.3. **Establishing Information on Erosion and Degradation**

To develop a policy towards soil erosion it is necessary to have information on its actual and potential extent. Obtaining such information on a European and world scale presents a major challenge. One reason why this is difficult is because soil erosion and degradation mapping presupposes that definitions of erosional systems exist and that soil erosion and degradation can be quantified in a way that allows assumptions to be tested. This Section will consider some of the problems at a European scale and then discuss an attempt to evaluate degradation at a world scale.

12.3.1. **Erosional systems**

The loess belt of north-west Europe is, as already mentioned, an area of high erosion risk. By way of example, the following main erosion systems in this area have been described (de Ploey, 1989). These systems can be used to characterize mapping units at a particular moment in time but they also form temporal sequences.

1)   *System I*.   High aggregate stability, low susceptibility to crust development, and erosion limited to rainsplash.

2)   *System II*. Low aggregate stability (consistency C 5-10 index below 2.0). The topsoil is prone to liquefaction which leads to sheetwash and to colluvium deposition in valley bottoms.

3)   *System III*. Slopes are greater than 5-6 per cent and the material is cohesive, although the consistency index is below 3. Rill erosion is high on such slopes and the rills end in colluvial cones which expand from the rill apex.

4)   *System IV*. As for III with respect to soil material but the slopes are less than 5 per cent. Inter-rill erosion predominates. Runoff production on long slopes may lead to ephemeral gullies where temporary creeks are generated along the valley bottoms with peak discharges in the order of cubic meters per second.

Variants of these four systems are related to the relative importance of linear elements in the landscape (tracks, roads, ditches and lynchets) which produce specific erosion hazards such as pipes and gullies. A linearity index expressing the effect of such features needs to be developed.

Such a classification for the loess belt would have to be adapted to take account of other processes. In mountainous regions, for example, the balance between various forms of erosion and mass-wasting will need to be considered. Further, it can be based on the elaborated topology of both these main process groups. Consequently, geomorphological maps can provide information for defining erosional systems at a regional scale. The classification will also have to include erosion hazards from expanding urbanization and tourism, covering, for example, areas of ski-grounds, forest degradation and deforestation.

One of the major aims of elaborating erosional systems is to help identify options for erosion control. In doing this, basic questions could be asked concerning man's impact and the inevitability of having to accept some slope denudation. In this context Bork (1988) noted that periods of intensive soil erosion in the fourteenth and eighteenth centuries were the result of frequent, extreme rainfall events and not simply due to man's activity.

12.3.2. **Erosion control: options and problems**

The main problem of soil erosion in Europe is the progressive structural degradation of the soil. This is an increasing problem which leads to widespread surface sealing, crusting and soil loss. It is particularly serious in the loess belt, the Alpine forelands and in the Mediterranean. There are several causes, notably the reduced use of manure, reduced soil biological activity and the increased use of heavy machinery. The decline of mixed farming has led to a reduced availability of organic manure. Countering this trend by using polymers (known as soil conditioners) to improve soil structure is not yet a practical general solution.

There is evidence that slight modifications of the tillage system will not completely solve the soil erosion problem (for example, contour plowing, different cultivation procedures and other machinery). To improve soil structure and reduce erosion, it would seem that in many cases conservation tillage will be necessary. Under such tillage the topsoil is kept compact and the surface protected by a mulch. The most extreme application of this approach uses no-till (de Ploey, in press) which is exceedingly effective for cereals. Root crops, however, produce lower yields under this system. Sowing catch crops, which either overwinter or are frost-killed, followed by direct drilling of sugar beet, may offer a solution (Buchner, 1988).

A basic distinction should be made between the effect of such techniques on sediment production and on runoff. Although no-till may prevent soil loss, the compact soil can sometimes produce large amounts of runoff which may activate channel erosion at offsite locations. The erosion hazard may be shifted from the slopes to the valley bottoms. This is serious because of the large increase in the potential sediment transport capacity when high flow velocities and low sediment loads are combined. In planning soil conservation measures such effects should be considered.

Geomorphological studies have shown that slope instability has existed in Europe since the Holocene and that man's impact should not be overestimated. The exact extent of this impact has not been clarified yet. In some forested catchments mass-wasting and valley channel erosion have always occurred. Many of the badlands in the Mediterranean region are thousands of years old and were not produced by massive deforestation by man. Today the rates of erosion are less than visual evidence suggest (Yair et al., 1980; Wise, 1982).

A complicating factor when planning soil conservation methods is that for them to be effective they need to be socio-economically acceptable (Napier, 1988). Chisci (1986), for example, analysed different factors influencing soil loss. He concluded that for some catchments the only solution would be the conversion of the steeper arable land and vineyards to forests and/or pasture. At the same time it is realized that this would encounter major socio-economic and political objections. It is necessary, therefore, to find conservation options that require minimal interventions in terms of required changes in crop rotation and land use. On steep vineyards in the Beaujolais region of France tests with mulching and grass offer good perspectives for erosion control, and the same is true in Hungary (Gril and Canler, 1985; Goczan et al., 1987). The success of these and many other options will depend, in the final analysis, on their financial feasibility.

In controlling one form of erosion another could be encouraged. Erosion control to prevent wash on steep slopes could lead to more infiltration and to a greater risk of mass-wasting. Thus, it may be preferable to accept a certain amount of wash rather than to promote a more serious hazard. Alternatively, it could mean that mass movements in some areas are unavoidable and that protection requires a delicate strategy of passive defence as is the case for the migration of huge longitudinal dunes into a Saharan oasis.

Where the state of erosion is already extreme, as in the case for some degraded regions in the Mediterranean, there may be little soil still remaining above the lower slopes. Here controlled colluviation (de Ploey and Yair, 1985) could be a more achievable alternative goal for erosion control.

Finally, from a methodological point of view, there are good reasons for greatly increasing the number of experimental sites for measuring erosion in Europe. Slopes should, however, be studied in their total length rather than within plots. The complex hydrology of slopes requires that they be considered in their entirety. Such studies need to be conducted for a period of at least ten years because of the return period of the most important erosive meteorological events.

12.4. **Land Degradation: The World Problem**

The future perspectives for soil erosion control in Europe have been considered against the background of basic relationships between land use and erosion and in the light of current concepts of conservation.   One suggestion was that basic data could be provided by mapping and surveying. When the issue is considered on a global scale the major problem is to obtain information that can be correlated between different regions.

Particularly in developing countries, increasing pressure on the land, indiscriminant destruction of forests and the loss of productivity resulting from land degradation make the availability of information on erosion a major priority. There are strong arguments for:

i)   strengthening the awareness of decision- and policy-makers on the dangers resulting from inappropriate land and soil management;

ii)   improved mapping and monitoring of land resources;

iii)   development of an information system capable of providing the necessary information on land resources.

In response to this need a major project has been initiated to provide a global assessment of soil degradation (GLASOD Project UNEP, 1988; ISRIC, 1988), the major outputs of which will be a global assessment of the status of human-induced soil erosion, with a map of scale 1:10 million; and a similar assessment at a scale of 1:1 million, using a soil and terrain digital data base, for a pilot area in adjacent areas of Argentina, Brazil and Uruguay.   At both scales the objective is to delineate areas where the resistance of the terrain to degradation has been disturbed by man.

The forms of soil degradation that these surveys include are:

i)   erosion of soil material by wind and water, including offsite effects;

ii)   chemical (nutrient loss, pollution, acidification and salinization), physical (soil structure decline and

compaction) and biological soil degradation (microbiological imbalance). An indication is also given of
the kinds of human activities responsible for the
above forms of degradation (for example, deforestation, overexploitation of vegetation, overgrazing,
contamination by (bio)industrial wastes).

A major problem faced is that of standardization. Consequently,
guidelines have been prepared (ISRIC, 1988) and are being tested in the
first pilot area.   These are now being used by local scientists who are
collecting the information for the regional soil degradation status maps
which are to be used for the compilation of the global map.

The world soils and terrain digital database, which is to be used for
the pilot survey in Latin America, has a scale of 1:1 million; it is
compatible with other environmental databases. It can be easily updated
or purged of unwanted information and is accessible to a wide range of
users. The database uses ORACLES as a relational database management
system. In particular, it is transferable to developing countries who might
wish to extend it for more detailed development. In the pilot area in Latin
America, water erosion and soil compaction appear to be the most serious
problems. Algorithms will be developed for assessing these problems as
quantitatively as is possible.

The project is being extended to include other pilot areas in West
Africa (adjacent areas of Benin, Burkina Faso, Niger, Ghana, Nigeria and
Togo) and along the border of Canada and the USA. When the project
becomes operational, it will enable an objective global assessment of the
world soil degradation problem to be made.

12.5. **Soil Erosion and Climatic Change**

12.5.1. **Introduction**

The potential effect of climatic change on soil erosion needs to be
considered against the background of changing land use patterns in
Europe. This is because the hazard of soil erosion is closely associated
with land use. Changes in land use will in the future as in the past
produce changes in the areas at risk to erosion. Locating areas of future
potential risk is possible because the distribution of soils which are
inherently vulnerable to soil erosion is generally known. These are soils
with inherently poor physical properties. However, even soils with relatively
good physical properties can suffer from degradation and erosion under
inappropriate management.

The effect of climatic change on erosion will be superimposed upon
the effects produced by shifts in land use boundaries resulting from the
introduction of new cultivation techniques, economic changes and

technological innovations. Some potential shifts in land use boundaries are explored in Chapter 4. Climatic change will affect soil erosion both indirectly, for example, via land use changes and economic development, and directly through the effect it has on the various processes of erosion. Future land use changes and new technologies could lead to either positive or negative effects on erosion and these will be difficult to isolate from the effects of climatic change.

In the following Sections the potential effect of climatic change on erosion will be considered for locations in two regions where soils are particularly vulnerable: in The Netherlands and in Spain. In one region climatic change will probably have little impact; in the other region it could be very serious.

When the distribution of potentially erodible soils, land use and erosive rainfall are considered together, two regions emerge as being particularly sensitive to erosion. These are the loamy or silty soils of northern Europe developed in loess and soils low in organic matter and clay in Mediterranean regions. Some potential effects of climatic change will be considered for soils from these two regions. In both regions a very important factor is the potentially poor physical structure of the soil.

12.5.2. The loess belt of The Netherlands

12.5.2.1. Present erosion on loess

Soil erosion is being experienced increasingly in the loess region of South Limburg and in adjacent areas of the loess belt in Belgium and Germany. The increase in erosion and damage by flooding and sedimentation is associated with large increases in the area under maize, with an increase in the asphalted and built-up area and with the increased mechanization and scale enlargement of agriculture. The cultivation of maize began in the region in 1975 and this crop now covers 26 per cent of the cultivated area. Erosion is confined to loess soils under cultivation. Surveys have been made of the areas affected by erosion and of the financial damage incurred (Schouten, 1987). Measurements of erosion on experimental plots at Wijnensrade by Kwaad (1988) confirm that nearly all soil loss occurs during high intensity summer rainstorms between April and August. Rainfall occurring at other times of the year is effective in producing runoff but does not involve sediment transport. From the data available it can be inferred that to be able to predict the impact of climatic change on soil erosion, information will be required on the magnitude and frequency of rainfall events during the critical period between April and August.

The erodibility of the loess soils in South Limburg has been investigated under different types of vegetation cover by Eijsden and Imeson (1985), by Roelse and de Jager (1988) and by van Andel (1988). These studies showed fluctuations in erodibility during the growing season which can probably be explained by the cyclical effect of the soil micro-fauna during periods of rainfall following dry periods on the production and breakdown of soil stabilizing substances (Lynch, 1984). Soil microbiological effects mean, therefore, that the soil stability will be influenced by soil temperature and by fluctuations in soil moisture. Consequently, in addition to the erosive high magnitude rainfall events, the frequency of low magnitude rainfall events also needs to be considered in the doubling of atmospheric CO_2 scenario. Low magnitude rainfall events are also important because of the way in which loess soils respond to wetting.

When a dry Ap horizon (the plowed layer) of a tilled loess soil is suddenly moistened, it responds by slaking. Slaking is a process whereby under wetting a large soil ped or aggregate breaks down into small constituent microaggregates. This process is resisted by organic matter in the form of very fine roots, fungal hyphae and possibly polysaccharides. The rate of wetting and the moisture content of the soil influence the severity of slaking. Wetting and drying can lead to partial slaking or mellowing whereby cracks develop in aggregates which become progressively weakened (Utomo and Dexter, 1981). Consequently, the moisture regime of the soil is extremely important in determining soil erodibility. Laboratory experiments (de Ploey, 1989) have also shown that moist loess soils in Belgium resist rill forming more than dry soils. Similarly, it was found that whereas rills and pipes did not form in a slightly moist soil in which the structure of the Ap horizon is reconstituted in a laboratory experiment, they did form when the soil was dry. Whether or not rills develop during a rainfall event is very important because rills increase sediment transport by an order of magnitude and transport gravel sized fractions.

12.5.2.2. The impact of climatic change

The impact of climatic change is approached by considering the implications of predictions from an analysis by Bultot *et al.* (1988) of the impact of a doubling of atmospheric CO_2 on the water balance of three small drainage basins in Belgium, and then by considering the implications of small changes in climate on erosional processes studied in the field.

The changes in precipitation and temperature in Belgium indicated by Bultot *et al.* (1988) are shown in Table 12.1. For the critical period, precipitation is shown as increasing by about 10 mm in March and April but as decreasing by a few millimeters each month between May and August. The hypothetical temperature increases range from 3.4°C in March to 2.3°C in August and September. It is assumed that the frequency of rain days remains unchanged under the CO_2 doubling scenario. Of particular interest for soil erodibility is the implications of these changes for soil moisture levels. At Dyle in Belgium where the PE (Potential Evapotranspiration) is similar today to that at Maastricht Airport in Limburg, the hypothesized change in PE is greatest in June (7.4 mm) and

Table 12.1. A comparison of monthly regimes of air temperature (T, C), precipitation (P, mm) potential evapotranspiration (PE, mm) and soil water content in the aeration soils (SM, mm), under present conditions and for a doubling of atmospheric carbon dioxide, for the river Dyle drainage basin (645 km 2), Belgium.

Present conditions:	March	April	May	June	July	August	September
T	5.4	8.2	12.4	14.9	16.7	16.5	14
P	59	56	66	73	74	77	65
PE	32	60	89	92	89	74	57
SM	134	134	141	146	145	140	135

Climatic change due to CO_2 doubling (difference)

	March	April	May	June	July	August	September
T	+3.4	+3.1	+2.8	+2.7	+2.5	+2.3	+2.3
P	+10	+10	-1	-3	-2	-2	0
PE	+6	+6	+5.5	+7	+6	+5	+1.5
SM	-0.8	-0.6	-2.3	-6.5	-8.9	-9.6	-7.4

(Source: Bultot *et al.*, 1988.)

ranges between 5 and 6 mm during the other summer months. During the autumn the predicted PE values will be only 1 or 2 mm higher than today. Changes in the effective evapotranspiration are indicated as being about 5.5 mm in March but only 2.1 and 1.5 mm in July and August respectively. Simulated soil moisture contents in the aeration zone show a decrease of only 0.6 mm in April and 0.8 mm in March but of between 9 and 10 mm in July and August.

The predictions indicated by Bultot *et al.* (1988) suggest that the changes that might be expected in soil moisture are rather small. Between June and August when the decreases are at a maximum these are only about 5 per cent lower than those calculated for the present time. In fact, when monthly rainfall records from South Limburg are compared for wet or dry years, these show changes of an order of magnitude greater than those considered above. When evaluating the significance of the predicted changes for erosion, the assumption that the frequency of rain days will remain unchanged and the lack of information on rainfall intensity pose restrictions. It is the time between showers and the intensity of rainfall events that are of major interest and about which little can be inferred. The climatic changes will, however, mean that the sowing of row crops such as maize, potatoes and sugar beet can be advanced and that these will provide a protective cover several weeks earlier than is the case today.

This, together with the more active soil fauna, may more than compensate for the increase in precipitation. The date at which aggregate stability begins its seasonal increase might also be several weeks earlier. There is, however, no evidence to suggest any large change in the rate of erosion.

The implications of climatic changes in the order of magnitude of those suggested in the preceding paragraph can also be considered against the background of data obtained from field measurements. When rainfall is applied above a particular threshold intensity, it will begin to pond on the surface after a certain time, or after a certain amount of rainfall has been applied. The time to ponding (tp) is an important parameter because ponding represents one of the threshold conditions which have to be reached for overland flow and rainwash erosion to occur. The relationship between tp and rainfall intensity is known as the infiltration envelope. Such curves suggest for the loess soils tested that ponding points are likely to be exceeded regularly each year. This is more likely to be the case for soils under row crops. When loess soils are very dry sudden wetting can cause slaking; crust formation and ponding may be very rapid. On slightly moist to wet soils ponding points are reached sooner with increasing moisture content. Consequently, except on dry soils, the wetter the soil, the greater the chance of runoff.

From a large number of field measurements it can be deduced that the infiltration envelopes of many of the loess soils that suffer from poor structure are so unfavorable that runoff production is likely to remain high irrespective of the small changes in precipitation predicted under the CO_2 doubling scenario. The data also clearly indicate that changes in the type of cover or the adoption of different tillage methods are likely to have a larger effect on erosion than climatic change. In this context it should be stressed that climate usually influences soil erosion indirectly through vegetational effects.

12.5.2.3. Conclusion

Cultivated, loess soils in north-west Europe are highly vulnerable to erosion. There is no reason for expecting this situation to either improve or deteriorate on the basis of the data examined. Wherever loess soils occur in north-west Europe, in different climatic zones, they are usually reported as being sensitive to soil erosion. Soil erosion can, however, be minimised by adopting different tillage techniques (de Ploey, 1987), or by growing different crops. As soon as information is available on the magnitude and frequency of showers and on the length of dry periods under a CO_2 doubling scenario, it will be possible to predict the effect of climatic change on soil erosion. Should loess soils, or other soils with low clay content sensitive to erosion, be taken out of cultivation and put under grass or woodland in order to improve soil physical properties and to decrease erodibility? Today, new techniques are being developed to improve the physical properties of soils by means of synthetic polymers. These can prevent soils from forming crusts and can improve water retention so that erosion and runoff are greatly reduced. However, their application is still subject to economic and practical limitations.

12.5.3. **Mediterranean Spain**

The potential impact of climatic change on erosion in the Mediterranean region has recently been reviewed by Imeson *et al.* (1987) and by Imeson and Emmer (1988).

Although there appears to be little information about the changes in precipitation that are expected under the CO_2 doubling scenario, it is predicted by global climatic models that temperature will increase by about 3°C and potential evapotranspiration will increase by about 400 mm (Table 12.2). This is considerably greater than for the example from Belgium described above. If there is no increase in precipitation it could result in a potentially large increase in aridity, especially in the spring and autumn.

In a very general way, soil degradation and erosion can be hypothesised as being likely to increase because of climatic change irrespective of precipitation changes, because of the effects that the altered water balance will have on soil organic matter (amount, composition and form) and on water soluble salts. In addition, indirect impacts on erosion can be expected, for example, those resulting from more frequent forest fires. Seasonally high exchangeable sodium percentages (ESP) in combination with low electrolyte concentrations will lead to an extension of rill, gully and badland erosion. The resulting increase in dispersive soil conditions could lower infiltration rates by an order of magnitude (Agassi *et al.*, 1985). This will be most likely in regions slightly more humid than those subject to salinization today (Chapter 13). Higher rates of organic matter mineralization together with lower contents of organic matter in the soil could decrease the content of those soil organic matter compounds responsible for soil stability. This would most probably have a negative effect on soil erodibility.

The general effect of a lower organic matter content and an increase in dispersive soil conditions mentioned above should be seen as not impossible trends operating within complex ecosystems. Because of the specific soil conditions, soil structure is, in general, less resilient and more easily destabilized than in north-west Europe. Whilst the effects of the chemistry of the soil solution and destabilization of clay are rather general, the changes in the organic soil horizons that can be expected will be highly dependent on the prevailing ecological conditions (Sevink, 1988). Any negative effects will be greatest for sandy and silty soils. The highly calcareous soils found in Mediterranean regions are particularly vulnerable to lower organic matter levels (Imeson and Verstraten, 1985).

Table 12.2. Predicted changes in temperature (T, °C); precipitation (P, mm) and potential evapotranspiration (PE, mm) at Madrid for a GISS scenario for a doubling of atmospsheric carbon dioxide.

					Present conditions							
Mean Annual Average	D	J	F	M	A	M	J	J	A	S	O	N
T 14	6	5	7	10	13	16	21	22	23	20	14	9
P 444	48	39	34	43	48	47	27	11	15	32	53	47
PE 746	11	10	15	30	48	79	113	144	135	87	52	22

Climatic change due to CO_2 doubling (difference)

T +4	+4	+4	+4	+4	+4	+4	+4	+4	+4	+4	+4	+4
P +54	0	0	0	+6	+6	+6	0	0	0	+12	+12	+12
PE +312	+8	+8	+14	+32	+32	+32	+32	+32	+32	+32	+32	+26

The magnitude of the changes in erosion that might occur can possibly be deduced from the relationship between precipitation, temperature and erosion modelled by Kirkby and Neale (1986). The relationships they put forward are potentially important but field data to evaluate the predictions are scarce.

12.5.4. Specific field situations

Some of the implications of the general statements mentioned above can be considered for specific field conditions. This can be illustrated for two locations.

12.5.4.1. Highly calcareous soils in Alicante and Valencia

In south-east Spain extensive areas of highly calcareous soils occur on marls in a region receiving between 350 and 500 mm of rainfall per year. Where these soils are under cultivation, soil erodibility is very low (Imeson and Verstraten, 1985). About 60 per cent of the soil consists of silt sized calcium carbonate particles and, because clay contents are low, organic matter in the form of fungal hyphae, fine roots and polysaccharides are required for the formation of stable aggregates. Today, on tilled soils, organic matter contents are often so low that there is little possibility of further decrease should the climate change.

At the foot of slopes, material eroded in the past has accumulated as colluvium. Here, gentle slopes and stones reduce the erosion hazard and soils are slightly less erodible than on the marl slopes. At a few locations, water soluble salts have accumulated in the colluvial deposits and these are subject to gully erosion. An increase in the area of soils affected by water soluble salt concentrations could lead to extensions of the gullied areas. On the higher rocky slopes of the region former erosion has led to the removal of erodible material so that further erosion is unlikely. The main areas likely to be directly affected by climatic change are the sediments deposited during previous erosional phases at slope foot positions.

12.5.4.2. Soils on granodiorite in Gerona Province

Soil erosion in the Province of Gerona is locally severe. This erosion is common in the Selva region, in areas of granitic rock, on granodiorite and on the Pliocene-Pleistocene deposits derived from them. Soil erosion is especially prominent where soils have a texture-contrast or duplex profile. These soils suffer erosion where the Mediterranean forests have been replaced by Eucalyptus, when the soils are tilled on slopes and sometimes after forest fires.

Erosion on cultivated soils is a result of the poor structure of Ap horizons which develop in the albic horizons of the original soil, and because of the ease with which saturated overland flow develops on the impermeable B horizons in the winter even at low rainfall intensities. Erosion following fire is frequently the result of the construction of access roads or Eucalyptus terrace construction. Whilst low intensity rainfall events produce much erosion, so also do the infrequent high intensity summer thunderstorms. In the forest and burnt forest sites infiltration rates are very low due to water repellence.

Several types of data are being collected to examine the potential impact of climatic change on erosion. These include information on the relationship between weather and fire frequency, on the relationship between site aridity and the recovery of burnt forests, and on the infiltration characteristics of soils along climatic gradients. Several relationships between fire frequency and climate have been described by various workers in Gerona. It is known that fire frequency increases with drought and is influenced by humidity. It can also be seen that the regeneration of burnt forests is greatly influenced by the available soil moisture during periods of potential growth. On cultivated land, relationships between rainfall intensity and infiltration are being collected similarly to those described for the loess region. The data describing these relationships will be reported in full elsewhere.

12.5.5. Conclusion

It is concluded that the potential impact of climatic change due to a doubling of atmospheric CO_2 on soil erosion in the loess region of South Limburg is likely to be small. There are many strategies which could be developed to prevent erosion from increasing. In the Mediterranean the potential for an increase in land degradation exists. However, indirect effects which operate through vegetation or land use changes are likely to be greatest. The direct effects are limited by local on-site factors which will in the future, just as today, restrict erosion to specific areas.

REFERENCES

Agassi, M., Shainberg, I. and Morin, J., 1985, Infiltration and runoff in wheat fields in the semi-arid region of Israel. *Geoderma*, **36**, 263-276.

Ahnert, F., 1986, An approach to the identification of morphoclimates. In *International Geomorphology Part II* (ed. by V. Gardiner), 159-188. Wiley, Chichester.

Andel, C. van, 1988, Ontwikkeling aan het oppervlak, erosie processen in de tijd. Report, *Fysich Geografisch en Bodemkundig Laboratorium*, Universiteit van Amsterdam.

Balteanu, D., 1983, *Experimentul de Teren in Geomorfologie* (ed. by Ac. Rep. Soc. Romania), Bucuresti.

Bergsma, E., 1980, Provisional rain erosivity map of The Netherlands. In M. de Boodt and D. Gabriels (eds.), *Assessment of Erosion*. John Wiley, New York.

Bonnard, C., 1985, Détection et utilisation des terrains instables (Rapport École Polytch., Lausanne).

Bork, H.R., 1988, Bodenerosion und Umwelt, Landschaftsgenese und Landschaftökologie, TU Braunschweig, 13.

Buchner, W., 1988, Cultivation techniques for reducing soil erosion in the Rhineland. Agriculture, Erosion Assessment and Modelling. CEC Report EUR 10860, 283-298.

Bultot, F., Coppens, A., Dupriez, G.L., Gellens, D. and Meulenbergs, F., 1988, Repercussions of a CO_2 doubling on the water cycle and on the water balance - a case study for Belgium. *Journal of Hydrology*, **99**, 319-347.

Carrara, A., Sorriso-Valvo, M., Reali, C., 1982, Analysis of landslide form and incidence by statistical techniques, southern Italy. *Catena*, **9**, 35-62.

Chisci, G., 1986, Influence of change of land use and management on the acceleration of land degradation phenomena in Apennines hilly areas. *Soil Erosion in the European Community*, 3-16. Balkema, Rotterdam.

Eijsden, G.G. and Imeson, A.C., 1985, De relatie tussen erosie en enkele landbouwgewassen in het Ransdalerveld, Zuid Limburg. *Landschap*, **2**, 133-142.

Fournier, F., 1960, *Climat et érosion.* (Presses Univ. France, Paris.)

Gerard, C.J., 1987, Laboratory experiments on the effects of antecedent moisture and residue application on aggregation of different soils. *Soil and Tillage Research*, **9**, 21-32.

Goczan, L., Loczy, D., Szalai, L., 1987, Zalahalap. Hillslope experiments and geomorphological problems of big rivers. Geogr. Research Inst. *Hungarian Ac. Sc.*, 61-64.

Govers, G., 1986, Mechanismen van akkererosie op lemige bodems. Ph.D. thesis, Univ. Leuvan.

Govers, G., Everaert, W., Poesen, J., Rauws, G., de Ploey, J., 1987, Susceptibilité d'un sol limoneux à l'érosion par rigoles: essais dans le grand canal de Caen. *Bull. Centre Geom. CNRS de Caen*, **33**, 84-106.

Gril, J.N. and Canler, J.P., 1985, "Beaujolais": l'érosion attaque les vignobles. Cultivar, spécial Sols et Sous-Sols, **184**, 35-39.

Hansen, L., 1983, Soil erosion in Denmark. Report CEC Soil Erosion EUR 8427, 45-46.

Imeson, A.C., Dumont, H. and Sekliziotis, S., 1987, Impact analysis of climatic change in the Mediterranean region. Discussion paper, European Workshop on Interrelated Bioclimatic and Land Use Changes. Vol. F.

Imeson, A.C. and Emmer, I.M., 1988, Implications of climatic change for land degradation in the Mediterranean. *UNEP Mediterranean Action Plan Symposium.*

Imeson, A.C. and Verstraten, J.M., 1985, The erodibility of highly calcareous soil material from southern Spain. *Catena*, **12**, 291-306.

ISRIC, 1988, Guidelines for a General Assessment of the Status of *Human-induced Soil Degradation* (ed. by L.R. Oldeman). ISRIC/GLASOD Working Papers. International Soil Reference and Information Centre, Wageningen.

Kirkby, M.J. and Neale, R.H., 1986, A soil erosion model incorporating seasonal factors. In *International Geomorphology, Part II.* (ed. by V. Gardiner). Wiley, Chichester.

Kwaad, F.J.P.M., 1988, Part 2, South Limburg, *Excursion Guide of the 4th Benelux Colloquium on Geomorphological Processes.* Report, *Fysich Geografisch en Bodemkundig Laboratorium*, Universiteit van Amsterdam.

Langbein, W. and Schumm, S., 1958, Yield of sediment in relation to mean annual precipitation. *Trans. Am. Geophys. Un.*, **39**, 1976-1084.

Loczy, D. (ed.), 1987, Hillslope experiments and problems of big rivers. Excursion Guide. Publication 43 Geographical Research Institute, Hungarian Academy of Sciences.

Lynch, J.M., 1984, Interaction between biological processes, cultivation and soil structure. *Plant and Soil*, **76**, 307-318.

Morgan, R.P.C., 1985, Effect of corn and soybean canopy on soil detachment by rainfall. *Trans. Am. Soc. Agric. Engnrs.*, **28**, 1135-1140.

Morgan, R.P.C., Martin, L. and Noble, C.A., 1987, Soil erosion in the United Kingdom: a case study from mid-Bedfordshire. Silsoe College Occasional Paper No. 14. Silsoe College, Bedfordshire.

Napier, T.L., 1988, Socio-economic factors influencing adoption of soil erosion control practices in the United States. Agriculture, Erosion Assessment and Modelling. CEC Report EUR 10860, 299-328.

de Ploey, J., 1979, Landslides in the Serra do mar, Brazil. *Catena*, **6**, 111-121.

de Ploey, J., 1981, The ambivalent effects of some factors of erosion. *Mem. Inst. Geol. Louvain*, **31**, 171-181.

de Ploey, J. and Yair, A., 1985, Promoted erosion and controlled colluviaton: a proposal concerning land management and landscape evolution. *Catena*, **12**, 105-110.

de Ploey, J., 1989, Erosional systems and perspectives for erosion control in European loess areas. Soil Technology Series No. 1, 93-102.

Roelse, I. and de Jager, A., 1988, Twee technieken voor de bepaling van de aggregaat stabiliteit in loessleembodems. Report, *Fysich Geografisch en Bodemkundig Laboratorium*, Universiteit van Amsterdam.

Schouten, C.J., 1987, De omvang en gevolgen van erosie en wateroverlast in Zuid Limburg. Report, *Geografisch Instituut*, RU Utrecht.

Sevink, J., 1988, Soil organic profiles and their importance for hillslope runoff. In A.C. Imeson and M. Sala (eds.), *Geomorphology in Environments with Strong Seasonal Contrasts. Catena, Supplement* **12**.

Utomo, W.H. and Dexter, A.R., 1981, Tilth mellowing. *Journal of Soil Science*, **32**, 187-201.

Wise, S.M., Thornes, J.B. and Gilman, A., 1982, How old are the badlands? A case study from south-east Spain, Badland (eds. R. Bryan and A. Yair). Geo-Books, Norwich, 259-278.

Yair, A., Bryan, R.B., Lavee, H. and Adar, E., 1980, Runoff and erosion processes and rates in the Zin Valley badlands, Northern Negev, Israel. *Earth Surface Processes*, **5**, 205-225.

CHAPTER 13

SALINIZATION POTENTIAL OF EUROPEAN SOILS

I. Szabolcs

13.1. Introduction

The process of salt accumulation in soils, poses a very real threat to the land user. Salt accumulation is a global problem; in some countries the problem is localized whereas in others, 40-50 per cent of the total land surface is already salt-affected. Historically salinization has caused the decline of whole civilizations (Whyte, 1961; Evenari, Shanan and Tadmor, 1971) and has long been the subject of study in many countries (Kelly, 1951; UNESCO, 1961). Salinization suppresses agricultural productivity and may even make crop growth impossible (Arnon, 1972) due to the inhibitory effect of high concentrations of salts (Szabolcs, 1987). High salt concentrations in water courses and storage bodies limit their utility as irrigation and domestic water sources since beyond certain threshold values salt concentrations are toxic to humans, animals and plants (Kovda, 1947).

At present approximately 10 per cent of the World's land surface is salt-affected. This proportion is rapidly increasing, in part because of natural processes but mainly because of anthropogenic activity such as irrigation, deforestation, overgrazing and desertification (Szabolcs, 1979).

According to recent research the extent of potential salt-affected soils is at least two to three times the area of existing salt-affected soils. Intensive agriculture and increasing demand on water supplies, for agricultural and domestic uses, are accelerating the salinization process and will therefore exacerbate the problem (Dregne, 1976).

Where salinization has reached extreme proportions expensive mitigating actions can be taken in an attempt to reclaim soils. However, much less attention has been paid to the actual prediction and prevention of salinization (Alekseevsky, 1971).

13.2. Salinization of Groundwater and Soils

The accumulation of electrolytes in the upper horizons occurs as a result of various processes acting under a range of environmental conditions. Salts may accumulate as a result of:

F. M. Brouwer et al. (eds.), Land Use Changes in Europe, 293–315.

i)   weathering of rocks and transport of salts by
        wind and water;

ii)   circumstances hindering the removal of salts
        from a system, such as improper drainage.

Salt-affected soils with varying pedological, physical, chemical and biological properties are formed according to the particular environmental conditions of the site (Kovda, 1980). Table 13.1 presents a simple grouping system of the various types of salt-affected soils. A more detailed description can be found in the technical literature (Szabolcs, 1979). The term salt-affected soils encompasses a wide range of soils spanning various degrees of salt concentrations, pH and morphologies. Different types of salt-affected soils have different effects on land productivity.

Table 13.2 lists the main cations and anions that play a decisive role in the salinization and alkalization of the environment. Elements in ionic form behave according to the particular weathering process in operation. In arid and semi arid areas the weathering processes produce water soluble compounds which, due to the lack of precipitation, are not removed from their place of formation.

Table 13.3 represents a sequence of ion exchange during weathering and shows the relative mobility of constituent ions. The sequences with the largest energy coefficients have the lowest mobility. The energy coefficient is calculated on the basis of lattice energies of inorganic salts (Fersman, 1934). Energy coefficients are closely related to:

i)   the sequence of extraction of ions from minerals;

ii)   the rate of migration of ions; and

iii)   the ability of the ions to accumulate in
        sediments and soils.

The elements that play a dominant role in salinization and alkalization are found in phases I and II of the extraction sequence; these phases contain elements that are capable of intensive migration.

13.3. Major Aspects of Soil Salinity

The extent of salt-affected soils in Europe is in excess of 50 million hectares. Table 13.4 illustrates the regional occurrence of salt-affected soils and indicates the broad range of climatic and altitudinal characteristics within which salinization occurs. The three countries most adversely affected are Hungary, Spain and the USSR. These

Table 13.1. Grouping of salt-affected soils.

Electrolyte(s) causing salinity and/or alkalinity	Type of salt-affected soil	Environment	Main adverse effect on production	Method for reclamation
Sodium chloride and sulfate (in extreme cases also nitrate)	Saline soils	Arid Semi-arid	High osmotic pressure of soil solution (toxic effect)	Removal of excess salt (leaching)
Sodium ions capable of alkaline hydrolysis	Alkali soils	Semi-arid Semi-humid Humid	Alkali pH Effect on water physical soil properties	Lowering or neutralizing the high pH by chemical alterations
Magnesium ions	Magnesium soils	Semi-arid Semi-humid	Toxic effect High osmotic pressure	Chemical alterations Leaching
Calcium ions (Mainly $CaSO_4$)	Gypsiferous soils	Semi-arid Arid	Acid pH Toxic effect Induration	Alkaline alterations
Ferric and aluminium ions (mainly sulfates)	Acid sulfate soils	Sea shores and lagoons with heavy, sulfate containing sediments	Strongly acidic pH toxic effect	Liming

Table 13.2.   Chemical elements and compounds critical in soil salinization and alkalization.

Cations	Anions
Ca^{2+}	Cl^-
Mg^{2+}	SO_4^{2-}
Na^+	HCO_3^-
K^+	SiO_3^{2-}

Table 13.3.   Sequences of ion extraction during weathering.

Sequence of extraction	Ions	Energy coefficient
I	Cl^-	0.23
	Br^-	0.23
	NO_3^-	0.18
	SO_4^{2-}	0.66
	CO_3^{2-}	0.77
II	Na^+	0.45
	K^+	0.36
	Ca^{2+}	1.75
	Mg^{2+}	2.10
III	SiO_3^{2-}	2.75
IV	Fe^{3+}	5.15
	Al^{3+}	4.25

Table 13.4. Climatic conditions and altitude of regions with salt-affected soils in Europe. (Mean annual temperature in °C, mean annual precipitation in mm, altitude in m.)

Country, region	Temperature	Precipitation	Altitude
Austria			
Pulkantal (Illmitz)	9.2	566	187
Seewinkel (Apetlon)	9.8	623	140
Bulgaria			
Danube Valley	11.6	585	50-100
Maritza, Tundja,			
Strema Valley	12.3	500	100-200
Czechoslovakia			
South Moravia	9.0	500-550	200
Danubian Lowland	9.0-10.0	500-600	100-150
East Slovakian Low-			
land	9.0	600-650	100
France			
Atlantic sea coast	10.5-11.5	700-800	0-100
Mediterranean sea			
coast	14.0-15.0	550-650	0-100
Greece			
Ionian sea coast	17.5-18.5	650-750	0-100
Thessaloniki Plain	14.0-15.0	500-600	0-50
Hungary			
Hungarian Plain	10.0-10.5	524-585	80-120
Italy			
Northern Italy	13.7	744	0-100
Southern Italy +			
island region	17.6	478	0-100
Portugal			
Atlantic sea coast	14.0-18.0	660-1,400	0-100
Romania			
Black Sea shore -			
Danube Delta	11.0-11.3	359-439	0-50
NE Romanian Lower			
Danube Plain	9.6-11.1	400-515	0-100
Western Romanian			
Lower Danube Plain	10.6-11.5	480-570	100-200
Tisza Plain	10.7-10.8	558-620	80-100
Moldavian Table-			
Land	9.0-10.5	399-588	200

Table 13.4 (continued)

Spain			
Southern region	17.6-18.3	535-651	30-100
South-western region	17.0-17.6	295-419	0-60
Basin depression of the Ebro	14.6-15.1	324-378	118-380
Central plateaux	11.6-14.2	389-403	700
European part of the USSR			
Ukraine	7.0	500-550	100
Oka-Don Plain	7.0-8.0	500-550	100-200
Preazovian Plain	9.0-10.0	300-550	100
Prevolgian Plateau	6.5-7.5	500-600	200-300
Precaspian Lowland	6.0	150-250	0-100
Transcaucasian plains	14.0	200-250	100-200
Armenia	11.3-14.0	200-220	500-700
Transvolga region	4.0-5.0	300-400	100-200
Yugoslavia			
Vojvodina	10.0-10.5	550-600	100
Macedonia	11.2	652	100-200
Adriatic sea coast	14.5-15.5	500-900	0-50

countries alone account for more than 75 per cent of the salt-affected soils in Europe. The most extensive areas of salinized soils are found in the semi-arid steppe and forest-steppe regions of the USSR, on the lowlands of the river Danube in Czechoslovakia, Romania, Hungary and Yugoslavia and in parts of Spain. The present extent of salt-affected soils is summarized in Table 13.5. Comparing Tables 13.4 and 13.5 shows that salt-affected soils develop under both semi-arid and semi-humid climatic conditions and at altitudes from less than 100 m to over 500 m above sea level.

The potential for salt accumulation in European soils has not, as yet, been extensively investigated. Only a few studies have focused on assessing possible future changes in the areal extent of salt-affected soils (Szabolcs, 1974). Table 13.6 presents an assessment of the present and predicted distribution of such soils in selected countries. The assessment is based on local investigations of physico-geographical factors taking into account a projected increase in irrigation practices.

Irrigation is the main anthropogenic factor influencing salt accumulation. However, other factors which have direct or indirect effects must not be overlooked.

To accurately characterize the likelihood of future salt accumulation in soils due to changes in environmental and socio-economic conditions it is necessary: i) to gain a detailed knowledge of

Table 13.5. The major extent of salt-affected soils in Europe (area in 1,000 ha).

Country	Saline soil	(Sodic) Alkali soil without structural B horizon	(Sodic) Alkali soil with structural B horizon	Total
Austria	0.5	-	-	0.5
Bulgaria	5.0	-	20.0	25.0
Czechoslovakia	6.2	7.5	7.0	20.7
France	175.0	-	75.0	250.0
Greece*	-	-	-	3.5
Hungary	1.6	58.6	325.9	386.1
Italy	50.0	-	-	50.0
Portugal*	-	-	-	25.0
Romania	40.0	100.0	110.0	250.0
Spain*	-	-	-	840.0
USSR	7,546.0	1,616.0	20,382.0	29,544.0
Yugoslavia	20.0	50.0	185.0	255.0

*For Greece, Portugal and Spain only information was available about the total coverage of salt-affected soils.

Table 13.6.   Coverage of present existence and the projected potential
increase of salt-affected soils (as a result of a doubling of irrigation in
agriculture) in a few European countries (area in 1,000 ha).

Country	Salt-affected soils	
	Present	Potential
Austria	0.5	2.5
Czechoslovakia	25.0	80.0
Hungary	740.0	885.0
Italy	50.0	400.0
USSR	28,000.0	18,000.0

the present extent of salt-affected soils in Europe; ii) to examine the rate
of change of the processes involved; and iii) to explore the available
options, in the short and long term, for limiting or controlling the
processes. Figure 13.1 illustrates the present extent of salt-affected soils
in Europe. The future salinization potential will be discussed in the
following Section.

13.4. Future Potential for Salt-affected Soils in Europe

Three scenarios were selected to represent the most important processes
involved in salinization occurring in different parts of the continent:

Scenario 1 -   a potential increase in the extent and degree of salt-
affected soils in relation to an increase in irrigation
practices;

Scenario 2 -   a potential increase in the extent and degree of salt-
affected soils in relation to a projected change in
climate;

Scenario 3 -   a potential increase in the extent and degree of salt-
affected soils in relation to a projected rise in sea
level.

Different regions have been selected for each scenario. Figure 13.2
shows the representative areas. For Scenario 1 regions experiencing
continental and semi-arid climates were chosen. Here, the main
components affecting secondary salinization are in operation and
irrigation practices are expanding.   The Mediterranean region was
selected for Scenario 2 where climatic changes such as increasing

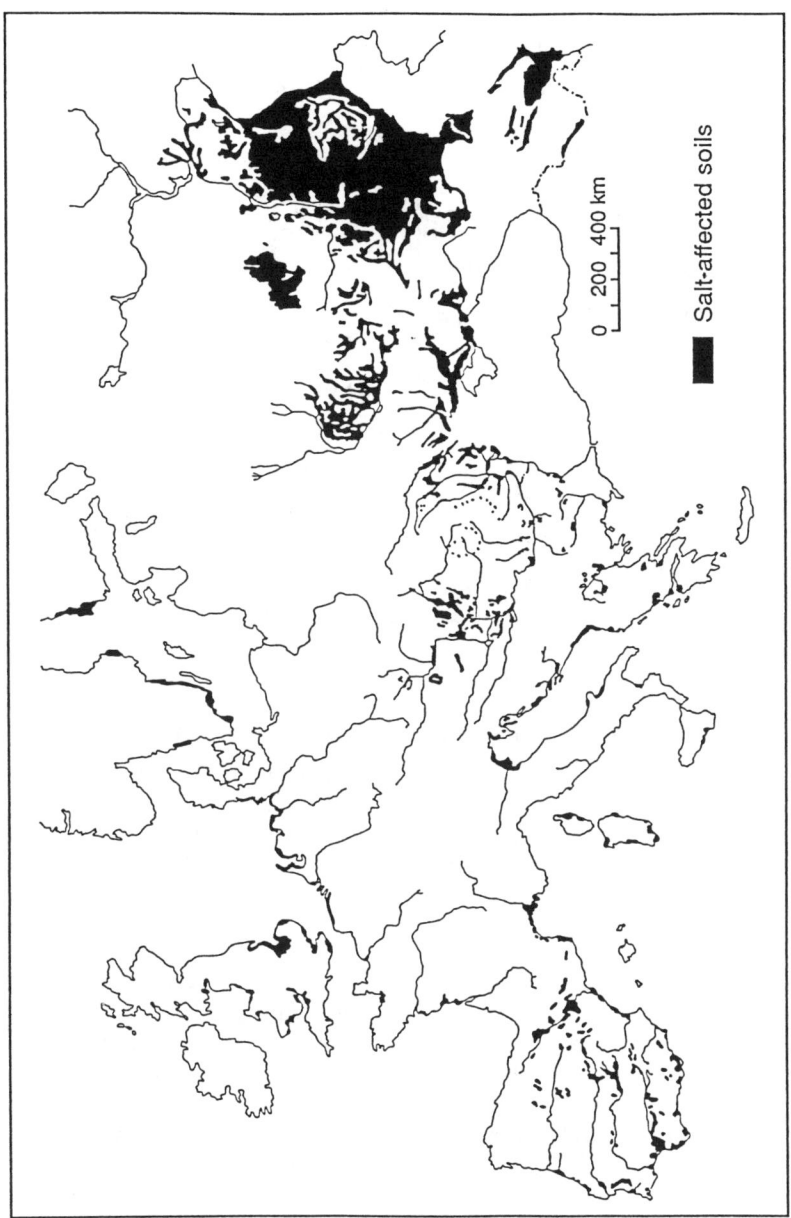

Figure 13.1. Present occurrence of salt-affected soils in Europe.

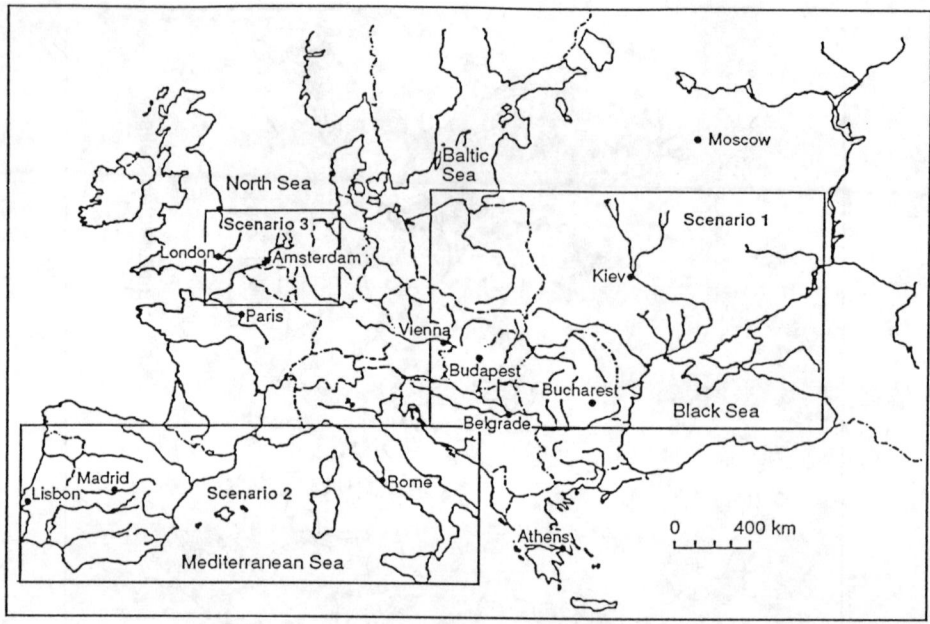

Figure 13.2.   Areas included in Scenarios 1, 2 and 3.

temperatures, due to changes in the atmospheric CO_2 balance, are likely
to be experienced.   *Scenario 3* focuses on the coastal regions of north-
west Europe where a projected sea level rise is likely to cause a major
increase in the area of salt-affected soils resulting from salt intrusion in
groundwater and surface deposition.

13.4.1. Scenario 1

The area selected for *Scenario 1* has a long history of irrigation due to
suitable economic conditions and the prevailing climatic conditions. The
area encompasses more than 50 per cent of the total irrigated land in
Europe and the distribution of the existing salt-affected soils correlates
highly with this agricultural practice.

Extension of irrigation practices is envisaged for most countries in
Scenario 1. This is particularly likely in those countries which already
experience low annual rainfall such as the USSR, Bulgaria and
Romania. Countries with higher annual rainfall are obviously less
dependent on irrigation, for example, Austria and Poland. The expected
increase in the area of irrigated land will, therefore, vary from country
to country. However, it is unlikely that the proportion of irrigated land
will more than double over the next 50-70 years. One of the major
factors limiting the extension of irrigation is the shortage of good quality
irrigation water.

Irrigation can lead to the development of salt-affected soils via two main processes:

i)   salt accumulation in the soil profile due to a high salt concentration in the irrigation source;

ii)   salt accumulation in the soil profile caused by a high salt content in rising groundwater levels.

The countries concerned have regulations governing the required quality of irrigation water. Consequently only a small proportion of the existing salt-affected soils have developed as a result of poor quality irrigation water. Soils are mostly threatened by rising groundwater tables on irrigated land. Taking into consideration climatic, environmental and agricultural conditions, a map demonstrating the salinization hazard has been produced for *Scenario 1* (Figure 13.3).

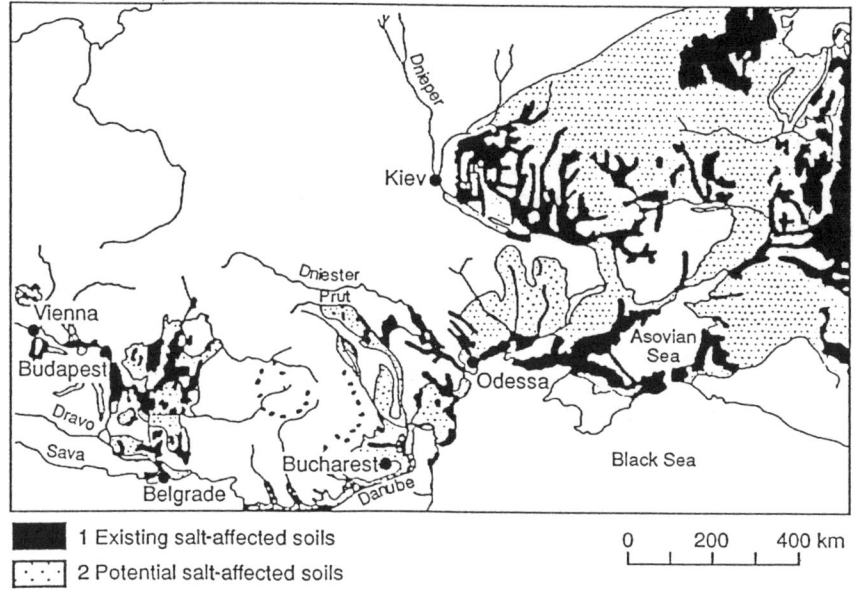

Figure 13.3.   Present coverage of salt-affected soils and the potential increase as a result of an increase in irrigated land.

The following assumptions were made:

-       the percentage of irrigated land will increase over the
        next 50-70 years (maximum increases will be double the
        present proportion of irrigated land);

-       there will be no substantial changes in climate and no
        natural disasters will occur;

-       the existing standards for irrigation water quality will be up
        held;

-       the overall hydrological system will be unchanged.

Scenario 1 covers approximately 20 per cent of the total land
surface of Europe.   From Figure 13.3 it is evident that the area
threatened by salinization over the next 50-70 years greatly exceeds the
area of existing salt-affected soils. Table 13.7 presents the areal extent
of existing and potential salt-affected soils for *Scenario 1*.

Table 13.7.   Areal extent of salt-affected soils in *Scenario 1*.

Total land area Scenario 1	= 2,080,000 km^2
Area of existing salt-affected soils	= 218,000 km^2 10% total area
Area of potential salt-affected soils	= 418,000 km^2 20% total area

The map, Figure 13.3, shows that soils in the vicinity of existing salt-
affected soils are most susceptible to salinization. The presently patchy
distribution of salt-affected soils is likely to become more uniform over
the next 50-70 years. The river basins and corridors of the Danube,
Prut, Dneiper, Don and Volga rivers are especially endangered.
Salinization will diminish the ecological value of an area, lower the
biological potential of soils and may render land entirely unproductive.
There will be associated secondary impacts on water supplies; both
drinking water and irrigation sources are likely to be severely affected.
Figure 13.4 illustrates, in detail, the present and future situation for
Hungary as predicted by Szabolcs (1987).

13.4.2. Scenario 2

The existing salt-affected soils are mainly concentrated in the Iberian
peninsula and are found to a lesser degree in southern France, Italy,
Sicily, on Sardinia and on Corsica as well as on the Dalmation coast of
the Balkan peninsula. The total land area included within *Scenario 2* is
approximately 885,000 km^2.
        It was assumed for the purpose of this study that as a

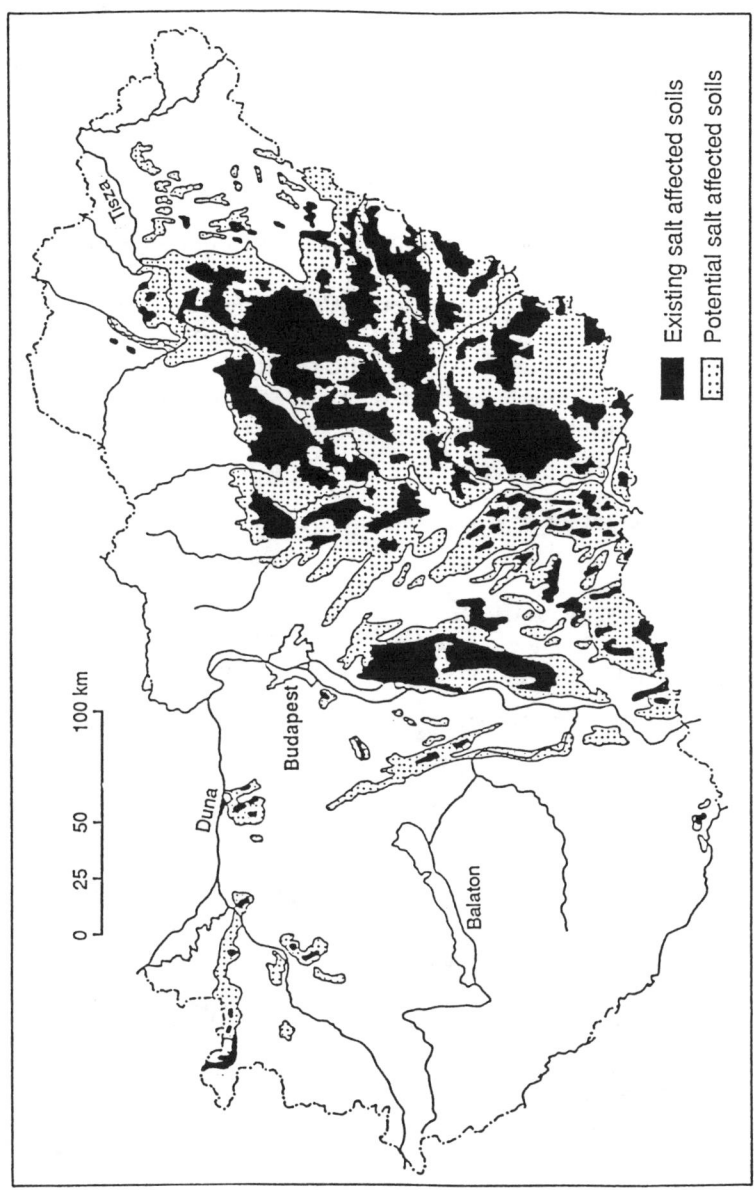

Figure 13.4. Present existence and future potential for salt-affected soils in Hungary.

consequence of CO_2 accumulation the average annual temperature of the area would increase by approximately 1°C over the next 50-70 years. Concomitantly the aridity index would increase, prompting progressive salinization of marginal areas. The following assumptions were made:

- the areal extent of irrigation will not change substantially;
- no natural disasters or tectonic changes will occur;
- the hydrology of the area will not change substantially.

Figure 13.5 shows the present extent of salt-affected soils and the future potential in relation to the projected climatic change. The proportion of salt-affected soils in this area is significantly less than in *Scenario 1*. However, similar trends are apparent in that the area threatened by potential salinization is substantially greater than that area already affected (Table 13.8).

Table 13.8. Extent of salt-affected soils in Scenario 2.

Total land area in Scenario 2	= 885,000 km²
Area of existing salt-affected soils	= 56,000 km² 6% total area
Area of potential salt-affected soils	= 122,000 km² 14% total area

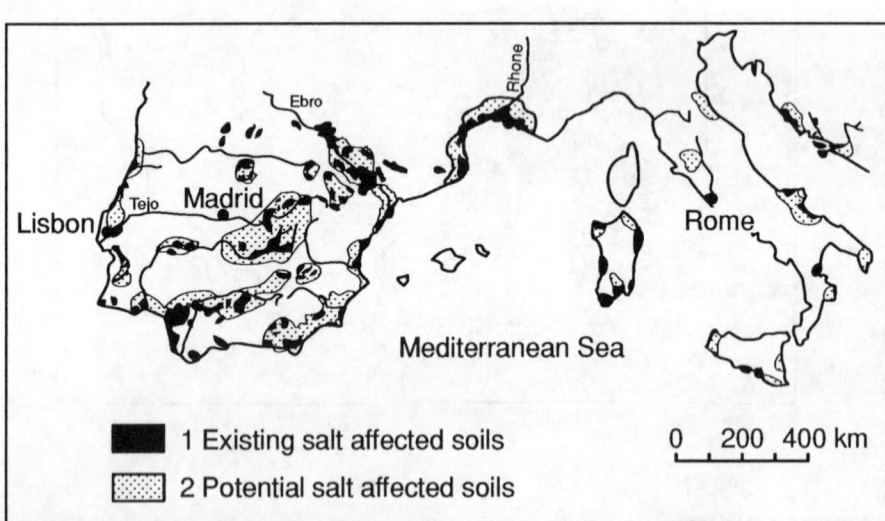

Figure 13.5. The existing and potential distribution of salt-affected soils resulting from climate change.

Aridity depends on effective precipitation, that is precipitation minus evaporation. The ratio of rainfall/temperature (r/t) has been employed as an index of precipitation effectiveness; the ratio r/t was proposed by Lang in 1915; where r = mean annual rainfall in mm and t = mean annual temperature in °C, such that r/t < 40 is considered arid and r/t > 160 is considered humid (Barry and Chorley, 1982). The estimates presented in Table 13.9 for selected locations in Scenario 2 clearly show that the r/t index is expected to decrease by about 10 per cent due to a 1°C increase in mean annual temperature. The dry areas of the Iberian peninsula such as Castile, the Ebro valley and south-west France are particularly exposed to the threat of potential salinization due to their marginal status with respect to likely increases in aridity.

Table 13.9. Aridity factor (r/t), relating mean annual precipitation to mean annual temperature for Scenario 2 under present conditions, and a scenario for climate change with an increase in mean annual temperature of 1°C.

Location	Present	Future
Central part of the Iberian Peninsula (Madrid)	31.27	29.18
Southern part of the Iberian Peninsula (Iaen)	35.22	33.27
Southern France (Nimes)	50.76	47.46
Corsica (Ajaccio)	46.10	43.14
Sardinia (Sassari)	37.27	35.17
Italy (Terni)	61.27	57.42
Southern Italy (Catanzaro)	60.22	56.63
Yugoslavia (Dubrovnik)	79.00	74.38

Potential salinization threatens fertile agricultural land and must be anticipated so that necessary measures to prevent and mitigate this environmental hazard can be developed. Potential salinization threatens river valleys and estuaries such as Ebro, Neretva, Rhone, Guadalquivir and Tajo, as well as plains and plateaux such as Castile, Aragonea in Spain and Umbria in Italy.

13.4.3. Scenario 3

The area chosen for *Scenario 3* includes part of north-west Europe (Figure 13.2). The area was selected on the basis of i) the major impact

on soil quality of the area that would occur as a result of a projected
rise in sea level, and ii) the information available concerning the present
extent of salt-affected soils.

The total land surface covered by Scenario 3 is 225,000 km^2
including south-east England, western region of The Netherlands,
Belgium and north-east France. The Scenario mainly encompasses
coastal soils high in sodium chloride. Figure 13.6 shows the existing
and likely future salt-affected areas. 5 per cent of the total land surface
(19,500 km^2) is presently salt-affected whereas the area threatened by
potential salinization comprises 11 per cent (25,000 km^2) of the total
land surface.

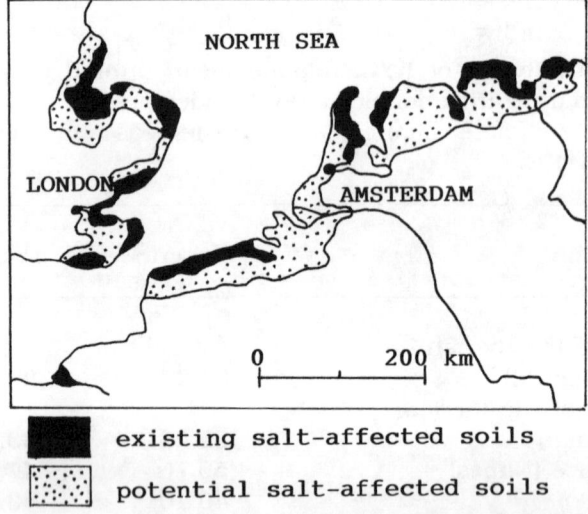

existing salt-affected soils

potential salt-affected soils

Figure 13.6. Salt-affected soils and future potential for salinity as a
result of sea level rise.

Scenario 3 is based on a rise in sea level of 1 cm yr^{-1} and the
consequent intrusion of salt. If the sea level rise exceeds the projected
1 cm yr^{-1} then associated expansion of salt-affected soils will not show
a linear relationship but probably an exponential one. The assessment
for likely impacts of Scenario 3 is based on several assumptions:

- irrigation practices will not change;

- no natural disasters or tectonic change will
  occur;

- the hydrology of the area will remain
  unchanged.

To summarize, the three scenarios elucidate the main processes leading to secondary salinization in Europe. However the study is by no means exhaustive; it does not investigate other factors such as changes in cropping patterns, intensive use of chemicals or other such changes in farm management. Neither does the study consider a combination of events; it concentrates on steady state phenomena.

To predict more accurately the likely adverse salinization processes, appropriate comprehensive survey methods should be developed.

13.5. **Prevention and Monitoring of Secondary Salinization**

Reclaiming salt-affected soils involves difficult and expensive processes. It is therefore highly desirable to explore methods of preventing the salinization processes. Some prevention options are discussed in this Section. In the past research has concentrated on salinization as a result of irrigation; very little work has been done regarding the control and prevention of salinization resulting from climatic or sea level changes.

It is important to recognise the interdependency of the apparently individual factors contributing to salinization problems. For example, a decrease in precipitation will probably necessitate an increase in the use of irrigation. The salinization problem which may have been initiated by climatic change can then be exacerbated by irrigation. A full assessment of the complex interlinkages between contributory factors is beyond the scope of this Chapter.

The primary objective of irrigation is to remove water stress from crops and hence improve productivity. The threat of irrigation-associated salt accumulation is one of the major obstacles limiting the development of irrigation systems. It is necessary to intensify theoretical and technical research into secondary salinization before the use of irrigation is expanded further. In the past, the importance of geochemical and hydrogeochemical processes has been underestimated. There has been an abundance of studies on irrigation water quality whereas the role and importance of salt content in groundwater and its impact on soils has been ignored.

13.5.1. **Preliminary Study**

A comprehensive study should be undertaken before planning new irrigation systems or expanding existing ones. Such a study should include a detailed investigation of both the surface and subsurface hydrology of the area (Darab and Ferenz, 1969).

The preliminary study should determine first, the critical depth of the groundwater table. The critical depth is that below which, owing to natural or irrigated conditions, leaching prevails while above this level salt accumulation occurs in the soil profile. Thus the salt regime of a given location is in equilibrium at the critical depth. Based on the determination of the critical depth it is possible to elaborate a specific "turning-point-scenario" enabling the monitor or surveyor to trace and record salt accumulation in the profile from groundwater sources.

To establish the salt balance of a given location it is necessary to measure the total amount of soluble salt both at the beginning and at the end of an observation period so that the change in salt concentration can be found. The salt balance may be in equilibrium and therefore stable; or the rate of incoming salts may exceed the leaching rate so that salts accumulate; or the leaching rate may exceed the rate of input so that the salt content of the soil is depleted. The salt balance of soils depends on the combined effect of numerous factors: the depth and chemical composition of groundwater, the type of irrigation method employed and the influence of relief are particularly relevant components.

13.5.2. Calculation of the salt balance

The salt balance can be determined using the following equation:

$$b = a + d + \frac{cv}{Mt_{fs}} \times 10^{-5}$$

b = soluble salt concentration of the soil at the end of the observation period (mg/100g soil)

a = soluble salt concentration of the soil at the beginning of the observation period (mg/100g soil)

c = salt concentration in irrigation water source (mg/100g soil)

v = amount of irrigation water applied during the observation period (m^3/ha)

M = thickness of soil layer for which the salt balance is being measured (m)

t_{fs} = bulk density of the soil

d = salt regime coefficient - changes in the salt content of the soils during the observation period are expressed in the salt regime coefficient.

13.5.3. Determining the salinity hazard

To monitor the possibility of salinization or alkalization in irrigated areas the following factors should be taken into account:

Climatic factors: temperature; rainfall; humidity; vapour pressure; evapotranspiration; and their temporal and spatial variation.

Hydrogeological factors: drainage patterns; fluctuations of the water table; direction and rate of horizontal groundwater flow; salt content and general chemical composition of groundwater and parent material.

Edaphic factors: soil texture/structure; saturated and unsaturated hydraulic conductivity; soluble salt content and salt profile; cation exchange capacity; pH.

Agrotechnical factors: land use patterns; crop type; cultivation methods; machinery.

Irrigation practice: application method/rate; irrigation interval or frequency; water quality; drainage conditions.

The above-mentioned factors determine the objectives and methods necessary for the preliminary survey to define the present and future extent of salinization and alkalization. Elucidation of soil and water properties to determine the salt balance conditions facilitates decision making regarding recommendations for optimal irrigation methods or strategies which will most effectively avoid secondary salinization of the environment.

Table 13.10 presents related factors to be considered in the preliminary survey and in the post implementation monitoring phase.

Failure to thoroughly investigate the potential hazard of salinization and alkalization can have drastic consequences. For example, the Hungarian experience dramatically illustrates the implications. After World War II irrigation became an important agricultural practice on large areas of the Transztisza (eastern) part of the Hungarian plain. The Tisza-1 (Tiszalok) Irrigation Project in 1953 was one of many schemes put into operation during this time of expansion. In the 1950s the harmful processes that could occur in association with irrigation were known. However, such processes were largely ignored in the technical planning, territorial installation and

Table 13.10. Recommendations to control salinity and alkalinity in irrigated areas (with a preliminary survey on landscapes and irrigation practice before cnstruction of an irrigation system, and a monitoring phase during irrigation).

Preliminary survey

Environment	Planned irrigation
climate	water quality and quantity available for irrigation
hydrology	groundwater depth and quality
hydrogeology	technology of irrigation
geomorphology	cropping pattern tolerance

Monitoring

salinity and alkalinity of soils and the level of groundwater

chemical composition of groundwater

chemical composition of irrigation water

physical soil properties

toxic elements in soil and water

---

operation of irrigation projects. Consequently 120,000 ha of land were affected by undesirable soil processes such as peat formation, salinization and alkalization.

Such disastrous effects have prompted the evolution of maps illustrating the findings of preliminary surveys of geological, hydrological, climatic and agrotechnical characteristics. Figure 13.7 is a map indicating the possibilities for low risk irrigation implementation or extension in Hungary. The map can be used as a useful tool for decision making in irrigation and agriculture. Thus a proper projection of potential salt-affected soils should be based on reliable evaluation of the natural and man-made factors involved.

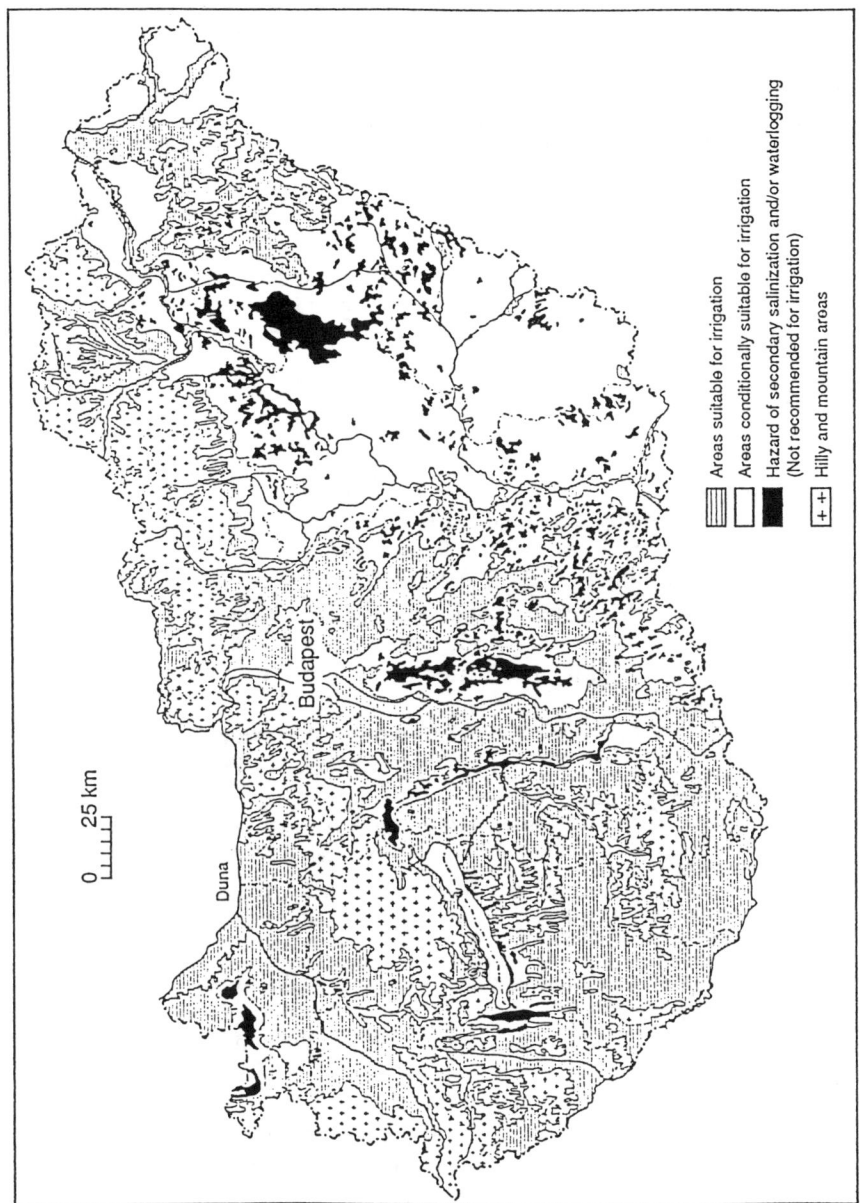

Figure 13.7.   Possibilities of irrigation in Hungary in respect to prevention of salinization.

Having determined "low risk" or "safe" regions for irrigation implementation a carefully organized and structured monitoring system should be devised. The monitoring system should be conducted during the operation of irrigation schemes so that any changes are recorded and preventive actions can be taken in good time. The particular methods and schedule will obviously depend on local conditions.

Visually illustrating the results of preliminary and "operation phase" surveys provides a comprehensible display of soil and water conditions and gives guidelines for the proper management of land. Such approaches can and should be employed by decision makers, farmers, irrigation engineers and economists in order to stop or mitigate secondary salinization in the praxis of irrigated agriculture.

REFERENCES

Alekseevsky, E.E., 1971, *Irrigation and Drainage of the World.* Isdatelstvo Kolos, Moscow. (In Russian.)

Arnon, I., 1972, *Crop Production in Dry Regions Volume 1: Background and Principles.* Leonard Hill, London.

Barry, R.G. and Chorley, R.J., 1982, *Atmosphere, Weather and Climate* 4th Edition. Methuen, London and New York.

Darab, K. and Ferencz, K, 1969, Soil mapping and control of irrigated areas. National Institute for Agricultural Quality Testing, Budapest.

Dregne, H.E., 1976, *Soils of Arid Regions.* Eslevier, Amsterdam.

Evenari, M., Shanan, L. and Tadmor, N., 1971, *The Negev. The Challenge of a Desert.* OUP, London.

Fersman, A.E., 1934, *Geochemistry.* Leningrad. (In Russian.)

Kelly, W.P., 1951, *Alkali Soils.* Reinhold, New York.

Kovda, V.A., 1947, Origin and regime of salt-affected soils. Volumes I and II. Izadatelstvo Akademii Nauk, SSSR. Moscow. (In Russian.)

Kovda, V.A., 1980, Problems of combating salinization of irrigated soils. UNEP, Nairobi.

Polynov, B.B., 1956, Selected papers. Izdatelstvo Akademii Nauk, SSSR. Moscow. (In Russian.)

Szabolcs, I., 1947, *Salt-Affected Soils in Europe.* Martinus Nijhoff, The Hague, The Netherlands, and Research Institute for Soil Science and Agricultural Chemistry of the Hungarian Academy of Sciences, Budapest.

Szabolcs, I., 1979, *Review on Research of Salt-affected Soils.* UNESCO, Paris.

Szabolcs, I., 1987, The global problems of salt-affected soils. *Acta Agronomica Hungarica,* **36**(1-2), 159-172.

Szabolcs, I. and Darab, K., 1982, Irrigation water quality and problems of soil salinity. *Acta Agronomica Hungarica,* **31**(1-2), 173-179.

Szabolcs, I., Darab, K. and Várallyay, G., 1969a, The Tisza irrigation systems and the fertility of the soils in the Hungarian Lowland. II. The "critical depth" of the water table in the area belonging to the irrigation system of Kisköre. *Agrokémia és Talajtan*, **18**(3-4), 211-218.

Szabolcs, I., Darab, K. and Várallyay, G., 1969b, The Tisza irrigation systems and the fertility of the soils in the Hungarian Lowland. III. Methods of the preparation of 1:25,000 scale maps indicating the possibilities and the conditions of irrigation. *Agrokémia és Talajtan*, **18** (3-4), 221-231.

UNESCO, 1961, *Salinity Problems in the Arid Zones*. United Nations Educational, Scientific and Cultural Organisation, Paris.

Whyte, R.O., 1961, Evolution of land use in south-western Asia. In *A History of Land Use in Arid Regions*. UNESCO, Paris.

Chapter 14

CHANGES IN RATES OF WEATHERING AND EROSION INDUCED BY ACID EMISSIONS AND AGRICULTURE IN CENTRAL EUROPE

T. Paces

14.1. Introduction

Integrated effects of human activities have changed chemical and mechanical weathering and erosion rates in Europe. Environmental acidification by SO_2 and NO_x from sulfur dioxide (SO_2) and nitrogen oxides (NO_x) deposition and modern agricultural practices accelerate these rates, deplete soils of exchangeable calcium (Ca^{2+}) and magnesium (Mg^{2+}), and enrich them with aluminium (Al^{3+}). Such changes are associated with dieback of forests in Europe (Ulrich, 1981; Paces, 1985, 1986).

The mass balance of eleven chemical elements, based on studies in small catchment areas in Czechoslovakia since 1975, indicates that present agricultural practices in the region increase the rate of weathering by a factor of about 3 and the industrial acidification by a factor of about 4. Acidification is harmful because it depletes soils of Ca^{2+} and Mg^{2+} and mobilizes Al^{3+}. Data on the present input of SO_2 and NO_x indicate that harmful acidification of soils will continue. Reversibility or irreversibility of environmental acidification of soils is a crucial point in planning future land use patterns. This Chapter shows how land use patterns may lead to a depletion (in quantity and quality terms) of forest as well as agricultural soils.

This Chapter presents data on the rates of weathering and erosion in four representative catchments in Czechoslovakia (Figure 14.1). Methods concerning the measurements and the theoretical models of weathering and erosion which enable the calculations of the rates are presented in previous publications (Paces, 1985, 1986).

14.2. Results and Discussion

The influence of different types of land use on weathering of bedrock and chemical and mechanical erosion has been studied in small hydrological basins (catchments) in Czechoslovakia since 1975. Major characteristics of the catchments are summarized in Table 14.1. The calculated rates

317

F. M. Brouwer et al. (eds.), Land Use Changes in Europe, 317–323.
© 1991 Kluwer Academic Publishers. Printed in the Netherlands.

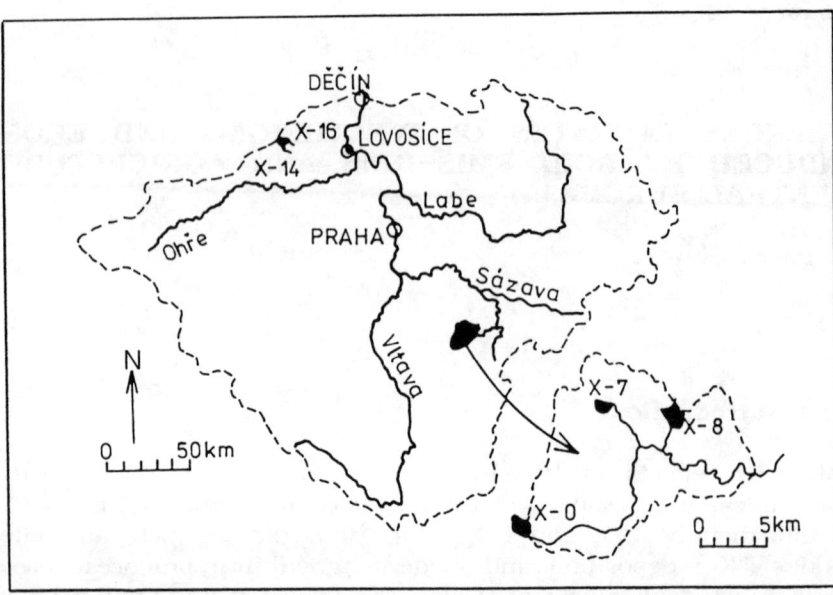

Figure 14.1. Location of representative basins in the watershed of the upper Elbe River, Czechoslovakia.

of chemical and mechanical erosion are presented in Table 14.2. The environmentally significant properties of soils are given in Table 14.3. The location of the representative catchments are shown in Figure 14.1.

Weathering rates in forests, before man appeared on the continent of Europe, have been estimated to correspond to the consumption of hydrogen ions. A mean rate of 0.6 kmol ha^{-1} yr^{-1} has been derived (Mazzarino et al., 1983). During the Holocene period this rate increased to 1.9 kmol ha^{-1} yr^{-1}. Results from the Bohemian-Moravian Highland give a rate of 1.1 and 1.5 in forests, depending on slope (3.8 per cent and 13.3 per cent respectively), 4.1 in fertilized fields and 6.1 (all values in kmol ha^{-1} yr^{-1}) in forests of the Ertzgebirge where spruce has disappeared due to acid emissions. These results suggest that the present agricultural practices in Czechoslovakia increase the rate of weathering by a factor of about 3, and industrial acidification due to emissions of SO_2 and to a lesser extent NO_x increases the rate by a factor of about 4.

The weathering process can be subdivided into three components: mechanical weathering, primary chemical weathering (mainly the dissolution of rock-forming minerals and precipitation of secondary minerals such as clay minerals) and secondary chemical weathering (mainly desorption of exchangeable cations from soils). The primary weathering is a useful process; it consumes acidity and produces soils. However, the secondary chemical weathering is a harmful process since

Table 14.1. Major characteristics of small catchments under different influence of man.

Catchment	X-0	X-8	X-7	X-14
Location (Czechoslovakia)	Bohemian-Moravian Highland			Ertzgebirge Mountains
Type of countryside	Agricultural			Industrial
Type of catchment	Forest	Forest	Fields	Forest
Slope (%)	3.8	13.3	4.9	18.0
Bedrock	Biotite gneiss and quartzites			Biotite-muscovite
Soil	Brown earth, Acid brown earth, Brown earth gleyic			
Air:				
SO_2 (microgram per m^3)	8.3	8.3	8.3	112
NO_x (microgram per m^3)	11.2	11.2	11.2	33
Area of catchment (km^2)	0.98	1.68	0.59	2.7
Forested area (%)	100	100	1.3	30
Arable land (%)	0	0	98.7	0
Vegetation after decline of Spruce (*Picea abies*)	0	0	0	70
Annual precipitation (mm)	781	685	736	812
Period of monitoring	1975-88	1945-88	1975-88	1978-83

Table 14.2.   Rates of weathering and erosion (in kg ha^{-1} yr^{-1} if not given otherwise.

Catchment	X-0	X-8	X-7	X-14
Mechanical and chemical weathering of gneiss	420	732	1,700	1,510
Mechanical erosion of weathered layer (regolith)	374	732	1,991	1,402
Chemical erosion of gneiss	64	109	160	318
Consumption of hydrogen ions by chemical weathering of bedrock and soil (in kmol (H$^+$) ha^{-1} yr^{-1})	1.1	1.5	4.1	6.1
Depletion of exchangeable Ca^{2+} and Mg^{2+} from soil (in kmol (H$^+$) ha^{-1} yr^{-1})	0.34	0.75	0.85	4.3
Separately				
Ca^{2+}	-7.1	-11	+6.9	-58
Mg^{2+}	-0.66	-2.4	-15	-17

it depletes nutrients such as Ca^{2+} and Mg^{2+} and mobilizes potentially toxic metals such as Al^{3+}. Results from this study indicate that depletion of Ca^{2+} and Mg^{2+} is accelerated in areas with industrial acid emissions (mainly SO$_2$ and NO$_x$). The present rate of depletion of these two cations from acid brown earth soils in the Bohemian-Moravian Highland (where the concentration of SO$_2$ and NO$_x$ in air is 8.3 and 11 µg m^{-3} respectively) is equivalent to the hydrogen-ion consumption of 0.34 and 0.75 kmol ha^{-1} yr^{-1} in forests and 0.89 kmol ha^{-1} yr^{-1} in fertilized fields. The rate of depletion reaches 4.3 kmol ha^{-1} yr^{-1} in the damaged forest in the Ertzgebirge (where concentration of SO$_2$ and NO$_x$ in air is 112 and 33 µg m^{-3} respectively). These cations are replaced by Al^{3+} mobilized from hydroxide by acid rain.

According to Ulrich (1981) the length of branch rootlets of Norway spruce (*Picea abies*) decreases when the mole ratio of Ca^{2+}/Al^{3+} in soil decreases below 1.0. This is supported by the data from the catchments monitored in this study (Table 14.3) where the dieback of *Picea abies* coincides with a Ca^{2+}/Al^{3+} mole ratio of approximately 0.1 and an Mg^{2+}/Al^{3+} mole ratio of approximately 0.03. First signs of the decline of

Table 14.3. Soil properties indicating environmentally significant changes in catchments.

Catchment	X-0		X-8		X-7		X-14	
Soil horizon	A	B+C	A	B+C	A	B+C	A	B+C
pH_{KCl}	3.0	3.6	3.1	3.8			3.4	3.8
Humus content (%)	34	0.71	38	1.5	1.9	0.7	18	1.6
Cation exchange capacity mmol (p$^+$) 100g^{-1}	16.7	4.6	13.8	5.9			12.1	5.4
Mole ratio								
Ca^{2+}/Al^{3+}	0.62	0.91	0.81	0.84			0.13	0.10
Mg^{2+}/Al^{3+}	0.18	0.53	0.13	0.46			0.03	0.02

spruce have been reported from the catchment X-8 (Figure 14.1) where the ratio of Ca^{2+}/Al^{3+} in soil is 0.8 and Mg^{2+}/Al^{3+} is 0.1 in the A horizon of the soil.

The mass balance measurements indicate that the decrease in the Ca^{2+}/Al^{3+} and Mg^{2+}/Al^{3+} ratios continues even if pH_{KCl} of soils is buffered at values between 3 and 4. Thus, the inputs of environmental acids are consumed by reactions with aluminium salts. Such reactions mobilize inorganic aluminium ions which are harmful to fish.

The supply of essential exchangeable cations to plants is not limitless. For example, the supply of exchangeable Ca^{2+} and Mg^{2+} in the A horizons of the studied soils is equivalent to 90 kmol (H$^+$) ha^{-1} and the supply for the total weathering zone including A, B and C horizons is equivalent to 2,200 kmol (H$^+$) ha^{-1}. At the minimum rate of leaching (that is, 0.34 kmol (H$^+$) ha^{-1} yr^{-1}) the A horizon would be depleted of Ca^{2+} and Mg^{2+} after 265 years and the whole profile after 6,500 years. However, the time to complete depletion would be reduced to 21 years in the A horizon and to 510 years in the whole soil profile if acidification proceeds at the maximum rate (4.3 kmol (H$^+$) ha^{-1} yr^{-1}).

Research at the Ertzgebirge and the Bohemian-Moravian Highland sites indicates that an acidification in excess of 1.1 kmol (H$^+$) ha^{-1} yr^{-1} is harmful to forest and should be reduced. This rate of acidification is equivalent to an input of 18 kg ha^{-1} yr^{-1} of S-SO$_2$ or 15 kg ha^{-1} yr^{-1} of N-NO$_2$. Inputs of this magnitude are common over large areas of Europe and will not be sufficiently reduced even if a 30 per cent reduction of emissions is reached. Hence, the data suggest that the acidification of soils and the depletion of biologically essential metal cations will continue. Liming with dolomite is an expensive remedy. However, this treatment will not restore the original status of soils. The relatively high exchangeable sorption complex characteristic of acid brown earths is not readily developed under liming. Thus, the reversibility of environmental acidification of soils (or the irreversibility) is a crucial point for consideration in the planning of future land use practices.

Agricultural utilization of land has also caused excessive acidification. The acidification, caused by intensive application of fertilizers and increased biological consumption of cations, is generally combatted by periodic liming of fields. The mass balance in the agricultural catchment X-7 (Figure 14.1) in the less acidified Bohemian-Moravian Highland indicates that mechanical erosion of the weathered layer has increased by a factor of 1.5 when compared to the rates in forests of the same region. The rate of chemical erosion is much less than is observed in the damaged forest catchment in the mountains of the Ertzgebirge. Here, the chemical erosion has been increased by a factor of 2.9. The increase in the mechanical erosion due to deforestation and intensive agricultural practices in the Bohemian-Moravian Highland is from 732 to 1,967 kg ha^{-1} yr^{-1}. This corresponds to an increase in the region of 10 kg ha^{-1} yr^{-1} per 1 per cent of arable land. This increase is greater than that measured by Ryding (1984) in southern Sweden (where the increase is 7 kg ha^{-1} yr^{-1} per 1 per cent of arable land). The discrepancy may be due to the steeper slopes characteristic of the Bohemian-Moravian Highland. The major problem associated with the intensive exploitation of agricultural soils and their excessive mechanical weathering is the depletion of humic substances from the topsoil.

14.3. Conclusions

The long-term monitoring of the mass balance in the four representative catchments in Czechoslovakia quantifies the changes in weathering and erosion rates due to industrial and agricultural practices in central Europe. The present land use pattern leads to a depletion of the quantity and quality of forest as well as the degradation of agricultural soils. The harmful acidification of soils will continue.

If areas are subject to altered climatic conditions that would result in a change in the potential for the growth of crops and natural and semi-natural vegetation, the man-induced limitations outlined here will continue to constitute part of overall environmental conditions to which the new assemblages of plant species will need to adapt. The migration of the boundaries of plant distribution will not only be determined by altered climatic conditions but by the interaction of these with the prevailing localized effects resulting from the kind of altered rates of weathering and erosion brought about by the acidic depositions from the causal emission sources. This may make predictions of the new land use capability more difficult and complex but, nevertheless, they are factors of which account needs to be taken. The use of models is important to anticipate how the modifying effects of acidification vary across Europe.

REFERENCES

Carlsson, B., 1987, Analysis of trends in acidification by an integrated hydrological and hydrochemical conceptual model. In B. Moldan and T. Păces (eds.), *International Workshop on Geochemistry and Monitoring in Representative Basins.* Geological Survey, Prague. 145-147.

Christophersen, N. and Neal, C., 1987, Some results important for further development of hydrochemical models describing freshwater acidification. In B. Moldan and T. Păces (eds.), *International Workshop on Geochemistry and Monitoring in Representative Basins.* Geological Survey, Prague. 139-141.

Mazzarino, M.J., Heinrichs, H. and Folster, H., 1983, Holocene versus accelerated actual proton consumption in German forest soils. In B. Ulrich and J. Pankrath (eds.), *Effects of Accumulation of Air Pollutants in Forest Ecosystems.* Reidel Publishing Company, Dordrecht, The Netherlands. 113-123.

Neal, C. and Whitehead, P., 1987, Long term modelling of streamwater acidification in the British uplands. In B. Moldan and T. Păces (eds.), *International Workshop on Geochemistry and Monitoring in Representative Basins.* Geological Survey, Prague. 74-76.

Paces, T., 1985, Sources of acidification in Central Europe estimated from elemental budgets in small basins. *Nature,* **315**, 31-36.

Paces, T., 1986, Weathering rates of gneiss and depletion of exchangeable cations in soils under environmental acidification. *Jour. Geol. Soc.,* **143**, 673-677.

Ryding, S.O., 1984, Erosion of agricultural land - a growing water quality problem in rural areas. In E. Eriksson (ed.), *Hydrochemical Balances of Fresh Water System,* Uppsala Symp., IAHS Publ. 303-311.

Ulrich, B., 1981, Destabilisierung von Waldökosystemen durch Akkumulation von Luftverunreinigungen. *Der Forst- und Holzwirt,* **36**, 525-532.

Chapter 15

POTENTIAL FOR ACIDIFICATION OF FOREST SOILS IN EUROPE

M. Posch and L. Kauppi

15.1. Introduction

Over the last thirty years concern about acid rain and its consequences in terrestrial and aquatic ecosystems has grown in conjunction with increasing public awareness of environmental issues in general. The first signs of acidification were observed in vulnerable lake ecosystems in Scandinavia (Odén, 1968). About ten years later soil scientists in Central Europe reported detrimental changes in the chemistry of forest soils (Ulrich et al., 1980). The central role of soils as a link between acid rain and lake and groundwater acidification as well as between acid rain and forest damage was recognized quite early on, for example in the Norwegian SNSF-Project (Overrein et al., 1981).

Soil acidification plays a significant role in lake acidification since all water entering lakes has been in contact with soil, except the direct precipitation on to the lake surface. If the buffering capacity of the soil is exceeded, the acid load will reach the water. Depending on the retention time of the lake, as well as its internal alkalinity production, there is a time lag before the lake becomes acidified. Linkages of soil acidification to lake and groundwater acidification have been described by physico-chemical models (for example, Cosby et al., 1985a, 1985b; Kämäri et al., 1985).

For the phenomenon of forest damage, the link to soil acidification is not so well understood. A decrease in the buffering capacity also implies changes in nutrient ratios which might affect plant growth due to an imbalanced supply of different elements. When acidification proceeds far enough, dissolution of aluminium results in concentrations that can be toxic to plants. However, it is presently understood that forest decline is a multifactor phenomenon and soil acidification is only partly to blame.

In the following discussion only forest soils are considered. Agricultural soils are managed intensively with lime, fertilizers and other chemicals; these practices alter and control the soil chemistry and acidification can be counteracted.

F. M. Brouwer et al. (eds.), Land Use Changes in Europe, 325–350.

15.1.1. The concept of soil acidification

Soil acidification has been defined as a decrease in the acid neutralizing capacity (ANC) of the inorganic fraction of the soil including the solution phase (van Breemen *et al.*, 1984):

$$ANC_m = B_m - A_m$$

where B = basic components (the cations contributing depend on the reference pH chosen), A = strongly acidic components (anions of strong acids), and m = mineral soil.

de Vries and Breeuwsma (1987) redefined the concept of van Breemen *et al.* (1984) by distinguishing actual and potential soil acidification. They differentiated the acid neutralizing capacity (ANC) and the base neutralizing capacity (BNC). ANC is defined as the sum of the basic components and BNC as the sum of the acid components of the soil:

$$ANC_s = B_m + B_o$$

$$BNC_s = A_m + A_o$$

where A = strongly and weakly acidic components, s = solid and solution phase (total soil), and o = organic phase. Actual soil acidification was then defined as a decrease in ANC_s and potential soil acidification as an increase in BNC_s. Thus, actual acidification is manifested by leaching of cations from the soil, regulated by the mobility of major anions. Potential acidification is primarily due to accumulation of atmospherically derived nitrogen and sulfur (de Vries and Breeuwsma, 1987). The weakly organic acids are also included in the definition of BNC_s. An increase in exchange acidity caused by accumulation of acid organic matter increases the BNC of the soil, and should therefore be considered as potential soil acidification. Potential acid threat is realized by mineralization processes after the removal of vegetation.

15.1.2. The spatial extent of forest soil acidification in Europe

As yet, there have been no systematic surveys on the extent of forest soil acidification in Europe. However, a rough estimate of the areas where acidification proceeds can be obtained by comparing the acid stress due to air pollutants with the weathering rate, which is the only counteracting process. Only areas remote from emission sources, such as parts of the Nordic countries, the southern countries and the British Isles, receive such a low level of acid deposition that weathering should be able to counteract it. However, it does not mean that no soil acidification occurs because factors other than deposition, such as biomass accumulation and forestry practices, may also contribute to acid stress.

Results of soil studies from some European countries confirm the above. Berdén *et al.* (1987) made an extensive literature survey on the observed changes in pH or exchangeable cations. Information was available from the Federal Republic of Germany, Sweden, Austria, Czechoslovakia and the United Kingdom. Most of the studies showed an acidification trend. The only exceptions were those where calcareous dust deposition or liming occurred. The degree of acidification, however, varied remarkably among the different sites.

There are only a few observations relating to the acidity of forest soils in the past. In southern Sweden the acidity of forest soils was first measured in the 1920s and the same sites were resampled in 1982-1984 (Hallbäcken and Tamm, 1986). A pH decrease of about 0.9, 0.2 and 0.1 units was found in A0-, A2- and B-horizons respectively. Falkengren-Grerup (1987) also observed increasing acidity in forest soils in southern Sweden in 1984 compared to earlier measurements from the period 1949/50. The pH change was 0.5-1.0 unit.

The time periods covered in studies from other European countries are much shorter. Generally, the earliest observations date back to the 1950s or 1960s. Recent measurements normally show increasing acidity over the last 30-40 years (Berdén *et al.*, 1987). The majority of the forest soils investigated had already lost their cation exchange capacity; their pH was below 4.0.

15.2. Causes of Soil Acidification

The occurrence of soil acidification depends on ecosystem and soil processes. Berdén *et al.* (1987) mention the following characteristics that could affect the acidification of soils:

i)   increased deposition of acid or potentially acidifying compounds;

ii)   decreased deposition of acid-neutralizing compounds; increased primary productivity (and/or biomass harvest);

iii)   increased rate of nitrification or sulfur oxidation; changes in land use, for example afforestation, introduction of acidifying species and changes in forest management;

iv)   reduced decomposition rate of litter and soil organic matter;

    v)   increased production and vertical movement of
                organic acids.

Most attention has been paid to the acidifying effects due to the deposition of (anthropogenic) sulfur. Annual sulfur emissions in Europe increased from about 10 million tonnes in 1900 to some 27 million tonnes in 1980, most of the increase occurring after 1950. The emissions of nitrogen oxides have increased continuously, while sulfur emissions have leveled off and even decreased during the 1980s (Figure 15.1).

The spatial distribution of sulfur deposition is very uneven. The highest sulfur deposition rates (>10 g m^{-2} yr^{-1}) are observed in the heavily industrialized regions of central Europe. Most of central Europe as well as parts of the United Kingdom and USSR receive more than 5 g m^{-2} yr^{-1} of sulfur (Figure 15.2).

Nitrogen deposition comprises the deposition of nitrogen oxides and ammonia (see Figures 15.3 and 15.4 for their spatial distribution in 1980). Nitrogen deposition in the Benelux countries and in a large part of the FRG and GDR is greater than 3 g N m^{-2} yr^{-1}. In some areas, for example, The Netherlands, ammonia emissions from agriculture play an important role.

Atmospheric deposition of these potentially acidifying substances leads to actual soil acidification to the extent that they are not taken up by vegetation or immobilized into soil organic matter, whereas potential acidification is manifest by the accumulation of organic N and S in the vegetation and soil.

Most forest ecosystems are nitrogen-limited (Cole and Rapp, 1981; Ågren, 1983). As long as these conditions prevail, no actual soil acidification due to nitrogen deposition occurs. In the long run, however, forest ecosystems might become saturated by nitrogen. According to Ågren and Kauppi (1983) the system can be defined nitrogen saturated when the nitrogen concentration in the leaf biomass is the maximum attainable without tissue injury. Using the model developed by Ågren (1983) the time scales for nitrogen saturation in European forests were estimated. Assuming that the geographical distribution of nitrogen deposition is equal to one-third of the sulfur deposition in 1974, and that the deposition rate of nitrogen increased by 4 per cent a year, the typical time to saturate a pine forest on medium soils in central Europe is about thirty years, whereas in northern Scandinavia over a hundred years is required. With spruce forests the time required to reach saturation was estimated to be typically 50 per cent longer. The actual rate of acidification would depend on the balance between the input and the output of NH_4^+ and NO_3^-.

Also, if the vegetation is removed, mineralization is no longer balanced by uptake, leading to potential acidification. According to de Vries and Breeuwsma (1987) this effect is not usually important in

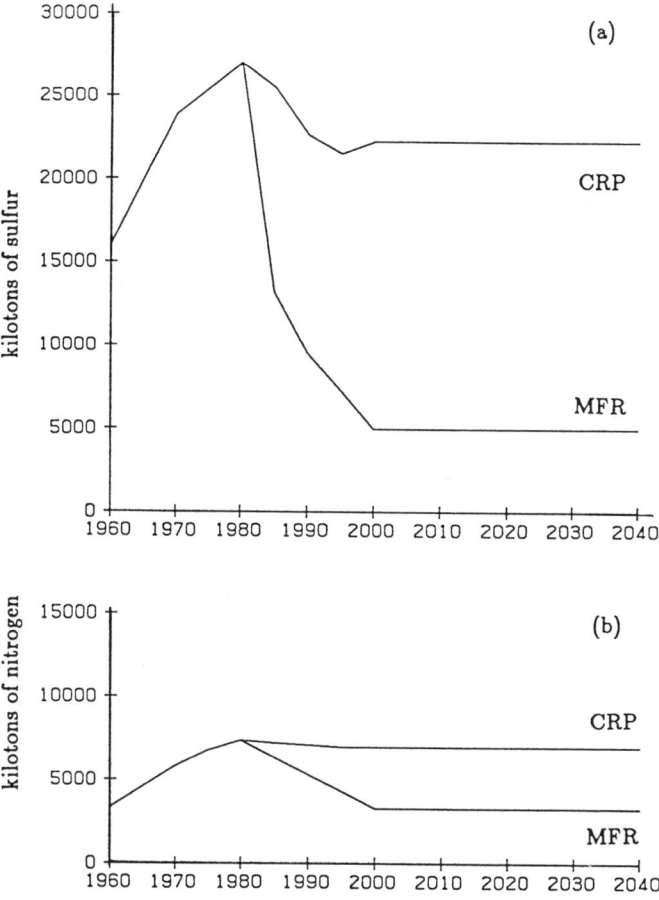

Figure 15.1. a) Total European SO_2 emissions (in kt yr^{-1} of sulfur) and b) total European NO_x emissions (in kt yr^{-1} of nitrogen for the period 1960-2040. From 1980 to 2040 two scenarios are displayed: Current Reduction Plans (CRP) and low Maximum Feasible Reductions (MFR).

forest soils because clearing is not frequent and vegetation generally regrows rapidly.

The uptake of sulfur by vegetation is considerably lower than that of nitrogen. Therefore, the acidifying effect of sulfur deposition is usually high compared to that of nitrogen deposition (van Breemen *et al.*, 1984). The mobility of sulfate can also be affected by soil adsorption or by precipitation. Thus, the annual rate of actual acidification caused by sulfur transformations is equal to the difference

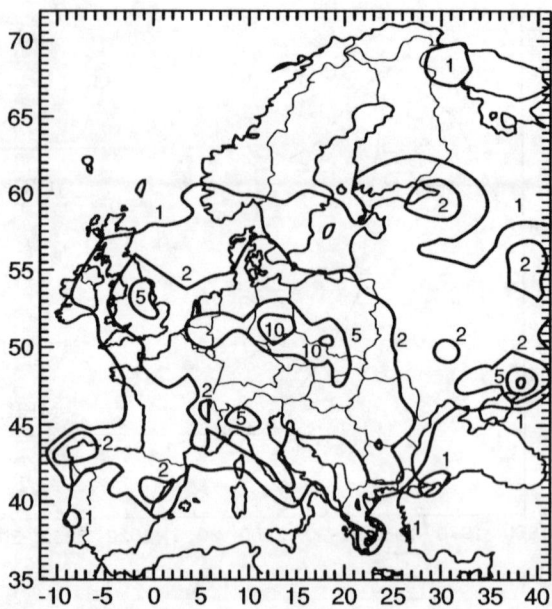

Figure 15.2.   Sulfur deposition (g S m^{-2} yr^{-1}) in Europe in 1980.

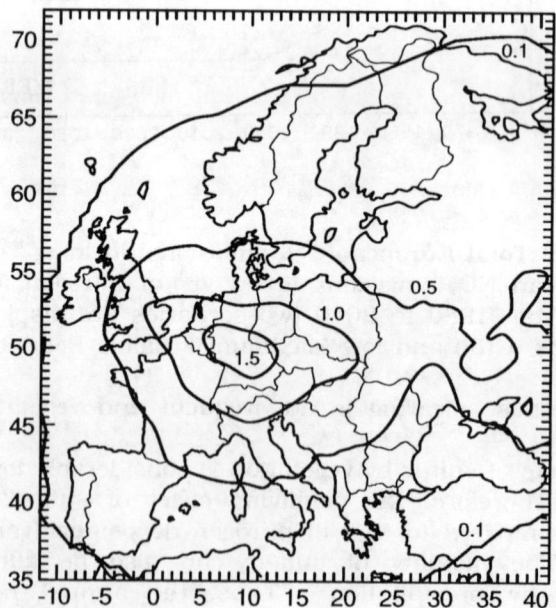

Figure 15.3.   NO$_x$ deposition (g N m^{-2} yr^{-1}) in Europe in 1980.

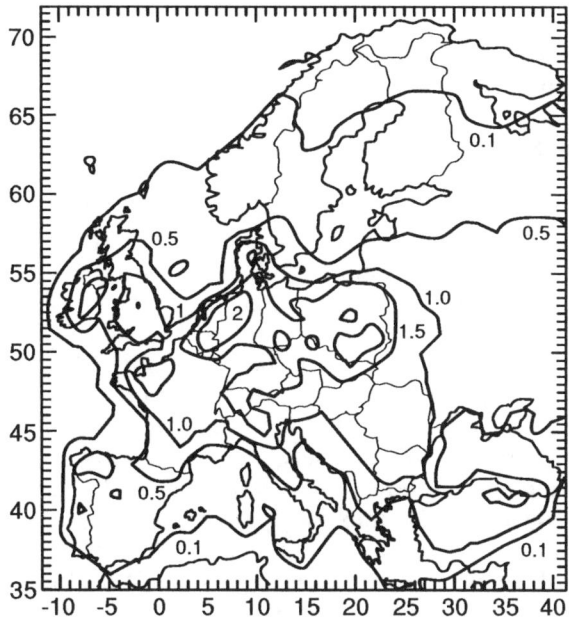

Figure 15.4.   Ammonia deposition (g N m^{-2} yr^{-1}) in Europe in 1980.

between the input and output of sulfate.   Potential acidification occurs in the case of sulfur retention.

Mineralization, weathering and desorption of cations neutralize the acid production induced by the production of anions in mineralization and oxidation processes.   Removal of cations thus results in acid production in the soil.   In northern Europe intensified forestry contributes to the removal of cations from the soil via increased biomass growth and harvesting.

The relative significance of the different factors in actual soil acidification varies depending on the region.   Berdén et al. (1987) concluded that acid deposition is of great importance in many cases, especially in southern Scandinavia and central Europe.   In northern Europe where low temperatures retard biomass decomposition, natural accumulation of biomass contributes to acidification by removing cations from the mineral soil.

Based on the synthesis of major element cycles (carbon, nitrogen, sulfur and cations) de Vries and Breeuwsma (1987) concluded that natural soil acidification, that is the dissociation of weak acids (carbonic and organic), is most important in calcareous soils.   Acidification rates measured in these soils vary between 7 and 13 kmol ha^{-1} yr^{-1}, whereas in podzolic soils the rate is only 0.1 to 0.7 kmol ha^{-1} yr^{-1}.   In non-calcareous agricultural soils land use largely determines acidification.

The rate of acidification is approximately 8 kmol ha^{-1} yr^{-1}. Finally, in non-calcareous forest soils, acid precipitation is the most important cause of acidification, the rate varying between 1 and 6 kmol ha^{-1} yr^{-1} (de Vries and Breeuwsma, 1987).

15.3. Modeling Forest Soil Acidification

Information on the long-term effects of acid deposition on soils is very important for the formulation of policies for emission reductions. In this respect models provide an important tool to assist decision makers in evaluating the effectiveness of abatement strategies. Therefore, at the International Institute for Applied Systems Analysis (IIASA) a Regional Acidification Information and Simulation model (RAINS) has been developed that analyses environmental impacts on a European scale for different emission scenarios. Predictions are based on quantitative descriptions of the linkages between emissions, deposition and environmental impacts such as soil acidification and effects on terrestrial and aquatic ecosystems (Alcamo et al., 1987).

Within the overall framework the soil acidification submodel forms an important link between atmospheric deposition and effects on forests (as a risk indicator), surface waters and groundwater. The RAINS regional soil model includes the impact of SO_2, NO_x and NH_3 deposition, as well as the effects of natural soil acidification due to the dissociation of CO_2. The model predicts the major components of the soil solution, that is H^+, Al^{3+}, BC^{2+}, NH_4^+, NO_3^-, SO_4^{2-} and HCO_3^-, and not only pH, since the ratios of aluminium and/or ammonium to divalent base cations (represented by BC^{2+}) are more sensitive parameters with respect to effects on forests.

15.3.1. Modeling approach and model structure

The model structure is based on the anion mobility concept by incorporating the charge balance principle (Reuss et al., 1986). Apart from the precipitation surplus which affects all ion concentrations except HCO_3^-, the concentrations of SO_4^{2-}, NO_3^- and NH_4^+ are completely determined by the net element input; the Al^{3+} concentration is controlled by chemical interaction with the soil (mobilization and cation exchange) and the concentration of BC^{2+} is regulated by both. The HCO_3^- concentration is based on an equilibrium approach with CO_2 and regulated by both the CO_2 pressure and the pH. The equilibrium approach has also been used with respect to base cations in calcareous soils and with respect to aluminium in non-calcareous soils. Consequently, the mobilization rate of BC^{2+} and Al^{3+} from carbonates and hydroxides respectively is not determined by kinetic constants, such as silicate weathering, but by the leaching of these elements with the precipitation surplus. The pH is determined by the acid load, which is induced by the external input of strong acid anions (SO_4^{2-} and NO_3^-) minus base cations (BC^{2+} and NH_4^+) plus the internal generation of weak acid anions (HCO_3^-), and the acid neutralization due to weathering and exchange reactions.

H^+ transfer in the soil is influenced by numerous reactions (de Vries and Breeuwsma, 1987). In order to minimize data input general assumptions have been made.

a) The soil solution chemistry is only determined by the net element input from the atmosphere and geochemical interactions (weathering and cation exchange) in the soil. Apart from the net uptake of nitrogen and base cations in harvests and net immobilization, affecting the net element input, the influence of the nutrient cycle (foliar exudation, litter fall, mineralization and uptake) is not taken into account. This does affect model predictions in the upper unsaturated zone (the upper 25 cm). Consequently, in a regional soil acidification model developed for The Netherlands (RESAM), nutrient cycling processes have been included (de Vries, 1990; de Vries and Kros, 1989). However, this requires a large amount of data which is not available on a European scale. Furthermore, the runoff and leachate concentrations that are important for surface water and groundwater quality are almost unaffected by this assumption.

b) Sulfate output is in equilibrium with sulfur input. Uptake, immobilization, reduction and adsorption of sulfate is considered negligible. Soils that are high in hydrous oxide minerals can strongly adsorb sulfate, thus reducing cation leaching from soils (Johnson, 1980). However, element budget studies in northern and western Europe (Rosén, 1982; van Breeman et al., 1984; Nilsson, 1985) indicate that sulfate adsorption is generally negligible.

c) Biological fixation of nitrogen and denitrification are negligible. Both assumptions are reasonable for most forest ecosystems (Klemedtson and Svensson, 1988). Notable exceptions are red alder with a high rate of nitrogen fixation (van Miegroet and Cole, 1984) and extremely wet forest soils with a high rate of denitrification (Tietema and Verstraten, 1989).

d) In acid forest soils (pH<4.5), natural soil acidification is negligible. Organic acids can have a significant impact in these soils, but the aluminium that is mobilized by these acids is non-toxic (Ulrich and Matzner, 1983) and has therefore been neglected.

e) The weathering rate of base cations from silicates is independent of the soil pH. This agrees with a thermodynamic analysis of published laboratory experiments, which shows that the weathering rate of pure feldspar minerals is pH-independent between 3 and 8 (Helgeson *et al.*, 1984).

f)   The soil is considered as a homogeneous box with a constant density. However, it is possible to divide the soil into several layers. The soil depth affected by acid deposition is variable, but generally a thickness of 50 cm is assigned since most biogeochemical interactions occur in this zone.

g)   The element input mixes completely with the soil. This is a reasonable assumption for most forest soils which are well drained and poorly structured (sands, loamy sands and sandy loams). However, in well structured soils with channels and cracks this assumption does not hold.

h)   The water flow is stationary on a yearly basis, which equals the time step of the model. The reason for this time step is the focus on long-term effects. Seasonal, monthly or even daily fluctuations in soil water chemistry are potentially very important but the model only predicts trends of increasing probabilities of crucial events.

i)   The water flux percolating from the top 50 cm of the soil equals the precipitation surplus which equals precipitation minus interception evaporation minus evapotranspiration. This implies that forests obtain all of their water from the top 50 cm of the soil. This is a reasonable assumption since most of the fine roots responsible for water and nutrient uptake occur in this soil compartment.

The conceptual basis of the model, as outlined above, is similar to various surface water acidification models (for example, MAGIC (Cosby *et al.*, 1985a, 1985b)). It is assumed that the soil water chemistry is mainly governed by atmospheric deposition as the major driving variable and chemical equilibrium reactions as the buffering processes.

In calcareous and slightly acid soils (pH > 5) acidity is generated in the soil water by the formation of bicarbonate from dissolved CO_2 and water according to:

$$CO_2 + H_2O \rightleftharpoons HCO_3^- + H^+ \tag{1}$$

In calcareous soils the free hydrogen ion, produced by this mechanism and by acid input, is neutralized by the dissolution of calcite (carbonate buffer range) according to:

$$CaCO_3 + H^+ \rightleftharpoons Ca^{2+} + HCO_3^- \tag{2}$$

Combination of equation (1) and (2) gives:

$$CaCO_3 + CO_2 + H_2O \rightleftharpoons Ca^{2+} + 2\,HCO_3^- \tag{3}$$

In non-calcareous soils the hydrogen ions can be neutralized by exchange with divalent cations (exchange buffer range) according to:

$$2H^+ + \overline{BC} \rightleftharpoons 2\overline{H} + BC^{2+} \tag{4}$$

where the bar denotes the adsorbed phase. Organic matter has a high affinity for hydrogen ions, leading to a strong decrease of the effective cation exchange capacity (CEC) with the pH of the soil solution (Helling et al., 1964).

Upon further acidification, due to a decrease in base saturation, H^+ will start to react with aluminium (aluminium buffer range). This becomes important below pH values of about 4.5 (Ulrich, 1981). In the model we assume that the concentration of Al^{3+} in soil water is in equilibrium with the solid phase of $Al(OH)_3$ according to:

$$Al(OH)_3 + 3H^+ \rightleftharpoons Al^{3+} + 3H_2O \tag{5}$$

In this pH range cation exchange will not be limited to H/BC exchange, since the exchange sites of the soil matrix have a high affinity for trivalent aluminium cations, leading to Al/BC exchange according to:

$$2Al^{3+} + 3\overline{BC} \rightleftharpoons 2\overline{Al} + BC^{2+} \tag{6}$$

However, the amount of aluminium hydroxide is not infinite and might be depleted from the topsoil over the next decades in areas with high acid deposition rates. If this occurs, iron hydroxides may start interacting with H^+ (iron buffer range), but the mobilization rate of Fe^{3+} is extremely small (de Vries, unpublished laboratory experiments) and has therefore been neglected in the model. In this stage of acidification the Al^{3+} concentration is only determined by congruent weathering from primary silicates, while there will be temporarily a strong buffering of H^+ by H/Al exchange according to:

$$3H^+ + \overline{Al} \rightleftharpoons 3\overline{H} + Al^{3+} \tag{7}$$

The model consists of a set of mass balance equations, which describe the soil input-output relationships for the cations (Al^{3+}, BC^{2+},

NH_4^+) and strong acid anions (SO_4^{2-}, NO_3^-), and a set of equilibrium equations, which describe the equilibrium soil processes. An explicit mass balance for the ions H^+ and HCO_3^- is not necessary, since these ions have diffuse sources and sinks (dissociation of water, dissolution of CO_2). Their concentration is determined by equilibrium equations and the concentration of the other ions by the charge balance principle. A complete overview of the various process formulations and the solution methods are described in de Vries *et al.* (1989).

15.3.2. Scenarios and data

The application of the RAINS soil acidification model on a European scale requires a large amount of data. Presently, such a data base containing all necessary information on input parameters and driving forces is not available. Therefore, the analysis is restricted to two major forest soil types. The goal of this exercise is to show the type of information that can be obtained from the model, and to demonstrate the applicability of the model on a large regional scale for analysing the impacts of various deposition reduction scenarios.

The Soil Map of the World (FAO-UNESCO, 1974) and the World Forestry Atlas (1975) have been used to digitize the spatial distribution of soil types and forests in Europe on a grid with 1° east-west and 0.5° north-south extension. From this data base we find that about 33 per cent of the forests grow on Orthic Podzols (Po) and about 8.5 per cent grow on Dystric Cambisols (Bd). The spatial distribution of these two soil types is displayed in Figure 15.5. These figures show that the podzol covers most of Fennoscandia, while the Dystric Cambisol is scattered over central Europe.

From the existing soil data base, which has been used for the old RAINS soil acidification model (Kauppi *et al.*, 1986) and which contains information for 80 different soil types, the following soil parameters for Po and Bd are taken (Table 15.1):

Table 15.1. Soil parameters for Orthic Podzols (Po) and Dystric Cambisols (Bd).

Variable	Po	Bd	Unit
Cation exchange capacity (CEC)	156	220	eq m^{-3}
Volumetric water content	0.22	0.27	m^{3}m^{-3}
Initial base saturation	0.10	0.15	fraction

The weathering rate of base cations, which counteracts acidification, depends on the parent material of the soil. Ulrich (1983) reports a range of variation in European soils from 0.02 to 0.2 mol m^{-3} yr^{-1}. The International Geological Map of Europe and the Mediterranean Region (UNESCO, 1972) has been used to determine the dominant parent material of the soils in each grid square. For Po (mostly in Fennoscandia) the values centered around 0.05 mol m^{-3} yr^{-1}, whereas for Bd (Central Europe) the values are about 0.1 mol m^{-3} yr^{-1}.

Also, precipitation and evapotranspiration vary significantly over Europe. In the RAINS data base, derived from thirty-year climatic mean statistics of 253 stations in Europe, the Soviet Union and northern Africa (Müller, 1982), the precipitation surplus (that is, precipitation minus evapotranspiration) ranges from almost zero to about 2,500 mm. In Figure 15.6 the (area weighted) distribution of the precipitation surplus on the soils Po (mean 464 mm) and Bd (mean 314 mm) is displayed. For the other model parameters, such as equilibrium and exchange constants, no data are available on a European scale. Therefore, reference values had to be used (de Vries et al., 1989).

As driving forces for the soil acidification model, scenarios for the different atmospheric pollutants are required, that is, scenarios for the deposition of sulfur (SO_2) and nitrogen compounds (NO_x and NH_3).

Sulfur deposition: For the years 1960 to 2040 the sulfur emission scenarios are taken from the RAINS model (Alcamo et al., 1987). Between 1900 and 1960 the European sulfur emissions are taken from Fjeld (1976). The resulting deposition patterns are computed with the aid of source-receptor matrices derived from the EMEP-II long-range transport model (Eliassen and Saltbones, 1983, and Lehmhaus et al., 1986). In Figure 15.2 the SO_2 deposition pattern over Europe due to the 1980 emissions is displayed.

NO_x deposition: NO_x depositions are computed by the IIASA NO_x model (Alcamo and Bartnicki, 1988), which is a simplified version of the "Harwell model" for the long-range transport of nitrogen species (Derwent, 1986; Derwent and Nodop, 1986). The spatial coverage of the Harwell model has been extended to cover all Europe. Between 1960 and 2040 NO_x emissions are again taken from the RAINS model. Between 1900 and 1960 depositions are linearly interpolated between an assumed background level of 0.01 eq m^{-2} yr^{-1} and the 1960 values. In Figure 15.3 the NO_x deposition pattern resulting from the 1980 emissions is displayed.

NH_3 deposition: Ammonia calculations use a model developed by Asman and Janssen (1987). European NH_3 emissions for the year 1980 are estimated at about 10,000 kt (expressed as N). Since neither historical emission figures nor future emission scenarios for NH_3 are available, the following procedure has been adopted: between 1900 and 1980 the NH_3 deposition is assumed to increase linearly from a background value corresponding to 0.01 eq m^{-2} yr^{-1} in 1900 to the

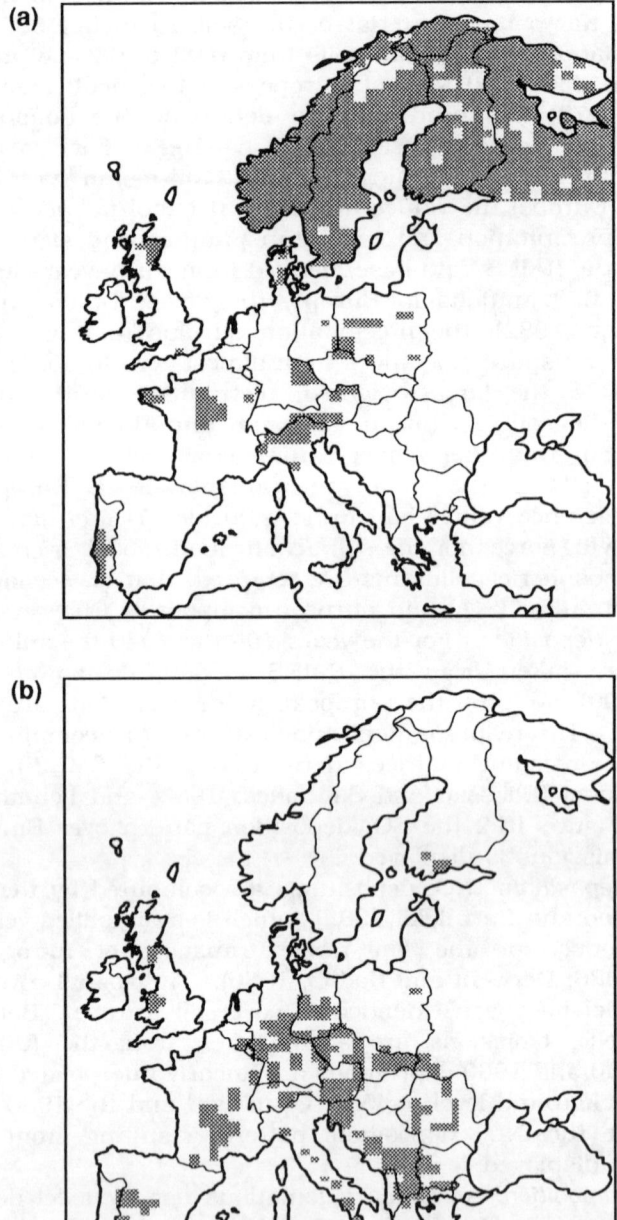

Figure 15.5.   Distribution of a) Orthic Podzol (Po) and b) Dystric Cambisol (Bd) on forest soils in Europe.

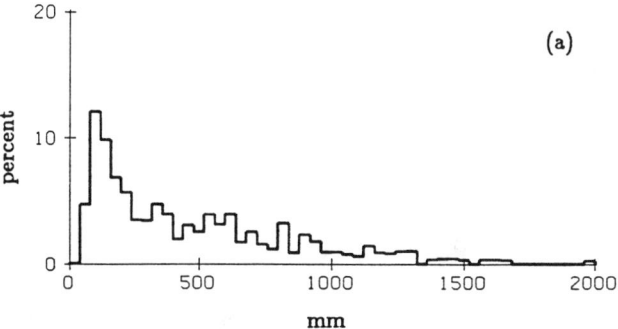

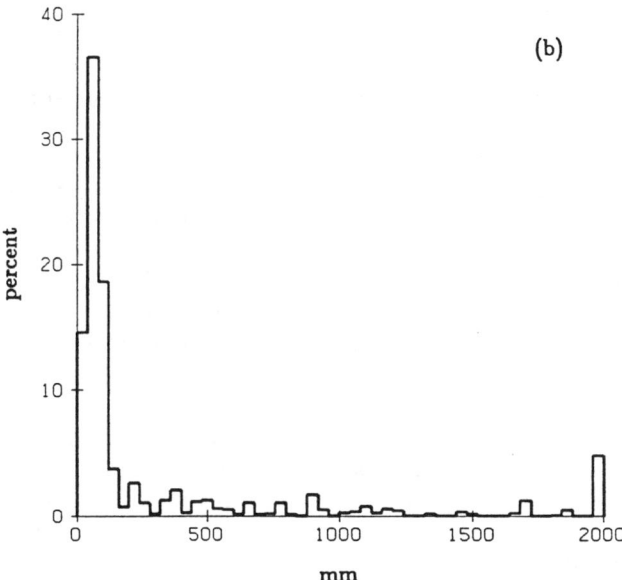

Figure 15.6.   The (area weighted) and distribution of the precipitation surplus (that is, precipitation minus evapotranspiration) on a) the soil Po (mean 464 mm) and b) the soil Bd (mean 314 mm).

deposition values computed with 1980 emissions; and from 1980 onwards the emissions (and depositions) remain at the 1980 level.   In Figure 15.4 the NH_3 deposition pattern resulting from the 1980 emissions is displayed.

Base cation deposition: No atmospheric transport models for base cations are currently available, and even reliable emission estimates are lacking. Therefore, it can be assumed that base cation deposition equals base cation uptake by plants (trees).

15.3.3. **Results and discussion**

Running the soil acidification model from 1900 to 1980 with the historical deposition patterns described produces the current state of acidification for the two soil types Po and Bd. The spatial distribution of this state of soil acidification is displayed in Figure 15.7. The cumulative distributions of the forest area on Po and Bd soils are shown in Figure 15.8 (thick lines). It shows that all Po forest soils have pH values between 3.4 and 5.0 with about 70 per cent between 4.5 and 5.0, whereas Bd soils have pH values between 3.7 and 4.8 with only 25 per cent above pH 4.5. By taking the derivative of these cumulative distributions, it can be seen that the frequency distribution for Po is unimodal with a maximum value around pH 4.7; the frequency distribution for Bd soils has two maxima, one around pH 4.1 and a second smaller peak around pH 4.7.

Next the model is run with two different deposition Scenarios: the *Current Reduction Plans* (CRP) Scenario containing the current plans to reduce SO_2 and NO_x emissions as announced by the individual European governments. The second Scenario, called *Maximum Feasible Reductions* (MFR) is based on official energy consumption figures and assesses the impacts of a radical decrease in emissions (mostly SO_2). It is assumed that all potential emission reductions achievable with today's pollution control technology are realized, but no measures are taken for energy conservation and fuel substitution. The time development of the total European sulfur and nitrogen emissions due to these two scenarios is displayed in Figure 15.1a and Figure 15.1b. The results of the model runs with these scenarios are not illustrated in a map format because of the difficulties in making comparisons; for the year 2040 the results are displayed as cumulative graphs in Figure 15.8. One can see that for Po both Scenarios yield an overall increase in the soil pH; however, the increase for the CRP is only small. For Bd the CRP further decreases the pH by the year 2040, while the MFR increases the soil pH substantially. For example, the Bd soils with pH below 4.0 comprised about 30 per cent of the area in 1980, whereas this area goes down to about 10 per cent for this low Scenario. It should be borne in mind that Po soils are located in northern Europe where deposition is much lower than in central Europe where most of the Bd soils are located.

The concentration of hydrogen ions in the soil solution (pH) is not the only variable to describe the state of a soil in response to the deposition of acidifying compounds. Therefore, the Al/BC ratio, the ratio

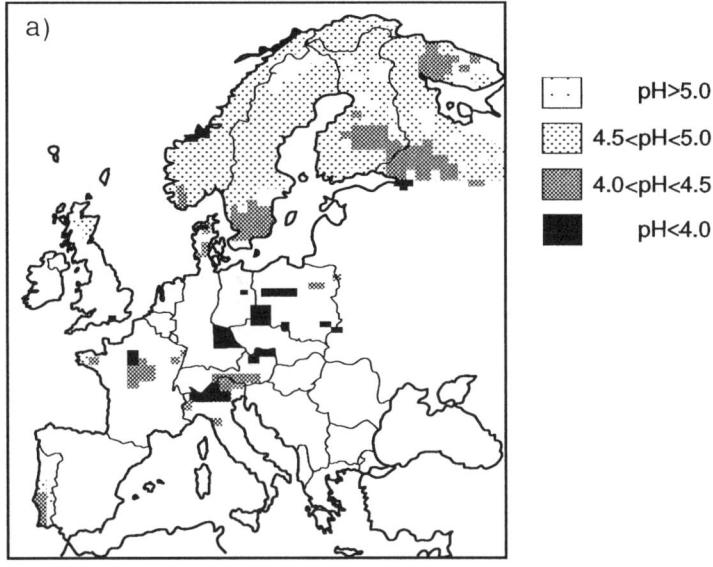

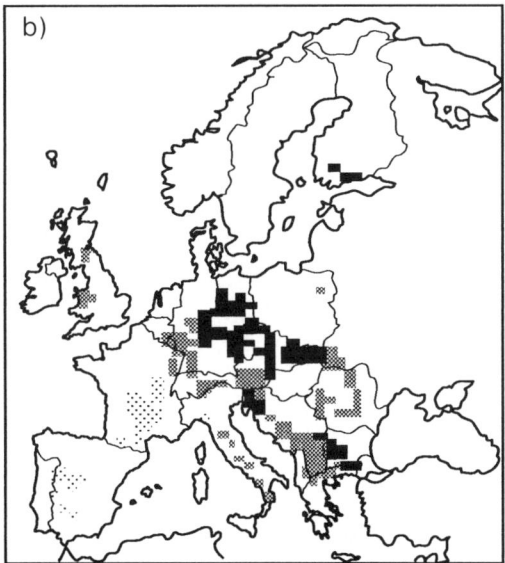

Figure 15.7.   The spatial distribution of the acidification of a) Po in 1980, and b) Bd in 1980 according to the simulation with the soil acidification model.

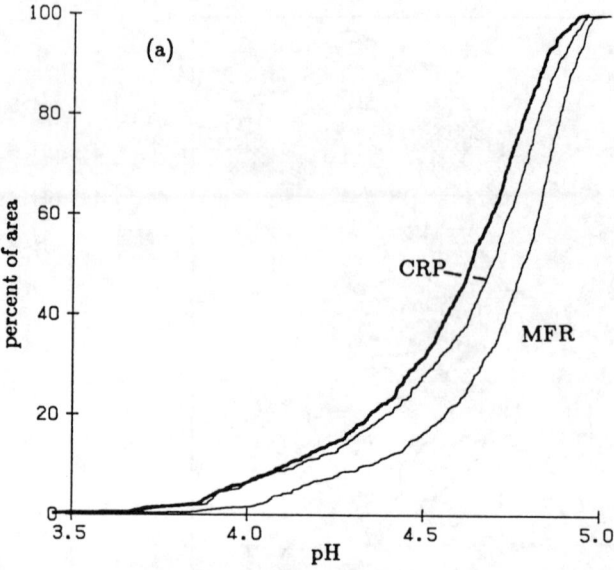

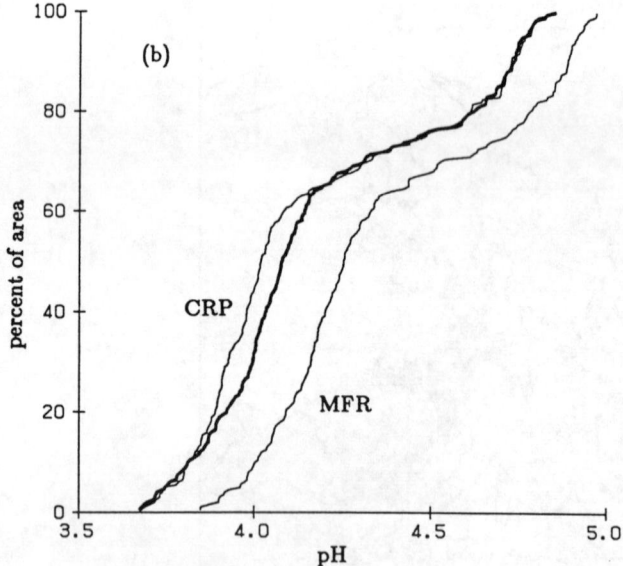

Figure 15.8. The cumulative distributions of pH for a) Po soils and b) Bd soils. The thick lines refer to 1980; the other lines refer to the pH of the soils in 2040 due to the Current Reduction Plans (CRP) and the Maximum Feasible Reductions (MFR).

of the aluminium concentrations and the concentration of divalent base cations (Ca^{2+} and Mg^{2+}) is used as an indicator of the state of the soil. In Figure 15.9 the cumulative distribution of the Al/BC ratio for the Po and Bd soils, respectively, are displayed for 1980 (thick line) and for the two deposition Scenarios in 2040. It shows that in 1980 about 93 per cent of the Po area has an Al/BC ratio below the critical value of 1.0; and this percentage does not change much under the two Scenarios (about 90 per cent for CRP and about 95 per cent for MFR). For the Bd soil about 80 per cent of the area is below 1.0 in 1980, but the CRP is not sufficient to maintain this value; the fraction of the area with an Al/BC ratio smaller than 1.0 drops to about 60 per cent in 2040. However, with the drastic MFR it is possible to maintain the 1980 percentage even in the year 2040.

Another variable characterizing the state of a soil is the base saturation (ƒBC), that is, the percentage (or fraction) of the exchangeable base cations. Figure 15.10 shows graphs of the cumulative base saturation distribution for Po and Bd respectively.

The cumulative graphs shown in Figures 15.8 - 15.10 give a good overview of the state of a soil for a particular Scenario, but the graphs do not readily indicate *changes* in acidification. The changes of the three soil variables (pH, Al/BC ratio and base saturation) due to the two different Scenarios are presented in Table 15.2, Table 15.3 and Table 15.4. The Table entries show the changes between 1980 and 2040. Note that a 50%-50% entry means that there is no overall change; nevertheless, the soils might change locally. Also note that pH<0 and ƒBC<0 means a deterioration, while Al/BC<0 means an improvement in the soil.

The tables show that for soil Po, which is located in areas where the dose is already low, both Scenarios yield an increase in pH (almost) everywhere, while base saturation hardly changes for CRP and improves on average by 1 per cent in about 25 per cent of the area for the MRF. The Al/BC mol ratio decreases for the low MFR Scenario but increases on average for the CRP Scenario; however, the area where it increases is much smaller than the area where it decreases. For the Bd soils, which are located mostly in areas with a high acid load, the pH decreases in many places for the high CRP Scenario, while it increases everywhere for the low MFR Scenario. The base saturation of this soil type decreases for both Scenarios, whereas the Al/BC ratio mostly decreases for the low Scenario, while it increases almost everywhere for the high Scenario.

The question whether the long-term soil responses estimated by the model are accurate projections of the real system cannot be answered satisfactorily. The model could be verified by comparing simulations of past soil chemistry with historical observations, but only few records (for example, Falkengren-Grerup, 1987) are extensive enough to be suitable for rigorous model testing. However, it can be concluded

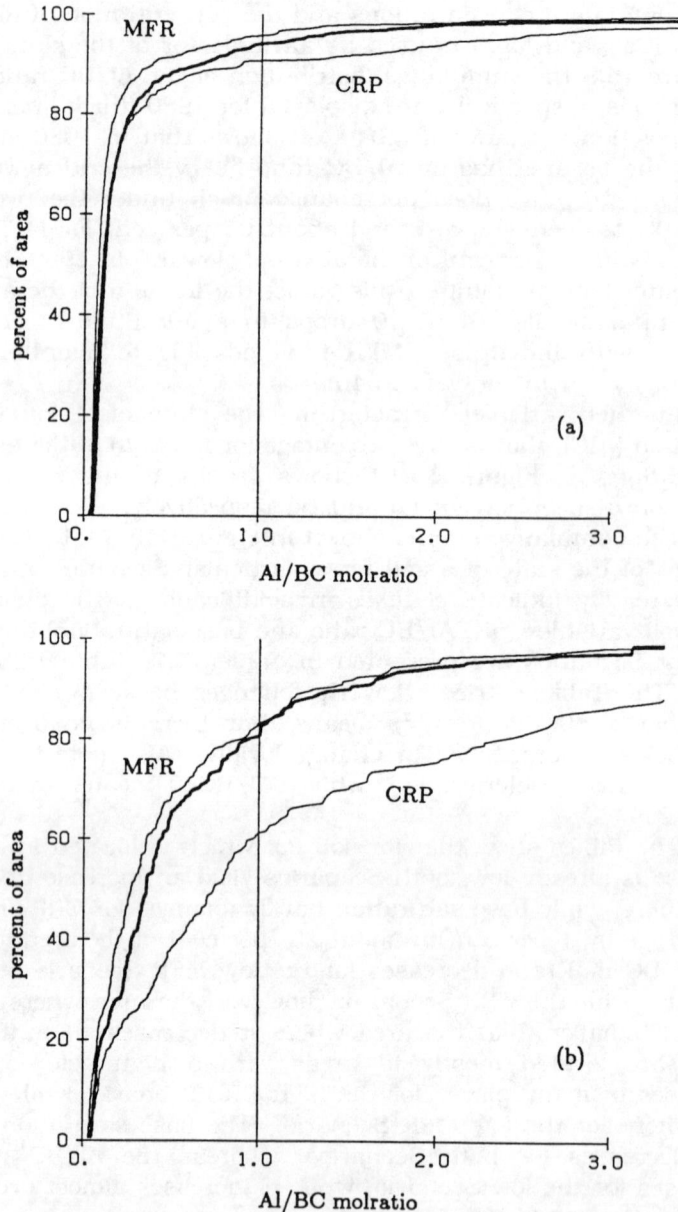

Figure 15.9. The cumulative distribution of the Al/BC ratio for a) Po soils, and b) Bd soils. The thick lines refer to 1980; the other lines refer to the Al/B ratio in 2040 due to the Current Reduction Plans (CRP) and the Maximum Feasible Reductions (MFR).

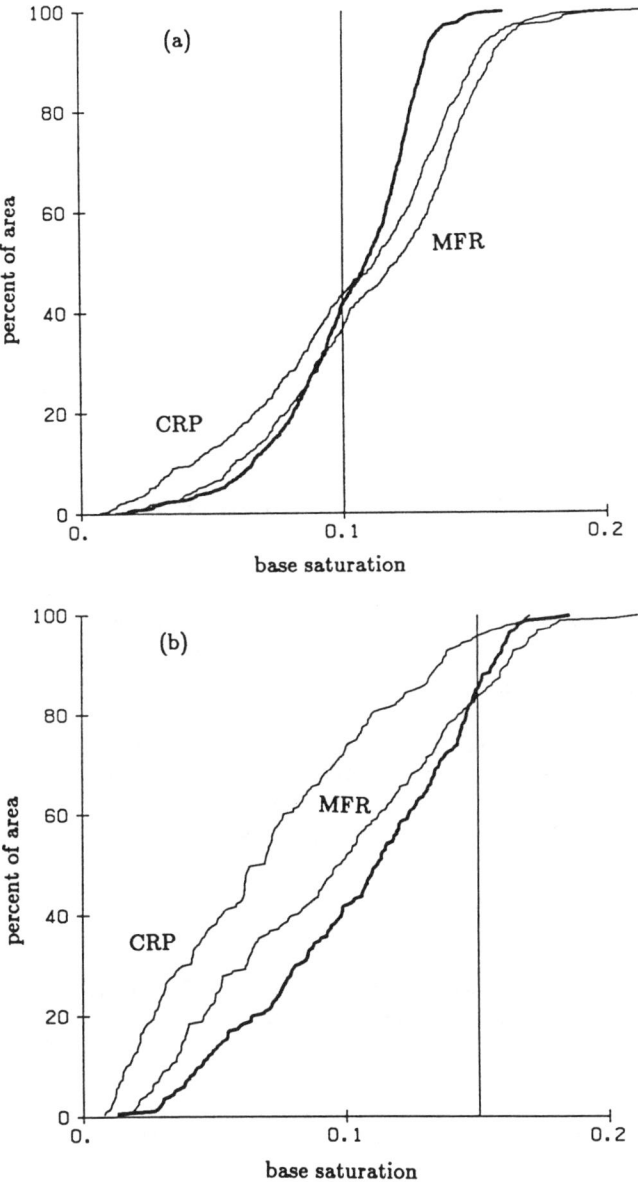

Figure 15.10. The cumulative distribution of the base saturation for a) Po soils, and b) Bd soils. The thick lines refer to 1980; the other lines refer to the base saturation in 2040 due to the Current Reduction Plans (CRP) and the Maximum Feasible Reductions (MFR).

Table 15.2.   Changes in pH for Orthic Podzols (Po) and Dystric Cambisols (Bd) for both scenarios between 1980 and 2040.

Soil	Scenario	ΔpH < 0 (%)	ΔpH > 0 (%)	$\overline{\Delta pH}$
Po	CRP	3.2	96.8	0.05
Po	MFR	0.0	100.0	0.14
Bd	CRP	67.3	32.7	-0.04
Bd	MFR	0.0	100.0	0.18

Table 15.3.   Changes in the Al/BC ratio for Orthic Podzols (Po) and Dystric Cambisols (Bd) under both scenarios between 1980 and 2040.

Soil	Scenario	ΔAl/BC < 0 (%)	ΔAl/BC > 0 (%)	$\overline{\Delta Al/BC}$
Po	CRP	69.5	30.5	0.06
Po	MFR	97.8	2.2	-0.07
Bd	CRP	2.8	97.2	0.93
Bd	MFR	91.6	8.4	-0.06

Table 15.4.   Changes in base saturation (ʃBC) of Orthic Podzols (Po) and Dystric Cambisols (Bd) under both scenarios between 1980 and 2040.

Soil	Scenario	ΔʃBC < 0 (%)	ΔʃBC > 0 (%)	$\overline{\Delta ʃBC}$
Po	CRP	43.0	57.0	0.00
Po	MFR	24.1	75.9	0.01
Bd	CRP	99.1	0.9	-0.04
Bd	MFR	75.0	25.0	-0.01

that the soil acidification model, including processes that are thought to be most important in influencing soil responses to acidic deposition and using parameter values within ranges appropriate for natural soils, produces plausible results. Even though it cannot be strictly verified for its ultimate use of long-term predictions, it provides a conceptual understanding of long-term soil responses and projections for policy-makers.

REFERENCES

Ågren, G.I., 1983, Nitrogen productivity in some conifers. *Can. J. For. Res.*, **13**, 494-500.

Ågren, G.I. and Kauppi, P., 1983, *Nitrogen Dynamics in European Forest Ecosystems: Considerations Regarding Anthropogenic Nitrogen Depositions.* IIASA CP-83-28. Laxenburg, Austria.

Alcamo, J., Amann, M., Hettelingh, J.-P., Holmberg, M. Hordijk, L., Kämäri, J., Kauppi, L., Kauppi, P., Kornai, G. and Mäkelä, A., 1987, Acidification in Europe: a simulation model for evaluating control strategies. *Ambio*, **16**, 232-245.

Alcamo, J. and Bartnicki, J., 1988, *Nitrogen Deposition Calculations for Europe.* IIASA WP-88-25, Laxenburg, Austria. 25.

Asman, W.A.H. and Janssen, A., 1987, Long-range transport model for ammonia and ammonium for Europe. *Atm. Environ.*, **21**, 2099-2119.

Berdén, M., Nilsson, S.I., Rosén, K. and Tyler, G., 1987, *Soil Acidification - Extent, Causes and Consequences. An Evaluation of Literature Information and Current Research.* National Swedish Environmental Protection Board, Report 3292, Solna, Sweden.

van Breemen, N., Driscoll, C.T. and Mulder, J., 1984, Acidic deposition and internal proton sources in acidification of soils and water. *Nature*, **307**, 599-604.

Cole, D.W. and Rapp, M., 1981, In *Dynamic Properties of Forest Ecosystems* (ed. by D.E. Reichle). IBP Report 23, 341.

Cosby, B.J., Hornberger, G.M., Galloway, J.N. and Wright, R.F., 1985a, Modeling the effects of acid deposition: assessment of a lumped parameter model of soil water and streamwater chemistry. *Wat. Resour. Res.*, **21**, 51-63.

Cosby, B.J., Wright, R.F., Hornberger, G.M. and Galloway, J.N., 1985b, Modeling the effects of acid deposition: estimation of long-term water quality responses in a small forested catchment. *Wat. Resour. Res.*, **21**, 1591-1601.

Derwent, R.G., 1986, *The Nitrogen Budget for the United Kingdom and North-west Europe.* ETSU Report 37. ETSU, Harwell, Oxfordshire, U.K.

Derwent, R.G. and Nodop, K., 1986, Long-range transport and deposition of acidic nitrogen species in north-west Europe. *Nature*, **324**, 356-358.

Eliassen, A. and Saltbones, J., 1983, Modelling of long-range transport of sulphur over Europe: a two-year model run and some model experiments. *Atm. Environ.*, **17**, 1457-1473.

Falkengren-Grerup, U., 1987, Long-term changes in pH of forest soils in southern Sweden. *Environ. Pollut.*, **43**, 79-90.

FAO-UNESCO, 1974, *Soil Map of the World*. Vols. I and V, FAO-UNESCO, Paris.

Fjeld, B., 1976, *Forbruk av fossilt brensel i Europa og utslipp av SO_2 i perioden 1900-1972* (in Norwegian). NILU Teknisk Notat Nr 1/76, 6.

Hallbäcken, L. and Tamm, C.O., 1986, Changes in soil acidity from 1927 to 1982-1984 in a forest area of south-west Sweden. *Scand. J. For. Res.*, **1**, 219-232.

Helgeson, H.C., Murphy, W.M. and Aagaard, P., 1984, Thermodynamic and kinetic constraints on reaction rates among minerals and aqueous solutions II. Rate constants, effective surface area and the hydrolysis of feldspar. *Geochim. Cosmochim. Acta*, **48**, 2405-2432.

Helling, C.S., Chesters, G. and Corey, R.B., 1964, Contribution of organic matter and clay to soil cation exchange capacity as affected by the pH of the saturating solution. *Soil Sci. Am. J.*, **28**, 517-520.

Johnson, D.W., 1980, Site susceptibility to leaching by H_2SO_4 in acid rainfall. In *Effects of Acid Precipitation on Terrestrial Ecosystems* (ed. by T.C. Hutchinson and M. Havas). Plenum, New York.

Kämäri, J., Posch, M. and Kauppi, L., 1985, *A Model for Analysing Lake Water Acidification on a Regional Scale. Part I: Model Structure*. IIASA CP-85-48. Laxenburg, Austria.

Kauppi, P., Kämäri, J., Posch, M., Kauppi, L. and Matzner, E., 1986, Acidification of forest soils: model development and application for analysing impacts of acidic deposition in Europe. *Ecol. Modelling*, **33**, 231-253.

Klemedtson, L. and Svensson, B.H., 1988, Effects of acid deposition on denitrification and N_2O emission from forest soils. In *Critical Loads for Sulphur and Nitrogen* (ed. by J. Nilsson and P. Grennfelt). Miljørapport 1988:15, Nordic Council of Ministers, Copenhagen, Denmark. 343-362.

Lehmhaus, J., Saltbones, J. and Eliassen, A., 1986, *A Modified Sulphur Budget for Europe for 1980*. EMEP/MSC-W Report 1/86.

van Miegroet, H. and Cole, D.W., 1984, The impact of soil acidification leaching in a red alder ecosystem. *J. Environ. Qual.*, **13**, 586-590.

Müller, M.J., 1982, *Selected Climatic Data for a Global Set of Standard Stations for Vegetation Science.* Dr. W. Junk Publ., The Hague.

Nilsson, S.I., 1985, Why is Lake Gardsjön acid? An evaluation of processes contributing to soil and water acidification. In *Lake Gardsjön. An Acid Forest Lake and its Catchment* (ed. by F. Andersson and B. Olsson). *Ecol. Bulletins,* **37**, 311-318.

Odén, S., 1968, *The Acidification of Air and Precipitation and its Consequences in the Natural Environment.* Ecology Committee, Bulletin No. 1. Swedish National Science Research Council, Stockholm, Sweden.

Overrein, L.N., Seip, H.M. and Tollan, A., 1981, *Acid Precipitation - Effects on Forest and Fish.* Final Report of the SNSF Project 1972-1980, Oslo, Norway.

Reuss, J.O., Christophersen, N. and Seip, H.M., 1986, A critique of models for freshwater and soil acidification. *Water, Air and Soil Pollut.,* **30**, 909-930.

Rosén, K.O., 1982, *Supply, Loss and Distribution of Nutrients in Three Coniferous Forest Watersheds in Central Sweden.* Reports in Forest Ecology and Forest Soils **41**, Swedish University of Agricultural Sciences, Uppsala.

Tietema, A. and Verstraten, J.M., 1989, *The Nitrogen Budget of an Oak-Beech Forest Ecosystem in The Netherlands in Relation to Atmospheric Deposition.* Dutch Priority Programme on Acidification, Report 04-01.

Ulrich, B., 1981, Theoretische Betrachtung des Ionenkreislaufs in Waldökosystemen. z. Pflanzenernähr. *Bodenk.,* **144.**, 647-659.

Ulrich, B., 1983 Soil acidity and its relation to acid deposition. In *Effects of Accumulation of Air Pollutants in Forest Ecosystem* (ed. by B. Ulrich and J. Pankrath). D. Reidel Publishing Company, Dordrecht, 127-146.

Ulrich, B., Mayer, R. and Khanna, P.K., 1980, Chemical changes due to acid deposition in a loess-derived soil in central Europe. *Soil Science,* **130**, 193-199.

Ulrich, B. and Matzner, E., 1983, *Abiotische Folgewirkungen der weiträumigen Ausbreitung vol Luftverunreinigungen,* Umweltforschungsplan der Bundesminister des Inneren, Forschungsbericht 10402615, BRD.

UNESCO, 1972, *International Geological Map of Europe and the Mediterranean Region,* Bundesanstalt für Bodenforschung, Hannover, UNESCO Paris.

de Vries, W. and Breeuwsma, A., 1987, The relation between soil acidification and element cycling. *Water, Air and Soil Pollut.,* **35**, 293-310.

de Vries, W., 1990, Philosophy, structure and application methodology of a soil acidification model for The Netherlands. In *Impact Models to Assess Regional Acidification* (ed. by J. Kämäri). Kluwer Academic Publishers, Dordrecht.

de Vries, W. and Kros, J., 1989, The long-term impact of acid deposition on the aluminium chemistry of an acid forest soil. In *Regional Acidification Models: Geographic Extent and Time Development* (ed. by J. Kämäri, D.F. Brakke, A. Jenkins, S.A. Norton and R.F. Wright). Springer, New York, 113-128.

de Vries, W., Posch, M. and Kämäri, J., 1989, Simulation of the Long-term soil response to acid deposition in various buffer ranges. *Water, Air and Soil Pollut*, **48**, 349-390.

Weltforstatlas (World Forestry Atlas), 1975, Paul Parey, Hamburg and Berlin.

Chapter 16

HISTORICAL LAND USE CHANGES AND SOIL DEGRADATION ON THE RUSSIAN PLAIN

N.A. Karavayeva, T.G. Nefedova and V.O. Targulian

16.1. Introduction

The European portion of the USSR covers more than 500 million hectares, an area approximately half of the total European continent. A major part of European USSR comprises the Russian Plain which stretches from the Urals in the east to the Baltic in the west and the Black Sea, the Caucasus and the Caspian Sea in the South but excludes the northern regions falling within the Arctic Circle (Figure 16.1). This area has always been the economic and social center of the USSR and this is reflected by the development trends of industry, agriculture and transportation networks, as well as in population trends:

i) approximately 75 per cent of the gross national product originates from this region;

ii) about 70 per cent of the national population lives on the Russian Plain;

iii) the population density is approximately 32 people km^{-2}, almost three times the national average.

The central and western parts of the Plain have the longest history of land use (about 1,000 years) whilst human activities in the remaining portions have developed mainly over the last 100-150 years.

  This Chapter discusses land use changes over the last three hundred years, covering some of the major changes in socio-economic, demographic and political conditions. Changes in soil quality in response to shifting land use patterns will also be discussed.

F. M. Brouwer et al. (eds.), Land Use Changes in Europe, 351–377.

Figure 16.1. The Russian Plain.

16.2. **Socio-economic and Political Factors Determining Land Use Changes**

Economic conditions which have determined land use patterns include
i) the level of regional development such as the growth and spatial
distribution of transportation and communication networks and other
infrastructural facilities; ii) the regional economic structure in terms of
the use of land for activities such as agriculture, forestry, mining and
industry; iii) the efficiency of agriculture.

Important social conditions include i) demographic characteristics
such as population density and age distribution; ii) urbanization trends,
for example, the emergence and development of cities. Factors such as

the political system or prolonged wars have also affected land use patterns on the Russian Plain (Emeljanov, 1987).

More than 90 per cent of the Russian Plain is presently used as agricultural land and for forest production, with only small areas being used for other purposes such as industry, urban centers and transportation networks. The increasing demand for timber and food accordant with the population increase has directly affected the nature of the land use and the processes employed. The increased demand has necessitated the introduction of new technologies with an emphasis on chemical-oriented activities in agriculture and on lumbering efficiency in forestry (Alayev *et al.*, 1990).

The major socio-economic changes which have led to the transformation of land use patterns can be linked to three different and consecutive political systems:

i)   *FEUDALISM* until the decay of serfdom in 1861;

ii)   *CAPITALISM* between 1861 and the Russian Revolution of 1917;

iii)   *SOCIALISM* post 1917 (the socialist period is divided into the pre-World War II and post-World War II period).

16.2.1. The feudal period (until 1861)

The total population of the Russian Plain was approximately 17 million during the first part of the eighteenth century. Moscow was the main urban center. The taiga zone in the northern part of the country was only locally developed and the steppe zone in the south was gradually colonized as the threat of nomad invasions receded. Some of the major characteristics of the socio-economic conditions and population trends of the feudal period can be summarized as follows:

a)   by 1700 about 4 per cent of the total population lived in urban areas (600,000 people in 253 cities);

b)   by the end of the eighteenth century, there were about 30 branches of the metallurgical and chemical industry, although industrial development was still slow;

c)   cartage and waterways were the major means of transportation; the first main railroad between Petrograd (Leningrad) and Moscow was not constructed until the middle of the nineteenth century.

Agriculture was the most important economic activity during the feudal period, and this remained so until the 1930s. Feudal relations inhibited agricultural development; there was a lack of interest in achieving higher yields. Table 16.1 shows some trends in the yields of cereals during the feudal period in three zones of the Russian Plain (Zonn and Chikishev, 1976).

Table 16.1. Yields of cereals on three zones of the Russian Plain between 1700 and 1850 (in t ha^{-1}).

| Zone | Period | | |
	1700-1750	1750-1800	1800-1850
Northern (taiga)	0.40	0.49	0.43
Mixed forests	0.44	0.44	0.41
Steppe and forest steppe	0.51	0.56	0.45

The yields were low and even showed a decreasing trend during the first half of the nineteenth century, consequently the increasing food requirements of the growing population could only be met by increasing the area of cultivated land. The increase in the utilization of arable land was mainly achieved by the cultivation of forested areas and other types of natural vegetation. Soil degradation on agricultural land was not widespread and organic fertilizers were widely applied. The major environmental stress observed during this period was depletion of the forest sector since wood was used for fuel, the construction of houses and as a raw material for industry.

16.2.2. The capitalist period (1861-1917)

This period was marked by structural changes in society. Demographic trends altered with industrial development and urbanization. The urban population made up 6 per cent of the total population in 1861 compared with 18 per cent of the total in 1913.

Heavy industry was becoming more important in society. Table 16.2 illustrates the rise in industrial production between 1860 and 1913. During the first decades of the twentieth century Russia was the fifth largest industrial center in the world and the fourth largest in Europe. The rising demands of the growing population and the growing

Table 16.2.   Trends in industrial production between 1860 and 1913 (production in million tons).

Product	1860	1900	1913
Coal	(n.a.)	16.9	33.3
Iron	0.3	2.9	4.6
Steel	0.2	2.2	4.2

industrial sector could only be met by extending the transportation networks.   The length of railroads increased by 1,500 km per year between 1860 and 1870 and by 2,500 km per year after 1890.   The total railroad network was approximately 65,000 km by 1913.   Water transportation facilities were similarly improved during this period.

Agricultural practices were still dominated by extensive farming. The average cereal yield was approximately 0.8 t ha^{-1}, while countries such as Germany, Belgium and The Netherlands had already achieved yields about three times higher.   The difference was mainly due to technological conditions, the availability of machinery and of fertilizers. The western and south-western part of the Russian Plain, that is, Byelorussia and part of the Baltic region, were the most productive. Soil degradation was observed on about 70 per cent of farms, especially in the south-eastern regions.   This was mainly due to overexploitation of the Chernozem soils.   Soil depletion resulted in considerable yield decreases towards the end of the nineteenth century (Dulov, 1983; Alayev et al., 1990).

Major changes were made in the forest sector; 28 million hectares of forest were cut during this period and the land subsequently turned over to agriculture.   Timber exports increased dramatically and wood was used to fuel the ever increasing number of steam engines (Tsvetkov, 1957).

16.2.3. **The socialist period (since 1917)**

Some of the important socio-economic and demographic changes that have taken place since 1917 are reflected by the following trends:

i)   the total population of the Russian Plain increased from 17 million in the eighteenth century to 126 million in 1913 to a level of about 161 million in the 1980s; the urban population has increased by more than 250 per cent over the past 70 years and today represents 68 per cent of the total population;

ii)   the rural population has decreased correspondingly both in relative
        and absolute terms;

iii)   industrial production has increased by a factor of 70 since 1913
        and now the European part of the USSR produces about half of
        the raw material required for domestic energy production, and is
        responsible for at least 60 per cent of the total national energy
        production; twenty out of twenty three of the USSR's nuclear
        power stations are located on the Russian Plain;

iv)   the railroad network now comprises 100,000 km and new means
        of transport have emerged, including roads, airlines and pipelines;
        concurrently transportation is becoming increasingly associated
        with land, water and air quality deterioration.

Agricultural   productivity   slumped   during   the   period   of
collectivization following the Russian Revolution in 1917 (the 1913 cereal
yields were not matched again until 1940). The Second World War also
greatly affected socio-economic conditions in the country.   The
population had recovered to pre-war levels by 1955. The area of utilized
agricultural land took slightly less time to recover and reached pre-war
dimensions in 1950.   Between 1947 and 1951, average cereal yields
were 0.6 t ha^{-1} in the central part of the Russian Plain, 0.7 t ha^{-1} in the
central Chernozem region and 0.5 t ha^{-1} in the Volga region.   The
development of the agricultural sector after the Second World War can
be divided into two main periods (Agricultural Atlas of the USSR, 1960;
Zonn and Chikishev, 1976).

i)   pre-1960: the period before 1960 showed an increase in the
        utilization of agricultural land; land productivity was improved
        only on a limited scale through the application of chemical
        fertilizers;

ii)   post-1960: the period after 1960 showed a trend towards
        intensification of agriculture, with large improvements in land
        productivity due to an increasing use of mineral fertilizers and
        machinery; the average yield of cereals in 1986 was about 1.6 t
        ha^{-1}, and the productivity of livestock reached a level of about
        2.5 t yr^{-1}; however, the increasing application of chemical fertilizers
        and the construction of large cattle breeding farms is accelerating
        environmental deterioration.

There are considerable regional differences in agricultural productivity on the Russian Plain both in space and time. Before the 1960s,
land productivity was primarily determined by biophysical conditions and
the highest yields (1.5 t ha^{-1}) were obtained in the Chernozem regions of

the Ukraine and the northern Caucasus, compared with an average level of between 0.5 and 1.0 t ha^{-1} in Byelorussia and the Baltic republics. Over the last twenty years, with the introduction of fertilizers and improved management yield increases of about 1.0-1.5 t ha^{-1} have been observed in the Baltic republics, Byelorussia, western Ukraine, as well as in the areas around Moscow and Leningrad. The actual yields in these areas are now approaching their maximum potential as defined by natural soil and climatic limitations (Figure 16.2). A considerable increase in land productivity has also been achieved in the eastern part of the Ukraine and the northern Caucasus region.

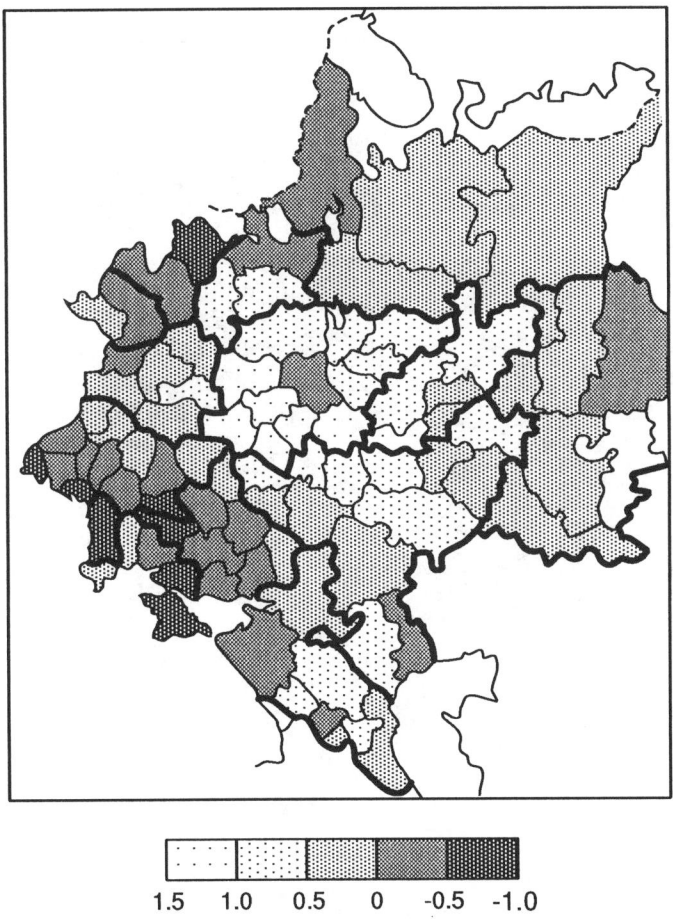

1.5   1.0   0.5   0   -0.5   -1.0

Figure 16.2. Difference between potential and actual yield levels (averages for the 1981-1984 period) (in t ha^{-1}).

Figures 16.3 and 16.4 show present yields of the major cereal
crops and meadow. The highest yields (more than 2.0 t ha⁻¹) are
observed in the region stretching from Karelia in the north-western part
of the Russian Plain, including the Leningrad region, the Baltic republics,
Byelorussia and the Ukraine across to Krasnodar in the northern
Caucasus region. The Moscow region exhibits similarly high yields, as a
result of the local socio-economic conditions and the advanced level of
farming coupled with the high level of investment to promote improved
agricultural practices and infrastructural facilities. The productivity of
meadows (Figure 16.4) shows a similar regional pattern with highest
yields attained in the north-western part of the Russian Plain.

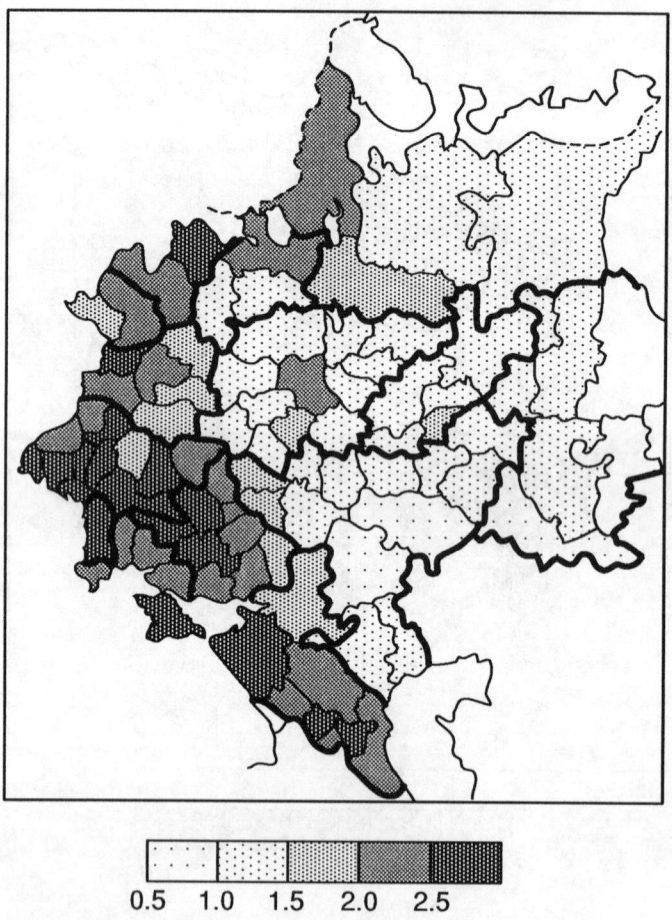

Figure 16.3. Annual average yield of cereals during the period 1981-
1984 (in t ha⁻¹).

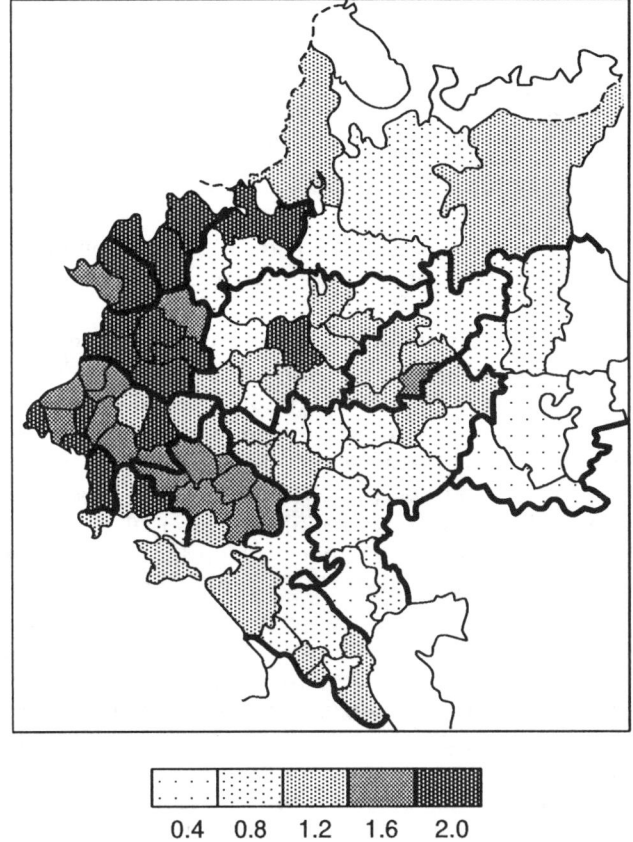

Figure 16.4.   Annual average yield of meadows (in t ha⁻¹).

The peripheries in the southern, eastern and south-eastern parts of the Russian Plain have not been able to achieve a substantial increase in land productivity over the last twenty years and consequently a migration from rural to urban areas has been observed. Most of the regions outside the Chernozem area are affected by soil degradation due to the increased use of chemical fertilizers and heavy machinery. The central Chernozem and Volga regions have been extensively farmed for a long period of time and therefore also experience soil degradation problems. Such problems are likely to be the cause of differences between the potential and actual yields observed in the central part of the Russian Plain. The increasing rates of soil degradation in relation to changing land use patterns in different parts of the Russian Plain will be discussed in Section 16.4 (Dolrovolsky and Levin, 1978; Emeljanov, 1987).

16.3. Land Use Changes since 1700

16.3.1. Introduction

Two major land use trends can be recognised for the Russian Plain over the past three hundred years (Zonn and Chikishev, 1976):

i)   an increase in the use of land for agricultural purposes;

ii)   an increase in the use of land for the development of industry, urban centers and transportation networks.

Figure 16.5 presents the relative changes in land use for four major vegetation zones. The major land uses by the end of the nineteenth century were forestry in the northern and broad-leaved forest zones, arable in the forest-steppe zone and meadow and pasture in the steppe zone. Forestry remained the major land use category throughout, although by 1887 10 per cent of the total land resource had been transformed to arable (Tsvetkov, 1975; Alayev *et al.*, 1990).

Table 16.3 shows the changes in land use patterns for the Russian plain between 1887 and 1966: arable land more than doubled in less than a century; this increase was accompanied by considerable increases in the area of meadow and pasture. At the end of the nineteenth century forestry was the largest land use category, occupying more than 40 per cent of the European USSR. This percentage fell to approximately 30 per cent by 1966. During this time much of the remaining land resource was taken into cultivation (Dulov, 1983).

During the first part of the twentieth century the area of tilled land decreased on the central part of the Russian Plain. This was primarily because of the abandonment of land as farmers moved to cities or to more productive areas in the south. In general, however, the area of tilled land showed an overall increase on the Plain until the 1960s and then stabilized at the present level of approximately 210 million hectares. Details of the land use pattern in 1966 are presented in Table 16.4.

The nature of rural areas has changed considerably in the recent past with various parts of the Plain experiencing different trends and different rates of change. Land has been transformed from agricultural land to alternative uses particularly in the northern part of the Plain. The transformation of such land was prompted by the perpetually low rates of return from the agricultural sector, forcing the rural population to migrate. Natural pasture was abandoned and even arable land was left fallow in some cases. The nature of the rural environment has also changed due to increased pressures from recreation and from the urban population infiltrating in this area. However, the poor quality of the transport infrastructure continues to be a major obstacle to rural development.

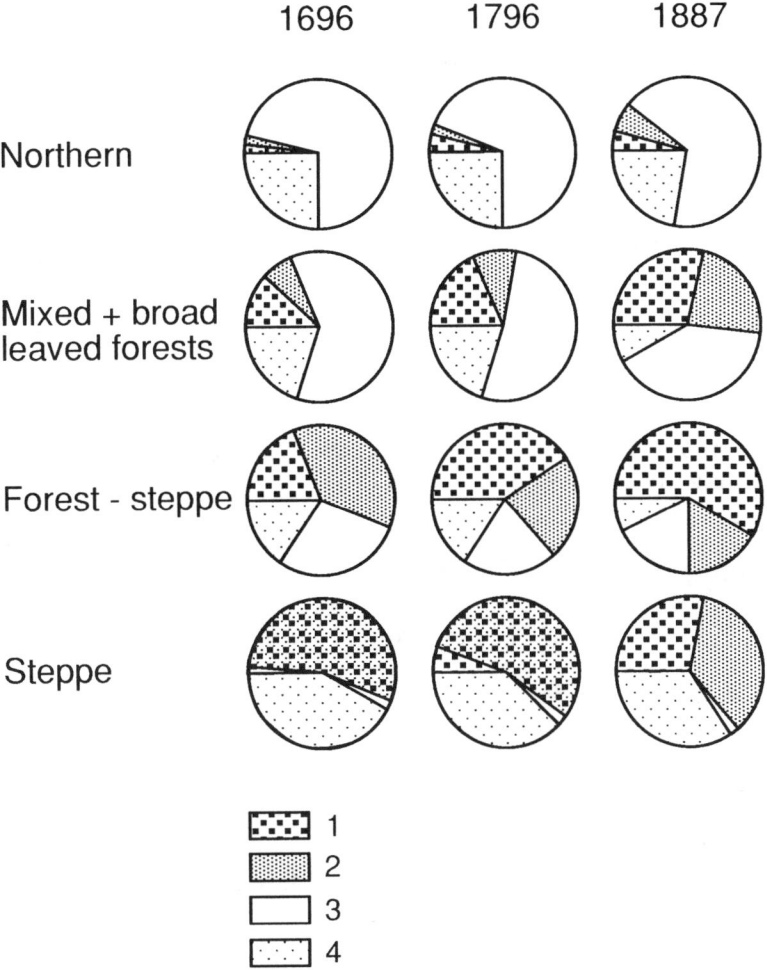

Figure 16.5. Land use patterns between 1696 and 1887 in four major vegetation zones of the Russian Plain; 1 = arable land, 2 = meadows and pasture, 3 = forests, 4 = other.

The distribution of forests (Table 16.5) has shown a steady decrease over the past three hundred years, from about 52 per cent at the end of the seventeenth century to approximately 32 per cent in 1966. About 15-30 per cent of tundra forests have been transformed as well as 30-50 per cent of the taiga forest (Tsvetkov, 1957; Alayev *et al.*, 1990). Between 40-60 per cent of mixed and broad-leaved forests have been affected along with 60-90 per cent of the steppe and 50-70 per cent of the semi-desert area. In addition, forests in the vicinity of

Table 16.3.   A comparison of land use distribution in European USSR
in 1887 and 1966 (percentage of total area).

Land use category	1887	1966
Arable land	20.0	41.4
Meadows and pasture	14.6	22.0
Forests	42.3	30.2
Other land	23.1	6.4

Table 16.4.   Land use distribution on the Russian Plain in 1966
(percentage of total area).

Agricultural land			56.8
	arable	39.4	
	meadow	5.7	
	pasture	11.7	
Forests			28.2
Settlements and roads			3.3
Water bodies (rivers)			3.4
Man-made water bodies			1.3
Other			7.0*

(*including bog, sand, scrub and eroded land).

rivers were destroyed in the 1950s and 1960s to make way for
engineering projects.
        The forested area in the southern and western parts of the region
have recently begun to show an increasing trend primarily because of
the reforestation of abandoned agricultural land. These areas are being
developed by the State Forest Fund. Forests, in general, remain under
stress from exploitation, industrial activity and recreation. Figure 16.6a
and Figure 16.6b show the areas under threat in the 1980s.
        Approximately 5 per cent of the land resource is, today, being
used for industrial activity, settlements and transportation.   Some
400,000 ha have been made available for such purposes in recent
decades; the transformation has been particularly noticeable in the
southern region of the Russian Plain. Land least suitable for agriculture
was earmarked for this conversion. As industrial activity becomes more
intensive environmental degradation will become more widespread and
consequently more land is likely to be classed unsuitable for agriculture.

Table 16.5.   Area covered by forest on the Russian Plain in several natural zones in 1966 (million ha). The percentage of each zone that is covered with forests and the percentage of the total forest coverage within each zone are given in brackets. Tundra also includes part of the northern taiga.

Natural (sub) zones	Total forested area in zone		Total area of zone	
	(mln ha)	(%)	(mln ha)	(%)
Tundra	25.9	(38.9)	66.6	(14.4)
Central taiga	52.3	(59.0)	88.6	(29.0)
Southern taiga	31.7	(55.8)	56.8	(17.6)
Mixed forests	31.4	(36.4)	86.3	(17.5)
Broad-leaved forests	16.1	(28.5)	56.4	(8.9)
Forest-steppe	8.1	(13.5)	60.1	(4.5)
Steppe	3.6	(4.9)	72.5	(2.0)
Semi-desert	0.1	(0.8)	12.2	-
Mountain regions	11.2	(20.4)	54.9	(6.1)
Total	180.4	(32.5)	554.4	(100.0)

16.3.2. **Farming systems on the Russian Plain**

Various farming systems are present on the Russian Plain. Each alters the landscape and soil conditions in a different way. Farming practices have developed through three basic stages over the last three hundred years.

1) The *primitive system*. This system was employed in the forest and steppe zones where cut-and-burn and forest-field methods prevailed. A piece of land was farmed for a period of ten years with shallow plowing (10-14cm) and without fertilizer. The land was cultivated intensively and then abandoned when yields began to fall. Farmers simply moved to new land leaving the abandoned land to recover fertility. Shallow plowing played an important role in assisting nutrient cycling and helped to sustain fertility to some extent.

2) *Fallow farming*. An alternating system of crop and fallow. The introduction of a fallow period allowed a three- to fourfold increase in the amount of land that could be farmed at one time. However, this system is prone to soil degradation especially if the fallow period is shortened because of economic pressure.

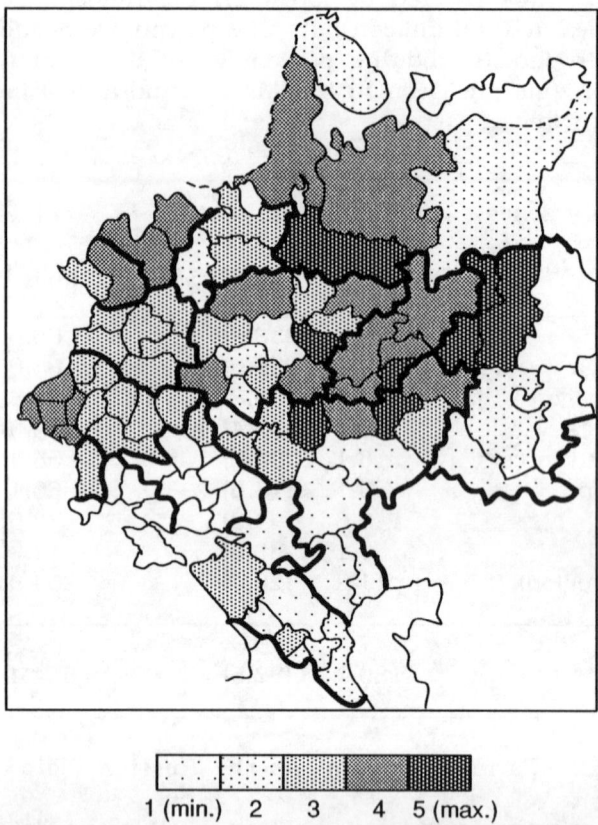

Figure 16.6a. The extent to which forests are affected by exploitation for industrial activities. (1 = minor influence; 5 = severe stress.)

3) *Grass-arable farming.* This is the present system based on crop rotation. A typical example is a crop rotation system of 5-9 years (with crop tillage rotation or grain-arable rotations). This is the most intensive of the three systems requiring deep plowing (up to 35 cm) and large, frequent fertilizer applications. Poor land management and fertilizer shortages may result in rapid soil degradation. Crop rotation systems incorporating grass crops are least damaging since such crops do not significantly alter the nutrient status of the soils and provide good ground cover. Rotations that comprise solely tilled crops are most likely to instigate soil degradation.

The fallow system was common around 1700 and examples could still be observed up until the 1930s in combination with primitive farming systems. At the end of the nineteenth century and the beginning of the twentieth century all three of the systems described

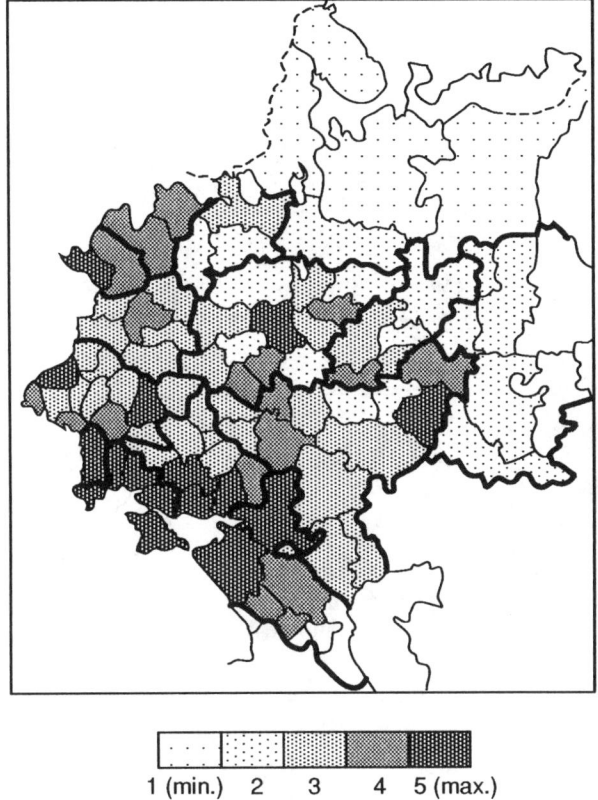

1 (min.) 2 3 4 5 (max.)

Figure 16.6b. The extent to which forests are affected by recreation (long-and short-term recreation).

above were in operation on the Russian Plain. Changes in farming systems have been initiated by advances in technology particularly in the fields of fertilizers, drainage and irrigation. These advances will be explored in the next Section.

16.3.3. The advancement of fertilizer use, drainage systems and irrigation in agriculture

Fertilizers have been applied, in one form or another, since the eleventh century. Manure and ash were used originally and peat and compost were introduced on a small scale in the eighteenth century. Mineral fertilizers have only been used since the 1950s and 1960s (Figure 16.7). Presently as much as 250-300 kg ha^{-1} fertilizer is applied in the Baltic Republics and Byelorussia and as little as 30-50 kg ha^{-1} is applied in the Volga region. Thus the spatial economic discrepancies are reflected

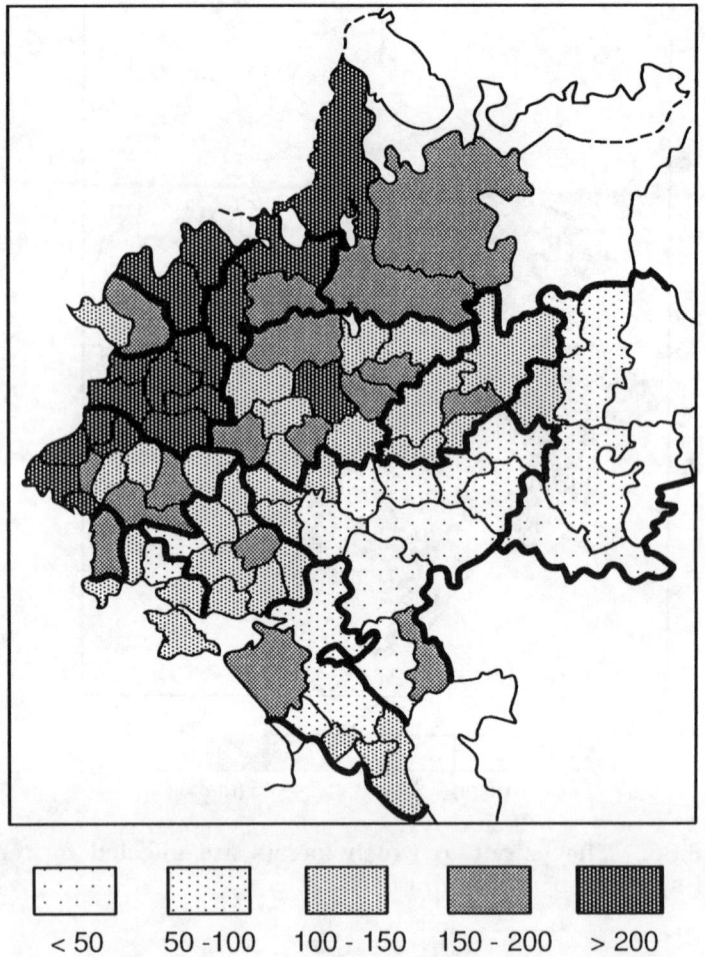

< 50   50 -100   100 - 150   150 - 200   > 200

Figure 16.7. Application of mineral fertilizers in 1985 (kg ha⁻¹).

in agricultural development on the Russian Plain. Major problems of
soil degradation are observed on the Chernozem soils of the eastern and
south-eastern part of the Plain as a result of intensive use without
nutrient replenishment in the form of fertilizer (Alayev *et al*, 1990).

The practice of drainage was implemented even before 1700, but
its purpose was mainly to accomodate urban runoff. At this time
drainage in agriculture was only applied on a local scale. Land had not
been recognized as a finite resource; there was still plenty of land
available for cultivation. Those agricultural areas that were drained
were generally used to grow rye. At a later stage drainage was used to
improve pasture and hay fields (Tables 16.6 and 16.7).

Table 16.6. Development of drainage and the major objectives of land drainage on the Russian Plain until 1917 (areas in ha).

Period/region	Area	Objective of drainage
Eighteenth century		
Leningrad region	2,400	urban areas, pasture, hay fields
Forest lowland around Moscow	(unknown)	Urban areas, pasture, hay fields, water transport
1800 -1861		
Baltic region	(unknown)	Pasture, arable land, hay fields
Western regions	5,000	Improvement of forests and hay fields
Central regions	85,000	Improvement of forests and hay fields
1861-1917		
Russian regions	891,000	Improvement of forests, pasture, hay fields, agriculture and urban areas
Byelorussia	295,000	Peat extraction, hay fields and improvement of forests
Ukraine	44,000	Hay fields and pasture
Baltic region	64,000	Agriculture, pasture and hay fields

Land drainage of farmland became increasingly important during the nineteenth century when swamps were drained for use as pasture, for cereal cultivation and for peat mining. Tile drainage was introduced in the middle of the nineteenth century. Post 1917 improvements in drainage practice were mainly aimed at regulating river runoff, however, this required deep drains to depths of up to 4m. Drainage for agricultural purposes was widely implemented after the 1960s and wetlands are still being drained today (Figure 16.8).

Between 25 per cent and 50 per cent of cultivated land in the western part of the Russian Plain has been reclaimed. This mainly includes the area around Leningrad and the Baltic region. Large areas of land beyond the Chernozem region in the south have been reclaimed by drainage since the 1970s (Alayev *et al.*, 1990).

Table 16.7. Development of soil drainage on the Russian Plain between 1913 and 1980 (in 1,000 ha).

Region	1913	1921-23	1928	1932	1967	1980
Russian Federation	800	420	1,430	2,000	1,640	3,860
Byelorussia	290	110	290	380	290,000	6,400,000
Ukraine	44	20	73	160	537	681
Lithuania	64	-	-	-	970	2,180

Irrigation has played an important part in the agricultural development of the Russian Plain. Before the nineteenth century, irrigation was only used locally in the southern part of the Plain. The average field size of irrigated land was around 2 to 4 ha. The amount of land irrigated showed a gradual increase during the nineteenth century. Irrigation was particularly concentrated on the south-eastern region of the Plain and later spread to the west. The size of irrigated fields was beginning to increase by the early part of the twentieth century. Irrigation originally involved flooding or basin techniques; since the eighteenth century methods have gradually improved through furrow irrigation to gravity fed systems. The development of irrigation on the Russian Plain is detailed in Tables 16.8 and 16.9.

16.4. Soil Degradation and Changes in Land Use Patterns

The major changes in the quality and quantity of the land resource have been associated with land transformation in agriculture and forestry. The following have been the main changes in soil conditions:

i)   changes in the biological cycle of soil nutrients; the extraction of nutrients from the soil exceeding input;

ii)   the application of fertilizers and soil amendments causing changes in the composition and ratio of the soil nutrients;

iii)   stratification of soils, primarily brought about by the use of heavy machinery;

iv)   changes in the hydrological and thermal regimes of the soils, that is the microclimate of soils; land reclamation causes a major transformation of the hydrological regime followed by transformation of the thermal regime.

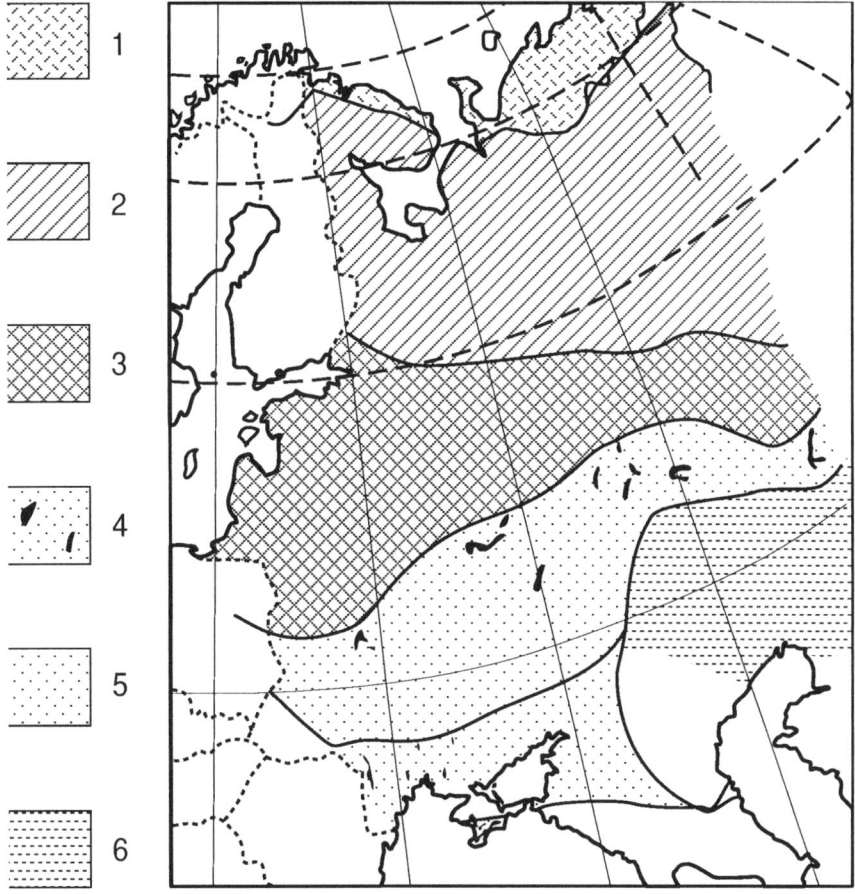

1 = tundra and boggy pasture;
2 = taiga (coniferous young forests and maturing and
      small-leaved secondary forests;
3 = forest-meadows (mixture of coniferous, coniferous
      broad-leaved and small-leaved forests, meadows
      and plowed land);
4 = combination of almost completely plowed land
      and gully used for pasture; predominantly dry
      agriculture;
5 = predominantly irrigated agriculture;
6 = semi-desert and pasture, development of irrigated
      agriculture in some parts.

Figure 16.8.   Major biomes on the Russian Plain.

Table 16.8. Development of irrigation on the Russian Plain between the eighteenth century and 1917 (in 1,000 ha).

Region	Size	Irrigation method
Eighteenth century		
Volga delta	fields of 2-3 ha	Flooding, basin irrigation
Nineteenth century		
South-eastern part	1.8	Basin irrigation
1861-1917		
Western coast of		
Caspian sea	160	Flooding, furrow irrigation
Volga and Ukraine	240	Flooding and gravity irrigation

Table 16.9. Development of irrigation on the Russian Plain after the 1960s (in 1,000 ha).

Region	1960	1970	1980
Russian Federation	1,600	5,600	5,000
Ukraine	700	600	600
Volga region	300	300	1,300

Figure 16.9 shows the major differences in biological cycles between natural vegetation (forest and meadow) and cultivated crops (rye, barley, potatos and Winter wheat). The Figure shows that the *total living phytomass* (t ha^{-1}) is the major component of the biological cycle of forests and the *volume of the biological cycle* (t ha^{-1} yr^{-1}) is the major component for meadows. All components of the biological cycle for cultivated crops are much diminished in comparison with natural vegetation.

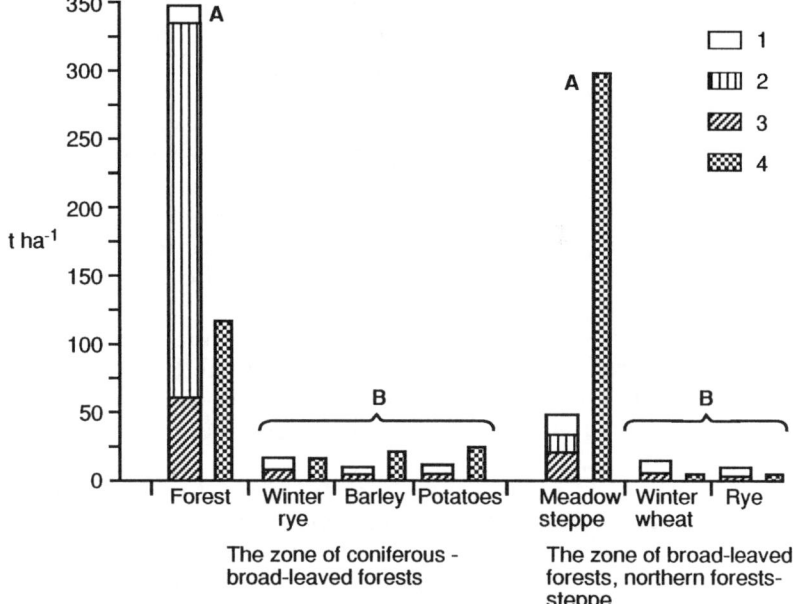

Figure 16.9. Some components of the biological cycle for natural vegetation and cultivated crops in the broad-leaved forest zones with coniferous forests and in the northern forest-steppe zone. 1 = increment of phytomass (t ha^{-1} yr^{-1}). 2 = total living phytomass (t ha^{-1}). 3 = total dead phytomass (t ha^{-1}). 4 = volume of biological cycle (t ha^{-1} yr^{-1}).

The increase of arable land over the past three hundred years has been concentrated in three zones on the Russian Plain:

i)   the taiga-forest zone on sod-podzolic soils (Eutric Podzoluvisols);

ii)   the broad-leaved forest and forest-steppe zone with gray forest soils (Albic and Orthic Luvisols);

iii)   the steppe zone with Chernozem soils.

16.4.1. Changes in the sod-podzolic soils of the taiga forest zone

Land management and fertilizer applications largely determine any changes in the chemical and physical characteristics of Podzols. The humus content may decrease considerably under poor management and nutrient deficiencies. Such changes increase the potential for soil erosion. There is a trend of increasing podzolization (or acid eluviation). The podzolization associated with the cultivation of land is a much more rapid process than the development of podzols under natural conditions: the more rapid the process the more destructive for the soil mass (Galeeva and Kurok, 1986; Dolrovolsky and Levin, 1978).

Cultivated Podzols are less prone to erosion than degraded land but more so than natural soils. Soil degradation of podzolic soils is more severe on arable land than on pasture and hay fields. Soil transformations mainly occur in the top horizon because of the changes in the biological cycle and due to compaction. Proper land management should be based on regular shallow plowing, moderate application of fertilizers and planting of grass to improve the floral composition and to control the biological cycle.

Improvements of the taiga forest zone by drainage are evident on peaty and peaty podzolic Gleysols and on peat land with excess soil moisture. Podzolic Gleysols resemble sod-podzolic soils after drainage, exhibiting similar trends under cultivation over time.

16.4.2. Changes in gray forest soils of the broad-leaved forests and forest-steppe zone

Tillage practices affect the basic properties of gray forest soils. The induced modifications are similar to those found on sod-podzolic soils. Differences between the two soil types are affected by:

i)   a moisturization coefficient of around 1 on gray forest soils may potentially accelerate the development of steppe after a period of deforestation;

ii)   extensive exploitation of plowed land may cause the development of xerophytic and steppe conditions; such processes will primarily affect the biological cycle and humus status of the soils; the conversion of natural vegetation to cereal crops can decrease the organic matter input four- to six-fold; tillage of land intensifies the microbiological decomposition of organic matter and can increase the salt content, resulting in the removal of humus, nitrogen and absorbed calcium from the plowing horizon; changes in the humus content and in the available humus reserves observed on gray forest soils under arable crops over the past decades are presented in Table 16.10.

Humus decline in the gray forest zone shows an increasing trend when moving from north to south because of the higher salt content of soils and probably because of the higher intensity of land exploitation in the south. Humus losses aggravate physical soil properties and may intensify soil erosion processes.

Table 16.10. Changes in the humus content and the humus reserves on gray forest soils (Luvisols) and Chernozem soils used for arable land use in the Tatar Soviet Socialist Republic between the 1940s and 1980. (Humus content in percentages and humus reserve in t ha^{-1} of a 0-20 cm soil layer.)

Soil type	Subcategory	Humus content			Humus reserves	
		1940s	1950s	1970-80	1940s	1970-80
Gray forest	light-gray	3-4	3.4	3.0	91	77
	gray	4-6	4.9	4.5	130	94
	dark-gray	5.5-6.5	6.3	5.5	144	116
Chernozem	leached	7-10	8.4	7.6	204	154
	typical	9-14	7.9	7.2	276	158

16.4.3. Changes in the Chernozem soils of the steppe zone

Plowing of the Chernozem soils began during the nineteenth century, transforming forest-steppe to steppe Chernozems. Between 82 and 94 per cent of all arable land is presently on Chernozem soils. Continued extensive use has resulted in major changes in both humus content and reserves over the last century. An assessment of the status of natural and cultivated soils in the 1880s was conducted by Dokuchayev. Table 16.11 illustrates the changes in humus content and reserves observed over the last century (Kovda and Samoilova, 1983). The largest decreases of humus content are observed in the eastern part of the Russian Plain with decreases in the Volga and Ural regions from 13-16 per cent to 7-10 per cent and even down to 4-7 per cent.

Chernozem soils are largely affected by irrigation since the groundwater level is likely to rise and the soils become increasingly moist. Irrigation alters the hydrological regime of Chernozem soils. The salinization potential of soils can also be increased under irrigation (Chapter 13). Irrigation-induced secondary salinization of Chernozem soils is primarily a local although common phenomenon. As long as the water table remains at a depth of more than 5-6 meters then the risk of secondary salinization is minimal. However, even in the absence of salinization, undesirable changes in the chemical content of the soil can occur, for example calcium may be replaced by sodium, such changes may result in alkalization.

Chernozems are drastically transformed by rice cultivation. Inundation changes the redox regime and the chemical properties of the soils. In addition, the inundation may exacerbate compaction problems

Table 16.11. Humus content and humus reserves of the arable soil layer (0-30 cm) of the Chernozem soils on the Russian Plain during 1881 and 1981. (Humus content in percentage, humus reserves in $t\ ha^{-1}$, and decrease in humus reserve as percentage.)

Subcategory of Chernozems	Region	Humus content 1881	Humus content 1981	Humus reserve 1881	Humus reserve 1981	decrease
Forest-steppe Chernozem						
Typical Chernozem	Central part	10-13	7-10	300-330	210-300	23-30
Typical Chernozem	South-eastern part	13-16	7-10	390-480	240-300	38-39
Leached Chernozem	Caucasus	7-10	4-7	221-315	150-263	20-34
Leached Chernozem	Volga region	13-16	4-7	390-480	120-210	56-69
Steppe Chernozem						
Typical Chernozem	Central part	7-10	4-7	221-315	150-263	17-32
Typical Chernozem	Moldavia	4-7	2-4	126-221	75-150	32-40
Typical Chernozem	South-east	9-11	6-8	270-330	180-240	27-33

and may induce humus loss.

To summarize the Chernozem soils have exhibited two development trends over the last century.   Under dry farming, without irrigation, Chernozems might become degraded due to poor management and insufficient fertilizer applications.   Bad farming practice can cause a decrease in humus content, soil acidification and the deterioration of physical properties.   Under irrigation Chernozems follow two basic patterns i) soils become increasingly prone to salinization, alkalization and compaction, or ii) the soils are converted to a degraded Gleysol, eluvial Solonetz soil as a result of inundation for rice cultivation.

16.4.4. Soil erosion on the Russian Plain

Soil erosion was first observed around 1700 in areas covered with broad-leaved forests and also in the northern forest-steppe, where it resulted in decreased yields and the siltation of small rivers and reservoirs.

About 1,200 t ha^{-1} of top soil has been lost over the past one hundred years due to erosion of sod-podzolic soils.   During that period substantial losses of humus, potassium, nitrogen and phosphorus were also incurred.   Erosion of arable land in the taiga forests is observed on between 2.5 per cent and 14 per cent of the cultivated area (Zonn and Chikishev, 1976).

In the areas with broad-leaved forests and in the northern forest-steppe zone, erosion has resulted in major changes in relief.   Gullies have grown irrespective of measures intended to contain and control the erosion.   Plowing during the autumn has slowed the erosion process to some extent by decreasing spring runoff.   Up to 60 per cent of the arable land is affected by erosion in some parts of these zones.

Plowing arable land in the steppe zone has intensified the gullying process in some cases.   In addition, wind erosion has become prevalent in this zone, a new denudation process which is not inherent to the natural steppe.   Eroded land is a common sight in the steppe zone.   It is observed on approximately 25 per cent of the Chernozem soils in the Ukraine; within this category 18 per cent is classed as having slight-severe erosion, 5 per cent as moderately-severe and 2 per cent as very severe.

The annual rate of humus formation is between 0.2 and 0.3 mm yr^{-1} for sod-podzolic and steppe Chernozem soils and between 0.35 and 0.45 mm yr^{-1} for gray forest and forest-steppe Chernozem soils (Table 16.10).   Long-term land productivity is endangered since the annual erosion rates exceed the rate of humus formation.

16.5. Summary and Conclusions

Major changes in land use patterns on the Russian Plain have taken place over the past three hundred years.   Such changes were initiated mainly in the central and western part of the Plain and later shifted to the south, south-east, east and north of the Plain. The transformations in land use have taken place in response to changes in forestry and agricultural practices, while additional major impacts on the land resource are due to changes in industrial processes. The two major trends in land use patterns include the increase of cultivated areas and the increase of land used for industry, settlement and transportation. The taiga zone has been least affected by these trends, while the coverage of arable land increased considerably in the forest-steppe and steppe zones. This was mainly due to the transformation of forests and pasture land to land for arable purposes.

The western, north-western and suburban regions of the Russian Plain are characterized by major economic progress but with serious deleterious environmental consequences. The extensive cultivation of the central Chernozem, the eastern and south-eastern regions has caused severe degradation of forests and farmland. A decrease of organic material and more severe and more frequent events of water and wind erosion have resulted in a reduction of land productivity in the forest and forest-steppe zones. Chernozem soils on steppe land are irreversibly transformed by irrigation practice into soils with reduced land productivity.

A series of issues and policy objectives are becoming increasingly important for land use planning on the Russian Plain; these include:

i)   the protection and conservation of forests;

ii)   the control and reduction of the rate of soil erosion;

iii)   the reclamation of land;

iv)   the conservation and the potential increase of natural areas suitable for recreation.

The history of land use patterns in the Russian Plain clearly illustrates that societal responses to negative environmental impacts of farming practices have been short-lived and only temporarily satisfactory. The present land use policies need to incorporate a long-term perspective in order to cope with the global scale of present environmental issues.

REFERENCES

This summary paper is based on a large number of historical documents; only a small number of the more accessible references are cited.

Alayev, E.B., Badenkov, Yu.P. and Karavayeva, N.A., 1990, *The Russian Plain.* Regional Review (in English) (in press).

Agricultural Atlas of the USSR, 1960, The Main Geodesic and Cartographic Administration (GUGK), Moscow, Volume 8, 309.

Dolrovolsky, G.V. and Levin, F.I. (eds), 1978, *Rational Land Use of Soils in the Non-Chernozem Zone of the RSFSR* (in Russian). Moscow State University Printing House, Moscow.

Dulov, A.V., 1983, *Geographical Environment and History of Russia. End of XV - Middle of XIX Centuries.* Publishing House "Nauka", Siberia Department. Institute of History, Philology and Philosophy, Moscow.

Emeljanov, A.G. (editor-in-chief), 1987, *Geographical Aspects of Rational Land Use* (in Russian). Transactions of Kalinin State University, Kalinin.

Galeeva, A.M. and Kurok, M.L. (eds), 1986, On *Environmental Protection: Collection of Documents Issued by CPSU and the Government 1917-1985* (in Russian). Political Literature, Moscow.

Kovda, V.A. and Samoilova, H.M. (eds), 1983, *Russian Chernozem: 100 Years After Dokuchaev* (in Russian). Nauka, Moscow.

Tsvetkov, M.A., 1957, Change of the forest cover. In *The European Russia from the End of the XVII Century until 1914*, 1957, Publishing House of USSR Academy of Sciences, Institute of Forests, Moscow.

Zonn, S.V. and Chikishev, A.G. (eds), 1976, *Natural Resources of the Russian Plain in the Past, Present and Future* (in Russian). Nauka, Moscow.

Chapter 17

HISTORICAL LAND USE CHANGES: THE NETHERLANDS

H.N. van Lier

17.1. Introduction

Land use patterns in The Netherlands have followed particularly
distinctive trends. These trends can only be clarified by an analysis of
social changes. This understanding of historical changes is important
for predicting future land use patterns. The management of present
land use activities and the planning of future land use patterns must be
based on ecological sustainability: to preserve natural resources for
future generations, it is imperative that they are used in the most
efficient and effective manner possible.

Some of the historical trends of land use patterns in The
Netherlands are discussed in this Chapter. The nature of changes show
definite differences over time, and therefore a distinction is made
between changes over centuries and more recent changes (over the past
few decades). Special attention will be given to recent trends in land
use planning on different spatial scales since this is a dominant factor
in managing the available land resources in The Netherlands.

17.2. Historical Trends of Land Use Patterns

17.2.1. Land use changes over the last thousand years

In about the year 1000, The Netherlands was a delta continuously being
reshaped by the sea. Land use changes over the centuries have been
a result of changes in natural conditions and then more recently of
human activities.

Changes in natural conditions over the past thousand years have
dramatically affected land use patterns. Of particular importance has
been the rise in sea level due to climatic change (excluding the
greenhouse effect). The rise in sea level has inflicted an increased
frequency of large scale floodings. Large areas of land have been lost to
the sea over the past 1,000 years as shown in Figure 17.1. This trend
of loss of land to the sea lasted until about the year 1500.
Technological developments in the construction of dikes and dams and

F. M. Brouwer et al. (eds.), Land Use Changes in Europe, 379–401.
© 1991 *Kluwer Academic Publishers. Printed in the Netherlands.*

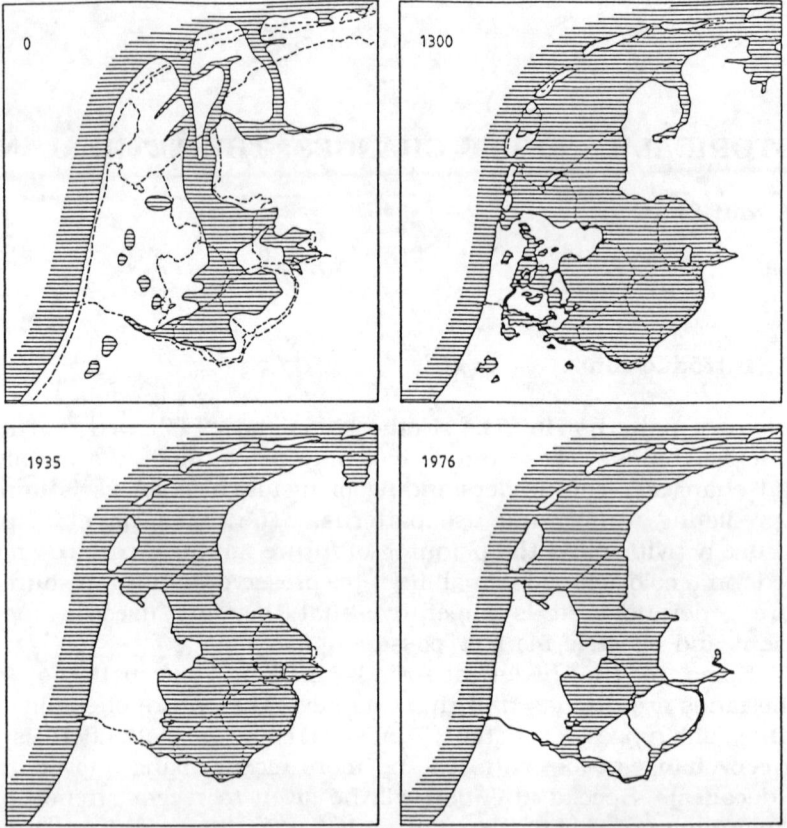

Figure 17.1. Development of the north-western part of The Netherlands over the past 2000 years.

the implementation of pumping stations have facilitated land reclamation since the sixteenth century.

Land reclamation affects land use patterns in two ways. It provides options for land use development on the newly available land (urban development, location of industry or highly intensive farming systems) and it may stimulate alternative development trends or reduce the need for land use transformations on the already available land.

The most important land use changes in The Netherlands include the reclamation of land from the sea; the cultivation of peat districts; and the cultivation of natural areas. Table 17.1 summarizes the scale of these activities over the last 1,000 years. It should be noted that about 520,400 ha of land became available through reclamation between 1500 and 1900, while about 566,600 ha was lost to the sea during the same period; the net result was therefore a loss of 46,200 ha. An additional 170,000 ha has been reclaimed during the twentieth century.

Table 17.1. Major transformations in the available land resource from different activities (reclamation from the sea and cultivation of peatland and heath) over the past 1,000 years (in ha).

Period	Water	Source Peatland	Heath
1200-1900	520,400		
1900-1987	170,000		
1000-1960		300,000	
1800-1975			775,000
Total	690,400	300,000	775,000

Major areas of land were transformed through the cultivation of natural areas. The cultivation of heath commenced only around the year 1800 whereas the cultivation of peatland has taken place since the year 1000.

Figure 17.2 shows the regions where land has been reclaimed, either by the construction of dikes or by land drainage and the development of polders. These actions have been mainly confined to areas bordering the North Sea. Figure 17.3 shows the natural areas which have been cultivated and transformed into agricultural land, primarily in the eastern and southern part of the country.

17.2.2. Land use changes in the twentieth century

At the beginning of this century, about 27 per cent of land was used as arable land, 37 per cent was pasture, 27 per cent of land was covered by forests, heath, dunes or marshes, 3 per cent was water and the remaining 6 per cent was used for urbanization, transportation, communication networks and other purposes. By this time approximately 64 per cent of available land was used agriculturally.

The major changes in land use patterns that have taken place during the twentieth century are: i) decrease in natural areas including forests; ii) increase in land use for human activities such as urbanization, industrial development and the construction of roads; iii) cultivated land showed a slight increase up until the 1960s but thereafter showed a decreasing trend. Consequently cultivated land takes up approximately 65 per cent of the available land resource of The Netherlands today.

Figure 17.4 schematically outlines the land use trends during the twentieth century. The Figure indicates trends away from natural areas to both i) cultivated land, and ii) other uses such as urbanization and socio-economic development, and also a shift from agricultural land towards urbanization and industrial development. The order of

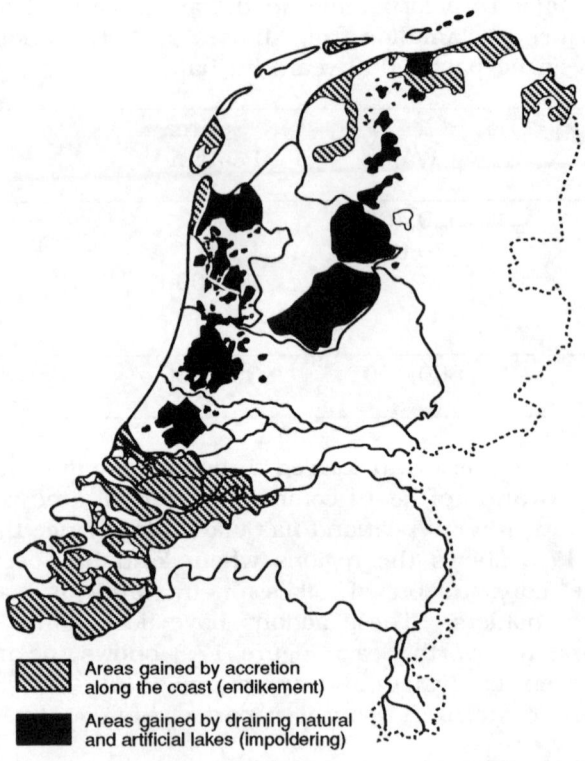

Figure 17.2. Extent of the areas in The Netherlands until 1980, where dikes and polders have been constructed to reclaim land. (Source: Held and Visser, 1984.)

magnitude of these land use changes are presented in Table 17.2 and Figure 17.5. The area covered by natural vegetation (including heath, forests and dunes) has dropped to approximately 12 per cent over the last 90 years. The total area of The Netherlands has shown a slight increase as a result of major land reclamation programs.

More detailed information is available on land use changes since the 1950s and Table 17.3 and Figure 17.6 summarize the major land transformations of this period. Additionally Figure 17.7 presents an outline of the mechanism of land use shifts in The Netherlands since the 1950s - major transformations have taken place from cultivated land and natural areas to forests, urban land, recreational use and industrial use. Arable land, pasture and natural areas show a decreasing trend - about 800,000 ha were transformed for the development of industry, urban areas and recreation. Forest areas have shown a major increase of some 50 per cent whereas natural areas have decreased by about 40 per cent.

Figure 17.3. Cultivation of peat areas and natural vegetation (heath) for agricultural use. (Source: Held and Visser, 1984.)

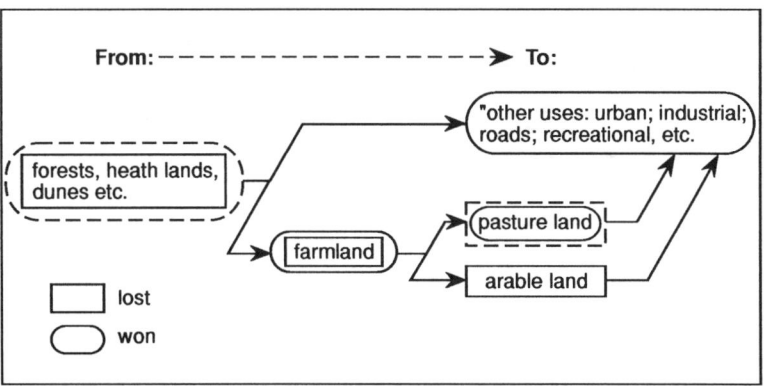

Figure 17.4. Outline of land use trends during the twentieth century.

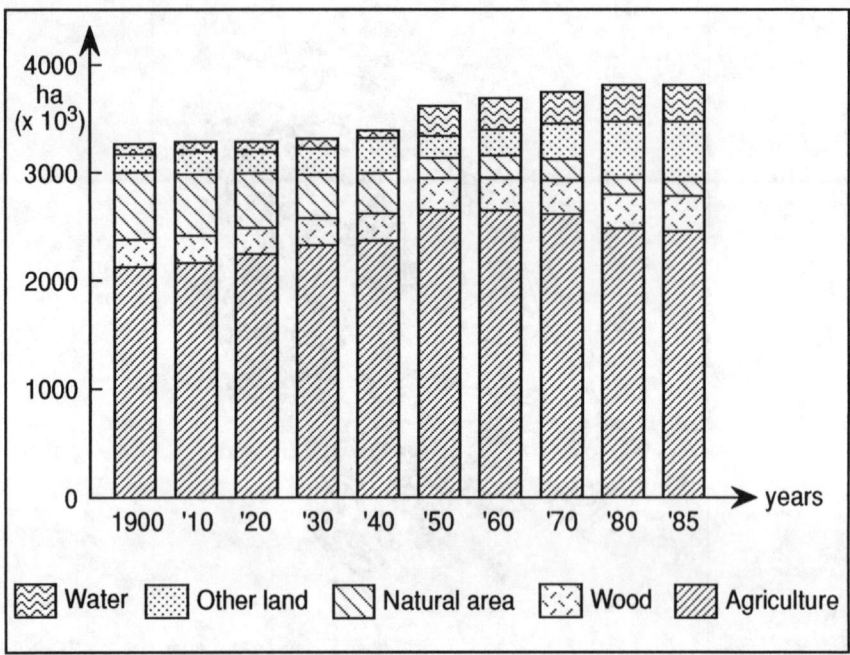

Figure 17.5. Trends in land use. (Source: de Groot *et al.*, 1988.)

Table 17.2. Land use patterns in The Netherlands between 1905 and 1985. (Land use categories in percent of total area. Total area in 1,000 km².)

Land use category	1905	1925	1955	1965	1975	1985
Arable land	27	29	32	29	26	28
Pasture	37	39	41	42	42	38
Forest, heath lands, dunes, marshes	27	22	13	13	13	12
Water	3	3	8	8	8	9
Other land	6	6	6	8	11	13
Total area	32.6	32.7	35.5	36.2	37.0	37.3

(Source: Held and Visser, 1984; de Groot *et al.*, 1988.)

17.3. Major Determinants of Historical Land Use Changes

17.3.1. Land use determinants for the period between 1500 and 1950

Changes in land use patterns in The Netherlands and changes in agricultural practice since the sixteenth century are characterized by three main stages of development (de Groot *et al.*, 1988), corresponding to the periods 1500-1880, 1880-1950 and 1950-1980s.

In the time between 1500 and 1880 the development differs according to geographical location; for example, depending on whether one observed the lowlands near the coast or the eastern part of the country predominated by sandy soils.

The main trends apparent in the coastal areas can be characterized as follows:

-   an increasing interdependency between the rural and urban economies which resulted in intensified use of the land, improved accessibility to the countryside through the construction of transportation networks, and a market-oriented agricultural system;

-   dairy products, arable and cash crops gaining importance;

-   an increase in capital investment in agriculture (for example, machinery) and the establishment of water management institutions; an increase in legislation regarding land use and governmental support for farmers to cultivate the land;

-   increasing dependence of the economy and society on international trade and development;

-   lack of agricultural development policy.

Development on the sandy regions in the eastern part of the country was relatively limited during the same period as a result of the following characteristics:

-   a peasant society with mainly subsistence farming;

-   insignificant capital investments in agriculture;

-   land was owned mainly by large land owners.

The period between 1880 and the 1950s began with an international crisis in agriculture and ended with the establishment of a strong international and national agricultural policy.

The most important changes in the agricultural industry during this period include: increased import of grain to Europe from North America causing a farm crisis; the increased importance of dairy farming;

Table 17.3. Land use trends between 1950 and 1985 (in 1,000 ha).

Land use	1950	1960	1970	1977	1985
Arable land	929	891	686	700	726
Pasture	1,717	1,327	1,335	1,250	1,164
Vegetable gardens	123	116	111	105	115
Greenhouses	3	5	7	8	9
Forests	242	269	299	309	362
Nature areas	264	230	190	164	156
Roads/dikes, etc.*	98	74	62	66	66
Water*	188	250	302	324	337
Outdoor recreation areas	-	-	-	64	78
Industrial area	-	-	-	39	48
Built-up area	-	180	-	219	265

(Source: CBS, 1985.)
*Differences are mainly caused by redefinitions.
- = No data available for that period.

increased diversity of arable crops; increased importance of horticulture; introduction of fertilizers; a slight decrease in the number of farms; the development of farm co-operatives; the differences in farm size and income between the coastal regions and the sandy eastern areas became much less significant; an increasing role for farm organizations and the development of an agricultural policy.

17.3.2. Trends in land use since 1950

A distinction should be made between i) changes in society which may indirectly affect land use patterns and farming policy (external influences) and ii) changes within the agricultural industry itself (internal influences).

i)   *Societal changes or external influences.* Table 17.4 summarizes the major changes that have affected land use patterns this century. Changes have taken place in terms of population trends; more people live in urban areas; the interests of citizens, changes in consumption patterns and recreational activities becoming more important; as well as in terms of the increasing role in transportation, urbanization and industrialization. A shift in perspective of land use policies has also played a key role in influencing land use patterns, this will be discussed in more detail in Section 17.4.

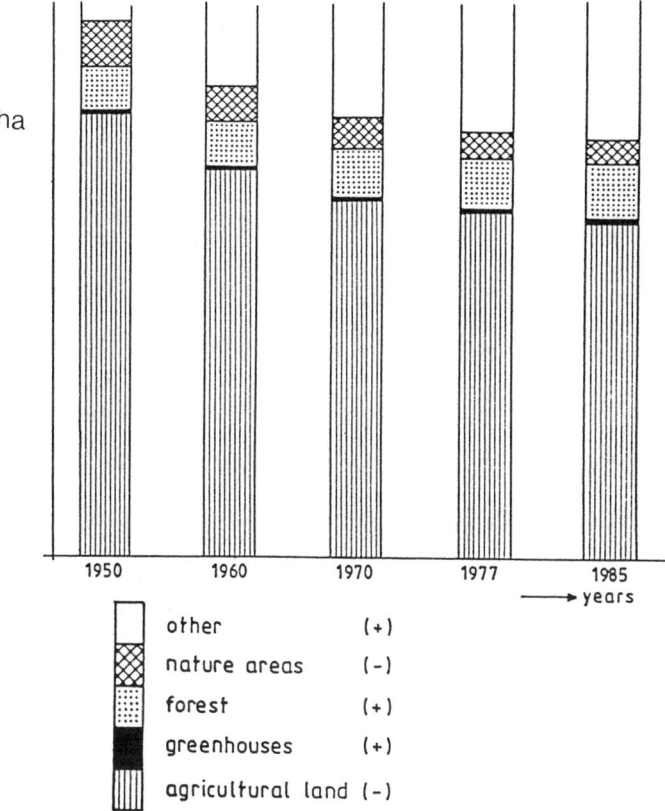

Figure 17.6. Land use patterns between 1950 and 1985.

ii)   *Changes within agriculture or internal influences.* Five periods
         of agricultural development can be distinguished (Held and
         Visser, 1984).

         1940-1950 -   food shortages, and food distribution measures
                           taken by central Government - an increase in
                           arable land and a reduction of land for dairy
                           farming;

         1950-1957 -   market balance of supply/demand, with first
                           signals of over production;

                    -      world market prices for most crops increased at
                           a higher rate than their cost price;

                    -      a trend towards free markets;

Table 17.4.   Major shifts in society which have been important for change in land use patterns.

Factor	Shift
Dynamics of population	
* population growth	1930: 2.6 million
	1988: 14.6 million
* living environment	rural urban
Citizen interests	
* income	redistribution
* life expectancy	1920-1980: 40% increase
* home ownership	1947: 28%; 1988: approximately 50%
* food consumption	flour/potatoes to meat/eggs/dairy
* outdoor recreation	increased demand
Transportation and mobility	
* car ownership	1930: 1 car per 120 persons
	1980: 1 car per 3.2 persons
* transport by car	tenfold increase between 1950 and
	1980 (in passenger kilometers)
* public transport	stable between 1950 and 1980
Land ownership	
* urban areas (15% of land)	mixture of private and public
* rural areas (85% of land	1985: 72% private
	28% by public/others
Urbanization and industrial-	
ization	
* number of people living in	1840: 20% of the population
cities >20,000 population	1980: 58% of the population
* labor participation in industry,	
service and governmental sector	1900: 67%; 1985: 95%
Land use planning	
* national level	national physical planning policy
* provincial level	regional plans
* municipality level	structure/allocation plans
* land consolidation	75% of the rural areas
	(1.5 million ha)
* land reclamation	available land resource has increased
	by 7%

1957-1963 -   development of the EEC market in terms of i)
                   removal of barriers between member countries,
                   ii) guaranteed prices for basic products, and iii)
                   a rapid increase in food production;

1963-1970 -   production of most commodities approached
                   overproduction levels;
         -         efficiency increase;

         -         major achievements in research and education;

1970-1980s-   selective growth;

         -         reduction of overproduction;

         -         formulation of new objectives for the rural areas,
                   in terms of landscape, ecological phenomena and
                   a focus on outdoor recreation.

   Changes in agricultural development over the past few decades are
reflected, *inter alia*, by rationalization, specialization, increasing scales,
mechanization and the establishment of agro-industrial complexes.
Some major trends illustrate these changes:

i)   the number of people employed in agriculture decreased from
       653,500 in 1947 to 273,000 in 1982, the total labor participation
       in the agricultural sector (not including employment in food
       processing industry) decreased during the postwar period from
       19.8 per cent in 1947 to a level of 5.6 per cent during the early
       1980s;

ii)   changes in wages - the average hourly wages in agriculture have
       increased at least fivefold since 1950;

iii)   agricultural production has increased threefold since 1949
       (measured in physical terms) and the increase in monetary terms
       has been more than fourfold;

iv)   number of farms decreased from 410,000 in 1950 to 139,700 in
       1982;

v)   the average annual rate of increase of production capacity shows
       a wide variation among the different types of farming, from
       relatively no change in open air vegetable farming (0.7 per cent)
       to as much as 5.9 per cent for horticulture, or 8.9 per cent
       intensive livestock industries;

vi)   farm size has increased steadily from an average of 11 ha in 1950
        to 17 ha in 1988;

vii)   income shows major differences among the farms; generally it is
        still less than the income level of other sectors, although it shows
        an increasing trend - the actual income from farming also shows
        large interannual variations (de Groot *et al.*, 1988).

viii)   increased use of farm equipment (machinery and barn systems) -
        the number of tractors, for example, has increased eightfold since
        1950;

ix)   the increase in the use of non-factor inputs, such as fertilizers,
        pesticides and insecticides - the level of nitrogen fertilizer use in
        1980 was more than ten times that of 1920.

Other factors also need to be mentioned such as:

1)   rise in farmland prices;

2)   increase in capital investments;

3)   increased imports of food products for intensive
        livestock production;

4)   introduction of sprinkler irrigation systems;

5)   recent introduction of computer-based farm
        management support systems.

Some of these trends in agriculture (employment, production and
wages) are presented in Figure 17.7. The total agricultural production
and the hourly salary costs (in real terms) both increased steadily after
the 1950s, while labor participation has decreased.

17.4. Land Use Planning

17.4.1. Physical planning

Physical planning is a tool used to manage and control land use
development. It is based on objectives for local, regional and/or
national development and provides legislation and political power to
enforce development trends in a participative planning system. Figure
17.8 shows that physical planning takes place at three spatial levels
(local, regional and national). Local and regional planning systems have
to comply with national objectives. Such regional and local systems
need to be consistent within the framework of the policies formulated for
the country as a whole.

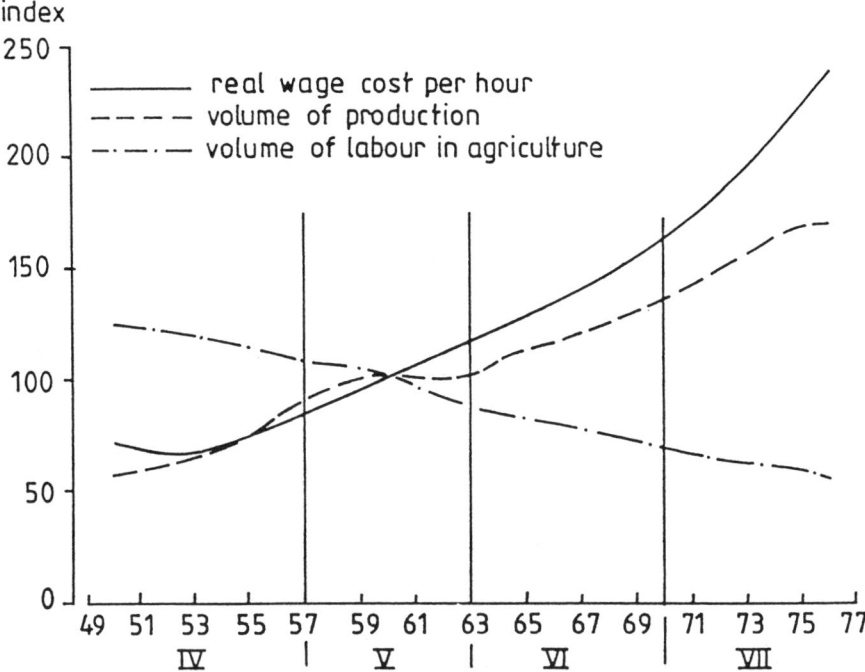

Figure 17.7. Trends in agriculture between 1950 and 1977, in terms of actual wages, agricultural production in physical terms and people employed in agriculture (1960=100). (Source: Held and Visser, 1984.)

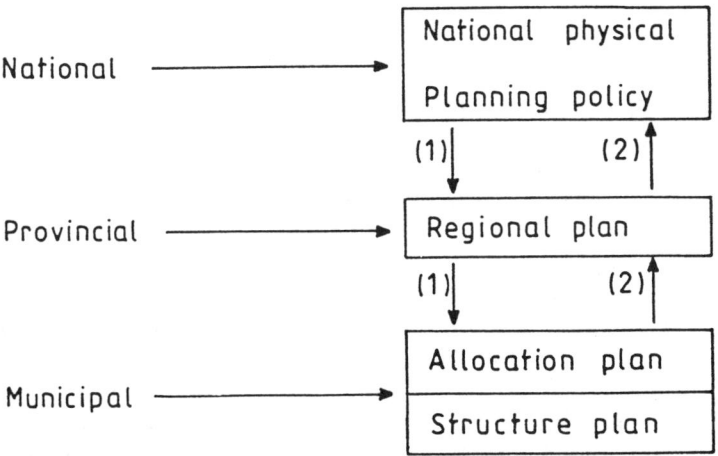

Figure 17.8. Three levels of physical planning in The Netherlands (1=establish and renewal; 2=approval).

The three spatial levels of physical planning include:

1)   a national policy with national objectives for
        land development, described in general terms;

2)   regional plans focused on regional development;

3)   detailed allocation and structure plans focusing
        at the municipality level.

Figure 17.9 gives an example of a national plan for rural areas in
The Netherlands. It shows the major functions which these areas might
engage. The areas where, for example, agriculture is considered to be
the basic function of land development, include the northern and south-
western part of the country as well as the reclaimed land of the
Ijsselmeer polders. The regional and municipal plans, as well as
reallocation and land reclamation programs, have to fit into the overall
objectives outlined in such a national plan.

17.4.2. Rural philosophies in The Netherlands

Private farming is the prevailing system in western Europe, although
many other forms exist (for example, kolkhoz, kibbutzim and large land
ownership). Thus private ownership of land is one of the basic
principles for the organization of agricultural production in conjunction
with freedom of management and production for the free market. In
theory prices of agricultural products are based on the balance between
demand and supply. Annual income of farmers is simply described in
the following way:

$$I = Q (P - C)$$

where: Q = total production of a farm.
            P = market price per unit of product.
            C = costs per unit of product (in terms of
                  labor, capital, fertilizer, machinery,
                  etc.).

Farm income can be increased through an increase of Q, a
decrease of C or a combination of both. An increase in production may
have advantages to farmers and society in general (such as enough food
to feed the population, wide diversity and high quality of products) as
well as disadvantages (overproduction problems of social stress and
environmental deterioration).

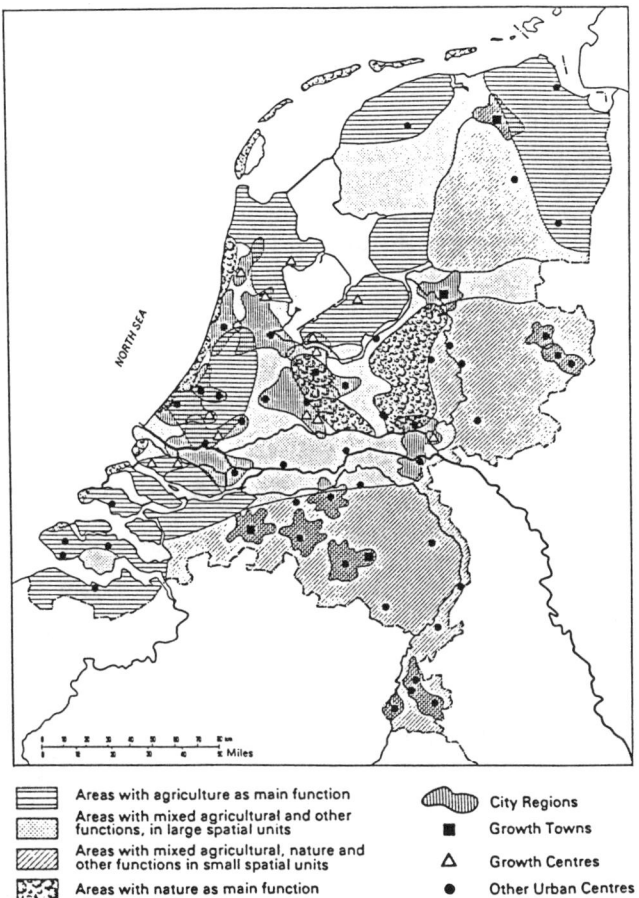

Areas with agriculture as main function   City Regions

Areas with mixed agricultural and other functions, in large spatial units   Growth Towns

Areas with mixed agricultural, nature and other functions in small spatial units   Growth Centres

Areas with nature as main function   Other Urban Centres

Figure 17.9. A framework for physical planning of rural and urban areas in The Netherlands. (Source: IDG, 1985.)

The increasing problems in present agricultural practice, at least in the western world, have led to alternative views on the planning of land use in rural areas. This changing perspective was described as "rural fundamentalism" by Wibberly (1981). The concept of rural fundamentalism is to remove any disadvantages of land use activities in the rural areas where possible, such as reducing the social and environmental stress.

17.5. Current and Future Trends

At least three alternatives are presently being evaluated in The Netherlands in order to cope with the environmentally detrimental effects of modern farming systems:

a)   organic farming;

b)   integrated agriculture.;

c)   multifunctional approaches to using rural areas.

Organic farming without chemical fertilizers and pesticides is practised in more than one form. The option of integrated agriculture requires a fundamental discussion on future agricultural policies and systems and is based upon alteration of the present agricultural system by inclusion of additional objectives to farming practices (WRR, 1984). Such objectives include:

-       an improvement in the supply/demand balance of agricultural production within the EEC;

-       a decrease in the competition with developing countries for markets;

-       a reduction in the use of non-renewable energy and other resources;

-       improvement of the environment, nature and working conditions.

Examples of multifunctional land use options are presented below.

17.5.1. **Land reallocation programs**

A land reallocation program is a collective action by farmers to improve farming conditions such as income and the working environment. The aim may also be to improve long-term perspectives for farmers. The process is generally implemented via Acts of Parliament and is guided by governmental support. The "external" production conditions are improved in a land reallocation plan, in terms of proper water management, soil quality improvements, field size and form, road systems, and the allocation of land and farmyards. Initially, it only focuses on improving farming conditions, but it has gradually changed into a reconsideration of the rural areas. The objectives of the most recent Act for land allocation in The Netherlands include:

-       a high, stable income level for the rural population;

-       proper conditions for living and working;

-       access to areas for recreation;

-       conservation of natural ecosystems;

-       improvement of the road system;

-       an emphasis on the preservation of the aesthetic and cultural value of the rural landscape.

Two new approaches have recently been introduced in land reallocation: management agreements and the establishment of an ecological infrastructure.   Management agreements are negotiated between Government and private farmers with the aim of farmers accepting restrictions on the management of their farm and on the farming system. Farmers receive an annual compensation in return, depending on the size of the farm. These management agreements are established on a voluntary basis.

Restrictions on the management of the farms may include conditions such as raising the water table, stabilizing field size and no alteration of soil conditions. Management agreements can also focus on farming practice and may involve major restrictions on the use of fertilizers and pesticides or minimizing the use of machinery.

It was initially estimated that between 500,000 and 700,000 ha (approximately a quarter to a third of all farmland) would be involved in these management agreements. Due to political, financial and practical reasons this was later reduced to 200,000 ha of which 100,000 ha were given priority.   The figure was further reduced to 86,000 ha. More than 1,000 agreements were established in 1988 encompassing over 10,000 ha of land, and over 5,000 ha of land have been converted to nature reserves.

Ecological infrastructure is the creation of a main corridor of "landscape-nature elements". Such an area would be part of existing or newly planned and improved land use patterns. Main corridors connect to several smaller linear elements to promote a network of herringbone structure. The basic principle is summarized in Figure 17.10. An important feature of the concept of an ecological infrastructure is that the main elements as well as the connecting smaller ones fit into an ecosystem as well as into the farming conditions. Locations on road banks, waterways and field boundaries are likely to be the most acceptable. The overall aim of an ecological infrastructure is to promote and maintain natural ecological value.

17.5.2. Land reclamation programs

Large areas of The Netherlands have been reclaimed from the sea by the construction of polders. One of the largest schemes is the Ijsselmeer project, where five new polders have been created since the enclosure dam was constructed in 1933 (Figure 17.11).

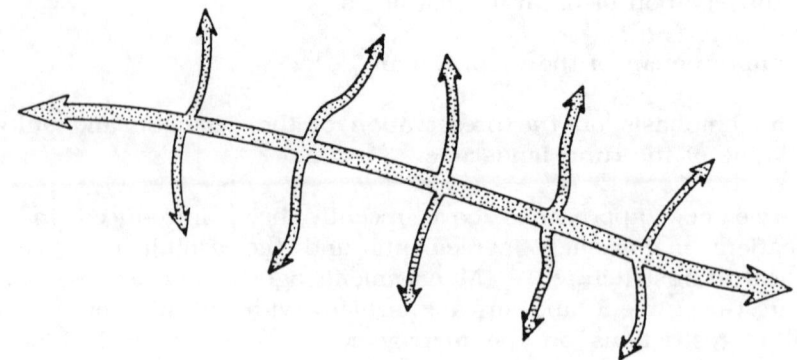

Figure 17.10. Basic principles of an ecological infrastructure. (source: Gelderblom *et al.*, 1985).

The original objective of these programs was to create new farmland, although other aims also played a role such as the protection of land against floodings, the provision of employment and reducing distance for travel. The first polders were predominantly used as farmland, with more than 80 per cent of the area being used for arable production. However, other land use types gradually gained importance during the postwar period, such as housing and infrastructure, recreation, forestry and especially the creation and preservation of natural ecosystems. This trend has resulted in a shift in land use patterns in the successive polders. Figure 17.12 shows the decreasing importance of land used for agricultural purposes, since forests, natural areas and urbanization cover larger areas in the newest polders. About 25 per cent of the land reclaimed by programs since the 1970s is used for forestry and nature preservation and only half of the total land coverage is used for agriculture.

A new approach to the planning of land use in the recently developed polders is the combination of farming practice with the preservation of nature by the development of a transition zone between farm land and land for nature conservation (Figure 17.13). Such a transition zone might be used for some agricultural purposes, but only under specific conditions. Figure 17.13 shows the outline of an area that is managed taking into account the principles of farming and nature preservation. The average size of such farms is between 25 and 40 ha, including a transition zone of up to 5 ha. Under such conditions farming can only be profitable if the farm is relatively large (Figure 17.14).

The results from Figure 17.15 are based on a linear programming procedure for agriculture under different conditions. It shows that farm income largely depends on farm size and also on the part of the farm in the transition zone which has special restrictions on farming practices.

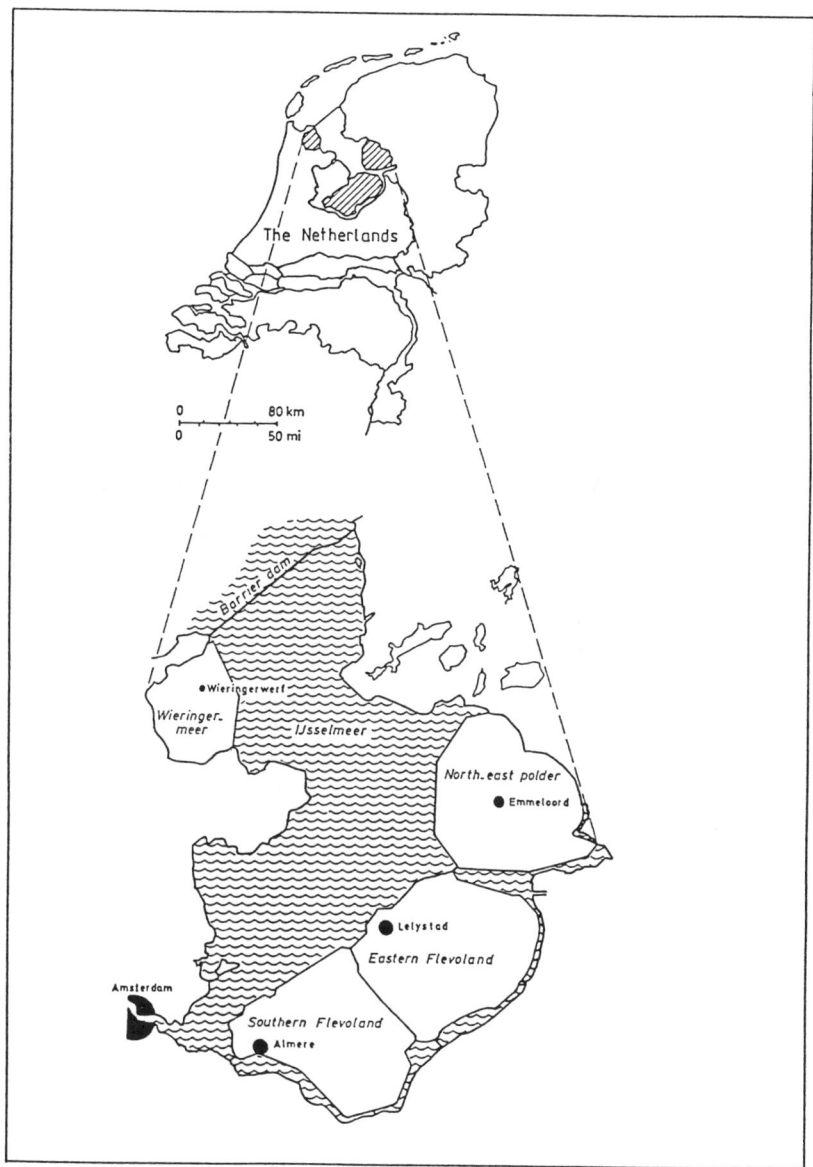

Figure 17.11. Land reclamation program in the Ijsselmeer area.

Projections on future land use patterns should therefore incorporate assessments of the direct effects on land uses, such as crop growth reactions (van Diepen *et al.*, 1987) as well as indirect effects through changes in socio-economic conditions. Scenarios on future land use patterns need to be based on, *inter alia*, relations between land use shifts and its related factors as observed in the past.

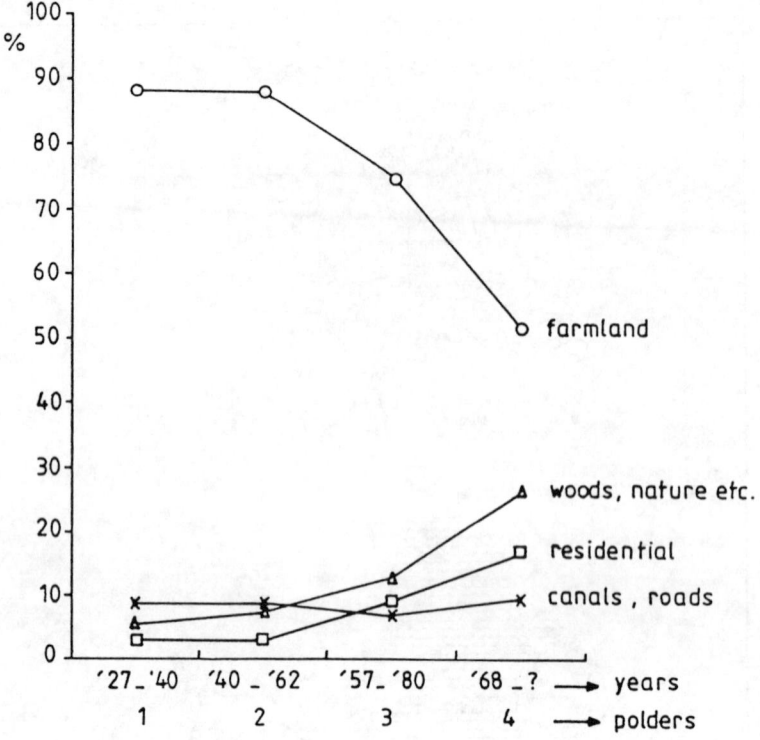

Figure 17.12. Changing land use patterns in four successive polders of The Netherlands.

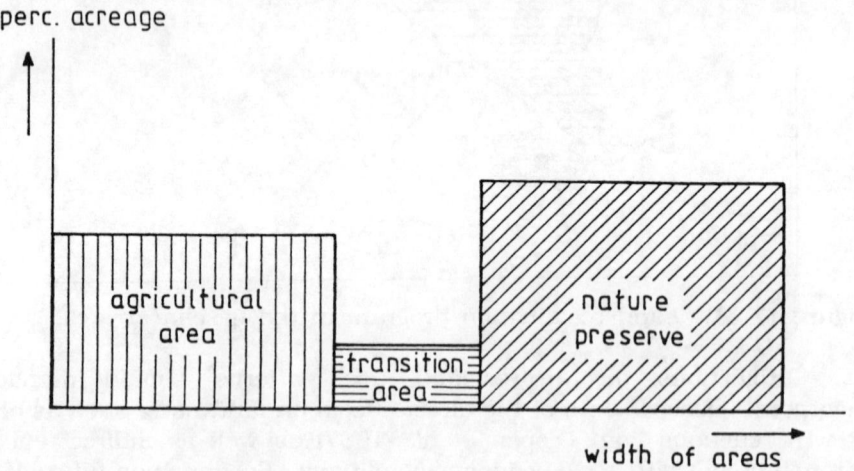

Figure 17.13. Outline of land use planning combining farmland with land for nature preservation.

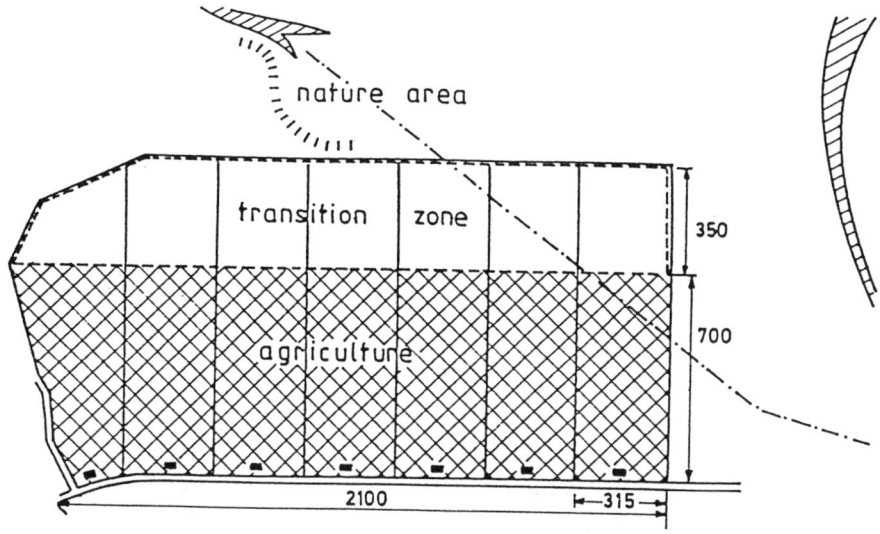

Figure 17.14. Outline of farmland, with a transition zone and nature reserve (size of zones in meters). (Source: van Heck *et al.*, 1985.)

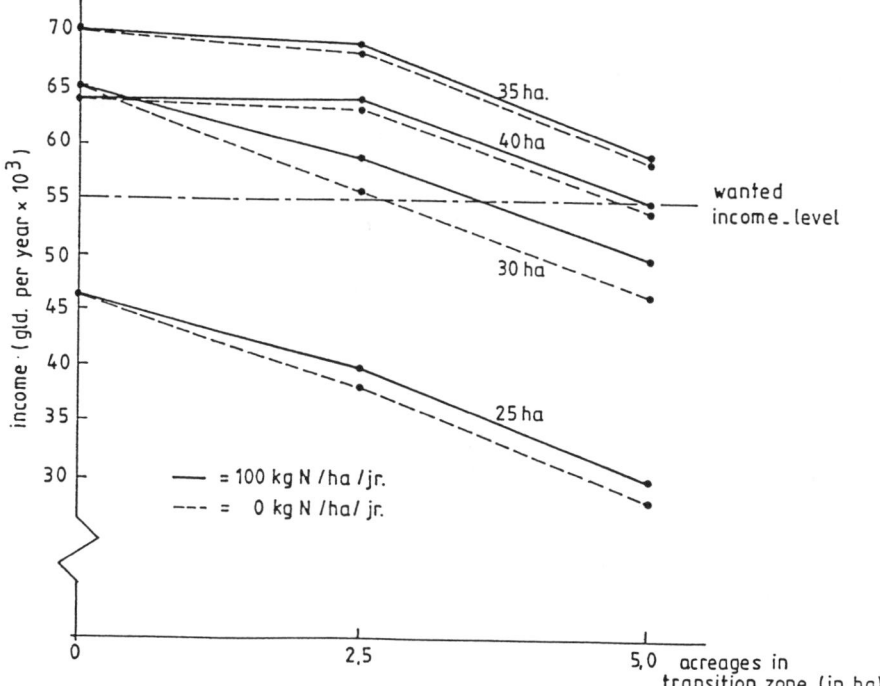

Figure 17.15. Income in relation to size of transition zone for four farm sizes. (Source: van Heck *et al.*, 1985.)

17.6. Conclusions

Physical as well as socio-economic factors have been the major determinants for the transformations in land use patterns in The Netherlands which have taken place in the past. Although the free market forces of land uses still explain the major shifts, recently developed systems of land use planning play an increasingly important role (for example, management agreements).

As the examples discussed in this Chapter demonstrate, land use patterns in The Netherlands no longer depend solely on the development of the free market but are strongly influenced by physical planning in general and specific land use planning programs in particular. This, however, does not mean that such programs do not incorporate socio-economic conditions. In fact many of the plans need to be revised over time, by redefining their objectives among other things. Whether or not land use changes take place through land use planning, they will also depend on societal changes. Even changes in physical conditions, such as climatic changes through the greenhouse effect, will probably affect future patterns of land use, primarily through complex reactions of society to such physical changes.

REFERENCES

CBS, 1985, Bodemstatistiek, 1985, National Statistical Office.   State Publishing Company, The Hague, The Netherlands (in Dutch).

van Diepen, C.A., Van Keulen, H., Penning, de Vries, F.W.T., Noy, I.G.A.M. and Goudriaan, J., 1987, Simulated variability of wheat and rice yields in current weather conditions and in future weather when ambient CO_2 has doubled.   Simulation Reports CABO-TT, Wageningen.

Gelderblom, L., Kuysters, R. and Meijssen, L., 1985, Ekologische infrastruktuur en landinrichting, Report 78.   Department of Land and Water Use, Agricultural University, Wageningen, The Netherlands (in Dutch).

de Groot, E.H.J.M., Kerkhof, H. and Veening, L., 1988, *Land Use Changes in The Netherlands: Description and Analyses of Developments in Land Use in the Past 40 Years*.   Report 124. Department of Land and Water Use, Agricultural University, Wageningen, The Netherlands.

van Heck, H., Hendriks, T. and van Lier, H., 1985, Operationele analyse.   Een toepassing in het Lauwerszeegebied, *Landbouwskundig Tijdschrift*, **97**(11), 23-25.

Held, R.B. and Visser, D.W., 1984, *Rural Land Uses and Planning. A Comparative Study of The Netherlands and the United States*.   Series Isomul, Volume 6A.   Elsevier Publishing Company, Amsterdam, The Netherlands.

IDG, 1985, Compact geography of The Netherlands. *Information and Documentation Center for the Geography of The Netherlands.* University of Utrecht, The Netherlands.

RYP, 1976, Room at last.   The Ijsselmeer polders described and illustrated.   Ijsselmeer Polders Development Authority, Lelystad, The Netherlands.

Wibberley, G., 1981, Rural planning in Britain.   Its development, problems and conflicts.   Bakaskog Conference on *The University and Rural Resource Development - the Road between Theory and Practice,* Sweden.

WRR, 1984, *Bouwstenen Voor een Geïntegreerde Landbouw* (Building stones of integrated agriculture).   Report V44.   Scientific Council for Governmental Policy, The Hague, The Netherlands (in Dutch).

Chapter 18

HISTORICAL LAND USE CHANGES: SWEDEN

S. Anderberg

18.1. Introduction

Swedish society has undergone great changes over the past few centuries. These are characterized by a considerable increase in population, industrialization and urbanization as well as the modernization of agriculture and forestry and the establishment of a modern welfare society. Such development trends were accompanied by major increases in mobility and energy consumption. They have resulted in fundamental changes in the use of natural resources and intensified the impact of man on the environment. The development of national and international markets, national policies and pollution also have an increasing influence on the landscape.

The objective of this Chapter is to discuss the major changes in land use patterns in Sweden over the past three hundred years and to present some of the major contemporary issues and the possible future development of the Swedish landscape. The focus will be on arable land and woodland.

The large areas of forests, the large number of lakes, the low population density and the small share of agricultural and urban land distinguish Scandinavia from continental Europe. Approximately 60 per cent of Sweden is presently forested, while farmland only occupies about 7 per cent. Meadows and grazing lands, once abundant, now cover less than 1 per cent; urban areas (including roads and airports) occupy 3 per cent of the land. The remaining land is primarily mountain areas and marshlands. There are 20 national parks covering 1.4 per cent of the land resource, and about 1,600 areas for nature conservation or the protection of birds and animals, comprising 3.4 per cent of the national territory.

Sweden has long-standing statistical records: the first national census originates from 1750 and some details of regional statistics of population and taxes (an indicator of agricultural production) are available from the sixteenth century.

F. M. Brouwer et al. (eds.), Land Use Changes in Europe, 403–426.

18.2. **Societal Development over the Past Three Hundred Years**

Some historical trends regarding the development of Sweden since 1650 are summarized in Figure 18.1. Population has increased from about one million to the present level of more than eight million. The largest increase of population took place around 1820 with the demographic transition from an agricultural society toward industrial development. Emigrations to America around the end of the nineteenth century reflect the problems of adjusting to the large increases in population. Despite this, Sweden witnessed a radical increase in the rural proletariat.

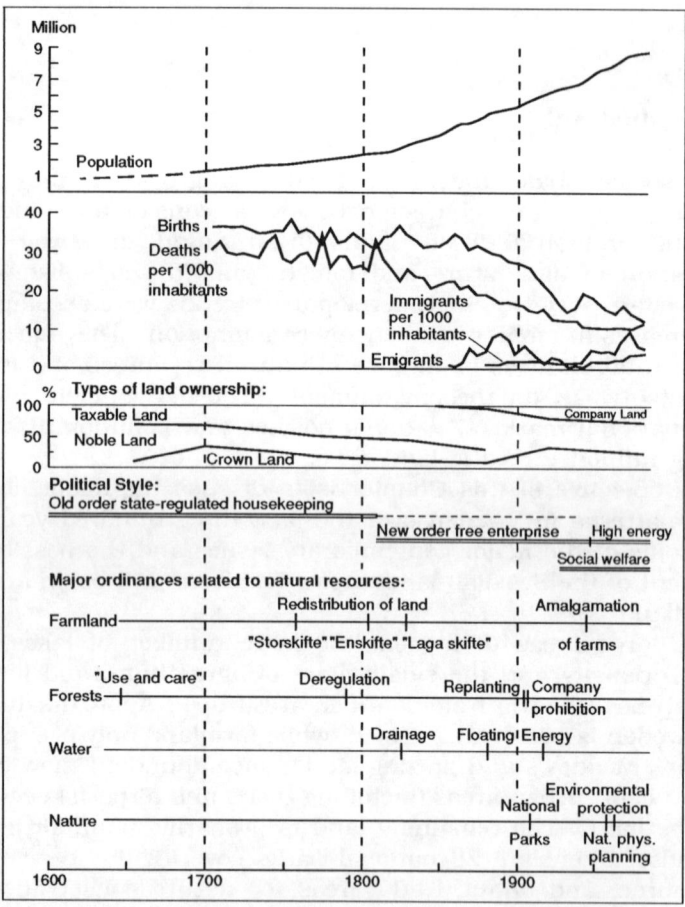

Figure 18.1. Total population, crude birth and death rates, rates of emigration and immigration (per 1,000 inhabitants); general trends in the development of ownership of land; periods with different political style; major ordinances related to the use of farmland, forests, water and natural resources in general.

Immigration to Sweden began after 1945, particularly from Finland and southern Europe.

The spatial distribution of population on a regional level has remained rather stable. The plains around the lakes in central Sweden and Skåne in the very south have always been the most densely populated regions. During the first stage of urbanization around the end of the nineteenth century, people moved only within small areas, and a network of small towns was developed. The relative importance of the metropolitan areas of Stockholm, Gothenburg and Malmö has increased since the last World War. The population of Norrland (in the north) was about one-eighth of the total population during the nineteenth century and early part of the twentieth century, although it covers about two-thirds of the whole country. The population has increased here with colonization and industrialization. But this region has also experienced economic decline. There have been several policies for the Norrland region, aiming to preserve the regional distribution of economic wealth and population.

Sweden became a centralized State in the sixteenth century, and a formal bureaucracy was gradually established. Three major eras of "political style" can be discerned. The ancient order was characterized by mercantilistic regulations with State-controlled housekeeping. These strict regulations of trade and industry were abolished during the nineteenth century and a capitalistic market economy evolved. The Welfare State was established during the twentieth century. This modern, engineered "system society" has much in common with the old order but the rules and regulations are much more effective than in the past. The State exerts strong influence on the economic sphere, but the market economy is still dominant.

The real estate unit, which the owner has the power to manage, traditionally has been the basic legal domain in Sweden, as well as in other western countries. Three types of land developed in the Middle Ages: i) Crown land; ii) noble land; and iii) taxable land. This distinction did not affect land use patterns in a major way. The peasants, for example, who worked on the land, either had to pay rent to the Crown, or to do day work and pay rent to the aristocracy, or to pay taxes to the State. Ownership became more decisive during the nineteenth century because of the varying entrepreneurial opportunities that followed the development. The large noble land estates often became centers of innovation. Regulations by the State or local community, imposed to reach specific goals, may include restrictions or compulsory actions, or incentives to direct land ownership.

A traditional and rather unique right for Swedish citizens is "every man's right" which allows the public to enter areas without major restrictions (with the exception of in the vicinity of dwellings) and to collect berries and mushrooms without damaging trees and animals. This tradition has strongly influenced the attitudes towards Nature.

18.3. **The Ancient Order**

The Swedish Baltic Empire was largest during the middle of the seventeenth century. A centralized State organization with an efficient tax system had been developed. The majority of the population were peasants living in small villages. Two-field or three-field farming was common practice on the plains and more than a third of the cropland lay fallow. Tilled land and meadows were divided into parcels and the remaining land was common woodland. The villages and the size of the cropland of the forested areas were smaller and the periods of fallow were generally shorter. The outfield forest was a necessary resource used for several purposes such as grazing and hunting and the provision of firewood, timber, berries, charcoal and tar.

The small communities were almost self-subsistent with limited connections or contact with other areas. Wood was the only fuel type, as well as the dominating material, and water played an important role as an energy source along the rivers and streams. Market conditions only played a major role in the mining areas in the southern and central part of Sweden. In these areas most of the socio-economic activities were related to the metal industry, and these products formed the basis of export in Sweden.

Food supply showed a large inter-annual variation. Statistics on food consumption show that farmers on the plains were able to manage after poor harvests, while many small farms in the woodlands struggled to produce enough food even during good years (Hannerberg, 1971). Crop failures were not unusual and often led to famines; a factor reflected in the population statistics up to the beginning of the nineteenth century.

18.4. **Land Use Transformations over the past Two Centuries**

Changes in the landscape were generally slow during the ancient order, but accelerated from the beginning of the nineteenth century with the population increase, societal development towards external markets and the increasing use of natural resources. The major changes in land use patterns during the nineteenth century include:

-       transformation of meadows to tilled land because of the larger food requirement; these transformations were concentrated during the agricultural expansion of the nineteenth century;

-       reclamation of land through water drainage projects was most intensive during the second half of the nineteenth century;

-       decrease of the forest areas as a consequence of the increase of agricultural land; the exploitation of the available forest resource also increased; peat burning, charcoal and tar production already put the forests under stress in some regions during the early development stages and cutting of forests increased largely with the growing population and industrial development.

The development trends have gradually changed during the twentieth century (Figure 18.2). The forest areas show an increasing trend, arable land is decreasing and the meadows are also decreasing. Urban areas have increased mainly on land which was originally for agriculture.

Land use patterns have changed in response to:

i)   intensification of agriculture and forestry (for example, by introduction of fertilizers and modern technology);

ii)   specialization in agriculture and forestry.

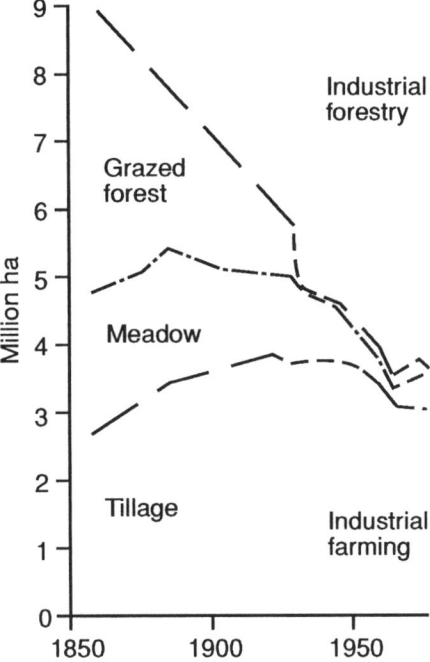

Figure 18.2. Changes in land use patterns in the rural areas of Sweden (meadows actually showed a larger variation than the figure shows). The developed grazed forest should be interpreted as a rough approximation only. (Source: Ingelög, 1981.)

18.5. The Agrarian Revolution during the Nineteenth Century

The hectarage of cultivated land more than doubled between 1800 and 1980, and production per hectare (Table 18.1) also doubled during this period (Hannerberg, 1971). Regional estimates indicate that the cropland area also increased during the seventeenth and the eighteenth centuries.

Table 18.1. Development of cultivated land, number of ha per 100 inhabitants, and yield levels in Sweden during the period 1800-1980.

Year	Cultivated land (000's)	Hectares per 100 inhabitants	Yield (t ha^{-1})
1800	1,500	64	0.8-0.9
1850	2,600	75	n.a.
1880	3,400	74	1.4-1.5
1930	3,800	64	1.7-2.0
1950	3,600	51	2.1-2.5
1980	3,000	36	3.3-4.3

n.a. = no data available.

The agrarian revolution started around the beginning of the nineteenth century. Agriculture expanded and was intensified because of increasing food requirements, the introduction of new technologies and major land reforms. The cropland area increased by transformation of meadows to cultivated land and reclamation through drainage projects. Crop rotation replaced the old field systems and intensified the use of cropland. Chemical fertilizers from abroad were introduced, although it was not until the 1930s that they were widely used and became a dominating factor of the increase in production.

18.5.1. Redistribution of land and the introduction of new technologies

A major factor in agricultural development in Sweden was the consolidation of land via three regulations:

i)   "Storskifte" during 1757: the major objective of this law was to link the scattered and narrow strips of a farm into larger blocks; its influence was mainly limited to the plains in the eastern part of central Sweden;

ii)   "Enskifte" during 1803 had a broader perspective and a wider impact than the former one; its goal was to gather all the strips to form one block and to move the farmsteads out to it; for topographical reasons such a radical program could only be realized in the plain region, and the landscape radically changed in these areas soon after the implementation of this policy;

iii)   "Laga skifte" of 1827 could be seen as a compromise; it permitted more than one parcel, but also required that the farms moved away from the village.

These reforms mainly originate from the physiocratic philosophies that dominated economic thinking during the eighteenth century. Agriculture was seen as the primary source of wealth; therefore, population and agricultural policies were implemented during this period.

The programs to reallocate the available land resource altered landscape, agriculture and the social life for the peasants. The ancient structure of settlements was changed and farms became dispersed. The peasants became independent individual farmers. The common outfields (grazing lands and forests) were also transformed into privately owned land. The quality of the soils was carefully assessed with the assistance of land surveyors to make the division as fair as possible. This reorganization process reached its climax during the middle of the nineteenth century, but continued locally until the twentieth century.

New forms of agricultural tools were introduced. The traditional wooden ards and plows were replaced by iron tools. Iron plows, often from abroad, were introduced during the eighteenth century but were not widely employed until local development of the English plow models began at the beginning of the nineteenth century. Iron ards, plows, harrows and rollers were generally in use around the 1870s. The substitution of wood by iron enabled improvements in the preparation of the soils. The plow depth doubled (from 10-12 cm to 20-25 cm), a prerequisite for both land reclamation and the cultivation of root crops. Traditionally, the ard was the common agricultural implement and plows were only in use in some regions, but this trend changed around the end of the nineteenth century. Towards 1900, threshers, horse hay-rakes and mowers were introduced. Mechanization has further revolutionized farming methods during the twentieth century.

18.5.2. **Reclamation of land**

The land reforms provided opportunities for farmers to introduce new technologies and crop varieties, as well as to take land into production. Reclamation and drainage were often necessary on the peripheral areas of the villages. Additional important factors behind the expansion of agriculture was the surplus of labor, taxes and rents based on harvests or yield from a certain acreage and increasing markets. A dramatic development, for example, took place in some districts when England abolished its barriers on corn imports during the 1840s, and Sweden became an important provider of oats (Fridlizius, 1957).

New arable land was seldom reclaimed for the growing rural proletariat. This group of landless people had to earn their living on small plots in the outfields, day work, berry-picking and handicraft production. In around 1880, when the rural population was largest, land for new settlement was a scarce resource in southern and central Sweden (Figure 18.3). Emigration of the landless rural population then started to increase.

Drainage of land continued until the middle of the twentieth century and has largely altered the hydrography of the country. Large areas were drained for agricultural purposes; by the year 1950, about 600,000 ha, or one-sixth of the total cropland area, was estimated to have been transformed from former lakes and wetlands.

The major objective of these efforts, besides reclamation of farmland, was to reduce the risk of frost damage to crops that had previously caused major difficulties in agricultural practice. Also the provision of employment to the rural areas with traditionally high unemployment must have been an important objective. The Government had supported these drainage programs for more than a century by the provision of loans and subsidies and the adjustment of legislation to make these projects possible. Expertise was available from abroad to ensure that the best available technologies were used. These projects were aimed at increasing the quality of the available land resource but were not always successful. Several small drainage projects had poor results because of the poor quality of soils, or because of unforeseen ecological side effects and, in some cases, surface shrinking also had a long-term impact (Persson and Lohm, 1977). No economic benefit analysis was required until 1918 and many undertakings were economic failures leaving farmers ruined or with heavy debts. Rather than having problems due to an excess of water, many areas became vulnerable to drought. Several formerly drained areas are now irrigated on a regular basis. Even if micro-climatic conditions had been influenced, there is no evidence of a reduction in the risk of frost damage (Sivertun and Castensson, 1989).

Only 10 per cent of the original wetland remains on the Isle of Gotland. The drainage of the mires which started during the 1820s has had a major impact on the landscape. These mires were originally used for fishing and the provision of construction material. A subsoil of pure lime was below the humus layer. The conditions for plant growth deteriorated when the running water that had brought necessary minerals from upstream was diverted (Romell, 1947).

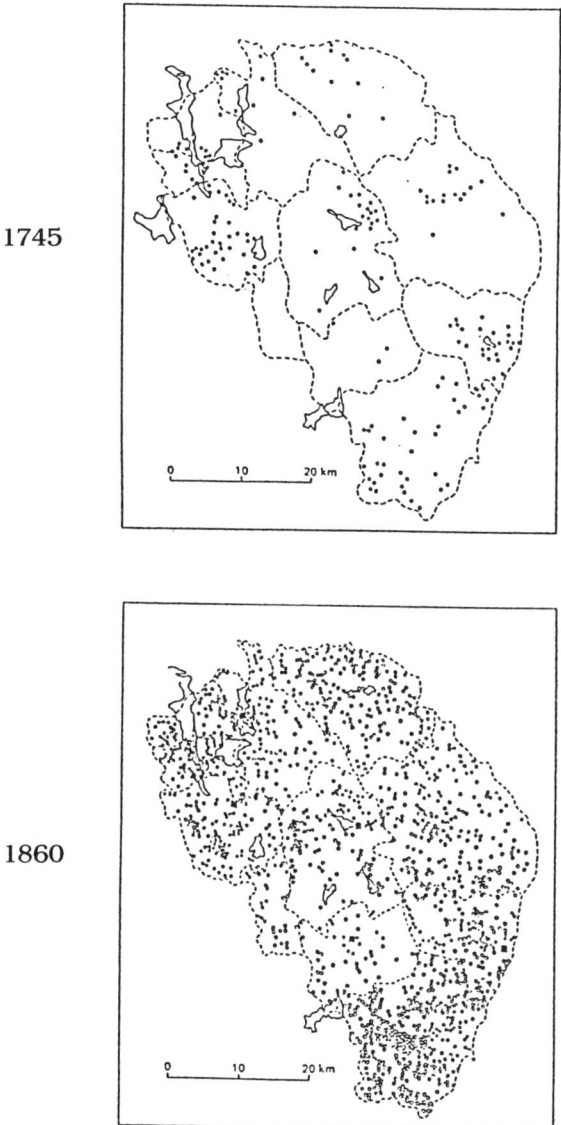

1745

1860

Figure 18.3. The large increase of crofts in a woodland area of southern Sweden between 1745 and 1860. (Source: Nordström, 1952.)

18.5.3. Intensification of agriculture

During the ancient order, land productivity was mainly maintained by fallow and by manure applications. Meadows were necessary in this system and the increase of cropland until the 1780s was accompanied by an increase of meadows. Land first became a scarce resource in the densely populated plain regions, and the close connection between meadow and cultivated land changed as new land was gained primarily from the former grazing lands. This was partially compensated by deeper plowing which intensified the circulation of nutrients and by converting the meadows on the best soils to crop production. However, gradually the need to substitute the natural fertilization of the land became a major challenge of the nineteenth century.

Marling was employed extensively during the 1840s in north-west Europe and reached the south-western part of Sweden. Marling resulted in good harvests but the benefits did not persist and increasingly negative results began to be experienced.

Guano from South America was introduced in 1844, and the use of artificial fertilizers increased with the construction of the railroad network in 1856. Local production of fertilizers started in 1862.

Swedish agriculture showed a period of major development after the 1850s as national and international markets for agricultural products evolved. The construction of new buildings, roads, wells, fences and ditches also provided employment to the rural population. This development was followed by a severe crisis in commercial farming around 1880 because of the increased competition on the international corn market between North America and Russia.

A system of crop rotation was introduced gradually, and this became very important after the crisis of the 1880s. The diversity in agricultural production increased with the growing importance of meat, dairy products and root crops; grazing and winter fodder were provided by the arable fields.

18.6. Agricultural Development during the Twentieth Century

Agricultural development during the first half of the twentieth century can be characterized from a biological point of view as fairly balanced (Mattson, 1985). Similar to development trends during the Ancient order, crop production and stocks of cattle and horses were integrated, and the major part of the plant nutrients circulated within the boundaries of the farm. The yields increased during this period because of improvements of strains, seed selection and tools. Only modest amounts of artificial fertilizers (mainly phosphate) were used.

Agriculture experienced severe economic problems during the 1930s as a result of decreasing prices on the world market. It became an economic activity protected by national policies. The preserved overcapacity was beneficial to Swedish society in the period of isolation during the Second World War. The regulations on food production during the 1930s were the beginning of the increasing influence of the Government. Present land use patterns probably would have been very different without the rules and transfers that were established in 1947.

The main objectives for the agricultural policy since then have been:

i)   self-sufficiency, a certain level of production must
        be maintained for reasons of national safety;

ii)   income of farmers and industrial workers should
        be comparable;

iii)   efficiency, aiming to provide food at the lowest
        price possible for consumers.

Agriculture was rationalized in terms of income and efficiency levels and farms had to be a minimum size in order to qualify for subsidies.

These guidelines have been the major driving force of the large changes that have taken place in agriculture since the 1940s. The number of farms showed a decreasing trend (from around 300,000 to 100,000) and farm size increased. The total area of land used agriculturally decreased considerably. The meadows disappeared to a large extent (Figure 18.2). The cropland areas decreased at a slower rate from 3.6 to 2.9 million ha, but the decrease was concentrated in forest areas with a low density of farmland. The agricultural population decreased from over 20 per cent in 1950 to a present level of less than 4 per cent. Depopulation trends of the rural areas began around the end of the nineteenth century, but the major changes in the rural areas of Sweden have taken place since the last World War.

Farming practice is now a highly specialized, mechanized and capital-intensive activity, largely dependent on inputs from abroad. The 600,000 horses of the 1940s, for example, have been replaced by 200,000 tractors, the fuel for which is imported from abroad. One-third of the farms for cereal production have no cattle, and are fully dependent on artificial fertilizers, while many of the farmers who concentrate on livestock breeding buy fodder and the farms produce more manure than can be disposed of. The use of artificial fertilizer (mainly nitrogen) has increased between four and fivefold during the period 1945 and 1970 (Chapter 10). Plant and animal breeding, the control of seeds and the use of pesticides have also contributed to the increase in land productivity. This increase has resulted in overproduction which has begun to create difficulties. The individual farmer depends on economic development because of the high capital costs required for modern agriculture.

Modern agriculture has severe impacts on the natural environment. The widespread use of pesticides creates problems such as decreases in bird populations as observed in the 1960s. The application of mercurial seed protectives has caused many environmental problems and the leaching of nutrients from the soils has caused eutrophication in lakes and rivers. Despite the fact that these latter problems were already well known in the 1950s, only in the 1980s was agriculture identified as a major cause of algal sea blooms and the mass death of fish and mussels in coastal areas.

Energy crops have been developed since the 1970s as an alternative in farming but these efforts have been unsuccessful partly because of the abundance of cheap energy and perhaps because Sweden has no long-term energy plan.

18.7. Woodland

18.7.1. The ancient order: regulated multiple use

By the beginning of the eighteenth century, woodland in the proximity of settled areas was already largely affected by man. However, in Norrland there were still large areas of primeval forests. Coniferous trees dominated in most areas of the country, but large areas of deciduous trees such as beech, oak and hazel were found in the south-western part of the country. In other regions deciduous trees were mainly found around urban areas.

The forest was primarily considered to be a self-replenishing resource but principles of sustaining the forests were already an important topic for discussion among intellectuals in the eighteenth century. Laws and regulations changed during the eighteenth century and new rules were introduced in order to prevent abuse, forest fires and secure the needs of the export industry and the military forces. However, the control of these regulations was very poor.

The woodlands of the mining areas were particularly under pressure and wood shortage was observed locally. The annual production of 50,000 tonnes of pig iron during the eighteenth century required about one million m^3 of wood for fire-setting in the mines and eight million m^3 for charcoal burning. The charcoal consumption was fairly stable during the eighteenth century and technical improvements during the nineteenth century (for example, the introduction of gun powder) made it possible to increase production without increases in wood consumption (Lindqvist, 1983). Firewood was also needed for glassworks, limeworks and saltpeter burning. Tar boiling and the production of potash were of particular importance along the coast of Norrland.

The most radical transformation of woodlands took place in the south-western part of Sweden where large areas of oak and beech forests became heather moorland. The mechanisms of this development are not completely known, but both the State and the farmers must have contributed. People in the area probably did not respect the rules which prohibited the cutting of oak and they allowed cattle to overgraze the land. As soon as the heather had established they started to burn it on a regular basis to provide fresh growth for cattle and sheep.

18.7.2. **The nineteenth century: deregulation and industrialization**

Forests could be exploited without major regulations around the end of the eighteenth century with the exception that oaks and pines suitable for masts were reserved for the Admiralty until 1830.

The sawmill industry in Sweden expanded during the 1840s as Britain decreased import taxes. The joint stock company law allowed larger units to be established and the development of the steam engine enabled sawmills to be located near rivermouths where sorting of floated timber, production and sea transport could be combined.

Exploitation of the forests in Norrland was limited at this time and this area was favored by the new export opportunities. The burning of charcoal still exceeded the consumption of wood in the mining districts of central Sweden and export opportunities in the more densely populated areas were limited due to the local use of firewood and timber. The construction of railroads and new settlements also required large amounts of timber and wood.

New areas were opened up in Norrland for the exploitation of forests. The sawmill companies first increased their influence by buying cutting rights from farmers and colonists and later on, when such rights were limited by a legislation in 1889, they stepped up their efforts to buy land.

The limitations of the available forest resource were already being discussed during the 1850s, along with the need to improve the management of the forests and develop reforestation programs. A central board of the State forests was established in 1859, and the State, having previously sold vast amounts of forest, once again started to try and increase the forested areas.

Peat burning was still practised, particularly in the forest areas of southern Sweden. This practice continued until the railroads introduced the timber market that made burning trees a non-economic activity (Larsson, 1980).

18.7.3. **Development during the twentieth century: restoration and cutting**

The proper management of Swedish forests started with a law of 1903. Timber was only allowed to be produced when it was also part of a reforestation program. The change from the traditional attitude to forests as a resource for multiple use into one where forests are primarily seen as a producer of wood also required some time to become effective.

The companies in Norrland were not allowed by law to buy more land. This law was subsequently applied to the whole country. Forest ownership was divided in the following manner: 25 per cent for State,

25 per cent for companies; and 50 per cent for farmers. The farmers held a privileged position as the major owners of forest and this reinforced their obligation to manage the forests in a sustainable manner.

The influence of the forestry boards has increased since that time. A kind of "negotiated order" with close co-operation between State, forest owners and industry has been established (Stjernquist, 1973).

Afforestation reached a peak around the year 1920 (Figure 18.4). A national assessment of the available forest resource was made for the year 1923. Despite the fact that harvests had doubled since 1850, the annual growth level still exceeded the rate of cutting (Figure 18.5). With the exception of the periods during the First and Second World Wars, yields remained stable at a level of 40-45 million m³ until the 1960s. The increasing demand for pulpwood was balanced by a decreasing use of charcoal and firewood.

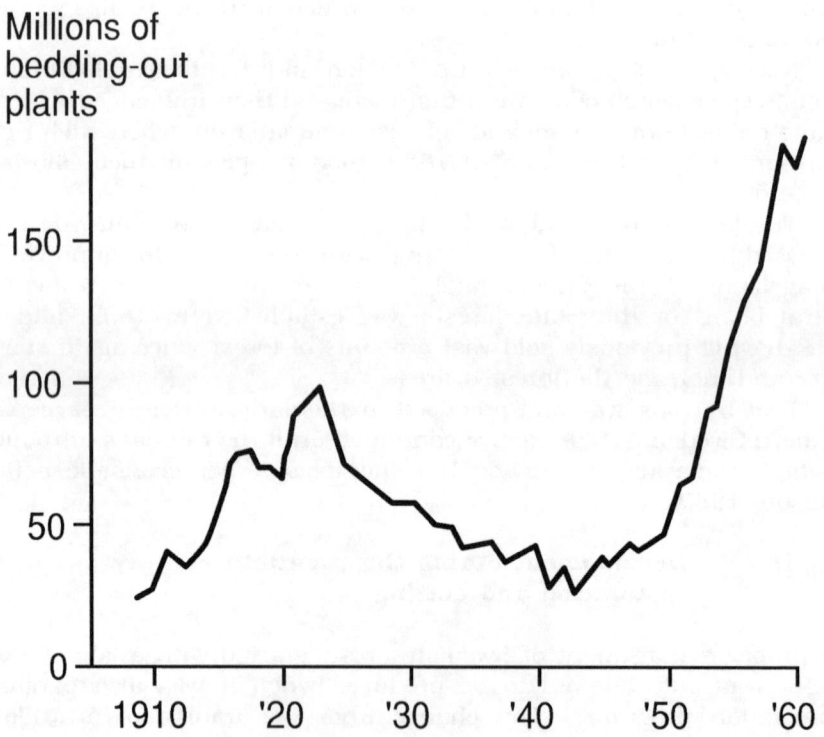

Figure 18.4. Number of forest plants distributed by the regional forestry boards from the beginning of the century up to 1960. (Source: Stjernquist, 1973).

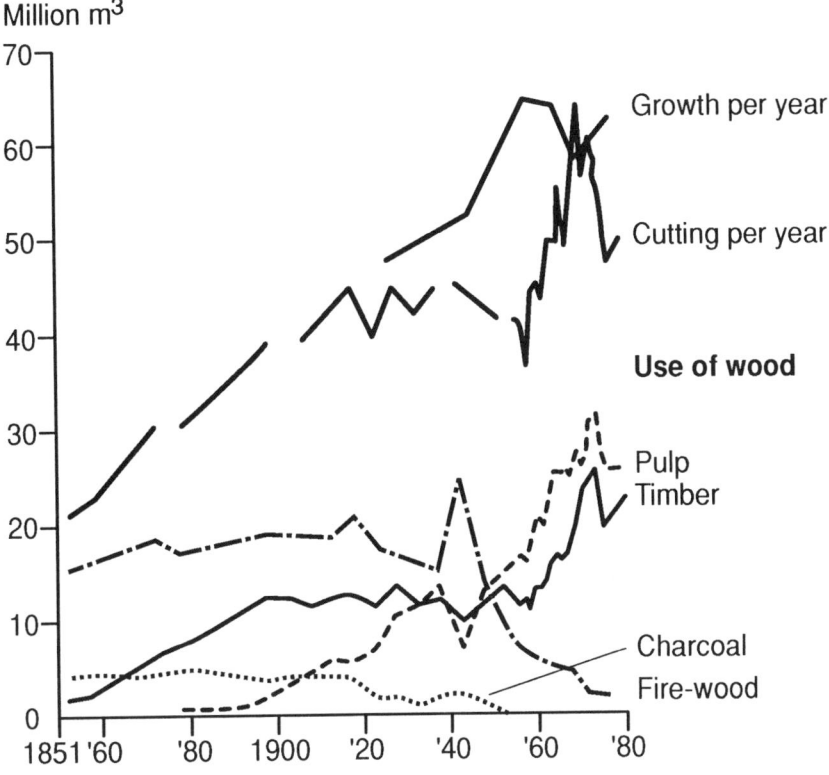

Figure 18.5. The Swedish forest budget from the 1850s up to 1980.
Note that the annual growth levels are only well known after the 1920s
when systematic field estimates were introduced. (Source: Stridsberg
and Mattsson, 1980.)

Major investments in the forest industry were stimulated by the
market conditions in Europe and the increasing gap between growth and
cutting. An increasing rate of the investments in the pulp industry took
place in the south because the regeneration time of the forests in
southern Sweden is less than half of that in the north (75 years versus
150 years) and the formerly dominating uses of wood decreased in this
region. This gradual movement decreased pressure on the woodlands
of northern Sweden but the supply in southern Sweden could not fully
meet the demand. Only about 10 per cent of the owners in the south
depend economically on the forests; thus, economic incentives are rather
ineffective. Cutting levels during the 1970s exceeded the annual growth.
The pulp industry then started to import raw material and also increased
their efforts to raise yields.

18.7.4. **Modernization of forestry**

New technologies in forestry have been introduced since the Second
World War. By 1950 the chain saw had replaced the timber saw and
the long distance transport of wood by floating was replaced by road
haulage. The tractor replaced horses during the 1960s and advanced
harvest machines were introduced. Many of the small forest owners
have formed co-operatives to cope with the modernization. Labor
productivity has increased greatly. Mechanization has also resulted in
larger clear-cuttings.

The forest composition has also changed. The stands of spruce
have increased and are now more important than pine, while the share
of deciduous and mixed forests decreased from a level of 40 per cent in
1920 to only 14 per cent in 1977. In the 1960s breeding of indigenous
trees began and foreign high yielding varieties from abroad were
introduced; for example, the North American *Pinus contorta* has been
predominantly planted in the State and company forests in northern
Sweden. About 35,000 ha are planted each year with *Pinus contorta.*
Approximately 8 per cent of the wooded area has been fertilized by
nitrogen, mostly in Norrland. The disappearance of cultivated patches
and grazing areas, particularly in Norrland, has resulted in greater
homogeneity.

18.7.5. **Present controversies on the management of forests**

State and industry have been working together on new methods and
technologies. Forestry is an important component of the national export
and national income. This implies a need to use all possible means to
increase the productivity of the forests and this tendency is, in fact,
supported by economic legislation. Modern forestry employs machines,
fertilizers and pesticides to maximize production of coniferous forests.

Shortage of wood is a major problem for industry and this
limitation has made it necessary for industry to improve its activities
and to incorporate a more strict regulation of forest management. The
sensitive and often primeval forests in the mountain areas are being
exploited at an increasing rate.

Forestry practices and the increasing wood demand of the forest
industry are causing environmental, conservational and recreational
problems. Forests are not only seen as a source of raw material to
industry. Forests are an essential part of the environment and
important for recreational activities. Uniform conifer stands with poor
ground vegetation is considered to be far from ideal. Multiple use of
forests has again been introduced on the political agenda and is
supported by changes in legislation. The conflicts between modern
forestry, conservation of flora and fauna and recreational interests are,
however, still unresolved.

18.8. The Impact of Increased Circulation of Chemicals in the Environment

The use of pesticides and artificial fertilizers in agriculture and forestry has had unforeseen side effects, such as eutrophication of lakes and rivers and poisoning effects on birds and fish. The runoff of nutrients from farmland and woodland has also been exacerbated by the construction of dikes and drainage programs.

The transformation that has taken place in energy consumption has had major impacts on the use of land and water and can be seen as an indicator of the increased circulation of chemicals from production and consumption processes (Figure 18.6).

The dominating energy sources during the nineteenth century were charcoal and firewood. Coal was mainly used for urban consumption in industry and rail transport. Coal gradually became the major energy carrier but energy consumption increased only slowly until the Second World War. Oil and water power was introduced on a small scale. A period of rapid economic growth after 1945, with the breakthrough of electrification and the welfare society, led to large increases in energy consumption. This was primarily provided by oil and water power. The substitution of timber floating and horses in forestry and agriculture by trucks and machinery enabled further development of water power, and about one million hectares of land used for hay production became available for alternative uses.

The use of oil has had a strong impact on the environment. Acidification and pollution of heavy metals have been recognized as serious problems since the 1960s. Acidification by sulfur and nitrogen oxides is regional and transboundary in nature. More than half of the SO_x and NO_x deposition in Sweden is estimated to come from abroad, so it is not only national development that has an impact on the Swedish natural environment. Lakes and soils of Sweden are sensitive to acidification. At the beginning of the 1980s, 18,000 of the 85,000 large- and medium-sized lakes of the country were classified as acidified. Four thousand were damaged seriously. Until recently soil acidification was only considered to be a long-term problem. But over the last ten years there has been a growing awareness of the threat of "forest death" from acid rain and air pollution. The issue is not as serious as in central Europe but many individual trees appear to have been affected, especially in the dominating wind direction from heavily polluting industries (Pleijel et al., 1987). The net annual increment of Swedish forests presently is larger than ever before. One explanation for this is the larger proportion of spruce (Tamm, 1987) but an additional factor probably is the increased flow of nitrogen which, besides its acidifying effects, also acts as a fertilizer. One of the hypotheses about the causes of forest death is that the nitrogen surplus might allow the trees to grow at a faster rate than the supply of other nutrients permits (Hinrichsen, 1986).

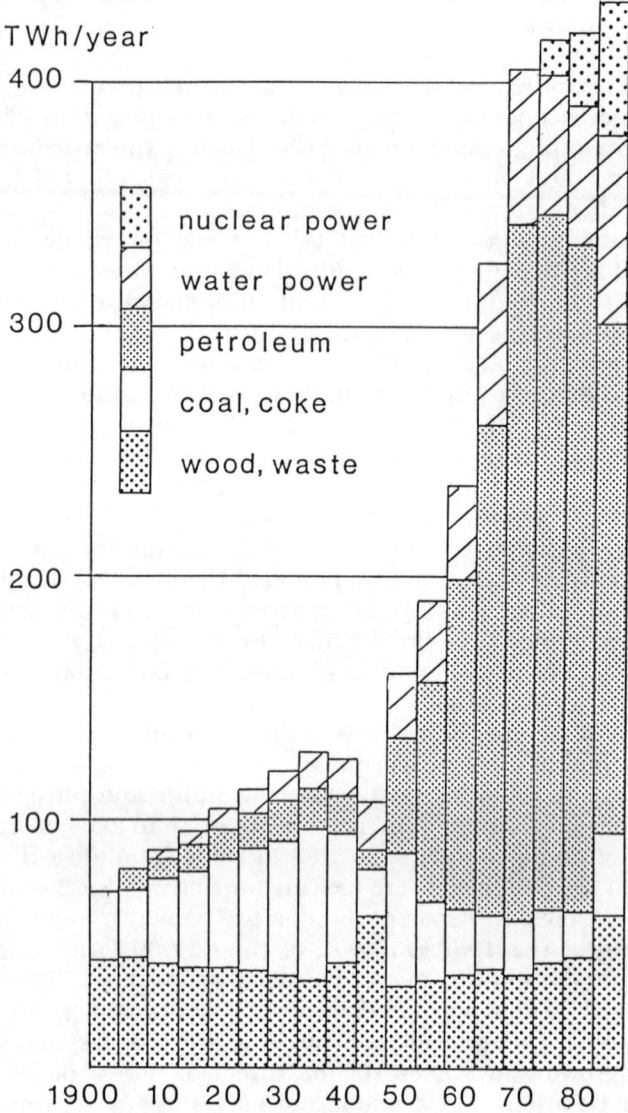

Figure 18.6. Trend in energy consumption and fuel types in Sweden between 1900 and the 1980s. Nuclear energy is planned to be phased out before 2010 according to plans adopted by Parliament. (Source: Swedish Official Statistics.)

Present environmental problems do not only result from recent development trends but also result from accumulation effects over long periods of time. An example of this is the mining district of

"Bergslagen" in central Sweden where attention was given in the eighteenth century to damages because of smoke fallout in the vicinity of the Falu copper mine.   Several lakes in this area are polluted by heavy metals that over the centuries have been leaking from mines and spoil heaps.   Various studies have shown that there have been other sources of pollution (Renberg and Wik, 1985).   The studies have examined natural phenomena, such as bird feathers, shells of mussels and sediment cores, found in museums or field sites to determine the accumulation of anthropogenic substances (Figure 18.7).   Despite the fact that the sample was taken about 600 km from the urban and industrial center of Sweden, the time series of soot sphere accumulation in varved lake sediments closely follows the development trends of coal and oil use in the country.   Approaches that use industrial and trade statistics to estimate emissions have recently been applied to attempt to complement the natural object studies (Anderberg *et al.*, 1989).

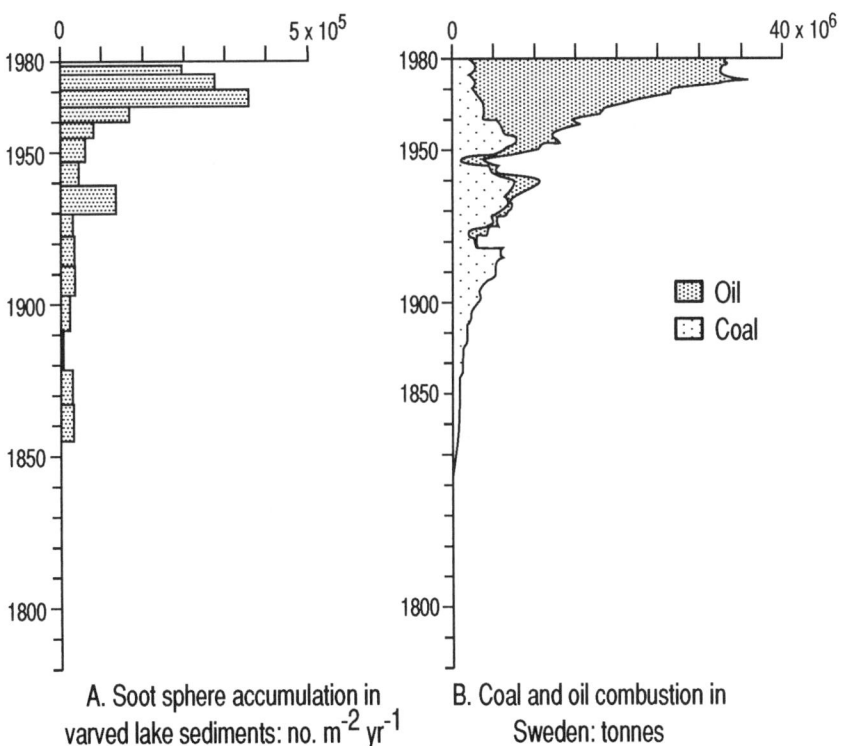

A. Soot sphere accumulation in varved lake sediments: no. m^{-2} yr^{-1}

B. Coal and oil combustion in Sweden: tonnes

Figure 18.7.   Accumulation of soot spheres over time in the sediments of hole Gråvästjärn, located in the coastal zone of Norrland, compared with combustion of coal and oil in Sweden corresponding years. (Source: Renberg and Wik, 1985).

The present decreasing trend of fossil fuel consumption is partly a result of the introduction of nuclear power but also reflects efforts to decrease the dependency on imported fuels as well as to decrease energy consumption. Oil is, however, still the most important energy carrier and the stagnation of the use of fossil fuels may be an episodic phenomenon. All nuclear power plants are projected to close before 2010 and the only energy alternatives presently available seem to be coal and natural gas.

Emissions may irreversibly change natural conditions. But environmental pollution has so far only had marginal effects on land use. Waste dumps and former industrial land that can no longer be used for biological production are probably the most dramatic examples. In the long run this continuous development threatens to radically change the biological and climatic conditions which, of course, will affect future land use patterns.

18.9. **Discussion and Conclusions**

Environmental conditions are largely affected in response to long-term changes in land use patterns, intensification and specialization of various land use types, modifications of water flows and the gradual change of the chemical environment.

There are several historical examples of environmental deterioration in relation to changes in land use patterns and societal development:

i)   overgrazing and cutting converted large areas of forest into heather moorland;

ii)   exploitation of charcoal and firewood caused local wood shortages in the mining districts during the eighteenth century;

iii)   overexploitation of forest resources resulted in a nation-wide shortage by the end of the nineteenth century;

iv)   the three decades of specialization in oats for the British market that was interrupted by the severe crisis of the 1880s and caused a deterioration in soil quality;

v)   drainage efforts that, despite many failures, continued for more than a century.

The present landscape results from a very complex long-term interaction between biophysical conditions and the socio-political, economic and technological development. Land use patterns develop in response to human interference. But the landscape is also affected by biophysical properties since they still limit the possible options of using the land, even if these limits have been altered by technological achievements.

The Government has played an important role to support development, with economic growth and increased efficiency as the dominating objectives. It has introduced several land transformation and drainage programs. The Government has increased its influence in both forestry and agriculture during the twentieth century.

The most important trait in the long-term is perhaps the development of a specialized and intensive production landscape. Rather than a varied landscape with a gradual transition from farmland to the deep forest used for multiple purposes, a "two-tone landscape" has been developed with sharp boundaries between the dense spruce forest and intensive crop production areas. The marshlands and the meadows in the open landscape have disappeared. Changes in agriculture and forestry have affected living conditions. Populations of domesticated animals and plants, including trees in the forest, have been promoted or suppressed. Noxious animals and weeds were controlled and the conditions for wild animals and plants have radically changed. The floral diversity of meadows and fields has been greatly diminished. The introduction of modern management of forests has gradually reshaped the varied indigenous forests which naturally provide favorable conditions for a diverse biotic community. The larger mammals are limited to the mountain areas while the elk and the roe have increased as their predators have been removed and they have also benefited from clear fell management of the forests.

So far, policies of environmental protection and nature conservation have been poorly integrated into industrial development programs. Physical planning has developed a spatial division with areas used for farming, forestry, industry, habitation, nature conservation, tourism and recreation. There is also a trend toward specialization of land use that may largely alter future landscapes. Activities tend to concentrate in certain areas which then focuses the environmental and social impacts.

The most likely scenario for the future would be a further development towards a landscape with separated areas of intensive specialized production or consumption of the various utilities such as wood, grain, recreation and values of culture or natural beauty. The conflicts of interest will be settled to some extent by separation, aiming to provide the various interests with space for their valued activities. Under this scenario "every man's right" will perhaps become restricted. Instead of general access to the landscape, people will be directed or at least expected to find recreation in more or less specialized recreational areas. Restrictions of access to national parks and reservations will probably also increase as outdoor recreation is concentrated and pressure on those areas increases. Agriculture and forestry will continue to modernize and intensify production. An increasing amount of agricultural land might be transformed into forest. Meadows might disappear completely with the exception of meadows within cultural reservations. This means continued international adaptation in economic terms of access to the landscape where Sweden traditionally has been exceptional.

There are, of course, alternatives to this development. The question is whether they are feasible without fundamental rethinking; for example, less emphasis on traditional economic goals and methods. Rather than specialization and separation, the goal must be to keep a varied landscape used for multiple purposes as far as possible. Different users must seek to develop practices that are compatible with other users and interests.

The environmental changes during the past centuries have been closely related to the development of transportation making it possible to use inputs from distant sources and to freight outputs to remote markets but also reflect the movement of people using Nature for different leisure time activities. Circulation can also refer to the increased movement of materials in the form of pollution. The new range of circulation has definitely resulted in new types of conflict because of the greater spatial overlap of interests.

The old order mainly generated problems at a local scale; at the level of farms and states. This type of conflict still appears but, in addition, one has now wide-ranging conflicts at larger scale, one example being the conflict between nations concerning atmospheric pollution. But modern farming, industry, forestry and the use of dangerous pollutants also threaten the quality of nature and the landscape in general. So far the differing opinions, attitudes and interests on this scale have been relatively poorly articulated and have had limited influence.

If the future landscape is not to become an arbitrary side effect of economic strategies and development, increased debates on the desirability of different alternatives are necessary. The environment must be viewed as a goal in itself, not only as a result, and be valued for its aesthetic and cultural aspects, not solely as an article of consumption. The most important challenge for the future is probably to find new institutional frameworks with an improved ability to manage the increasing conflicts, the high mobility and circulation and the complex changes of landscape and environment on various scales.

ACKNOWLEDGEMENTS

This Chapter is based on the major findings of the project *The Use of Natural Resources in Sweden since the seventeenth century*, which was directed by Professor Torsten Hägerstrand, Lund University, and Professor Ulrik Lohm, Linköping. Financial support was provided by the Bank of Sweden Tercentenary Foundation. The project also provided a major contribution to the Symposium *The Earth as Transformed by Human Action* at Clark University, October 1987 (Hägerstrand and Lohm, 1989).

REFERENCES

Anderberg, S., Bergbäck, B. and Lohm, U., 1989, Flow and accumulation of chromium as an example of a new approach to study environmental problems. *Ambio*, Vol. 18, No. 4, 216-220.

Fridlizius, G., 1957, Swedish corn export in the free trade era. Patterns in the oats trade 1850-1880. *Samhällsvetenskapliga studier*, **14**. Lund.

Hägerstrand, T. and Lohm, U., 1989, On the use of land and water in Sweden since the 17th century. In *The Earth as Transformed by Human Action* (ed. by B.L. Turner). Cambridge, Mass., USA (in press).

Hannerberg, D., 1971, *Svenskt Agrarsamhälle under 1200 År. Gård och Åker. Skörd och Boskap*. Stockholm.

Hinrichsen, D., 1986, Multiple pollutants and forest decline. *Ambio*, **5**, 285-265.

Jonasson, O., Höiher, E. and Björkman, T., 1952, *Jordbruksatlas över Sverige*. Stockholm.

Larsson, L.J., 1980, Svedjebruket i Småland. *Kronobergsboken 1979*. Växjö. **80**, 65-77.

Lindqvist, S., 1983, Naturresurser och teknik. Energiteknisk debatt i Sverige under 1700-talet. In *Paradiset och Vildmarken* (ed. by T. Frängsmyr), 82-108, Stockholm.

Mattson, R., 1985, Jordbrukets utveckling i Sverige. *Aktuellt från Lantbruksuniversitetet*, Uppsala.

Miljödatanämnden, 1982, Blågul Miljö. Miljödatanämnden, Stockholm.

Nordström, O., 1952, Relationer mellan bruk och omland i östra Småland 1750-1900. Lund.

Persson, T. and Lohm, U., 1977, Energetical significance of the annelids and arthropods in a Swedish grassland soil. *Ecological Bulletins*, **23**.

Pleijel H., Wastenson, B. and Wastenson, L., 1987, Luftföroreningar och naturmiljö i Göteborgs och Bohus län, Älvsborgs län. Göteborg.

Renberg, I. and Wik, M., 1985, Carbonaceous particles in lake sediments-pollutants from fossil fuel combustion. *Ambio*, **14**, 161-163.

Romell, L.G., 1947, Det gamla Gotland. *Ymer*, **67**, 108-126.

Schager, N., 1909, De Sydsvenska Ljunghedarna. *Ymer*, **29**, 309-335.

Sivertun, Å. and Castensson, R., 1989, Artificial drainage and its implications on the landscape ecology. A geographical analysis of the drainage activities in the Motala River basin, Sweden 1870-1950 (in press).

Stjernquist, P., 1973, Laws in the Forests. *A Study of Public Direction of Swedish Private Forestry.* Lund.

Stridsberg, E. and Mattsson, L., 1980, *Skogen genom Tiderna.* Helsingborg.

Tamm, C.-O., 1987, Tveksamt att använda ordet "skogsdöd". *Svenska Dagbladet.* 16/7.

Wester, E., 1960, Några skånska byar enligt lantmäterikartorna. *Svensk Geografisk Årsbok,* **36**, 162-180.

Chapter 19

HISTORICAL LAND USE CHANGES: POLAND

J. Okuniewski

19.1. Introduction

Most of Europe experienced a period of major economic progress during the nineteenth and the beginning of the twentieth century. At that time, Poland did not even exist on the political map of Europe; it was mainly part of the economic peripheries of surrounding countries such as Austria, Germany and Russia. This was reflected in agricultural practices which were primitive compared to the development trends in other parts of Europe. Similarly, the development trends of industry, trade and urbanization were also slow to occur.

Poland became independent in 1918 and integration of the whole territory was gradually achieved. This was complicated by the destruction of the country in the First World War. The development of the Polish economy was still behind that of other European countries. This is illustrated by the fact that about 70 per cent of the population lived in rural areas and that 60 per cent of the total population was dependent on agriculture. During the period between the two World Wars, Poland could be characterized as a typical agricultural society with low standards of living and slow economic development.

The population level increased rapidly in the period between the two World Wars, from a level of 27.2 million in 1921 to 35.1 million in 1939 (the average annual rate of population increase during this period was about 1.5 per cent). The demand for agricultural products also greatly increased during the early part of the twentieth century. The major increase in population, combined with slow economic development of the national economy, and shortages of facilities to modernize agriculture, resulted in an increase in arable land which enabled an increase in agricultural production to take place. The increasing population density in the rural areas, the low prices of agricultural products and the unemployment in the cities were also contributory factors. Arable land increased between 1924-1928 and 1934-1938 by almost 1.9 million ha (an increase of more than 10 per cent). Originally this land was primarily forest, meadows and pasture.

The total population decreased by about a third during the Second

F. M. Brouwer et al. (eds.), Land Use Changes in Europe, 427–440.

World War and more than 40 per cent of the national property was
destroyed during this period.   Major political and socio-economic
transformations have taken place since the Second World War.   These
changes have resulted in increasing industrialization and development rates
that have caused rapid changes in Polish society. Industrialization has had
a major impact on the development of the national economy, including
agriculture.   The demand for labor in industry, for example, increased
sharply and new jobs were created providing employment for many people
from rural areas. Only a small proportion of the people working in industry
actually moved to the cities, and many of them remained in the villages and
continued to farm on a part-time basis.   The high population densities in
the rural areas showed a decrease.   Industrial development also improved
the social and economic conditions for the majority of the rural population.
However, the migration trends which mainly involved young people leaving
the rural areas, over a period of more than 40 years, have caused a major
change in the age and sex distribution of the rural population in some
parts of Poland.
         The objective of this Chapter is to further analyse the trends in land
use patterns which have taken place in Poland during the twentieth century
as well as the major socio-economic and demographic changes which have
determined the changes in land utilization.

19.2. Changes in Economic Conditions and Land Use Patterns

The increased demand for food during the twentieth century has been
caused by the rapid growth of the population and the increasing demand
for employment outside the agricultural sector.   Prices of food were
relatively low and controlled; this has also contributed to high demand for
food.   In addition, wages and income have only increased slowly in Polish
society.   The income elasticity of the demand for food therefore remained
high, and this was particularly true of animal products.
         The trend of low-priced agricultural products persisted through the
first stage of industrialization and hampered the increase of agricultural
production.
         The amount of land used for agriculture has decreased over the last
30 years by about 1.6 million ha (7.5 per cent).   Forests increased during
the same period by almost 1.3 million ha (17.6 per cent).   Only part of the
land which was formerly utilized for agriculture was substituted by forest
during that period, although that was the main objective of the changes in
land use patterns.   Part of the arable land and afforested land was used for
industrial development, construction of roads and houses (Stola, 1978).
Three major trends have affected agricultural production and land
utilization:

         i)   increasing demand for agricultural products;

ii)   increasing yield levels of cereal crops, mainly due to advancements in agricultural sciences, and improved production tools;

iii)   increasing differences between the most favorable areas and the marginal areas, in terms of the efficiency of investments, and labor productivity in agriculture.

A regulated price mechanism of the increasing demand for agricultural products with decreasing prices is well known. This mechanism, however, has been used in Poland in an abnormal way since, on the one hand, food prices were kept at a low level, which increased demand, while, on the other hand, these low prices hampered an increase in agricultural production. This has been particularly true for cereal crops.

Policies of subsidizing food in order to keep prices low and demand high (particularly for animal products) have not been fully overcome yet. However, self-regulation mechanisms operate within the market for horticultural products to a wider degree inducing production increases and better adaptation to changes in demand. Present interventions by the Government to regulate agricultural prices (such as grain, meat, milk and industrial plants) are based on negotiations with farmers' organizations and also take the costs of production into account as well as market supply and demand fluctuations.

Despite a considerable backwardness in agricultural development during the first half of the twentieth century compared to other European countries, major progress has been achieved during the past few decades. The yield of cereals remained almost unchanged during the first half of the twentieth century, while it more than doubled between 1950 and the 1980s. Mean annual yields have increased during this period from 1.2 t ha^{-1} to a present level of 3.1 t ha^{-1}.

Such a considerable rise in yield could have taken place because of significant increases in the application of fertilizers and pesticides, the introduction of crop varieties which are more productive and through modernization of seed production. It should be emphasized that the largest increases in yield levels took place during the period 1966-1975 when the average annual yield increase was about 77 kg ha^{-1}. In spite of the slow growth in grain production the total area used for grain growth decreased by about 1 million ha between 1950 and 1965, and decreased by a further 0.6 million ha during the period 1965-1980. This trend was mainly due to the very low grain prices and the artificial relationship to other consumer products.

The decreases of grain yields and of cultivated area between 1976 and 1980 have caused a decrease in the total production and an increase in imports of grain. This has had a negative effect particularly in agricultural areas (in terms of decreasing income) and the national economy.

Changes in economic policy, improvements in plant productivity during the period 1982-1985, as well as progress in the application of high yielding crop varieties and the introduction of new technologies for crop cultivation have resulted in increasing yields and an increase of the cultivated area. This enabled an increase in production levels of 7.8 million tons during the early 1980s. It should also be emphasized that the yield increased by 770 kg ha^{-1} during the period between 1980 and 1987 and that the area for grain crops increased by 0.6 million ha during that period. The increase in grain cultivation was mainly achieved by decreasing the area of fodder plants and potato cultivation.

The area of cultivated land is continuing to show a decreasing trend. It has decreased by more than 300,000 ha during the 1980s. It should also be mentioned that the decreasing trends of cultivated land is a complex process which is caused by many factors whose strength of influence may vary in various time horizons. Some major socio-economic factors which have caused changes in land utilization in Poland will be discussed in the following Section.

19.3. **Major Socio-economic Determinants of Land Utilization**

The socio-economic factors which are important determinants of land utilization can be subdivided into four groups:

i)   *development of science and technology:*

- new means of production and technology;
- progress in biological and agricultural science;
- promotion of scientific achievements for implementation on farms;

ii)   *social changes:*

- population trends;
- changes in labor structure and migration;
- improvements of human skills and aspirations;
- forms and rate of generation exchange in agriculture;

iii) *economic conditions:*

- prices of agricultural products and their relation to means of production;
- effectiveness and rates of profitability in agriculture;
- alternative options for land use in the rural areas;

iv)   *political and institutional factors*:

- legal regulations concerning land ownership and land utilization;

- extension and modernization of technical and social infrastructure and multifunctional development of rural areas.

This framework does not incorporate changes in biophysical factors which also affect land use patterns such as changes in global climate.

The conditions of science and technology and its utilization may change at a medium-term level with a time scale of 5 to 10 years. Political and institutional factors change on a similar time scale.

Economic conditions may change over relatively short periods of time (one to two years). Alterations in crop structure, slight changes in land use in response to fluctuations in demand prices and profitability and the possibilities of obtaining employment outside the agricultural sector may take place over short periods of time. This might be described as considerable flexibility of production and land use in relation to the economic conditions or as *easy economic adjustment of agriculture*.

Development of agricultural and social sciences, and particularly the utilization of their achievements in agricultural practice, were comparatively slow during the first half of the twentieth century. However, new products and new technologies were introduced to agriculture between 1960-1975. Rapid improvements in mechanization of agriculture, progress in fertilization, dissemination of plant protection methods among others may serve as good examples.

Political and institutional determinants may change on medium- and long-term temporal scales. Legal regulations concerning forms of possession and directions of land use show major effects at a medium-term time horizon; similarly the promotion of the farmers' investment activities.

Extension and modernization of technical and social infrastructure as well as the creation of new jobs in rural areas, beyond agriculture, and the improvement of the agrarian structure require a long-term time horizon of at least 20 years. The influence of most social determinants also takes affect over a long period of time. Population trends and changing employment structure generally affect land use patterns over long periods of time, as they are linked to the growth of education and aspirations of new generations that will affect continuity of generations working in agriculture.

The decrease in the number of farms in Poland, the increasing acreage of the agricultural area, improvements in agricultural structure and the substitution of technical means for manpower also mainly take place over long time horizons. We observe here an interference or the influence of social - long-term, economic - short-term and political, institutional determinants as well as determinants of growth of knowledge and utilization of scientific achievements of medium time horizon.

19.4. **Regional Differences and Marginal Areas**

Land use patterns not only differ on a continental or country scale, but also at a regional scale. This is due to differences in: i) biophysical conditions, particularly the landscape ecological characteristics, quality of soils, climatic conditions and the availability and quality of water resources; ii) the level of industrialization; iii) technological achievements; and iv) technical and social infrastructure. Location with respect to urban centers as well as available infrastructure facilities and efficiency of the service sector also play an important role.

The development of industry and industrial areas and the increasing urbanization rates were the major characteristics of economic development in most of Europe, including Poland. The rural areas formed the basis for the provision of food as well as the supply of labor. The size of villages and the population density in the rural areas are the result of a long process of development.

The rural areas around the large urban and industrial centers are mainly characterized by intensive use of land for agricultural purposes and horticulture. The labor force is frequently employed in industry (in the urban areas) and concurrently runs a small farm. The expansion of towns, industry and roads in those regions curtail the areas used for agriculture. Only a small part of the population in rural areas (about 50 km radius around larger cities and industrial centers) depends entirely on agriculture and horticulture for its living.

There are four main types of agricultural practice and land use patterns in the rural areas of Poland that are not within the urban and industrial hinterland.

i)   Large socialized farms and large individual farms prevail in the western and northern part of the country; the number of people employed in this type of agriculture is small (up to about 12 persons per 100 ha of cultivated land).

ii)   A trend toward disintegration of farms is observed in the southern and south-eastern parts of Poland, with most farmers having a second income; the average size of a farm in the southern part of the country is only 2-3 ha, and the utilization of land for agriculture is very intensive. On the other hand, some areas are poorly utilized at least for some time since a proportion of these small farmers have no successors. Large villages prevail in the southern part of Poland, and the number of people employed in agriculture is more than 50 persons per 100 ha of cultivated land. High density of population, a proper transportation network and a considerable level of industrialization in this region result in the treatment of agriculture as a supplementary activity. Part of the southern provinces also have mountains and hills of major touristic and recreational importance; recreation is projected to increase further over the next few decades.

iii)   The average size of farms is about 6 ha in the central part of the country, and more than half of the cultivated land is used by farms which are larger than 10 ha; however, there are also many small farms with less than 5 ha of cultivated land and the owners of these farms usually have a part-time job elsewhere. The western and northern parts of the central region are characterized by a more favorable agricultural structure and a more advanced technical infrastructure. Land use management is also more advanced in this part of the central provinces; this is mainly due to a higher level of agriculture, more efficient functioning of institutions that serve agriculture and a more favorable agrarian structure. These characteristics have an historical origin; a properly developed network for processing agricultural products in the mid-west region is also of some importance, as this is rather underdeveloped in other regions.

iv)   The north-eastern part of the country has large areas exhibiting demographic trends which are unfavorable for long-term agricultural development; these phenomena, which have been observed over a long period of time, are also reflected in changing land use patterns, such as an increasing tendency to take land out of agriculture. These demographic development trends will be addressed in the following discussion.

Migration from the countryside to urban areas is increasing in the north-eastern part of the country, especially young women who do not return to the villages after graduation, since working and living conditions are more difficult on the farms than in the urban regions. The sex distribution in some parts of the region therefore is becoming unbalanced with a distribution of females to males in the age group 20-35 of about 50:100. Young male farmers, therefore, often leave their farms. This happens in many regions of the country, but its effects are strongest in the north-western part of Poland where regular development is endangered. Not only single farms but entire villages are becoming desolate and previously cultivated land becomes waste land or forest. There are several reasons for this phenomenon, with social conditions being most important. Socio-economic development trends in the country have provided new options for living and working and thus have abolished the need to remain on a farm; improvements in education on various levels have also provided new perspectives in life. Socio-economic development trends in the country are rather diverse. This is not only due to the dominating development trends of industries and urban areas compared to the trends in agriculture and the rural areas but also due to the lack of support to the so-called marginal areas which has resulted in increasing differences between regions, communes and even villages. Therefore, people are leaving the areas which do not offer improvements in living and working conditions. Less advanced rural areas with a poor infrastructure usually have a shortage of work in sectors other than agriculture. This holds particularly for the areas that are relatively remote from urban centers. In addition, modern agriculture requires less manpower and more means of production to increase land productivity as well as to improve working conditions and farming efficiency. Labor participation in agriculture therefore shows a decreasing trend. Finally, marginal areas are also subject to biophysical limitations, mainly consisting of land with poor soils with low natural fertility. These areas are therefore not able to sustain the average income standards because of production limitations.

Marginal areas in the south-eastern and south-western part of the country are also endangered by major demographic shifts although on a smaller scale than in the north-eastern part of the country. Deterioration of agricultural land is observed in some of the mountain areas when particularly difficult conditions for agricultural practice prevail.

A decrease in the extent of cultivated land is also observed in many other parts of Europe, most seriously in the marginal areas. This trend may increase within the next few decades.

19.5. Socio-economic Development Trends and Changing Land Use Patterns in Poland

Development trends of the Polish economy are affected largely by the foreign debt situation. This situation is seriously weakening the possibilities for the accumulation of capital and the modernization of industry as well as agricultural improvement and the technical and social infrastructure of the country. This problem is not likely to be overcome in a short period of time.

The present reform of the economic system, the increasing expenditure on research and the utilization of its results in industry and agriculture may improve economic development and the efficiency of the national economy. Structural changes in industry and major technological achievements in terms of a considerable reduction in energy and material consumption in industrial and agricultural production may occur during the period 1990-2000. This tendency is foreseen to gain momentum during the first decades of the twenty-first century.

Biological-agricultural research, as well as new means of production and new technologies for food production, may have a major impact on the development of agriculture, food production, living conditions in rural areas and land use patterns in general. The accelerating rates of yield increase for grain and other crops during the period between 1950 and 1980, and particularly during the 1980s, fully confirm the influence of genetics, plant breeding and new technologies. The major increase in crop productivity in the western and northern parts of Europe also shows that crop production can increase largely with the implementation of advancements in biological and agricultural science.

The genetic experiments on grain, which began in the 1970s, have not only brought new varieties with a potential yield level of 8-9 tons ha^{-1}, but also relatively new grain crops such as *Triticale*, capable of achieving high yields even on poorer soils. Scientific research and genetic engineering may result in a further rise in productivity and in improvements in grain quality, particularly in the increase of proteins and exogenous amino acids in fodder crops. Research into the possibility of growing genetically-engineered varieties of grain which will assimilate nitrogen from the atmosphere is ongoing. However, it is unlikely that any definite results will be achieved before the year 2010. The consequences could be much more far-reaching than indicated by Nielsen (1986). It would not only be possible to use lower rates of nitrogen fertilizer application and so decrease air and water contamination, but also to increase the production of protein in grain and reduce the import of soya and other high protein fodder components.

The development of biological and agricultural sciences will bring considerable changes in animal production. Further improvement of fodder composition as well as environmental conditions will make it possible, by improved animal conversion rates, to obtain animal products with a decrease in fodder consumption of 20-30 per cent. This will promote an increase in animal production in Poland with an increasing demand up to the end of the twentieth century and a decreasing demand after 2010. Social changes in Poland at the turn of the twenty-first century will be expressed by a further decrease in the birthrate. Table 19.1, for example, shows the population trends in Poland between 1950 and 1985 and a projection until the year 2030. This shows that population increased largely during the 1950s and 1960s with an annual increase of more than 400,000 people. Urban population more than doubled between 1950 and 1980, while rural population showed a slight decrease. Urban population is projected to increase to a level of 65 per cent of the total population by the year 2030.

Labor participation and demographic composition will change with the slowing rate of population increase and with economic development. Labor participation in agriculture is expected to decrease further over the next few decades (Manteuffel, 1987). The number of people dependent on agriculture has decreased from a level of 47 per cent in 1950, 30 per cent in 1970, to a present level of only 19 per cent. This may decrease further

Table 19.1. Population trends in Poland during the period 1950-1985 and a projection until the year 2030.

	1950	1960	1970	1980	1990	2000	2030*
Total population (in million)	25.0	29.8	32.7	35.7	38.3	39.9	43.0*
Population in urban areas	9.2	14.4	17.1	21.0	23.4*	24.9*	28.0*
in rural areas	15.8	15.4	15.6	14.7	14.9*	15.0*	15.0*
Annual increase (in 1,000)	474	445	279	321	260	160	100
Annual growth rate (number per 1,000)	19.0	15.0	8.5	9.0	6.8	4.0	2.4

Source: GUS Yearbook, 1987.
*Estimation made by the author.

to 10-12 per cent by 2000 and to 5-7 per cent by 2030. For some people agriculture will still constitute an additional occupation.

Future development trends regarding the relative distribution of the urban and rural populations are still uncertain (Okuniewski, 1988). The number of people living in urban areas increased between 1950 and 1980 at an annual rate of about 400,000, but only increased by 250,000 per year during the first half of the 1980s. Rural population, which showed a decreasing trend during the period 1950-1980, showed a slight increase of about 100,000 per year during the first half of the 1980s. This phenomenon is partly due to a setback in the development of industry, particularly in large industrial plants, and the establishment of small- and medium-sized enterprises in small towns and rural areas. This trend might continue in the near future. The rural population with an income from sources other than the agricultural sector doubled during the period 1950-1985 and at present comprises more than 50 per cent of the rural population. New jobs have been created in small towns and in rural areas during the last few years, especially in those areas with advanced technological infrastructure and touristic value. The combination of employment in agriculture together with other sectors is gaining importance not only on private farms in small areas but also on production co-operatives and state farms. A similar development trend is also observed in Israel (Pohoryles, 1985). The production of agricultural products may be combined with part-time employment in food-processing, service sector, production of construction material or in the industrial sector. Such a combination of jobs increases the income of rural people, enlivens small towns and promotes the development of rural areas. However, it also puts pressure on the local environment through problems of sewage and industrial waste disposal from small industrial plants. Experience shows that small- and middle-sized plants adapt much faster and more easily to the pressure of public opinion and respect the legal regulations concerning protection of the environment more than the large industrial centers.

Multifunctional development of rural areas means that agriculture and food production will be one of many possibilities for human activities in these areas. The next important socio-ecological function of rural areas is the regeneration of environmental concern as contact with the environment (forests, water, meadows, pasture, natural reservations and fields) is once again increased through increasing leisure time availability. This development needs encouragement before degradation of the environment has proceeded too far.

Changes in land use patterns of utilized agricultural land and forests since the 1930s are summarized in Table 19.2. Arable land covers more than 70 per cent of the utilized agricultural area, and it has decreased by about 2 million ha since the 1930s although the proportion of permanent grassland has remained constant. Arable land is projected to decrease to a level of less than 14 million ha by the year 2030. Forested land has shown a steadily increasing trend since the 1930s, and this is projected to continue over the next decades.

Table 19.2. Land use changes in Poland 1931-1986 and a projection until the year 2030.

	1931	1950	1980	1986	2000*	2030*
Utilized agri-cultural area (million ha)	20.8	20.4	18.9	18.8	18.3	17.5-16.8
Arable land**	16.8	16.3	14.9	14.7	14.3	13.7-13.3
Permanent grass-land	4.1	4.2	4.1	4.1	4.0	3.8-3.5
Forest area	6.8	7.4	8.6	8.7	9.3	9.6-10.8

Sources: Agricultural Yearbook of Statistics 1945-1965, GUS.
                   Agricultural Yearbook of Statistics 1971.
                   Statistical yearbook 1987.
   *Author's estimation.
**Including horticulture.

Agricultural transformations and changes in land use patterns until the year 2030 may vary from insignificant to very large. This will depend on two main factors:

i)   development of science, mainly biotechnological research, and the implementation of its findings in agricultural practice;

ii)   socio-economic development of the country.

There are three possible scenarios differing in terms of trends in science and economic development, in the rate of growth of crops and agricultural production as well as in the scale of changes in land use.

The first scenario is based on the assumption that a continuation of the existing trend of development will occur with minimal improvement in the efficiency of input and with reduction of cultivated land by 0.8 million ha by 2010, 1.0 million ha by 2030 and a total of 1.3 million ha by 2050 (7 per cent). It will be marginal land which will be mostly afforested and partly used for the construction of roads and infrastructural needs. Moderate increase of fertilization and diffusion of new technologies will enable an increase in crop yields up to about 4.6 t ha^{-1} in 2010, 5.5 t ha^{-1} in 2030 and 6.5 t ha^{-1} in 2050. This will secure a growth rate of agricultural production of 1.5 per cent; before 2010 the rate may be higher (1.8 per cent), and considerably lower (1.3 per cent) after 2010.

The second scenario is based on the consideration of a more rapid economic development and higher substitution rates of land and employment by technical means than would occur in the first scenario. This might result from increasing fertilization rates as well as higher efficiency of input as a result of wider implementation of science and technology achievements. Possible crop productivity to 5.0 t ha^{-1} in 2010, 6.2 t ha^{-1} by the year 2030 and 7.4 t ha^{-1} by the year 2050. This scenario would enable a decrease of cultivated areas of 1.0 million ha by 2010, of 1.5 million ha by 2030 and of 2.0 million ha by 2050, that is a reduction of 11 per cent by the middle of the twenty-first century.

The third scenario is based on a moderate productivity until 2010, but forecasts the possibility of genetic engineering solutions consisting of nitrogen fixation by cereals as well as further technological advance after 2010. This would allow a gradual decrease of nitrogen fertilization but would probably incur a slower growth of crops. However, energy inputs would be considerably lower and a significant reduction in soil, water and air pollution with nitrogen compounds would be achieved. To utilize this biological factory of nitrogen and proteins it will be necessary to maintain greater areas under cultivation. Nevertheless, one may expect diminution of cultivated land by about 0.8 million ha by 2010 and 1.0 million ha by 2030 and 2050, that is, by 5.5 per cent. Mainly marginal land will be taken out of production and might be afforested or used for recreational purposes.

Multifunctional development of rural areas will also require that some land presently used for cultivation will be deployed for the development of infrastructure, construction of housing estates and industrial and service objects, including recreation areas and wilderness parks.

19.6. Conclusions

This Chapter represents an attempt to identify possible paths and potential growth rates in agriculture and land use in Poland until the middle of the twenty-first century and has been based on the assumption that a faster rise in food production until 2010 is necessary, succeeded by a slower rise between 2010-2050. This results from demographic and socio-economic changes, decreasing birthrates and decreasing income elasticity of food demand.

Multifunctional development of rural areas has been instituted over the last decade and constitutes the major solution to counteract excessive migration from the rural areas to the urban centers. Such trends of development adapted to local conditions have been assumed probable in the long term. This means that due to the creation of new possibilities for employment in the villages and small towns there is no need for the rural areas to become desolate areas in spite of a considerable decrease of cultivated land, the number of farms and employment in agriculture.

Two main factors will influence the development of agriculture and food production: the progress in science and technology and the rate of socio-economic development of the country. Bearing this in mind, three possible scenarios of development were considered, assuring full accommodation of the food demand and an export surplus of 15-20 per cent, but considerably differing as regards to the area under cultivation and the range of substitution of land and manpower by means of production and efficiency of input, as well as their influence on the natural environment.

The fact that the degree of achievement is very similar in all three scenarios suggests that the real path of development may probably run between the first and the second. However, the third scenario is the most advantageous from the point of view of environmental protection and effectiveness of expenditure. There is, however, no certainty whether science will solve the issue of nitrogen fixation in cereals over the next few decades.

REFERENCES

Agricultural Yearbook of Statistics, 1971, Polish Office of Statistics, Warsaw.

GUS Agricultural Yearbook of Statistics, 1945-1965, Polish Office of Statistics, Warsaw.

GUS Yearbook, 1987, Polish Office of Statistics, Warsaw.

Manteuffel, R., 1987, *Filozofia Ronictwa.* Pánstwowe Wydawnictwo Naukowe, Warszawa.

Nielsen, M.H., 1986, Industrial Biotechnology in Europe, in D. Davies (ed.), *Industrial Biotechnology in Europe: Issues for Public Policy.* Frances Pinter, London.

Okuniewski, J., 1988, Strukturalne Przeobrażenia Wsi i Rolnictwa w Polsce - ograniczenia i mozliwości. Wiés i Rolnictwo, Nr. 1/58 1988. Państwowe Wydawnictwo Naukowe. Warszawa.

Pohoryles, S., 1985, *Introduction to Rural Planning and Development.* Israel Association for International Cooperation, Ministry of Agriculture, Centre for International Agricultural Development Cooperation.

Statistical Yearbook, 1987, Polish Office of Statistics, Warsaw.

Stola, W., 1978, Uzytkowanie ziemi. W. pracy pod red. J. Kostrowicki. Przemiany Struktury Przestrzennej Rolnictwa Polski 1950-1970. Warszawa.

Chapter 20

HISTORICAL LAND USE CHANGES: MEDITERRANEAN REGIONS OF EUROPE

H.N. Coccossis

20.1. Introduction

The present land use patterns and processes of change in the coastal areas of the Mediterranean part of Europe, the transformations of the environment and the prospects for future development in this area suggest the need to re-evaluate the assumptions and present context of land use policy.

The European regions that border the Mediterranean Sea have much in common in terms of the challenges and problems they face. This is reflected in development patterns and transformations which transcend administrative and political boundaries. Furthermore, differences in climatic and other geographical features (such as availability of water resources or geomorphology) coupled with cultural and historical influences (for example, Christianity) and differences in the level of development distinguish Mediterranean Europe from other countries bordering the Mediterranean Sea. There are several factors which should be taken into consideration in order to identify the geographical area of Mediterranean Europe in an objective manner, for example, the limits of olive trees or of hydrographic basins, rainfall regions or landscape. Each of these factors may lead to slightly different boundaries of the region. However, for an analysis of land use changes in relation to socio-economic conditions it would be convenient to use administrative units, as they encompass most physical boundaries, and socio-economic data are readily available for these areas (Figure 20.1).

The emphasis in this Chapter will be on the more recent past since this period shows evidence for a reversal of past trends and a rapid pace of change in land use patterns and environmental conditions. One of the theoretical postulates in regional planning which seems particularly applicable for a regional analysis of dynamics and change is the concept of ecological equilibrium (Coccossis, 1987). This concept states that a region has four sets of factors; population, resources, technology and institutions. These factors are constantly in a state of dynamic interaction. Regional changes are the result of changes in the equilibrium between these factors. A change in one generates changes

F. M. Brouwer et al. (eds.), Land Use Changes in Europe, 441–461.

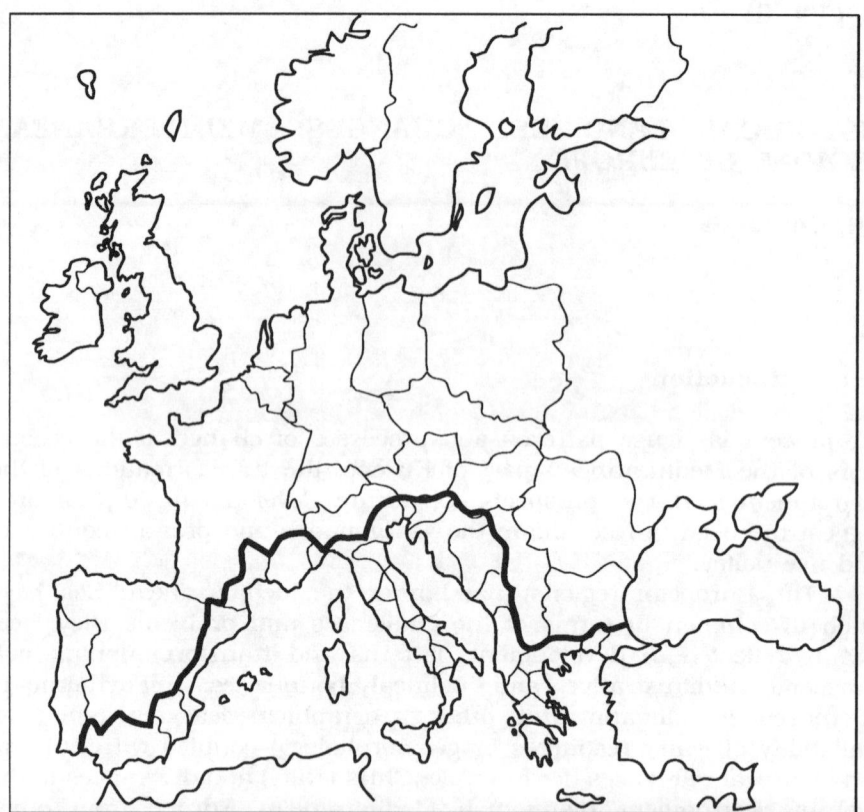

Figure 20.1.   European coastal regions in the Mediterranean.

in the others and the state of dynamic equilibrium shifts towards a new
level.   The most prominent contributors to the development of this
concept have been Gras (1922) for the relationship of the spatial
distribution of population to social organization (such as institutions);
McNeill (1976) for the relationship of population size to resources;
Rashevsky (1969) for the influence of the spatial distribution of
population on technological change; and Simon (1957) for the influence
of technological advances on the distribution of population in
geographical space.

In the above conceptual framework, changes in land use are the
result of changes in the size and distribution of population, technological
innovation and economic restructuring, social organization and policy.
A key feature in the concept of ecological equilibrium is urbanization or
the spatial distribution of population and this is directly linked to the
way land is used.   In fact, urbanization seems to have been the single
most instrumental process of change for the Mediterranean coasts of
Europe.

In the analysis particular emphasis will be placed on urbanization, regional development patterns and urban growth as these factors set the framework for changes in land use and the environment. Societal responses or policies aimed to affect the course of events will be also examined.

20.2. Historical Patterns of Development and Urbanization

"The Mediterranean is an urban region... The prevailing human order has been one dictated primarily by towns and communications, subordinating everything else to their needs. Agriculture, ... is dictated by and directed towards the town... The history and the civilization of the sea have been shaped by its towns" (Braudel, 1949)."

The development of the Mediterranean region has been marked by successive waves of urbanization and integration followed by abandonment and discontinuity. The earlier period of the expansion of Greek cities through colonization of the Mediterranean shores was succeeded by the sweeping conquests of Alexander the Great. At this time a large network of coastal and inland towns was established and reached a climax with the expansion and establishment of the Roman Empire which imposed an adminsitrative framework, with social organization supported by institutions over a wide area. The cultural and economic unity of the area was rapidly fragmented following the barbarian invasions and migrations (Gottmann, 1986). Pirates continually disrupted communications from the seventh century and this led to the economic decline and disintegration of the area until the tenth century when trade routes were re-established. Medieval times saw the rise of rich cities such as Genoa and Venice. The accumulation of wealth acceded the development and diffusion of ideas in the following period. The development of powerful sovereign states during the sixteenth and seventeenth century in the west and the east caused a decline in the importance of the Mediterranean in world affairs. The sixteenth century was an era of general prosperity and was marked by rapid population and urban growth, but also by destructive epidemics that decimated urban populations throughout the Mediterranean. The growth of the urban population in the next period (which lasted until the middle of the eighteenth century) occurred primarily in a few large cities already established (for example, Constantinople, Naples and Venice). Around the year 1700, approximately half of the largest ten cities in Europe were located in the Mediterranean. By 1860 only two of the largest ten European cities were in the Mediterranean and all of them were located outside the region by the beginning of the twentieth century. This pattern reflects wider-scale shifts of social and economic growth which occurred first in the resource-rich (coal) northern and western European regions. Technological innovation in certain key industries (textile, metallurgy, chemical, machinery and energy production) played a key role in the industrialization and modernization of western Europe. These processes shifted the centers of political and economic power away from the majority of the Mediterranean regions of Europe (Landes, 1969).

20.3. The Period after the Second World War

The population of the European countries bordering the Mediterranean increased from 141 million in 1950 to almost 183 million in 1980. This increase represents an average annual growth rate of 0.9 per cent (slightly above the European average). This figure, however, conceals the real magnitude of differences in population trends between the southern part of Europe and the rest of the continent, since it includes large countries with wide interregional differences such as France, Yugoslavia and Spain. Such interregional differences may lower the overall average rate for southern Europe. Recent trends during the 1980s also suggest a large divergence in growth rates between southern Europe and the rest of the continent. The rate of population increase during the 1970s in southern European countries (excluding France) was almost twice as much as in the rest of Europe indicating an upward trend in the growth rate for the south and a sharp decline in the growth rates for other areas (Table 20.1). The number of people living in the coastal regions of Mediterranean Europe was estimated to exceed 95 million in 1985, with an estimated 80.7 million in the level II regions of the EEC countries that border the Mediterranean (EEC, 1987), 2.7 million in Albania (United Nations, 1986) and approximately 11 million in the coastal republics of Yugoslavia (apportioned according to 1971 share of total population and adjusted for 1981 total population).

The population growth rates in Table 20.1 mask the structure and dynamics of population and, in the case of southern European countries, underestimate the potential pressures from the population on the resources of the area. A closer scrutiny of the structure of population growth in the period 1980-85 reveals that the natural increase of southern European countries is much higher than the corresponding figures for the industrialized northern and western countries (Table 20.2).

However, the differences between southern European regions and the rest of Europe are only partly due to demographic characteristics. The analysis of migration patterns and their implications for development is particularly important for southern Europe. Since 1875 the advantages of large cities for the location of industries has triggered a strong urbanization trend. The growing employment opportunities in central and northern European regions attracted migrants from the rural areas, mainly from Mediterranean Europe. As general health conditions improved, the ensuing population growth coupled with technological advancements in farming released a large amount of rural labor and led

Table 20.1. Population size and population growth rates in Europe (in millions) for 1950, 1960, 1970 and 1980, annual growth rates in brackets (%)

	1950	1950-60	1960	1960-70	1970	1970-80	1980
North	72.5	(0.5)	75.8	(0.6)	80.3	(0.2)	82.1
South	108.5	(0.8)	118.1	(0.8)	127.7	(0.9)	139.5
West	122.4	(0.9)	134.5	(1.0)	148.1	(0.4)	153.5
East	88.5	(0.9)	96.7	(0.7)	103.3	(0.6)	109.4
Europe	391.9	(0.8)	425.1	(0.8)	459.4	(0.5)	484.5

Source: UNCHS, 1986.
Notes: Total for Europe does not include the European part of the USSR.
          France is included in Western Europe.
          UK is included in Northern Europe.

Table 20.2. Population growth rates in Europe between 1980 and 1985. (Annual rates of increase per 1,000 people.)

	Growth rate	Birth rate	Death rate	Natural increase	Transfer rate
North	1.5	13.4	11.9	1.6	-0.1
South	4.4	14.0	9.5	4.5	-0.1
West	0.9	12.2	11.5	0.7	0.2
East	5.3	16.5	11.2	5.3	0.0
Europe	3.0	13.9	10.9	3.0	0.0

Notes: see Table 20.1.

to strong out-migration from the Mediterranean regions. The rural-urban shift is selective, hence the young labor force left the Mediterranean to search for employment, higher incomes and better living conditions, literally draining rural areas of key human resources. Urban employment increased at a higher rate than rural employment and an increasing number of people left the low income areas of southern Europe as income differences increased between industrialized and rural areas. This out-migration reached a peak in the 1950s and 1960s. Certain poor areas such as Mezzogiorno, Andalucia, Montenegro, Epirus and the Eastern Aegean islands were affected particularly badly by successive waves of out-migration. In general, the more disadvantaged areas (islands, mountain regions) and the less developed rural areas suffered most (EEC, 1981).

Even within each country there were strong population movements from rural areas to urban centers. The growth rate of the urban population in southern Europe was larger than in the northern and western European countries. In 1950 less than 45 per cent of the population in southern Europe lived in urban areas. By the year 1980 the urban population had increased to over 60 per cent, which was still lower than the average for Europe as a whole (Table 20.3). Urbanization was stronger in the 1950s and 1960s, while the 1970s experienced a decrease in the growth rate of urban population.

Southern European countries experienced rapid industrialization and economic growth during the 1950s and 1960s at rates higher than in many other countries. In spite of this, these countries never reached the development standards of other European regions due to structural weakness and the low level of development (reflected, inter alia, by the large share of labor force in agriculture, and production which was dominated by small enterprises.

Agricultural production has increased with mechanization and modernization of farming practices. Irrigation projects allowed intensification of agricultural activity in certain areas but at the same time increased pressures on local resources. Industrial growth has been largest in the intermediate goods sector (chemicals, metal-processing, electricity). Industry became concentrated in a few urban-industrial centers and coastal sites (due to the availability of water for processing and transport). Large-scale industrial complexes were established in most regions hoping to function as growth poles for their undeveloped hinterlands. Trade and transport also expanded which led to the development of key coastal port-cities (for example, Marseille, Barcelona, Genoa). One of the most important contributors to economic growth in this period, particularly for the Mediterranean regions, has been international tourism which has favored the southern regions. The number of tourists visiting southern European countries (France, Spain, Italy, Yugoslavia and Greece) increased within a decade by 150 per cent, from less than 20 million in 1960 to over 53 million in 1971 and over 80 million in 1982. Income from tourism also increased, for example, from US$ 4.7 billion in 1970 to over US$ 25 billion in 1982 (UNEP, 1986). Special regional development programs were prepared in some cases and these programs focused on the expansion of tourism (for example, Languedoc-Roussillon, and the Adriatic coast of Yugoslavia).

Table 20.3. Urbanization in Europe.

	Urban population (millions of people) and urban population growth (annual percent rates of increase in brackets)						
	1950	1950-60	1960	1960-70	1970	1970-80	1980
North	53.8	(0.8)	58.2	(1.3)	66.1	(0.5)	69.8
South	48.3	(1.9)	58.4	(2.0)	71.7	(1.6)	84.3
West	81.5	(1.6)	96.1	(1.6)	113.2	(0.7)	121.2
East	37.1	(2.2)	46.3	(1.8)	55.3	(1.6)	64.9
Europe	220.7	(1.6)	258.9	(1.7)	306.3	(1.0)	340.2

	Urban population as percent of total population			
	1950	1960	1970	1980
North	74.3	76.7	82.4	85.0
South	44.6	49.5	56.2	60.5
West	66.6	71.4	76.4	78.9
East	41.9	47.9	53.5	59.4
Europe	56.3	60.9	66.7	70.2

Notes: Total for Europe does not include USSR.
France is included in western Europe.
UK is included in northern Europe.

These growth rates in the Mediterranean region, which were larger than elsewhere in Europe, contributed, along with other factors, to a convergence of regional disparities between southern Europe and northern countries.

Regional development in southern Europe still lagged behind the development of the industrialized areas in the west and the north. The European south is therefore considered to be underdeveloped and peripheral compared to the rest of European regions (with some exceptions). Peripherality suggests lower development levels as compared with a European average. In 1977 for example, only one region (Provence-Alps-Côte d'Azur) had a Gross Domestic Product (GDP) per capita which was larger than the European Economic Community average, while Greece and southern Italy were below the average (EEC, 1981).

Even within each country, development was concentrated in a few areas, mainly large urban centers. This trend increased the role of urban centers in regional and national networks at the expense of smaller urban centers. Athens, for example, has evolved into an international center with a higher rate of population increase compared to other urban centers in Greece. The population of Athens increased from 1.8 million in 1961 to over 3 million in 1981. This city presently accommodates about one-third of the total population of Greece. Similar patterns can be observed in Barcelona, Marseille, Genoa, Salonika and other major urban centers along the Mediterranean coast.

20.4. Recent Development Patterns

Changes over the last decades present signs of possible reversal in the long-lasting trends and deserve special examination as they might foretell important shifts in the phenomena involved.

As already noted, population growth rates have declined but continue to be higher than in other European regions. Natural increase rates are also higher in southern regions and migration patterns in most cases have been reversed. Modest growth in population is therefore expected in the Mediterranean part of Europe.

The economic recession of the seventies had significant effects on industry and transport with serious repercussions for most southern European economies. Coastal regions were particularly affected as several large-scale industrial projects of the previous decades (production of iron and steel in Taranto in the south-eastern part of Italy and large industrial complexes in Fos-Marselle) faced serious problems of economic viability. This was particularly so for those that failed to develop linkages to other economic activities and failed to produce the expected spread effects of development in their hinterlands (Robert, 1981). Shipping, an activity that benefits coastal locations, was severely affected by the sharp reduction in international trade.

The rates of urbanization increase show a decreasing trend. Recent estimates for individual countries suggest slower rates of increase during the early eighties for southern countries (as well as for the rest of Europe) although the rates for the southern countries are still much higher than the rates for other European countries (Table 20.4). It is expected that southern Europe will continue to face increasing and intensive urbanization in the future. Although the above patterns of urbanization are expected to affect the whole Mediterranean region, the intensity of urbanization is not expected to be similar in all coastal regions.

Large urban centers still grow at faster rates than smaller ones, although the rates of increase were smaller than in the previous period.

Table 20.4. Urbanization in some European countries as percent of total population in urban areas during 1965 and 1985, and average annual growth rates of urbanization in brackets.

	1965	1985	1965-80	1980-85
Greece	48	65	(2.5)	(1.9)
Spain	61	77	(2.4)	(1.6)
Italy	62	67	(1.0)	(0.9)
France	67	73	(2.7)	(1.0)
Albania	32	34	(3.4)	(3.3)
Yugoslavia	31	45	(3.0)	(2.5)
FRG	79	86	(0.8)	(0.1)
UK	87	92	(0.5)	(0.3)
Denmark	77	86	(1.1)	(0.3)

Source: World Bank, 1988.

Some medium-sized centers have also been rapidly expanding. In contrast to several older industrial cities in the north and central European regions which face urban decline, most cities in the south are still growing. Almost all large urban agglomerations which experience growth problems (for example, Cagliari, Córdoba, Naples, Athens, Salonika, Barcelona, Murcia, Palermo and Taranto) in the EEC are located in Mediterranean regions (Cheshire *et al.*, 1987). Cities between 50,000 and 100,000 people are expected to experience dynamic growth and absorb a large share of urban population growth (OECD, 1981).

There are five Mediterranean regions within the EEC-12: Provence-Alps-Côte d'Azur, Liguria, Toscany, Emelia-Romagna and Veneto with income levels (in terms of per capita level of GDP) above the EEC average (EEC, 1987).

Most of the regions with average income less than 25 per cent of the Community average are located on the Mediterranean fringe (all are in southern Europe with the exception of Ireland). These regions account for one-fifth of the population of the Community. They generally have a low population density, a young and rapidly growing population and production oriented to agriculture (EEC, 1987).

Overall, within the EEC the economic recession of the seventies reversed these trends and regional disparities widened. This reversal can be explained by i) the structural deficiency of southern regions which makes them vulnerable to economic recession, ii) the cessation of out-migration and return of migrants home, and iii) general failure of regional policy in the previous periods to distribute the effects of development in rural areas (EEC, 1981).

These development trends reflect broader patterns of change in the spatial orientation of industry and in the preferences of people. They also result from technological innovation and institutional change. Technological advancements have had major impacts on industries and services throughout Europe leading to economic restructuring and changes in the locational patterns of economic activity. Increasing automation in production, for example, has reduced labor requirements, particularly the demand for unskilled employees. The availability of skilled labor has been an increasingly important factor in the decisions of industrial location. Technological improvements in transport and communication have reduced the importance of the cost of transport in the location of industry. Firms were able to move from the congested and expensive large urban centers of central European regions. Most of the employment growth has been in a few branches of industry, mainly those involved in research and development and high technology products which have been attracted to southern European regions offering better climatic and living conditions for their skilled employees. The preferences of employees have also changed. Standards of living rise with increasing income, and so do preferences of people to choose residential location. In the last decades, working hours have shown a steady decline, while the amount of time available for leisure and travel has increased. People become more apprehensive of the way to live with increasing income and leisure time. The problems of environmental degradation and crime, increasing commuting time and traffic congestion, the high cost of living, decline in the quality of services and increasing alienation in the large urban centers have shifted the orientation of employees, particuarly those who can afford it (highly skilled with high income) in favor of smaller communities with good climate and better living conditions. Southern cities could offer fewer problems, a better physical environment and better access to leisure facilities (for example, marinas, beaches). Thus, they have attracted a large number of people and firms.

Travel and tourism have grown very fast. The combination of good beaches, cultural attractiveness and a mild climate in southern European towns has attracted a growing number of national and international tourists. The Mediterranean area accounts for about one-third of the total world tourist business. The growth in tourist demand has led to an unprecedented increase in tourist services and increased employment opportunities not only in tourism but also in a range of other service activities.

The expansion of economic activities, the changing preferences of firms and people have induced growing pressures for development in southern regions.

20.5. **Development Patterns and Land Use**

Southern Europe is fragmented into many small coastal strips and valleys. Large plains are rare (Po Valley, Andalucia) and often inland of coastal mountain ranges. More than half of Greece, for example, has a slope of at least 10 per cent and a large part is also covered by shallow erodible soils.

Agriculture is the largest user of land accounting for 37 per cent of the total land area of coastal European countries (Table 20.5). Forests cover 27 per cent of the land, while pastures occupy 24 per cent of the total. Between 1969 and 1982 arable land decreased by 5 per cent, permanent crops remained stable, pastures and grasslands decreased by 5 per cent, while forest areas and other land use categories increased by 5 per cent and 10 per cent respectively. These figures underestimate the scarcity of land for the development of human activities in the region and diminish the problems faced since they include entire countries (Spain, France, Yugoslavia). There are only a few areas with soils suitable for agriculture, well drained, and with enough water available for irrigation. Actually, the land most suitable for urban development and farming (with an altitude of less than 500 meters) is narrowly confined in a 50 kilometer wide coastal zone.

Table 20.5. Land use distribution in 1974 in millions of ha.

	Total area	Arable and permanent crops	Permanent meadows and pastures	Forests	Other
Albania	3	1	1	1	-
France	55	19	14	14	8
Greece	13	4	5	3	1
Italy	30	12	5	6	6
Spain	50	21	11	15	3
Yugoslavia	26	8	6	9	3

Source: UNEP, 1986.

Despite being a scarce resource, land has not always been used properly. On the contrary, it has often been abused. Plato mentioned the depletion of forests in the fourth century B.C. in the Attica region around Athens. Wood requirements for heating, shipbuilding and early mining, as well as the destruction of forests during wars, through fire, overutilization but also abandonment of resources, have seriously undermined the survival of natural and man-made ecosystems throughout history.

Mediterranean areas have never been self-sufficient in food production. Over the past centuries, society has become increasingly dependent on agricultural imports and has shifted cultivation practices to a selected number of crops. Mediterranean areas therefore became increasingly integrated in the world economic system. Cultivation practices which evolved over centuries, incorporating sound management practices (such as crop rotation, fallowing/planting, soil protection through artificial stone walls and terraces) have been abandoned with severe impacts on the natural environment and society. Mediterranean agriculture is characterized by its ancient tradition in the cultivation of specific products, including olive oil, wine, citrus fruits, hard grain and sheep.

Around 1900, cultivated land covered 39 per cent of Spain, 46 per cent of Italy and 19 per cent of Greece (Braudel, 1949). The importance of the agricultural sector increased considerably in the following decades although this involved the intensification of cultivation rather than expansion. In 1970 the cultivated areas covered 42 per cent of Spain, 48 per cent of Italy and 31 per cent of Greece. No major changes have taken place over the past fifteen years (World Bank, 1988).

The intensification of agriculture has contributed to serious land management problems, although it has satisfied pressing needs and improved income in southern European regions.

The need to feed the local population has led to the extension of cultivation to marginal lands, often resulting in soil erosion and depletion. The mechanization of agriculture has enabled more intensive use of the land, reducing labor and rendering most of these marginal areas uneconomical for cultivation. As rural labor migrated to the large urban centers, marginal areas (uplands, steep slopes and islands) were abandoned. For example, arable land decreased by 7 per cent in the French part of the Mediterranean area between 1963 and 1977 (UNEP, 1986). More than half the land area in the southern European countries is considered to be prone to erosion, the intensity of the erosion being highest in Spain, southern France, Albania and Greece. The abandonment of lands disturbed the precarious balance attained through years of adaptation and the adoption of careful practices in cultivation and environmental monitoring. Soil erosion increased and intensified as a result of overgrazing practices (due to overstocking) leading to "desertification" (UNEP/FAO, 1977; Margaris, 1987).

The total coverage of irrigated land increased by almost 50 per cent (or 3 million ha) between 1965 and 1982. The annual increase of irrigated land in Greece during the 1950s and 1960s was about 25,000 ha. In some cases irrigation and drainage schemes have led to increasing soil salinization either through sea water intrusion (through excessive exploitation of underground water resources) or due to repeated irrigation and poor drainage (Chapter 13).

Overgrazing also accounts for a large share of land degradation problems. The overgrazing of bushes and young trees by free roaming herds of goats even in burnt out areas has been frequently cited as the main cause of environmental stress (Margaris, 1987) as it prevents forest regeneration.   Greece and Albania are typical examples of such deterioration processes. However, changing economic conditions and the depopulation of remote rural areas are expected to ease the pressures of overgrazing; this has already been observed in Italy and France (OECD, 1983).

Mediterranean ecosystems noted for their rich diversity have been subjected to periodic disturbances by fire. Fire is even considered an intrinsic natural regeneration mechanism. In the past few decades, however, forest destruction through fires has increased at an alarming rate.   The areas burnt by fire have increased dramatically for all southern European countries between 1960 and 1981. For example, about 500,000 ha of land were burned during 1979 in southern European countries (UNEP, 1986). Forest fires in Greece destroy each year an area which is slightly larger than the area replanted (OECD, 1983). These statistics underline important losses in fauna and habitats as well as increasing risks of soil erosion (Figure 20.2).

Although the level of urbanization in southern coastal regions (60.5 per cent of the total population) is lower than in other parts of Europe, the intensity of the phenomenon is much stronger than it appears. Urban development is concentrated in a few suitable sites along the coast and the pressures on land resources have been increasing (Figure 20.3). It is estimated that over 15 million inhabitants live in 10 large coastal metropolitan centers of southern Europe with more than 500,000 people, at gross densities that range from about 1,500 people km^{-2} to over 15,000 people km^{-2} (Barcelona). In some cases urbanization levels are already very high.   On the French coast between Marseille and Menton, for example, which is a stretch of 350 kilometers, coastal urbanization is estimated at 92 per cent (UNEP, 1986).

Spatial patterns of urban growth are not uniformly distributed. Since the early 1960s the region that encompasses Marseille and Nice has had the largest growth rate of all French regions.   The total population increase was about 1.1 million, and 50 per cent was concentrated in the coastal communities (Renucci, 1986). Coastal areas at the fringe of large metropolitan centers are becoming increasingly urbanized (Athens and Naples). In Greece about 70 per cent of the population, 90 per cent of tourist facilities and 80 per cent of industry are located within a narrow 50 kilometer wide coastal zone.   An examination of population growth by urban area in the decade 1971-81 for the EEC Mediterranean countries reveals that urban centers all along the coasts of Spain, France and Greece exhibited fast rates of growth (over 1 per cent) which persisted and in some cases increased further in the period 1975-81 (Seville, Málaga, Perpignan). Growth rates

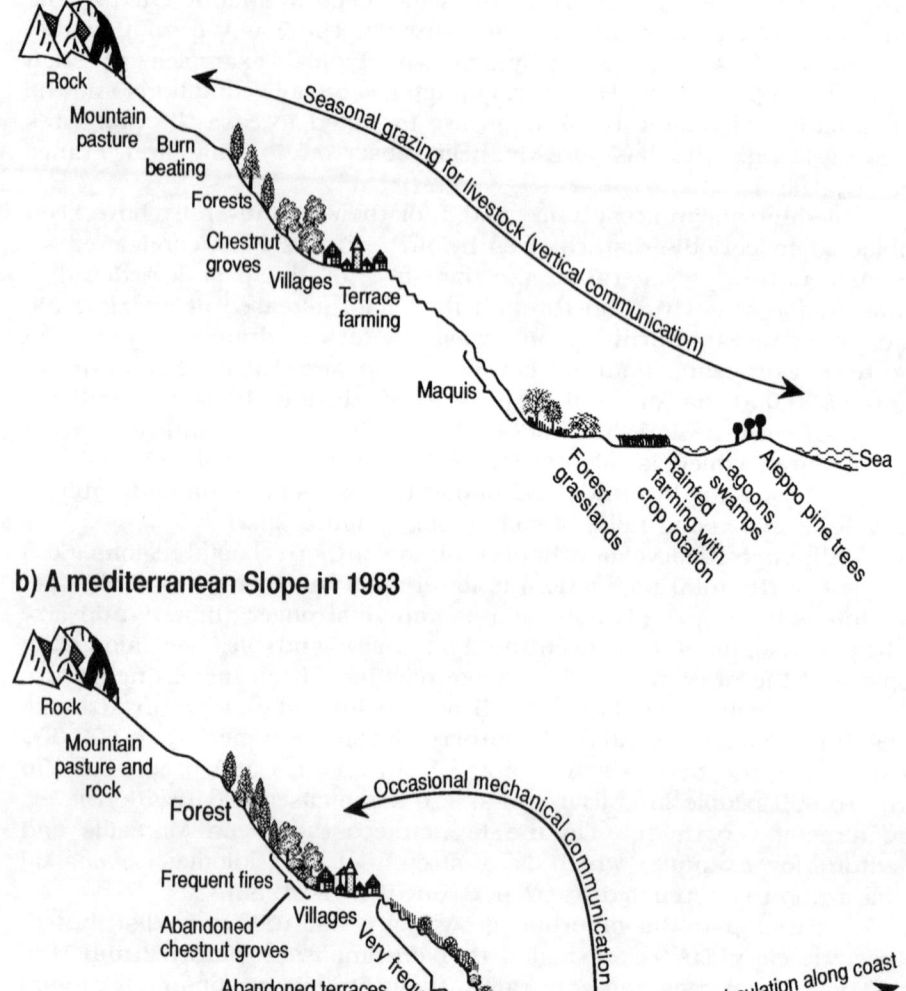

a) A Traditional Mediterranean Slope

Rock

Mountain pasture

Burn beating

Forests

Chestnut groves

Villages

Terrace farming

Seasonal grazing for livestock (vertical communication)

Maquis

Forest grasslands

Rainfed farming with crop rotation

Lagoons, swamps

Aleppo pine trees

Sea

b) A mediterranean Slope in 1983

Rock

Mountain pasture and rock

Forest

Frequent fires

Abandoned chestnut groves

Villages

Very frequent fires

Abandoned terraces and bush covered fallow land

Garrigues and degraded grassland

Occasional mechanical communication

Factories and warehouses

Intensive irrigated farming

Intense circulation along coast

Lagoon

Tourism

Sea

Figure 20.2. Changes in land use and environmental hazards. (Source: UNEP, 1986.)

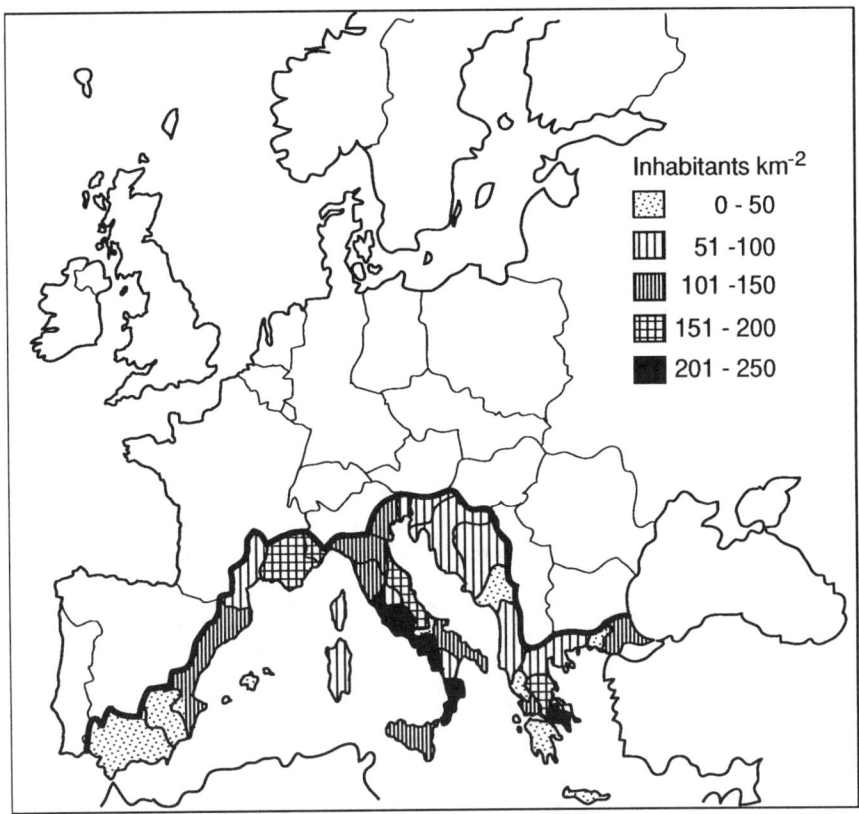

Figure 20.3.   Gross population densities in coastal regions.

have declined slightly in most Italian cities; Barcelona and Nice also exhibit this trend (Cheshire *et al.*, 1987).

Southern European shores are becoming increasingly attractive to industrial firms and enterprises as the development of infrastructure (such as roads, railways, airports, ports, industrial and office parks, communication networks) improves accessibility to the large international centers of the world and the range and quality of urban services provided.   The expansion of industry generates multiple increases in service employment and growth in urban development.   In some cases, zones of industrial and urban settlement have occupied the shores creating a coastal "megalopolis" (Doxiadis and Papaioannou, 1974).

However, the largest contributors to coastal urban development are recreation and tourism.   The expansion of tourism has led to the development of hotel complexes, camping areas, recreation and sports complexes, marinas, commercial facilities (for example, restaurants, bars, boutiques and souvenir shops) and a sprawling development of vacation houses, rental apartments and condominiums.   Along the southern coast of France (coastal departments of Provence-Alps-Côte d'Azur), for example, more than 160,000 new units were developed as permanent residences between 1975 and 1982 as well as more than 65,000 secondary residence units; almost 60 per cent of them were developed in the coastal communities representing a 15 per cent increase of accommodation (Renucci, 1986). Rural seaside communities are rapidly transformed into vacation areas with scattered holiday houses and summer resorts. The increasing mobility among European countries as a result of the integration of European markets will intensify such phenomena, as southern regions richly endowed in amenities are expected to attract a larger share of vacationers and pensioners and become increasingly urbanized.

Urban development patterns in coastal communities mask the actual pressures on coastal resources. Population density increases sharply during the tourist season. On the French Riviera tourist densities reach a level of as much as 1,800 tourists km^{-2}, since about four million people visit the coastal zone. Population figures may increase by an average 150 per cent during the tourist season (Renucci, 1986). Problems with the management of water resources increase in most coastal cities as consumption patterns of the local population change and the number of tourists increase. Urban infrastructure (sewer, water and electricity networks, telephone lines and roads) is heavily burdened by the summer peaks. The necessary investments often surpass the scale and costs justified by the size of local communities heavily taxing their natural, financial and human resources. These development patterns degrade the environment, mainly in the form of water pollution, urban sprawl and decline in the quality of life. High costs of services, crowding and traffic congestions threaten the once peaceful coastal communities (Figure 20.4).

20.6. **Policy Responses to Pressing Problems**

The growing pressures for development in southern European coastal regions and the conflicting demands on their limited resources, particularly land, have awakened local societies. The need for changes in policies is becoming a high priority. Policy responses address three sets of problems: i) regional development; ii) control of land use in urban areas and; iii) environmental protection.

Most countries attempted to restore their economies after the Second World War. Southern regions faced rural depopulation and abandonment of land, while a few metropolitan areas faced increasing pressures for development. Regional policies for rural areas in the period focused on the stimulation of national and local economies with support from regional development plans and projects. Programs first focused on the development of infrastructure, while the emphasis later shifted to the establishment of industrial complexes and tourist facilities.

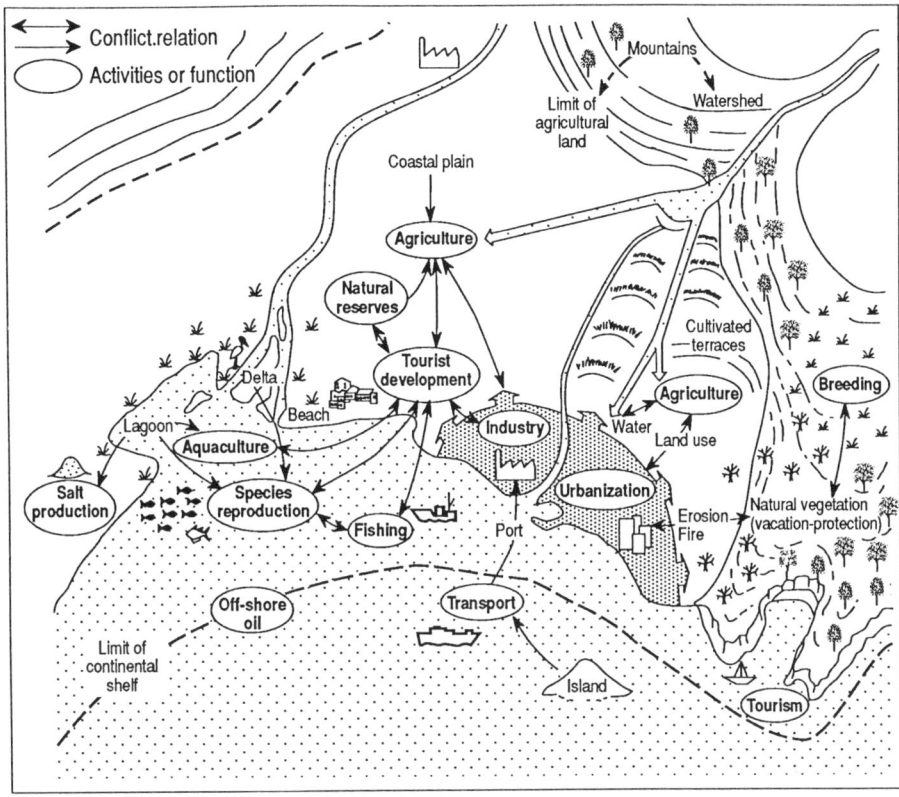

Figure 20.4. Land use conflicts in coastal areas. (Source: UNEP, 1986.)

The prevalent model of strategy in these plans was the stimulation of economic development through large-scale public investments, based on the "growth pole" concept, which were expected to contribute to the national economy, create employment opportunities and retain the out-migration of rural population to the large metropolitan centers.   The 1960s in particular was a period of intense policy making in regional development as countries realized the growing disparities between metropolitan regions and rural regions. Many of these programs focused on the Mediterranean   regions.   Among the   most characteristic examples are the programs for Mezzogiorno (southern Italy) and Languedoc-Roussillon (south-west Mediterranean coast of France).

Large-scale development programs even if well prepared and justified require long-term commitments in terms of financial and political support. These commitments generally exceed the time horizons of many politicians, require large amounts from national budgets, as well as the ability of the public administration to cope with complex operations. Substantial progress was achieved in the establishment of an administrative framework for regional development.   Often the assumptions of these programs did not reflect the complexity of regional development issues, nor incorporate the flexibility required to cover all the uncertainties involved. Subsequent changes in the economic growth prospects (energy crisis, recession and restructuring of world economies) and decreasing budgets have seriously undermined such programs and questioned their justification and survival. Several programs were drastically modified in scope or even abandoned. In the seventies and eighties regional policy turned away from grandiose development schemes adjusting its orientation to medium-term management of local resources reflecting the growing anxieties about the uncertainties of the future.

The control of urban development also attracted the interest of policy makers in southern European countries. Early policies were struggling to keep up the provision of basic urban infrastructure (roads, sewers, water, electricity) to the growing cities. Urban planning controls were adopted with varying success depending on political commitment and the efficiency of public administration. In most cases policies to contain and control urban growth failed to turn out the expected results. Urban sprawl in many southern cities continues to be a major problem. Some administrations have attempted to adopt strict policies to guide urban development away from certain areas (agricultural land, forests) with limited success, as the main emphasis was on physical planning control mechanisms (zoning in most cases). In some areas, urban policy was linked to regional policy in an attempt to exploit the local growth potential and guide development in desired directions (for example, "technology" and "research and development" parks in Barcelona and Sophia-Antipolis/Nice). As growth rates decline and standards of living rise the emphasis of urban policy shifts gradually to the expansion of the range of urban services provided. Urban policy increasingly reflects the concern for the quality of life and gradually merges with environmental policy.

Environmental degradation in urban and rural areas has led to a rise in public awareness of environmental issues and the adoption of environmental policy since the early seventies. Control of atmospheric and water pollution have been major concerns for the majority of urban areas in southern regions. In Athens, for example, no new industries are allowed to be established and expansion of existing industries is strictly controlled, traffic restrictions have been imposed and fuel improvements have been announced to control air pollution, while a large sewage treatment plant is under construction to control water pollution. Similar measures have been adopted or are under consideration in Barcelona, Nice, Genoa and Rome. A special lagoon management program was adopted in Venice to control the "sinking" problem of the city. Environmental protection measures in rural areas have been twofold: i) to improve, maintain and protect agricultural production which is still a major contributor to the local economies; and ii) to protect areas of environmental significance (wetlands, estuaries, lagoons) from the sprawling development of tourism. The diversion of a river has been promoted in Greece to provide energy and to irrigate the plain of Thessaly which faces salinization problems. The protection of the Adriatic coast (Jadran III project) has been implemented in Yugoslavia. A special program was adopted for the management of the Provence-Côte d'Azur region with the purchase of key sites by the Conservatoire du Littoral (a land banking institution) and the designation of natural reserve areas. This program aims to control the proliferation of tents, trailers and traffic and sprawling development which threatens the Camargue.

Most of the present problems in southern European regions, however, surpass the capabilities of individual local and national authorities, particularly since these problems are induced by larger scale changes. An increasing number of international bodies is becoming involved in issues which, until quite recently, were mainly of national and local concern. This is particularly the case with co-operation over environmental issues. The Mediterranean Action Plan is probably the most widely known program. Its special component, the Blue Plan, is of particular interest for coastal planning in the Mediterranean. The Council of Europe has repeatedly dealt with the problems of coastal areas through the Conference of Ministers of Regional Planning (Torremolinos, 1983) and the Conference of Ministers for the Environment (Athens, 1984). Various other organizations such as the Organization for Economic Co-operation and Development (OECD), the United Nations Economic Commission for Europe (UN-ECE) and UNESCO have also largely contributed to international co-operation. Special attention should be given to the European Economic Community which has particular interest for some of the southern European regions. In 1981 the European Parliament endorsed the European Coastal Charter which includes a special program of action. The Community has been actively involved in regional policy through the European Regional Development Fund and a special program for the assistance of southern regions to adjust in the restructuring of the European market (Mediterranean Integrated Programs).

20.7. Conclusions

The broad framework for the future of southern European coastal regions is set by increasing population pressure in terms of numbers (residents and visitors) and attitudes (preferences of lifestyle), rich endowment with natural (recreation related) resources and expansion in their accessibility to a widening number of European workers and pensioners, shifts in the locational orientation of economic activity, technological innovation and changes in social organization. The interaction between these elements sets the stage for changes in the way land is going to be used in the future, as it has done in the past.

The national economies and societies become increasingly integrated and interdependent. Patterns of change are much broader in scope and scale than they were previously; pervading and persisting.

> "...new orbits shape up...largely structured by networks of transactions, by communities of beliefs and interests between institutions and groups better identified by the cities and regions where they are based... The roots of modern problems and of basic philosophies are perhaps more obviously Mediterranean today than they were in recent centuries" (Gottmann, 1986).

Southern European coastal regions are expected to become more developed and more urbanized, to face increasing pressures to reconcile development and environmental protection and to change social responses, shape new attitudes and face the future.

Two major directions of policy changes appear to be promising for Mediterranean countries:

i)   an integration of scales in policy making as problems extend beyond administrative divisions; policies at the national level should be linked with regional and urban policies;

ii)   an integration in the scope of policy making as a reflection of the above through the integration of goals, consolidation of means and better co-ordination of efforts; policies for national growth regional development, urban growth, land use and environmental protection should be combined in order to face the interdependence and complexity of present problems in society.

REFERENCES

Braudel, F., 1949, *The Mediterranean and the Mediterranean World in the Age of Philip II.* (English edition 1973.) Collins, London.

Cheshire, P., Hay, D., Carbonaro, G. and Bevan, N., 1987, *Urban Problems and Regional Policy in the European Community.* CEC, Luxembourg.

Coccossis, H.N., 1987, Regional development and desertification in south Europe: from abandonment to overutilization. In R. Fantechi and N.S. Margaris (eds.), *Desertification in Europe.* Reidel, Dordrecht.

Doxiadis, C. and Papaioannou, J., 1974, *Ecumenopolis.* ACE, Athens.

EEC, 1981, *The Regions of Europe.* COM (80) 816 final. CEC, Brussels.

EEC, 1987, *Third Periodic Report from the Commission on the Social and Economic Situation and Development of the Regions of the Community.* COM (87) 230 final. CEC, Brussels.

Gottmann, J., 1986, Orbits: the ancient Mediterranean tradition of urban networks. *Ekistics*, **316-317**(1-2), 4-10.

Gras, N., 1922, *Introduction to Economic History.* Harper, New York.

Landes, D.S., 1969, *The Unbound Prometheus.* Cambridge University Press, Cambridge.

Margaris, N.S., 1987, Desertification in the Aegean islands. *Ekistics*, **323-324**(2-3), 132-136.

McNeill, W., 1976, *Plagues and Peoples.* Doubleday, New York.

OECD, 1981, *Urban Growth in OECD Mediterranean Countries in the 1980s.* Report UP/MG(81)1 draft. OECD, Paris,

OECD, 1983, *Environmental Policies in Greece.* OECD, Paris.

Rashevsky, N., 1969, *Mathematical Biology of Social Behavior.* University of Chicago Press, Chicago.

Renucci, J., 1986, Littoral "Hypertrophy" in the French "Provence-Alpes-Côte d'Azur" region. *Ekistics*, **316-317** (1-2), 74-81.

Robert, J., 1981, *Problèmes et Politiques d' Aménagement des Régions Côtières en Europe.* Paper MAT-6-HF 23 Conference Européenne des Ministres Responsables de l'Aménagement du Territoire. Council of Europe, Strasbourg.

Simon, H.A., 1957, *Models of Man.* Wiley, New York.

Spengler, O., 1928, *The Decline of the West.* Knopf, New York.

UNCHS (HABITAT), 1986, *Global Report on Human Settlements.* Oxford University Press, Oxford.

UNEP, 1986, *Overview of the Mediterranean Basin-Blue Plan.* RAC/BP, Sophia-Antipolis.

UNEP/FAO, 1977, Food, agriculture and the environment in the Mediterranean basin. *Ambio* **6**(6), 368-370.

United Nations, 1986, *World Population Prospects: Estimates and Projections as Assessed in 1984.* Population Division of the United Nations, New York.

World Bank, 1988, *World Development Report 1988.* World Bank, Washington D.C.

Chapter 21

LAND USE STATISTICS IN NATURAL RESOURCE ACCOUNTING SYSTEMS

A.M. Friend

21.1. Introduction

"The materials of all wealth originates primarily in the bosom of the earth; but it is only by the aid of labour they can truly constitute wealth. The earth furnishes the means of wealth; but wealth cannot have any existence, unless through industry and labour which modifies, divides, connects and combines the various production of the soil, so as to render them fit for human consumption." (Adam Smith, *Inquiry into the Nature and Causes of the Wealth of Nations*, 1776.)

21.1.1. Land: an economic perspective

The idea of sustainable development is increasingly evoked as a major theme in national and international progress and its quest viewed as an enlightened policy. The eighteenth century physiocrats and Adam Smith were also concerned with the very same issue with an agricultural bias. Thus, land was viewed as the ultimate source of wealth and, along with labor and capital, a factor of production. The finite nature of land resources was also considered a limiting factor to economic growth, a thesis developed by Malthus into an alarming specter of human impoverishment due to the differential growth rate between human reproduction and that of material/energy to sustain the population. Ricardo, another progenitor of modern economic thought, examined the origin of economic rent. The difference in the natural qualities of land was identified as the *source of rent* reflecting, in essence, the difference in productive potential between the most and least productive lands, such as marginal land. This concept is much in favor today for calculating an appropriate tax to encourage, through the price mechanism, more efficient uses of natural resources (public appropriation of natural rent).

By the mid-nineteenth century we begin to see concern about the destructive nature of economic growth and the upper limits for further progress. Again, nature is highlighted as the limiting factor. This theme is developed in the John S. Mill essay on *The Stationary State*, a kind

F. M. Brouwer et al. (eds.), Land Use Changes in Europe, 463–483.
© 1991 *Kluwer Academic Publishers. Printed in the Netherlands.*

of co-operative utilitarian state living within the means given by its natural resource base.   Marx, in his *Economic and Philosophic Manuscripts* said "The worker can create nothing without nature, without the sensuous external world. It is the material on which his labour is manifested, in which it is active, from which and by means of which it produces" (Capra, 1982). A mid-nineteenth century essay discussing the loss of nutrients from the soils by the removal of crops commented (Waring, 1857), "labor employed in robbing the earth of its capital stock of fertilizing matter is worse than labor thrown away. In the latter case it is a loss to the present generation; in the former it becomes an inheritance of poverty for our successors".

The turn of the century saw the development of the Conservation Movement in North America which expressed alarm at the rate of deforestation, degradation of soils and water, the rapacious exploitation and wasteful consumption of other natural resources, and the rapid disappearance of the wilderness.

In recent years a number of milestones have laid out an alternative path toward economic development in harmony with the capacity of nature to sustain the material and energy needs of the human population: the 1972 Stockholm Conference raised the alarm of global pollution and contamination caused, *inter alia*, by the rapid growth of industry, energy consumption, human settlements and the use of man-made toxic chemicals in agriculture; the publication of *The Limits to Growth* (Club of Rome, 1972) projected limits of the supply side of natural resources; and the Brundtland Report (WCED, 1987) focused attention on the potential of nature to satisfy the demand side of economic growth under constraints in the use of technology and re-evaluations of social and economic goals. The latter emphasized the instrumental role of governments in restructuring economic production processes in harmony with natural processes.

One of the conclusions that can be drawn from this recurring theme is that land, as a natural resource, plays a fundamental role in the theory of economic growth and the degradation of these resources is, in some sense, the Achilles' heel of human economic progress. Needless to say, neither analysis of the facts, nor interpretation of theory, will draw the analysis to the same conclusions or necessarily propose compatible solutions. Indeed, the current debate between (neoclassical) economists and ecologists serves to emphasize the different paradigms from which solutions (and policies) are proposed. It is difficult to make a case on the significance of the loss of soil productivity when much of the world is awash with food surpluses! In a similar vein the Club of Rome Thesis fell apart with the fall in oil prices.

In the past two hundred years the World's population has increased several fold accompanied by an even greater increase in economic production, consumption and capital accumulation. Does this not disprove the classical theory of land as the limiting factor to economic growth? Of course, progress in technology and settlement of new lands (including access to global natural resources) can be evoked as explanatory variables, which, in essence, may merely extend the time frame toward the inevitable "stationary state" or, in today's language, "the Steady State Economy" (Daly, 1973). Will it be a matter of a decade or two, a hundred years or the next millenium? Clearly, this is not an argument for those who are convinced that technology will master nature. A better argument is to *exact the price* of the "technology fix".

This Chapter attempts to address this question by outlining a framework for an information system which links the stock and flow of natural resources with the *System of National Accounts* (SNA), or with the equivalent in centrally planned economies, the *Material Product Balance System* (MPS).   By this means, it is hoped, that the trade-off between technology and nature can be raised to the level of macroeconomic performance or GDP. The need for this kind of accounting has been advocated in the numerous reports and analyses of environmental strategies, including the World Conservation Strategy and the Brundtland Report (WCED, 1987). A system of *Natural Resource Accounting* (NRA) is being proposed which:

i)   portrays national aggregates of the *stock and flow* of natural resources;

ii)   provides the physical (and parallel money) flow in the *input/output accounts*;

iii)   provides the statistical database to calculate the (monetary) value of natural assets in the *national balance sheets*;

iv)   provides the statistical database for the environmental component of *social benefit accounting* in the SNA such as satellite accounts.

Land use plays a special role in NRA; land mass and its qualitative attributes is a natural resource in its own right and land use mapping can be employed as a spatial integrator of natural resources. This Chapter describes in general terms the nature and the scope of NRA, and further explores the characteristics of land use databases in the context of constructing these accounts. Some of the salient points in favor of the NRA approach to support the information base for sustainable development policies conclude this Chapter.

21.2. Natural Resource Accounting

21.2.1. What is a system of NRA?

Statistical offices, in response to the demand for environmental and natural resource statistics, have considered various conceptual frameworks for their development. Three dominant approaches have emerged:

i)   the most straightforward is to compile data bases on the environmental problems as perceived by institutions concerned with environmental protection and conservation; this framework, sometimes referred to as the "media approach" tends toward sectorization of the environment into air, water, land, wildlife, forests and so forth;

ii)   a more abstract approach, referred to as "stress-response", defines the problem in terms of the dynamics of human (and natural) activity and associated environmental response; the key objective of the stress-response framework is to link statistics of socio-economic activity with indicators of environmental conditions (Rapport and Friend, 1979);

iii)   a third approach defines the problem as that of quantity and quality variables of stocks and flows of natural resources; an underlying objective is to develop a formal accounting system compatible with the SNA; Norway and France (Compte du patrimoine natural) pioneered in the development of NRA; these frameworks are described in United Nations (1984).

The difference between environmental statistical frameworks (i and ii) and NRA (iii) is on emphasis rather than substance. The latter is concerned first and foremost with quantity flows, whereas the former with qualitative states. Nonetheless, the content and data specifications, and *ipso facto* the supporting databases, are practically the same for both approaches. The key environmental questions raised, such as the capacity of natural resources to sustain economic development, the viability and integrity of ecosystems, and the carrying/assimilative capacity of the environment, are of sufficiently complex nature to require both stock/flow and quality assessments. It is generally agreed that NRA and environmental statistics are not alternative but essentially complementary approaches to environmental information systems. It should be noted that frameworks, along with concepts, definitions and classification systems, are the "meta language" for defining the boundaries and content of a statistical field. The users, on the other hand, are concerned with the relevance, validity and timeliness of data.

Figures 21.1 and 21.2 provide a diagramatic overview of NRA and its linkage with the SNA. The core of the system is the accounts of the *stock and flow of natural resources.* These measure the volume, weight and/or area of the accumulated natural resource stocks. The flow measures additions to, and subtractions from, stocks. That is the quantity of resources extracted or harvested by man, adjusted for natural (and human caused) losses and gains. These are recorded in *depletion and/or accretion accounts.* Inflow of biological resources is difficult to quantify from an accounting perspective, one of the reasons being the relatively slow incremental flow of biological growth rates in contrast, for example, to the catastrophic event of harvesting and natural mortality. Stock/flow balance of non-renewable resources is

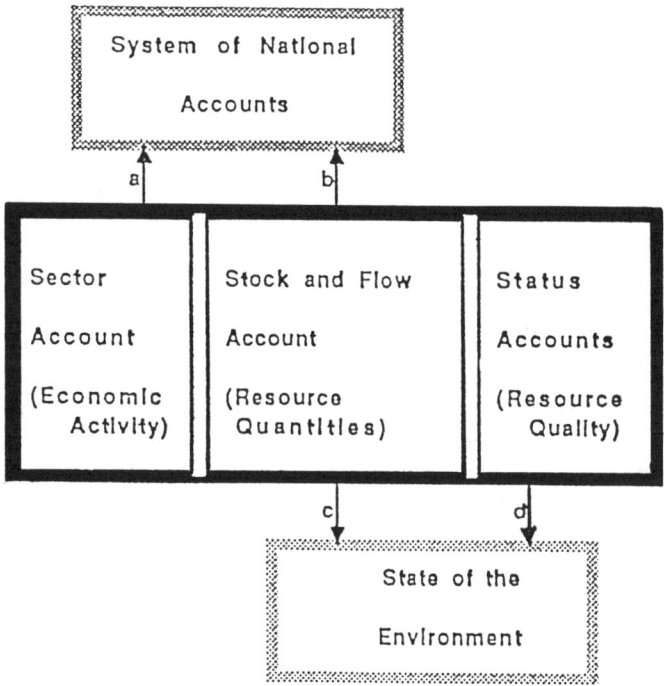

a.   Resource commodity flows into industry sectors, i.e., I/O Tables.
b.   Natural resource stocks in National Balance Sheets.
c.   Quantity measures of natural resource stocks, e.g., background and/or benchmark data.
d.   Quality measures of natural resource stocks.

Figure 21.1. Systems of natural resource accounts and linkage with the SNA and SOE.

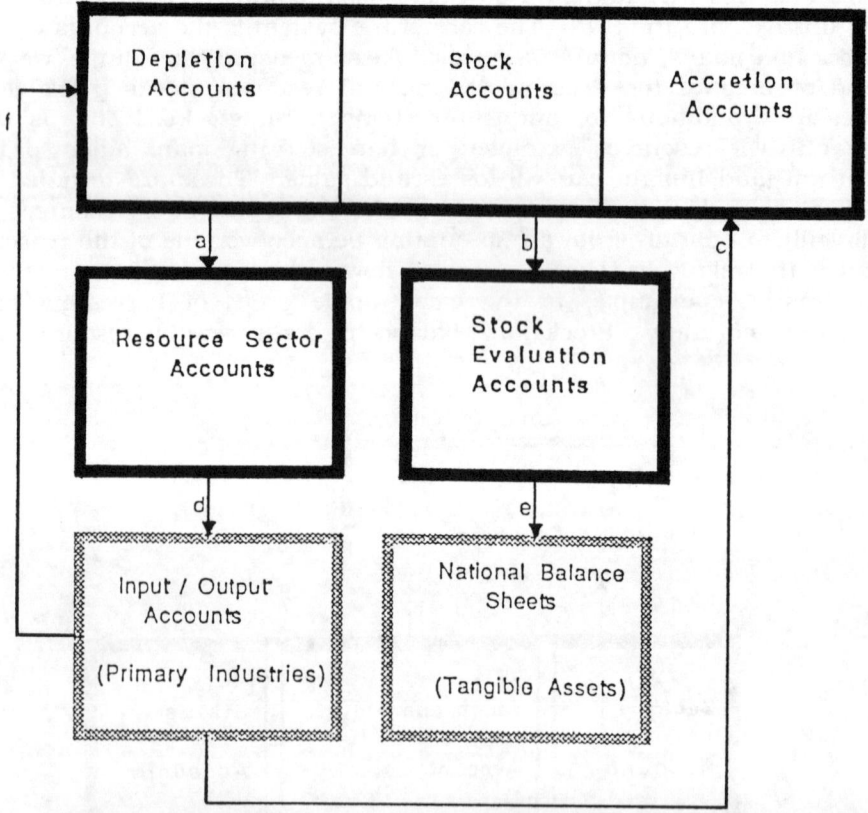

a.   Physical quantities of extracted/harvested natural resources.
b.   Physical stocks converted into monetary values.
c.   Economic activities contributing to augmentation and improvement
      of natural resource stocks, e.g., number/area of trees planted.
d.   Conversion to I/O coefficients, physical and monetary.
e.   Economic value of the national stock of natural resources.
f.   Pollution loadings depleting the stock of natural resources.

Figure 21.2.   Natural resource accounts linkage with the system of
national accounts.

generally better documented due to the conventions and practices
applied to the measurement of depletion and increases in mineral
reserves.   In areas where statistics on resource flows are poorly
documented or simply unattainbale, the flow is assumed by the
difference in stocks over two periods in time, that is:

$$\Delta S = S_{(t)} - S_{(t-1)}.$$

Figure 21.1 shows the intermediary role of NRA between the SNA and the State of the Environment (SOE). The *Sector Account* converts the part of natural resource flows that are treated as "economic commodities". This requires redefining natural resource flows (harvesting/extraction) to be compatible with the SNA "economic sectors". This depicts, in the first instance, the primary industry output such as agriculture, forestry, fishing and mining in physical quantities. These are then converted into intermediate inputs (in monetary values) of the secondary processing stream (food processing, sawmills, smelters) and natural resource commodity exports. These accounts also record the inflow of economic commodities which add to the stock of natural resources (for example, tree seedlings, livestock, industrial fertilizers). The "economic commodity" inflow/outflow is derived from the *I/O accounts* of the SNA (Figure 21.2). In addition to the intersector flows, there is a need to account for the intra-sector flows such as feed for livestock. In I/O accounting inter-industry transactions are recorded only as a "sale" or "purchase" within specifically defined sectors. Thus, pasturage, farm manure and internally produced seeds, for example, are excluded. The gross activity measures are particularly crucial in assessing land productivity and environmental carrying capacities and therefore should be included in NRA.

The *status accounts* record the qualitative dimension and spatial distribution of natural resources. Thus, just as the *sector accounts* transform natural resource data into economic data and vice versa, these accounts transform the stock/flow data into environmental statistics. Quality, *per se*, cannot be measured directly. Therefore, quantity parameters like average size of biota, yields per hectare and species diversity provide surrogates for quality. Assessments of quality contain elements of subjectivity and, more often than not, are judged in relationship to purpose. This is a complex process, particularly if more than one criterion is to be employed. For example, the quality of forests should be assessed not just by its economic/natural productivity potential, but also by attributes of species diversity, wildlife habitat, aesthetic and recreational values and by its uniqueness in relationship to other forests.

Figure 21.2 shows the relationship of NRA with the SNA. The linkage is achieved by two transition accounts where the data are reformulated to be compatible with the structure and concepts of the national accounts. The *sector accounts* are shown as an extension of the *input/output accounts*, and the *evaluation accounts* as an extension of *national balance sheets*. The purpose of the national balance sheets is to measure the national wealth as opposed to national income. Wealth is recorded in terms of financial assets and liabilities and as the monetary value of physical stocks, referred to as tangible assets. In practice the national balance sheets have included only natural resources for which there is a market. This is almost exclusively land

and livestock. NRA proposes to expand the wealth accounts to include the total stock of natural resources. Techniques such as shadow prices, willingness-to-pay surveys and methods to evaluate loss of environmental functions (for example, opportunity costs) can be applied to pricing assets where there is no actual or equivalency in "market transactions". Even in cases where there are recorded market values such as farmlands there remains some ambiguity in accounting for non-market functions. What is, for example, the economic value of hedgerow farmland in Britain? Economic evaluation techniques are clearly unsatisfactory when pricing permanent loss of natural assets, aesthetic and cultural values and, of course, higher-order functions associated with the long-term dynamics of ecosystems, and accounting for inter-generational transfers of natural resource stocks. Nonetheless, despite its limitations, economic evaluation of natural assets in the *national balance sheets* is useful in the context of NRA; it provides an explicit recognition that natural resources are a *means of production* and thus are treated as capital assets; natural resource stocks provide a stream of environmental services contributing to the socio-economic well being of the nation. In other words, they provide the incentive to develop the relevant databases for the environmental dimension of "social benefit accounting".

Economic activity which augments and/or enhances natural resource stocks is shown by the feedback loop (arc c in Figure 22.2) from the *I/O accounts* to the *accretion accounts*. This is manifest either through capital investment (environmental restructuring) or through current expenditures on resource management. These include major projects for environmental rehabilitation, soil conservation, afforestation and wildlife protection and those aimed at improving the economic productivity of existing resources, for example, irrigation projects, drainage of swampland and mineral/energy infrastructures. These latter, engineering type activities are often at the cost of the non-economic functions of the environment. In these cases a balancing entry should be made to the *depletion accounts*. Improvements, due to environmental management, are not as readily identified in an accounting matrix. It could be argued, on the other hand, that good management merely maintains the natural resource base on a sustainable basis whereas poor, or no management, results in drawing down of these resources.

Economic activities which degrade the environment indirectly are shown by the feedback loop (f) which links the *I/O accounts* with the *depletion accounts*. These can be classified under two headings: i) the generation of waste residuals and ii) permanent environmental restructuring. The latter comprises largely the part of resource depletion due to land use change caused by construction activities. New housing, non-residential construction and transport networks can be identified with specific land use conversions. Development of coefficients which relate the level of activity to depletion of natural resources are the basis for part ii) of the (f) loop. The type i) data are generated from pollution models that link pollution loadings to production levels (for example, pollution coefficients). An additional step is required, however, to specify the relationship of pollution loads to loss of natural resources.

To sum up, the NRA provides the conceptual framework for the development of a macro-statistical system for the management of natural resources and the environment. The core of the system is physical measures of the stocks and flows of natural resources. These are described by three distinct, although integrated accounts: *stock accounts; depletion accounts; and accretion accounts.* The system is linked to the SNA through the *sector accounts* and *evaluation accounts.* The latter document natural resource stocks in terms of the value of tangible assets in the *national balance sheets;* the former document the natural resource flows in the economic sector compatible with the national accounts, in particular with the industry/commodity classifications of the *I/O accounts.* Finally, the *status accounts* document natural resource stocks and flows by its spatial and quality dimension, that is, information on the state of the environment.

21.2.2. The major components of NRA

Land use statistics are contextual to the structure and conceptual framework of NRA. It should be pointed out that the specifications described here are not necessarily the same as those of other approaches to NRA. Nonetheless, the underlying logic of the dynamics of the environment and human activity suggests that the general lines described below need to be considered in any systematic approach to NRA. These are:

i)   *biological resources,* sometimes referred to as renewable;

ii)   *non-renewable resources,* composed largely of subsurface minerals and hydrocarbons;

iii)   *cycling systems of the biosphere,* composed of the atmosphere, hydrosphere and the lithosphere, these are sometimes referred to as "life support systems": solar energy is considered the most critical "life support" resource; it is measured in climate variables (atmospheric cycles) and plant growth rates (photosynthesis).

Biological resources are first and foremost characterized by their reproductive capacity and the growth cycle. In a general framework these can be specified in terms of i) *population* (numbers), ii) *habitat* (area) and iii) *biomass* (weight/volume). The biological parameters are linked to human activities by "economic production functions" (that is, volume and value of inputs and outputs) associated with agriculture, forestry, (commercial) fisheries and by the value of "environmental services". These latter include recreation potential, aesthetic enjoyment and the state of human health. The complex interdependency of economy, environment and human well being suggests a higher level function of "biological services" which maintain equilibrium conditions in ecosystems, chemical cycles and cycling systems. For example, deforestation has been blamed for accentuating floods and droughts and long-term climate change. From an accounting standpoint it is useful to distinguish:

i)   *composition* of natural resource stocks;

ii)   *spatial distribution* of natural resource stocks;

iii)   *gross natural production* of biological resources;

iv)   *net economic* or *harvested output* of biological resources;

v)   *stream of environmental services* obtained from the "inplace inventory" of biological resources.

Non-renewable resources in NRA record the stock/flow of minerals and hydrocarbons. In principle, the concept implies a one-way flow (depletion of stocks) leading eventually to complete exhaustion. In practice, these accounts do show additions to stocks based on new discoveries and re-evaluation of existing reserves based on commodity prices, new technology and reassessments. Proven reserves, particularly in minerals, often exhibit counter-intuitive patterns of growth in reserve stocks with increases in mineral output. This is explained largely by the additional investment made in mineral exploration associated with the growth in demand. The size and (economic) accessibility of reserve stocks are clearly considered as assets in economic accounting. Several proposals have been put forward on the treatment of mineral reserves in *national balance sheets* and methods to write-off depleted stocks against *national income*. A method to account for mineral depletions in the SNA was proposed by Ward (OECD, 1982).

Though non-renewable resources are generally thought of as sub-surface minerals and fossil fuels, there are also biological processes and physiological transformations which are, in essence, irreversible. The most obvious case is extinction of species (loss of genetic stock), but the same holds true for the maintenance of complex ecosystems which may never recover to their former state under severe disturbances. In the physical world land converted to urban uses, or flooded by dam construction, are for all practical purposes irreversible processes. Other resources may be treated as quasi non-renewable. Soil erosion is one of the best examples. Quite clearly, in many agricultural areas the rate of soil loss is often several orders of magnitude higher than the rate of soil formation.

Cycling systems accounts describe the change of stocks and dynamics of flows in the atmosphere, hydrosphere and lithosphere. In addition, chemical cycles of the biotic and abiotic environments, such as nutrient, nitrogen, oxygen, ozone and carbon cycles, should be considered in this domain. The global nature of atmospheric cycling systems and the geological time scales of lithospheric cycling systems place severe restraints on the capacity to describe them by statistical means. Climate variables are effective descriptors of atmospheric cycling. Hydrological cycling systems, on the other hand, lend themselves fairly well to statistical descriptions, for example, hydrometric measures of daily, monthly and annual flows of rivers. Data on chemical cycles are generally derived from minute behavioral observations of small-scale ecosystems. The translation of these scientific measures into global stock/flow balances have hardly progressed much beyond the heuristic stage. Nonetheless, despite these many limitations, there are compelling reasons for continuing to improve the monitoring of cycling systems; in part because of the fear that human activity could upset the critical balances resulting in climate change, impoverishment of soils and parched aquifers. The NRA would, of course, be selective with respect to the parameters which track the changing characteristics of cycling systems.

Although, in principle, it is possible to talk of additions to and subtractions from the accumulated stock of air, water and land, in reality this formulation makes little sense. A better concept is to apply the metaphor of the "Space Ship Earth". The concept of enclosed space is implied in the term cycling system, whereby materials are neither created nor destroyed but are continuously recycled. Data on the redistribution, and changing qualitative properties of these stocks, are relevant for environmental management and policies. The alternative term "life support system" suggests an approach which treats the statistical description of cycling systems as a stream of services drawn from stocks. This is also compatible with the economic concept of *capital stock as a means of production* valued as "service life". In general, biological stocks and mineral reserves are analogous to inventories, or goods available for consumption whereas cycling systems may be viewed as capital stock. This also allows for the measurement of the quality of services as a function of stock degradation/ enhancement, over/underuse and accessibility/location. The more apt analogy, however, is that of public assets rather than private capital stocks. The latter is almost exclusively viewed in terms of productivity (that is, capital/output ratio), whereas the former include other factors such as public accessibility, equitable distribution and contributions to social and cultural aspirations. In other words, multipurpose objectives need to be applied to the assessment of the "environmental services".

21.3. Role of Land Use Statistics in NRA

21.3.1. The need for land use statistics in NRA

Land use changes in Europe have been approached in this book from four basic perspectives:

i) *modeling* of the dynamic interrelationships to anticipate future trends in land use patterns;

ii) *mapping* to describe the spatial configuration of land use and its changes over time;

iii) *monitoring* to track the impact of human activities, in particular technology and waste residual generation, on the changing qualities of land;

iv) *historical analysis* based on archival records to portray the transformation of land use in relation to the dynamics of social, economic and demographic changes.

Land use statistics in the NRA framework provide the opportunity to integrate the above "analytical tools" in the context of "political accountability" at both national and regional levels. This approach implies that changes in land use become an "integrated variable" of the macro-economic performance indicators of production, consumption and capital accumulation recorded in the SNA. The legitimate question is whether we can still afford national assessments based on the evident dichotomy between environmental and economic analysis. The issue is complex. In the first instance, to bridge the gap between the economy and the environment requires recognition of the contribution of *natural production processes* to the economy and well being of the nation. Data collection activities generate massive, and detailed, statistics on the labor and capital inputs in production activities but practically ignore land (that is, the natural resource) as a factor of production. Secondly, there is a need for institutional commitment toward new approaches to national accounting, including the acceptance of scientific uncertainty in the measurement of the parameters of NRA. The perception that natural resource data is, by and large, "soft" has proven to be a major hurdle to institutional commitment. Although economic statistics such as the stock/flow of capital investment can be just as readily characterized as "soft data"; yet, uncertainty in many of the economic statistical series seems to have neither hindered their collection nor their use in assessing the state of the economy.

21.3.2. Land use statistics in NRA

Land use statistics play a central role in the development of NRA: primarily, by the treatment of the nation's *land mass* as a natural asset; secondly, the treatment of *land use* as a factor of economic and natural production; thirdly, *land use change* (flow) can be viewed as a change in the composition of the stock. Land use statistics describe the spatial distribution of natural resources (that is, mapping). Land use stock accounts are constructed by mapping i) human activities, ii) biological activities and iii) natural cycling systems. Figure 21.3 provides a general overview of the relationship of activities and resource mapping. The integrating role of ecosystem mapping by combining information from biological activity space, human activity space and cycling systems should be noted. Ecozones were employed as spatial analytical units in Canada's State of the Environment Report and Compendium of Environmental Statistics (Environment Canada, 1986; Statistics Canada, 1986).

In recent years multivariate analyses have been applied to produce synthetic spatial databases such as ecosystem mapping, climate mapping and intensity (or environmental loading) of human activity, for example acid deposition. Although the availability of natural resource maps is a prerequisite for the development of a land use component of NRA, this in itself is not sufficient. Additional analyses are required to fit into the NRA framework, with databases derived from resource maps.

The potential limits of the natural resource base are given by the unique geographical position in the biosphere; nonetheless, optimum productivity is a function of resource management techniques and practices. The concept of optimum production is complex. Over the past decade new kinds of analysis have emerged which once again place environment and natural resources on center stage. Georgescu-Roegan (1971) applied the concept of entropy to economic analysis. Boulding, Daly and Schumacher all take the position that income should be measured as a stream of services obtained from a sustainable stock instead of a reward for production.

The development of land use databases which distinguish different levels of natural productivity is a central problem of natural resource accounting. What is the appropriate measure of land productivity? What do we mean by quantitative in contrast to qualitative changes in land stock? In biological resource production, what part should be

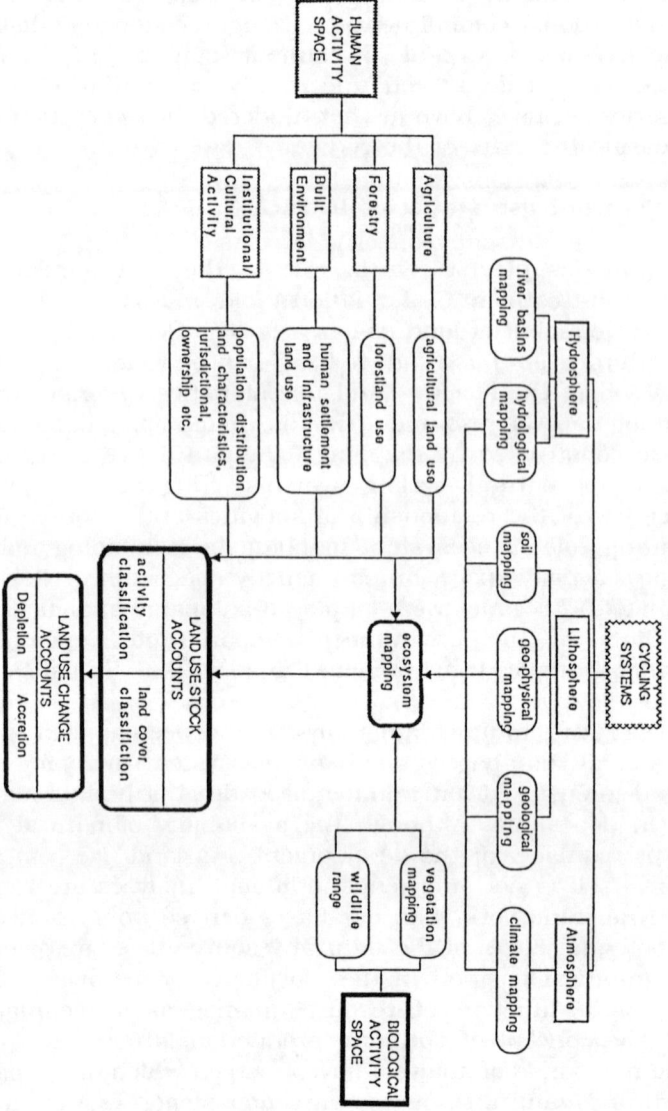

Figure 21.3.   The linkage of natural resource mapping and land use stock and flow accounts.

attributed to human intervention?   What part should be attributed to nature?   To what extent is there a need to modify the traditional land use classifications for the purpose of the NRA?   Methodological questions that need to be addressed are techniques to evaluate land resources in the national balance sheets, techniques for measuring natural productivity, and techniques for introducing "institutional attributes" of land which contribute to differences in productivity, such as ownership, zoning, environmental protection and policy instruments in different political/administrative jurisdictions.   (Some of these questions are discussed in Chapter 22.)

Answers (or at least working definitions) to these questions are expected to evolve from the research and it is possible, even at this stage, to provide some insight into the general direction of the role of land use statistics in NRA.   Some of these questions are being addressed in the current concern for assessing the state of the environment and natural resources.   These include development of capability classification of land, ecosystem mapping, and biomass (or other bio-productivity) measurements obtainable from satellite imagery. Area variables of the stock and flow of natural resources are largely obtained from abstracting information from the resource maps such as agricultural soils, forest stand mapping, and geological and hydrological surveys. The development of geographical information systems (GIS) has vastly improved the capacity of converting mapped data into digitized databases. Spatial analysis can be rapidly performed even with personal computers.   However, the production of digitized databases is still a tedious function, and it is this activity that is expected to absorb a large part of the resources for the development of NRA systems.   More difficult, however, are estimates of the natural resources stock in volume of weight.   These are largely estimated from models based on "ground sampling" of weight/volume variables multiplied by area. For example, the volume of merchantable timber (m^3) is estimated on forest stand information and a model of tree growth, as a function of species type, forest age, climate and soil.

In essence, land use statistics depict spatial activities. The end product, however, is determined by the definitions and classification system employed.   Current land use classifications tend to reflect a commodity production paradigm. This is due, in part, to employing the "standard industrial classification system" to define economic activities and, in part, to viewing resources in terms of commercial criteria, for example, forest rotation rates. Land use outside human settlements is generally described by its dominant natural features such as forests, deserts, heathlands, marshlands and glaciers. The mixing of activity and land cover classification (that is, vegetation and surface materials) also contribute to the inevitable confusion and inconsistencies. A multi-use classification of land provides a more realistic picture. Techniques of overlay mapping (efficiently performed by GIS) make it possible to derive synthetic, multidimensional land use databases (Figure 21.4). Natural resource accounting provides a different perspective to the traditional (cartographic) land use mapping. In the first instance, NRA databases are designed for macrolevel analysis of such questions as sustainable development, environment carrying capacity and natural productivity. This requires critical rating and evaluations of land use.

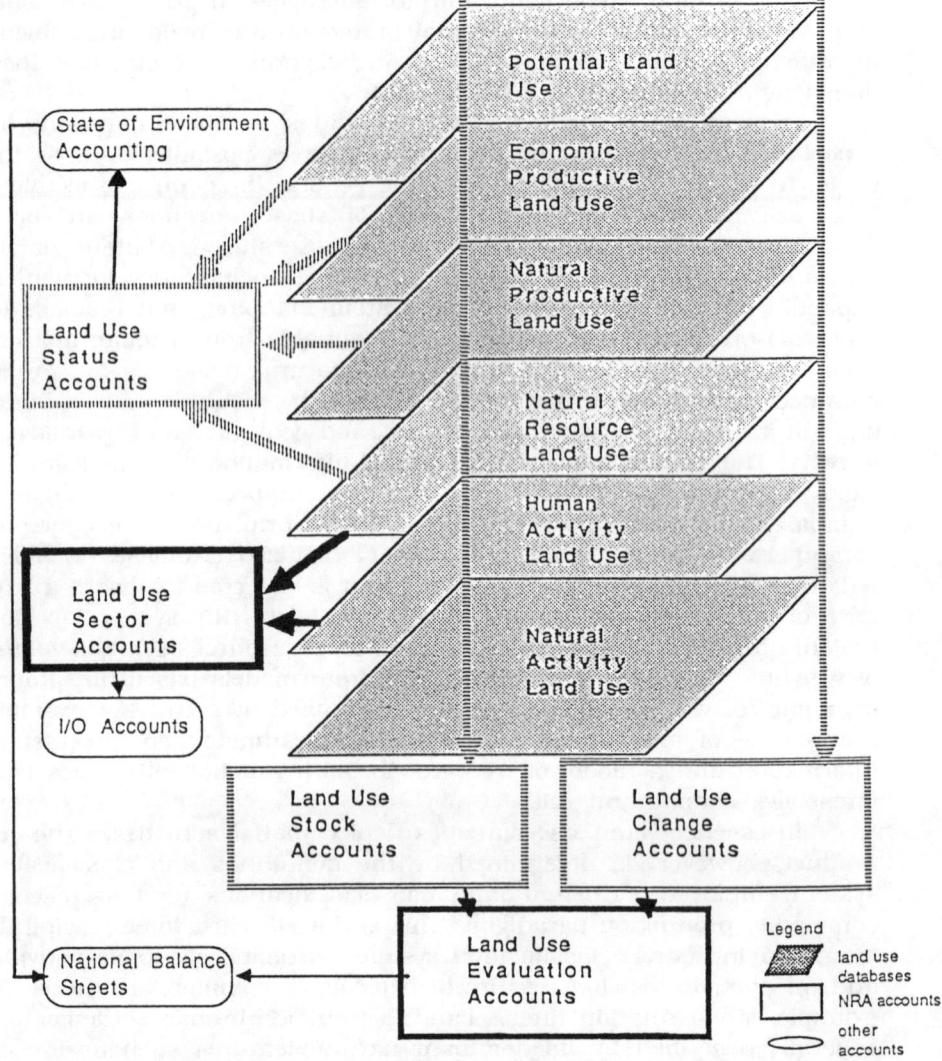

Figure 21.4.   Relationship of land use mapping and natural resource accounting.

The distinction of economic and natural productivity of land should be maintained in land use databases. Techniques and methods for measuring natural productivity at sufficiently sensitive scales to detect changes over time are still relatively underdeveloped. Research on the development of macrolevel ecological productivity indicators shows a promising approach to this problem (Rapport and Regier, 1980). Techniques have been developed to rate (or weight) a combination of variables for the purpose of identifying the suitability of land for different uses. For example, the Canada Land Inventory (CLI) was based on weighted variables of soil, climate and geophysical suitabilities for agricultural production, forestry, recreation and wildlife (ungulates and water fowl). The traditional "commodity paradigm" to land use classifications needs to be supplemented by process classifications. In the case of agriculture this would call for overlay mapping of "product" with "practice", such as small-scale mixed farming, large-scale monocultures and rotation systems. Similarly, the depletion/accretion accounts require specific focus on activities (both human and natural) which augment or diminish the resource base. Here, distinction should be made between permanent (or irreversible) processes and temporary (or recoverable) processes. The latter include harvesting, natural mortality and replantation of biological resources whereas the former focus on human-caused environmental restructuring (for example, land use conversion and large-scale engineering projects) and permanent changes due to natural phenomena (for example, climate change, geological, events, invasion of exotic species and natural population collapse).

Figure 21.4 illustrates the general relationships between land use databases and NRA. *Land use stock and flow accounts* form the core of the land use database. The capacity to produce these accounts depends, naturally, on the availability and suitability of existing land use maps, land use data from statistical surveys (for example, agricultural census) and the ability to abstract data from remote sensing imagery. Data inputs into *land use status accounts* are of a more complex nature. It would be desirable to include in these accounts natural productivity indicators (for example, average size of biota) and economic productivity indicators (for example, crop yields, land use intensity and land capability).

Data inputs in *land use sector accounts* are abstracted largely from natural resource mapping (for example, forest inventories, geological maps, soil maps). An important element is the linkage with the *input/output accounts*. This could be achieved by the development of coefficients which relate the economic production of goods and services to land use. This is, of course, particularly significant for land intensive production such as crops, livestock and timber. Land use coefficients can also be linked to mining, transport, construction and heavy industry. Service, household, institutional and governmental sectors can also be associated with land use change, for example, tourism, residential uses, military establishments and national parks. Several other avenues could be explored in developing land use databases in relation to economic activity. For example, income and lifestyle changes can be correlated to land use as can demographic variables such as population growth, migration patterns and age structure.

Table 21.1 provides a summary statement of the land use databases in Figure 21.4.

Table 21.1. Land use databases for NRA

Land use and spatial productivity mapping	Databases	Classification and integration
Natural activity	vegetation cover physiological type climate type	ecological classification
Human activity	agriculture forestry mining urbanization transportation industrial tourism hydrological infrastructure energy infra- structure	socio-economic classification
Natural resource mapping	soils (geological) forests (type) water minerals wildlife	resource classification
Natural productivity mapping	soil productivity forest productivity carrying capacity of habitat ecosystem productivity	rated by level of potential or capacity to produce biomass or indicators ecosystem health
Economic productivity mapping	agricultural harvest forest harvest fish harvest (inland) mineral extraction industrial (material) transport network capacity energy production population density	rated by level of output, by intensity of economic activity, by density of transport network and market mapping
Potential land use	agriculture forests wildlife renewable energy recreation human settlements	rated by soil, climate and other attributes

21.4. **Role of NRA in Policy Analysis and Public Information**

All governments are in the business of data collection, compilation and dissemination. A substantial amount of government generated data is a byproduct of administrative functions, although the collection of statistics is generally prescribed by formal legislation (Statistics Act). In its broadest formulation the purpose of government statistics is to monitor the state, and change of state, of the nation. The specification of the data collection mandate, however, refers almost exclusively to social, economic and demographic statistics. Therefore, environment and natural resource statistics are either perceived as being outside the domain of official government statistics or, at best, have an ambiguous and uncertain place in the development of national databases. Nonetheless, governments are deeply involved in the collection and organization of biophysical data. In some cases data collection is a major function of natural resource departments, for example, the Geological Survey, the Hydrological Survey and the Meteorological Office; in others these data are generated as byproducts of management and control functions. As has often been noted, the lack of a   formal statistical   framework in this area has led to fragmentation, inconsistency and gaps in natural resource databases (Environment Canada, 1986).

NRA offers a framework for organization and integration of natural resource statistics.   This alone might justify its development and implementation. It is difficult, at this stage, to say whether NRA would result in a net increase or decrease in government expenditures on natural resource data collection or, for that matter, what effective influence NRA would have on government policy.   Nonetheless, the underlying rationale would suggest potential benefits. Among these are:

-       the development of a macro database on natural resource assets and liabilities, and monitoring of their stock/flow relationships;

-       the provision of a framework for analysis of the adequacy and validity of current natural resource databases for the purpose of national environmental and economic policies;

-       the development of natural resource statistical indicators to assess the state of the environment and sustainable development goals;

-       the development of an accounting framework which complements the SNA.

The land use component of NRA could play a significant role in monitoring the spatial impacts of government policies, in particular those concerned with location of industry, human settlements, agriculture and resource development. Traditional land use impact analysis focused largely on land use change and its local effect on the environment and people, as well as the regional economy. NRA provides an opportunity to integrate "local impacts" into a national framework. A further step is to utilize land use statistics to explore the "spatial multiplier effect" of local activities. For example, the impact of a mining development goes beyond the immediate affected area. This, of course, requires the development of spatial models. Although NRA in itself is not a spatial model, it nonetheless is developed from databases which can be useful data inputs to calculate model parameters.

21.5. Summary and Conclusions

This Chapter presents the argument for natural resource accounting as a complement to national accounting. Although classical economic thought points to the importance of land in economic analysis, this factor of production was practically ignored in the "neoclassical model". The alarm of environmental degradation and resource supply limits have reintroduced land, expanded to include environment and "in-place" natural resources, in the economic analytical framework. The methods and techniques implied in the development of NRA are highly suggestive for integrative analysis of environment economy.

Important elements in this approach include the measurement, in physical parameters, of the stocks and flows on natural resources, the need to transform, both in physical and monetary terms, the portion that enters the economic production system, and the formal accounting of natural assets in the nation's "balance sheets". The quality of the environment is also considered in the measurement of the "status" of natural resources, in particular biological and cycling systems of air, water and land. Thus, NRA is viewed as the information link between economic performance and environmental change.

Land use statistics perform two important functions in NRA: these data are a spatial integrator of human (and natural) activity, and describe the spatial distribution of natural resources. The position taken in this Chapter is to view land use from the perspective of economic and natural productivity. The evaluation of land use productivity is complex. At the level of direct physical output from land, for example, crop production, tree harvest and minerals, the measure of (economic) land productivity is relatively straightforward. Environmental services of land is clearly more difficult to evaluate. Market prices might, in part, reflect the economic services, for example, desirability of location and, in part, evaluations may be obtained from opportunity cost analysis. Natural productivity, at a conceptual level, is well entrenched in the productivity models of ecosystems. Unfortunately, the transition of the model to real conditions presupposes, *inter alia*, agreed methods for qualitative ratings for ecosystems and species within ecosystems.

Despite the many unanswered questions the assessment of sustainable development ultimately depends on the juxtaposition of natural and economic production measures - a problem which may be expressed as the combined optimization of these two variables.

The inadequacy of national accounting aggregates to measure economic performance has also been raised by Nijkamp and Soeteman in Chapter 7. NRA does not, by itself, replace GDP, nor does it attempt to summarize human activities into one indicator. Rather it provides an organizational framework for the different dimensions of natural resource databases, including that of land use. Admittedly, the development of a sufficient statistical database to construct the different components of NRA is a long-term project. Nonetheless, Statistical Offices have recognized that sustainable development policies need statistics which monitor the availability and status of natural resources; they also have experience in developing systems of national accounts. The obvious conclusion is to marry the experience with the need.

REFERENCES

Capra, F., 1982, *Turning Point: Science, Society and the Rising Culture.* Simon and Schuster, New York.

Daly, H.F., 1973, *Towards a Steady State Economy.* W.H. Freeman, San Francisco.

Environment Canada, 1986, State of the Environment Report for Canada. Report EN 21-54/1986. Ottawa, Ontario.

Georgescu-Roegan, N., 1971, *The Entropy Law and the Economic Process.* Harvard University Press, Boston.

OECD, 1982, Accounting for depletion of natural resources in national accounts of developing countries. Report CD/R(82)3010, Paris.

Rapport, D.J. and Friend, A., 1979, Towards a comprehensive framework for environmental statistics: a stress-response approach. Report 11-510. Statistics Canada, Ottawa, Canada.

Rapport, D.J. and Regier, H., 1980, Ecological approach to environmental information. *Ambio,* **9**(1), 22-27.

Statistics Canada, 1986, Human activity and the environment. Report 11-509 E. Statistics Canada, Ottawa, Canada.

United Nations, 1984, Survey of environment statistics: frameworks, approaches and statistical publications. Statistical Papers, Series M, Vol. 73. United Nations, New York.

Waring, G., 1857, The agricultural features of the census of the United States for 1850. *Bulletin of the American Geographical and Statistical Society,* **11**.

WCED, 1987, *Our Common Future.* Oxford University Press, Oxford.

Chapter 22

ANALYSIS OF LAND USE DETERMINANTS IN SUPPORT OF SUSTAINABLE DEVELOPMENT

E.W. Manning

22.1. Introduction

There is a growing understanding that the fate of the environmental resource base is critical to the future of global societies. There is mounting evidence that human activity is altering global climate; the continuing use of components of the environment is causing reductions in land capability and waste and other byproducts of human actions are affecting the carrying capacity.

The planning and management of land use change today, for sustainable development tomorrow, requires knowledge of two basic features:

i)   the availability of land types;

ii)   the demands to be placed on the land resource.

Analysis of the current and future ability of the land base to serve the goals of society is dependent upon enhanced knowledge of the determinants of land use, particularly as these constitute opportunities and constraints to future land use options. This Chapter examines land supply and demand in terms of data requirements, useful analytical approaches and problem identification as well as solution development and implementation. These steps are central to the ability of planners to identify those factors (variables) most critical in the analysis of the ability of the land resource to serve the needs of society on a sustainable basis.

The World Commission on Environment and Development (Brundtland, 1985) supported sustainable development as a central goal of international environmental and economic planning. This position was echoed in the Proceedings of the 1986 Ottawa Conference on Conservation and Development (Jacobs and Munro, 1987). Sustainable land use can be defined as the best possible long-term product of the interaction of supply (generally defined in biophysical terms) and demand (described in socio-economic terms). Given known biophysical

F. M. Brouwer et al. (eds.), Land Use Changes in Europe, 485–514.

resources, the objective may be to take action in advance, ensuring that the land base will continue to serve the demands placed upon it whilst maintaining sufficient margins of safety.

In general, the carrying capacity of the land base can be defined in terms of specific biophysical variables such as temperature, water supply or soil type. These biophysical variables can be used within a supply/constraint modeling framework to analyse changes in the ability of the environment to support particular functions. In addition to the biophysical variables which set limits to resource use, there are many other social and economic variables, related both to past use patterns and to new needs and desires of users, which are important determinants of land use patterns both current and future. The biophysical variables and the socio-economic factors link to provide opportunities or constraints affecting the need, or ability, to respond in the face of changing circumstances. From a practical perspective these variables must be identified together with the relationships most critical to the ability of molding land use to future foreseen needs. These factors also apply to a rational assessment of possible significant disruptions (climatic changes, energy supply alterations, geopolitical changes) to the supply of, or the demand for, environmental resources with particular characteristics.

Fundamental to the problem of achieving long-term sustainable use of environmental resources is the fact that many of the most important future influences are unpredictable (Holling, 1986). Even where broad trends can be anticipated, the precise magnitude and the differential impact on different parts of the world can seldom be accurately predicted (Clark and Munn, 1986; Munn, 1987). Yet a fundamental precept associated with sustainable development is that problems should be foreseen and that actions should be taken to minimize or avoid problems before they occur. While it is impossible to anticipate all eventualities, actions can be taken to facilitate dealing with them when they occur or to eliminate, in advance, particular sensitivities of the system to a range of possible disruptions. In particular, analytical systems and data sets can be developed which support preventive actions or quick responses. Such anticipatory systems require that the relationships (linkages) between known attributes of the land base and likely environmental or ultimate human responses are clearly understood. In the case of land use, as will be addressed later, this may require a particular type of data and/or analytical procedure.

This Chapter examines, from a theoretical perspective, the way in which land attributes should be analysed in order to plan the achievement of sustainable development. It then identifies the key biophysical and socio-economic variables to support this approach. The considerations of a spatial framework are then addressed and a supply/constraint modeling procedure is proposed. The Chapter ends with an examination of the use of such an approach to test different scenarios within a land supply/demand framework.

The Chapter is intended to stimulate consideration of how to deal with future land use patterns and problems at a continental scale. It is based on the work of Lands, Environment Canada and on current work in Canadian, American and Australasian institutions. Observations are put forward that may contribute to the development of a program in Europe that ensures the land resource base is planned and managed to support sustainable development.

22.2. Analysis of Land in Support of Sustainable Development

The ability to analyse land use must be seen within a broader framework than just the collection and analysis of data. The identification of key determinants and the collection of supporting data are building blocks in a holistic approach to address the role of terrestrial resources in the achievement of sustainable development. Manning and McCuaig (1984) proposed a pyramidal structure as a conceptual framework describing an ideal program to support "wise land use". Subsequent work has developed this framework into a strategy for research in support of the goal of sustainable development (Figure 22.1). A precise definition will depend greatly on governments as they set priorities in terms of lifestyle and economic and social development. A range of demand modeling exercises could be undertaken wherein scenarios are developed and tested as alternative sustainable or non-sustainable futures.

In Figure 22.1 a strategy for sustainable development is presented; this is defined as the maintenance of the environmental resource base to sustain those functions which maintain life and socio-economic activity. In terms of land, this objective could be defined as the maintenance of an adequate quantity of land with the required qualities to support, indefinitely, the full range of societal demands upon it. These demands include not only the productive functions of the environment but also the support of a wide range of social, ecological and aesthetic values. The pyramid is designed from the top down to support a particular societal goal such as, in this case, sustainable development (level A). Each level of the pyramid is built upon lower ones to support the goal at the top. Level B of the pyramid, *implementation*, involves those activities necessary to modify the use of the land resource base in order to achieve the goal. This involves such steps as the planning and management of resources within our direct control, the use of other government powers to mobilize economic instruments in moving towards sustainability and the exercise of influence upon the actions of others. Implementation is built on practical *solutions* in the area of planning and management of resources (level C). Solutions are developed in response to *problems* which are known and/or foreseen (level D). At level C means of scanning for

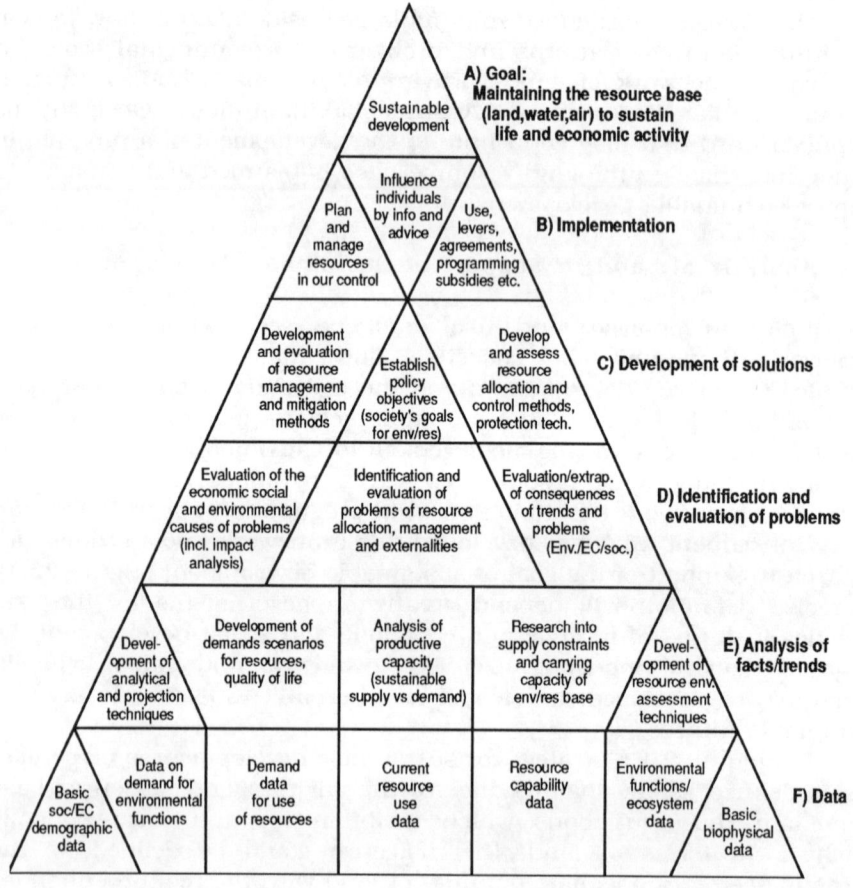

Figure 22.1.   Sustainable development: a strategy.

sensitivity of unforeseen but important disruptions can also be addressed.   The identification of specific current or foreseen problems, relative to measures of sustainability, will require the *analysis* of trends regarding land quality, land availability and land use (level E).   This involves the development and application of particular analytical procedures.   These analytical procedures, in turn, have certain requirements for *data* (level F).   Thus, by following the logic downward from the specific goal, the nature of data and analysis required to support the objective becomes more readily definable.   So, while the pyramid is *defined from the top down*, it is in reality *built from the base up* - commencing with data collection, then analyses, through to implementation.

The *raison d'être* for the pyramid itself is the center block: the *existence of problems*.   While we begin with awareness of problems (discontinuities) based on monitoring, scanning and analysis, the construction of this type of strategy requires data collection and analysis which will focus on the specific problems identified or anticipated in achieving the overall goal of sustainability.   To support the above strategy, much of the initial work would have to occur at the level of problem identification; this will imply the creation of scenarios, probable or possible, which have potentially significant impacts on either the resource supply or on the demands to be made on it.

A wide range of procedures can be used to create future scenarios. These range from single or multisector extrapolations and the modeling of changing relationships, to Delphi (consensus building) techniques. Because the range of possible future demands is so broad, it is essential to build an evaluative capability accommodating a wide range of potential demands and dealing with alterations or disruptions to the land supply. The identification of problems (unacceptable outcomes) will naturally depend on the ability of governments and society to clarify their own definitions of sustainability, in terms of their own goals for production levels, lifestyle, population numbers and environmental quality. Therefore, the decision of what is desirable or acceptable must precede any definition of what is unacceptable.

It is suggested that there are three distinct types of problems which can arise from changes in the land base supply (quantity and quality) or in demands on that base. These are:

i)   problems of *allocation* of land between users and user sectors (agriculture or wetlands, urbanization of prime resource lands);

ii)   problems of *management* of land once it has been allocated among sectors (agricultural land degradation, toxification, contamination);

iii)   problems of *externalities* or intersectoral impact involving disruptions caused to one user by others (downwind or downstream pollution).

All three types can be expressed in terms of a supply/demand equation for land.   In the case of allocation problems, the type of information necessary for analyses is generally quantitative as these refer to the amount of land with definable physical characteristics in any particular location.   With reference to management concerns, data requirements are more likely to focus on qualitative measures of the changes in those characteristics which can influence the productivity response for the functions for which the land is being managed.   To deal with issues of conflict, both quantitative and qualitative data concerning the attributes of the physical base may be required as well as socio-economic information. The juxtaposition of conflicting activities can create problems which influence land capability for a range of uses.

The nature of information (level F of the pyramid) required to address all three of these general types of problems should therefore comprise the following characteristics:

i)   variables identifying the quantity of land of different types and different capabilities;

ii)   variables identifying the quality of land with respect to certain biological and physical factors;

iii)   variables identifying the geographical location of the land;

iv)   variables identifying the current pattern of land use.

It is suggested that a spatial framework would be the most useful format for holding and presenting land data. The availability of land with a defined set of capabilities in any location could then easily be estimated. Information on demands for land and for the products of the land can then be related to particular sites or particular data units. If the information is analysed spatially, the impact of any alterations in the attributes of the land base in any data unit can be predicted. These impacts can be related to the provision of environmental functions within each spatial unit.

While the analysis of the biophysical resource base establishes, to a great extent, the long-term carrying capacity of the land base for different environmental functions, information on the present use of the land resource is also pertinent. This "socio-economic" information can be used to qualify further the availability of land with different characteristics to satisfy requirements (particularly in the short- and medium-term). Also this information can provide a baseline against which proposed alterations can be measured. Although current land use patterns are a reflection of past and present demands for the products and services of the land, these patterns may relate little to future demands. Time series analysis can permit the extrapolation of trends indicating possible future scenarios, but this is only one way of estimating future land use patterns. It may be more logical to treat the current biophysical and socio-economic determinants of land use as potential limiting or facilitating factors to the ability of the environmental resource to serve future demands (Manning, 1985, 1986b). If the information on land use determinants is handled in this fashion and if data on the key determinants can be held in a spatially compatible form, it becomes possible to test future scenarios (however developed) for sustainability, relative to the characteristics of the land base in each data unit.

22.3. Key Biophysical Variables

If we accept a supply/demand framework within which to analyse European environmental futures, the selection of biophysical variables can be focused on those which either i) directly *affect the capability of the land base* to serve environmental functions valued by society, or ii) *serve as indicators* of this relationship. de Groot (1986, 1987), Adamus and Stockwell (1983) and others have advanced the identification of the specific environmental functions served by the land base. Part of the problem is the selection of those which are most central to the objectives of society and that relate to the sustainable use of land. Harte (1979) and de Groot (1986) identified a very wide range of regulatory, carrier, production and information functions of the natural environment.   These serve particular human values, for example, production of food, or of minerals, or provision of information or space, or buffering of toxics.   Some of these environmental functions are difficult to evaluate because the linkages between the provision of functions and the market place are indirect or complex (for example, aesthetics). In other cases, links such as those between some of the production functions and marketable commodities are direct and make for easier measurement (for example, forest products). The selection of biophysical variables will therefore be dependent both on i) the particular environmental functions which are viewed as critical to supporting production and environmental quality in Europe (based upon priorities of European governments), and ii) data availability regarding the capability of the environment to support these functions.

Ideally, comprehensive data sets covering all of the chemical, climatic and physical parameters, which correlate well with productivity for each of the functions, would be developed, but this would be impossibly complex and expensive. Therefore, it may be wise to select specific variables which relate to *productivity for key natural resource products* and others which relate to *key measures of environmental quality* affecting human health or the overall usability of the environment.

Listed in Table 22.1 are representative environmental variables useful in assessing both gradual changes in supply/demand relationships for the land resource and evaluating the impact of surprise disturbances. These variables have the following characteristics:

i)   they have known relationships to productivity for key environmental products or functions (specific measures of forest productivity, crop productivity responses or the utility of the environment to provide key products or services valued by society);

ii)   they can be measured on a consistent basis both spatially and over time;

iii)   scenarios can be established wherein the quantity or quality of these variables are altered by physical changes, by human activity or by surprise events.

Table 22.1.   Selected biophysical variables for measurement of change in land use determinants.

Environmental functions (supply capability)	Environmental variable (measures of productivity potential)
Production of food	Soil texture, susceptibility to erosion Top soil depth Soil chemistry (for example, pH, nitrogen) Soil moisture Growing season, length, degree days Precipitation (in growing season) Slope
Production of wildlife	Vegetative cover Wetland incidence, quality Water table
Habitat open space recreation	Cover (measures of incidence, variety) Current land use designation Topography, shoreline quality Wildlife incidence
Regulation of toxics (Health, use ability)	Selected measurement of specific hazardous substances
Maintenance of genetic diversity	Biota counts Cover (incidence, percentage)

Notes:
a) Demand scenarios are independently developed by a range of means including projection, other models, Delphi, policy goals.
b) Selection of specific variables will depend on priorities established among valued environmental functions.

22.4. A Canadian Approach

Certain lessons regarding multisectoral data use at a continental scale can be learned from the Canadian experience in dealing with the national level databases. While baseline information for such factors as soil types (nomenclature classification) climate, geological and topographic maps has been in existence for several decades, existing data sets have proven very difficult to use within a framework for assessing the ability of the resource base to serve changing demands. During the 1960s, efforts were made to synthesize much existing biophysical information into standardized spatial units that could be analysed to help target economic development investments to areas where the return was likely to be greatest.   Therefore, climatic information (growing degree days, rainfall, frost free periods, snow cover) and physical data (such variables as slope, bedrock, soil type or salinity) were integrated into the Canada Land Inventory (Munn, 1986; Environment Canada, 1970, 1976).   For each of the sectors of agriculture, forestry, recreation and wildlife, a seven-level classification was developed relative to generalized production capability.   Class 1 land had no limitations for productivity for the particular sector, Class 4 had moderate limitations and Class 7 had no capability whatsoever. This was the first attempt to draw together a broad national database to influence the planning process.   Unfortunately, certain attributes of that database have made it inappropriate for the identification of changes in the physical land base in the 1980s (Manning, 1986a).   The multivariate data were synthesized to produce planning-level information; therefore, disaggregation to permit measurement of change in any one of the variables which caused it to be classed high or low is difficult. While certain relationships are known between productivity and land capability (for example, rye grass production on Class 1 land has proven to be approximately double that of similarly managed rye grass on Class 4 land), more sophisticated analysis using these data has not been possible.   While intersectoral trade-offs between land of high capability for one use versus another can be portrayed, the productivity implications of land allocation decisions are hard to analyse using this type of original data.   Analysis of the impacts of land degradation or climatic change fall outside the range of application of information held in aggregated classifications.

The problems in the analytical application of current data led to a new initiative mounted in the late 1970s to develop ecologically-based spatial units where data would be held on key biophysical variables. Measures of soil chemistry, key climatic variables (such as length of growing season, precipitation and degree days) and key physical parameters (such as depth to bedrock, soil texture, organic content) were collected for each unit.   These "soil landscape units" provide a baseline allowing measurement of change of individual biophysical variables (Coote and Shields, 1990).   The regionalization has been done on a multivariate basis; thus, other variables can, more or less, be held constant.   Work is now under way to develop empirical data on the productivity changes of important crops in response to changes in biophysical variables within each of the spatial units.

One of the lessons learned from this exercise is that the relevant variables may vary from zone to zone or from crop to crop.   However, for most agricultural products (food crops, grains, fodder crops) the same variables are pertinent in determining productivity response. The nature of the variables used (that is, climatic variables, soil depth, soil fertility, acidity, salinity, moisture holding capacity) suggest that they could be of considerable use in modeling forest crop productivity. Along with a few other variables, the utility of the land resource for different types of wildlife production, certain recreational pursuits and several other valued functions of the environment could be modeled. However, the empirical productivity-response relationships for the most important functions/products within the biophysical zones would have to be developed. The addition of certain measures of toxicity or certain indicator chemicals at the same scale might also be of considerable assistance in projecting or predicting environmental suitability for other production or life-support functions. The further addition of land use information for the same spatial units would allow estimates of remaining reserve capability, identification of some barriers to change and the estimation of minimum change responses for the achievement of particular production goals.

Based on the Canadian experience, some of the key considerations on the choice of variables which are most important for understanding what happens to the ability of the environment to serve critical societal functions under changed circumstances have been indicated above. It is possible to isolate the potential impact of a broad change (for example, climatic) down to the point where it can be shown to alter the attributes of particular sites or regions. Knowledge of the productivity response of key land functions to changes in factors such as the length of growing season or precipitation would permit us to estimate likely alterations in the ability of each data unit to support a particular crop or function. The choice of biophysical variables to be employed depends on which land uses are deemed to be most important. The base variables are summarized in Table 22.2. It would be useful if broader derived indicators of capability for important functions such as crop productivity potential could be obtained at an appropriate scale. The analysis of biophysical variables will assist in defining the opportunities and constraints of the environmental resource base at any point in time. The more variables held, the greater the versatility of the system and the greater the ability to handle an unforeseen range of changes which may involve new substances or new phenomena. Set against this is the increased cost and complexity of analysis which additional variables necessarily create. Based upon the variables held in the system, and the known relationships between the range of environmental uses or functions and these variables, it becomes possible to adjust the estimates of the carrying capacity of each part of the land resource base under any scenario involving changes in the biophysical base. The outline of an existing modeling procedure which could aid in achieving this objective is presented in Section 22.8.

Table 22.2.   The determinants of land supply and demand.

Determinants of land supply	Determinants of land demand
A. Biophysical (ecological) (determine the ability of the land to serve functions) Climate: growing degree days, frost free days, precipitation, Topography: slope, etc. Soils: depth, chemical composition, organic content, erodibility, etc. Current state of land resource (eroded, salinized)	**A. Economic** (influence the general market demand for products of the land) Sectoral demands for products (food fiber) Demands for space for specific valued functions (hunting, recreation, residence, factories)
B. Spatial Location, relative location Fragmentation	**B. Social** Changing perceptions of basic lifestyle requirements (housing, recreational demand) Changing demands for environmental quality (toxics, space) Demands for quality of life

<DECISION>

C. Socio-economic (affect the short-term ability to respond) Land: property size, field size, tenure, current value Labor: age structure, education, skills, expectations, productivity Capital: income, credit levels, mechanization Management: knowledge, technology, adaptability	**C. Policy** (specific demands can be established through Government policy) Conservation strategies, 5-year plans, sectoral goals, support programs
D. Political/Institutional (these constitute imposed limits or opportunities to supply) Political, institutional boundaries, subsidies Renewals, designations: military, single purpose (parks) Restrictive policies (zoning, regulated activities)	

22.5. **Key Socio-economic Variables**

If the land resource base was a "clean slate", the definition of carrying capacity in terms of biophysical variables would define the overall opportunities and constraints to change.   However, current and historical human activity provide further opportunities and constraints on land use.   Socio-economic factors are important to a society's ability to adjust land allocation and management, given new goals or opportunities or a changed biophysical base.   The situation with respect to the socio-economic variables which will influence societal response to future biophysical or political scenarios is at least as complex as the multitude of biophysical variables.   It is useful to divide these socio-economic factors into two groups: one pertaining to the land itself and another pertaining to current users and owners of the land.   No matter what the institutional system, the ultimate delivery of human response to changing environmental, economic or social situations is through individual action, influenced to varying degrees by those institutions.

22.5.1. **Current land use patterns as constraints and opportunities**

The following discussion is based primarily upon work from New Zealand (Manning, 1972) and North America (Beattie *et al.*, 1981; McCuaig and Manning, 1982).   This research indicates the range of variables with respect to individuals and property that are important in influencing the individuals' ability to respond to stimuli to alter land use.   European research is also generally supportive of these factors (Franklin, 1969; Galeski and Wilkening, 1987).

As for biophysical variables, it is important to limit the selection of socio-economic factors contained in any monitoring program or modeling exercise.   If we return to the concept of environmental functions, it becomes easier to see that some of the functions can be affected by changes in biophysical factors.   Most food and fiber production functions occur only via specific actions by property owners or users.   Realization of other functions (for example, habitat maintenance or genetic diversity) is less directly constrained or influenced by socio-economic factors.   With respect to the land itself, several variables related to land holdings are important.   The most obvious, that of property size, is a function of fragmentation and/or tenure.   The specific variables discovered in Canadian and New Zealand studies to be important in the decision process in response to external environmental or economic stimuli were the following:

i)   property size;

ii)   level of fragmentation of property holdings;

iii)   shape of property;

iv)   distance of property from home of owner/manager;

v)   tenure (owned or leased, or other encumbrances);

vi)   length of tenure (particularly with respect to leasehold or usufruct arrangements);

vii)   present level of infrastructure;

viii)   level of capitalization/debt;

ix)   land value.

The first four are direct characteristics of the land. The latter are characteristics of the ownership arrangements. The variables relating to the size, shape or fragmentation of the property unit relate to the ability of producers to adapt to changing economic or environmental situations. Larger property units are more versatile and open to fairly rapid change, whereas more fragmented patterns can provide severe logistical problems. It is clear that in order to temper any biophysically-based supply/demand modeling procedures, variables on the availability of land must also include size of property and fragmentation. Tenure is a further constraint or facilitating mechanism to action. For example, short-term leasing arrangements are not conducive to long-term investments in sustainable soil management or in products whose value will not be realized for decades. The nature of these property variables, however, is that they are not usually regionally homogeneous and decisions will have to be taken regarding the way in which the information is held. Mean size of property, for example, may be a very meaningful variable in regions where property size is relatively homogeneous. In areas of extreme variations it can be very misleading, particularly in terms of the ability of the land to adapt to changing economic or biophysical circumstances.

One significant socio-economic variable of considerable importance in short- to medium-term response to the need to alter land use is that of dedicated land areas. Many governments have made major subtractions from the land base to serve specific functions such as security (military bases), public recreation (parks), water supply (reservoirs), flood protection (flood plains) or wildlife reserves. These areas are, for most purposes, no longer available to serve most other functions (habitat, buffering and wildlife production are notable exceptions) and should probably be considered outside the land base for most production-related scenarios. Even so, under major distortions, or extreme stress, these barriers to change can also disappear.

22.5.2. **Characteristics of the individual owner or user**

Empirical research on the determinants of land use has highlighted the importance of landholder characteristics as they affect the ability of land-using systems to respond to needs to change (Manning, 1972; Mandale, 1984). What is becoming increasingly clear is that individual owners or managers of land resources vary greatly in their ability or willingness to undertake changes. Relative to continental-scale requirements to make fundamental changes to the way in which land is used, this factor may be temporal. At least in the short- and medium-term, however, it is important because it influences the nature and pace of response. It is clear that many individual factors relating to the structure of ownership or management of the land resource and relating to the individual decision-maker are important.

In the North American situation (McCuaig and Manning, 1982) a model was put forward identifying several factors as critical to the willingness or ability of individuals to change land use or management, given the requirement to do so due to economic or policy stimuli. This model is shown as Figure 22.2. It is suggested that the same individual characteristics shown in this model will often be operative as filters to individual response to major biophysical changes, particularly those which occur over short periods. Specific factors found to be important include age, level of education, level of capitalization and aspirations of the landholder. While it would be very difficult to model this type of information at a continental scale, broad generalizations regarding the nature of the population in given areas could be of use as a filter in the assessment of scenarios developed through the more biophysically-based supply/demand modeling. This socio-economic filter would be applied to the supply/demand model as a further step in evaluating the practicality of policy responses.

Some specific areas where information would be of particular use in evaluating the practicality of specific policies in certain situations are given below.

1)   *Level of education.* It would be useful to establish baseline information regarding the general level of technical education of land managers/owners in relation to crop production or other types of land management. More educated managers are more likely to be willing and able to adopt innovative technology or different crops/uses more quickly than those with less training in land management. Data pertaining to these qualities would allow general evaluation or assessment of any result heavily dependent upon rapid alterations by individual property owners or users.

2)   *Age.* In general, age structure or, as an alternative, family structure of the landholders or users is pertinent in terms of the

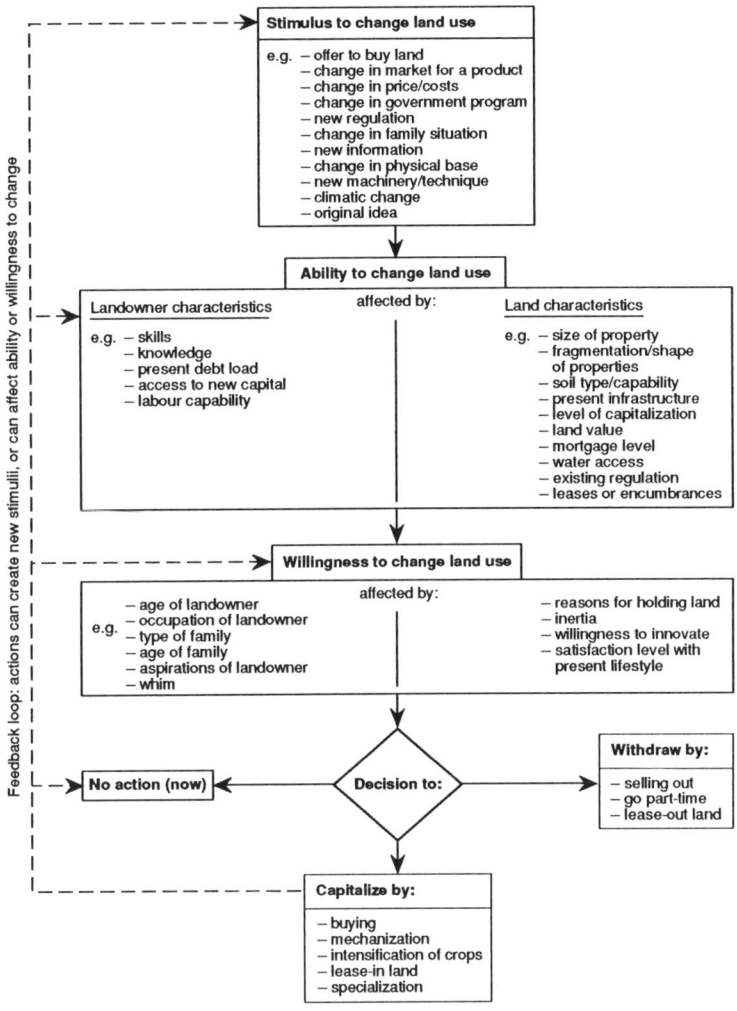

Figure 22.2. The decision process for rural land use change: a simple model.

ability and willingness to respond to the need for changes. In general, the nature of response relates very well to the level of continuity of landholding or management of the individual or corporate collective unit.   In the Canadian and Australasian context, research has demonstrated a clear relationship between the propensity to undertake investments which bring benefits in the longer term and the expectations of gain deriving from those benefits (Manning, 1972; McCuaig and Manning, 1982).   In general, younger owners, corporations and those with heirs who will continue to use the property were found to be most willing to make changes. With respect to the impact of personal factors on the ability to respond, it is clear that broad-scale anomalies or differences will act as filters to the overall response to the need for broad-scale change in the use of the land. For example, in an area where virtually all of the rural population is aged over 55, one would expect less direct response than in areas where the population is younger, or the land is held in corporate enterprises. At a continental scale, it may be difficult to generalize these variables, but major anomalies could be identified.

In Figure 22.2 other important factors at an individual or corporate scale are identified including income levels, present debt load or access to capital, occupation of owner and the aspirations or reasons for holding land. Whether any of these factors can be generalized to broad regional or continental scale differences pertinent to European futures would be worthy of further exploration.

From a holistic perspective a number of other social, infrastructural and policy variables are important both in terms of their influence on the development of current land use patterns and their utility as vehicles for/or impediments to change. Land use patterns can be seen as the product of the interplay of the biophysical variables and history. Research is ongoing to seek new variables which may be manipulated either to reduce risk in the future or to assist in directing land use patterns to serve current and foreseen societal goals. Therefore, the overall social, political and infrastructural situation becomes very important. The analysis of determinants of land use change would be incomplete without identifying the key Government activities which both directly and indirectly influence decisions on land use. In the Canadian instance these were categorized into several groups (Bond et al., 1982):

    i)   ownership and management of land;

    ii)   construction activities;

    iii)   regulation;

    iv)   financial policies;

    v)   sectoral support programs;

    vi)   regional development programs;

vii)   research, information and planning.

In a Canadian study (Bond *et al.*, 1986) over 100 key federal government programs were identified, comprising 25 per cent of discretionary federal government expenditures which significantly influence the use and management of the nation's land resources. These programs include those which act as important influences on current land use practices and decisions as well as those which could, in future, be manipulated in order to achieve broader goals relative to the land resource base (for example, Government of Canada and Government of the USA, 1985). Federal programs were complemented by an even greater number of provincial and local policies and programs which had direct and indirect impact upon the use of land resources (Audet and Le Hénaff, 1984; Ward, 1985).

In any given place the interplay of these policies and programs would be quite different, sometimes reinforcing trends towards sustainable development, sometimes blocking them or working at cross purposes (Manning, 1986a; Munton, 1987). In each country or area there is a fundamental framework of public policies and programs which, at least in the short-term, must be taken as "givens" within which changes will occur. A prime example is agricultural subsidization which has been a tenet of rural development for much of western Europe and North America for many years. The complex and comprehensive system of support to all aspects of agriculture can be taken as given within the decision process for land allocation and management. Any attempt to alter this system significantly will be disruptive because it will remove many of the assumptions on which past investments in land management have been made, both by individual landowners and by governments.

On a continental scale, if planning for sustainable development is to occur, it would be important to have information on the principal differences in policy from place to place. Most notable are those policies that are important in terms of infrastructure, direct and indirect price supports which protect particular lands for such functions as agricultural production, aggregate extraction or wetland habitat, or which favor particular sectors or regions. In any assessment of the impacts of change, these become important both in terms of influencing the changing demands on the resource base from one region to another and as levers by which land use change can be promoted or inhibited. Under most scenarios, many of these policy and program structures will have to be assumed to remain constant, with only incremental modifications at least in the short- and medium-term.

22.6. **Demands for Land**

The determinants of demand for land are diverse, reflecting all of the different functions the land must serve (Table 22.2). The estimation of future requirements for land can be done in a variety of ways. Sectoral demand projections have been commonly used to establish trends ranging from linear extrapolation of spatial trends through to more complex modeling of changing productivity. In the United States work at Ft. Collins, Colorado, has developed a linked set of sectoral models, beginning with forest sector demands, and carrying these through links to other sectors to examine sectoral impacts (Hoekstra and Joyce, 1988). Such approaches are workable in exposing the implications of demand trends but suffer from the inherent problem of linked models, since output estimates from one step become input into the next, usually yielding great (compounded) variability in the end product. Any predictive methodology is beset with uncertainty. Few multisectoral approaches improve on the Delphi approach wherein consensus futures are estimated by appropriate experts. For this reason this Chapter counsels an approach which does not depend on reliable predictions of future land demands.

This is even more essential given the concern for unpredictable events. It is suggested that one draws on existing goals, or develops scenarios which have some probability/possibility of occurrence. Work can occur to aid in deciding which of these scenarios are acceptable to people and their governments. Further effort will define the components of the chosen scenarios in terms relative to the demands they bring to bear on elements of the land resource. The literature on environmental functions, and work currently under way in North America on cumulative impact analysis (Peterson et al., 1987; Gosselink and Lee, 1987), give some indication as to how this can be accomplished. In research focusing on habitat preservation methodologies conducted under the North American Waterfowl Management Plan, specific links are being made to the various goals of society served by wetland habitat (Bardecki et al., 1988). This work shows the link between specific societal demands or goals and requirements for particular quantities of land with very specific characteristics. Therefore, the key to analysis of future demands placed on the land will be in establishing a range of scenarios, based upon explicit assumptions of lifestyle, levels of economic activity and population levels. The development of conservation strategies and national plans, inter alia, is also a goal setting procedure which defines demand scenarios. Scenarios which have implications for land supply, leading to a reduced or enhanced resource base, can be developed. Such scenarios can be based on ongoing predictive work or on "worst case" probable outcomes.

22.7. Selecting a Spatial Framework

Throughout this Chapter it has been suggested that a spatial perspective will be essential to permit productive analysis of the determinants of land use change. This implies a spatial framework for data collection and for any subsequent modeling. A key problem is identifying a suitable spatial framework that will permit the integration of information from diverse sources so that trends and surprises can be evaluated against the opportunities and constraints afforded by the land resource base.

The selection of a spatial framework should be closely linked to the needs and requirements of the users. However, it is also important to identify assumptions concerning resources, and access to geographic information systems (GIS) technology. The use of a powerful integration tool for discrete spatial data sets could reduce the importance of selection of a spatial frame (Chorley, 1987). There are several schools of thought with respect to spatial frameworks. One favors the use of geometric grids where data are synthesized and reported by geometric units, thus giving an evenness which supports mathematical analysis. If a 10 or 100 km^2 grid is developed, for example, all of the compromises in data are made at the stage of data input where information obtained through various routes such as census, field trials, political or ecological units is abstracted to the nearest point on the grid by using a range of subjective or mathematical procedures. This has the advantage of being neat and of not representing individual jurisdictions which may be sensitive to the portrayal of their data, particularly if they prove to be in the lowest cohort. Nevertheless, it has severe limitations, particularly for data which are not initially spatial (an attribute common to socio-administrative or ecological units) where homogeneity must be assumed. This occurs simply because the point may not be representative of the whole and, if larger grids were used, it may be quite unrepresentative of the locality. Thus, spurious spatial correlations can occur in terms of relationships that would drastically affect the testing of different scenarios dependent upon empirical relationships.

The use of political units (counties, districts) is another common way of data portrayal and integration. This is often easier to do than other means of portrayal simply because so much data tends to be collected on a jurisdictional basis. The key advantages of the use of political units, particularly fairly small ones (50-250 km^2 range), lie in the fact that they are often relatively physically homogeneous at this scale and reflect the levels of authorities which may be suitable to deal with the planning of land use. Further, at a continental scale, smaller units may produce data sets so large as not to be easily analysed. In Canada and the United States, county level data has frequently been used as a reasonable level to portray regional differentiation. Typically, counties in Canada and the United States are in the 1,000 to 10,000 km^2 range although there is some extreme variation. Variation in size is one of the drawbacks in the use of this type of spatial generalization, although the superimposition of ecumene boundaries in the Canadian case has been of considerable aid in resolving this problem. A further problem with the standardization of political units is their failure to correspond to environmentally homogeneous units in many areas, particularly in mountainous terrain or across natural boundaries. The principal advantage of the political units, particularly for output, is that the responsible authorities are clearly identifiable and when action needs to be taken the policy prescriptions can be phrased in terms that the current political structures find practical (Manning, 1986a, 1987a; Chorley, 1987).

A third type of unit is the ecological unit based on biophysical characteristics. The data acquisition and analysis of the significant biophysical variables is often easier, and far fewer spatial compromises need to be made with respect to physical information. The integrated analysis of the variables influencing the quality of the resource base for certain uses is therefore facilitated. Unfortunately, it is very rare that socio-economic information is available on the same basis. Further, if results are made available solely on the basis of biophysical units, experience has shown that it may be less easy to trigger political response as the units are unfamiliar and do not correspond to areas of jurisdiction (Manning, 1987b). It should be noted, however, that in some countries responsibility for conservation-related activity is sometimes based on biophysical regions (Conservation Authorities, River Basin Management Agencies). In these cases this type of reporting unit would be ideal.

An appealing compromise approach is the combining of well established political units with ecological units to form standardized hybrid units. Preliminary work using data for the Canadian prairies has demonstrated the utility of these "environomic" units to integrate, analyse and display both types of data at a regional/continental scale (Gélinas, 1988). A similar basis was used for the definition of modeling units by the University of Guelph in the development of a supply/demand response model for the Federal Government and the province of Ontario (Land Evaluation Project, 1982; Land Evaluation Group, 1985). A multisectoral application (agriculture, forestry, other) in New Brunswick shows some promise as a means of dealing with intersectoral competition for scarce resource lands under different scenarios of demand (Smit and Brklacich, 1985).

The improvement of geographic information systems (GIS) has made the integration of data from different units far easier although caveats still exist with respect to the way in which spatial data can be handled and integrated. Nevertheless, GIS make it possible to use a system of biophysical units and to integrate them with data from other spatial units permitting an output to be formatted on any geographical base or on hybrid units (Crain and MacDonald, 1984). The main concern is that the data held in the systems must be truly spatial and that this information must refer, to the greatest extent possible, to a homogeneous characteristic of all places within each spatial unit. Thus, extreme care must be taken when dealing with non-spatial data with the understanding that assumptions of averaging and homogeneity are inherent in any of the information used in spatial overlay. Nevertheless, the versatility of such systems can permit the generation of output on a wide range of spatial units, including biophysical units for analysis and political units for implementation. In the use of GIS the compromises need not be made at the input stage since data are held by the geographical units for which they were collected. They may, therefore, permit the data to be used to their limits of validity relative to other spatial data in quite sophisticated analysis.

The selection of specific variables to be held will be difficult. It would be folly to amass huge banks of data in its rawest form which may duplicate existing sectoral or national data banks but, at the same time, aggregation of data into very general indicators or very large compromise units is also inappropriate. The solution may well be in using an intermediate level of spatial aggregation for data holding, such as administrative units, drainage basins or subbasins, or ecological units, while maintaining the base variables in quantitative format. Ideally, one would use well established reporting units for which most of the required data are already available. Point data sources for other information such as extraction sites and soil quality monitoring points are obtainable. If GIS are used, then many means of integrating point data with spatial data are available.

22.8. Testing Scenarios within a Spatial Supply/Demand Framework

This Chapter has defined the types of information pertinent to the key variables determining land use which would contribute to a supply/demand modeling exercise for land. It is suggested that a supply/demand model would be the most useful framework within which to test changes either in the quality and quantity of land supply or in the nature and quantity of demands to be placed on the resource base. In this Section a particular type of model is proposed based on a number of modeling exercises which have been developed as policy support instruments in Canada and other jurisdictions (Land Evaluation Group, 1983a; Heckland, 1984). The particular framework put forward is, in the main, directly derived from the modeling exercise for the governments of Ontario and New Brunswick, through the University of Guelph. This model is a supply/demand model for land with a spatial framework (Smit, 1981). When tested, scenarios result in an output which gives not only the total requirements from the land base but shows where these land-based requirements for products or services can be satisfied across a large number of spatial units. This approach, therefore, permits the testing of scenarios and the valuation of their implications for each spatial data unit. This type of procedure can be called supply/constraint modeling. The supply/constraint model is dependent on a knowledge of approximately a dozen biophysical variables adequate to permit analysis of supply/response for the essential products or functions in question. In the Ontario experiment the spatial data units selected were combinations of political units (counties and townships) approximating to major biophysical zones within the province of Ontario. The selection of boundaries was done primarily through climatic data and broad physical differences. The units varied somewhat in size from one representing a single county with relatively unique biophysical features to another much larger zone representing several thousand km² of relatively flat, climatically favored land. The information necessary for the analysis of major changes in the biophysical base or in other political or socio-economic constraints is related to the quantity of land with particular qualities to be found within each of the data units (Land Evaluation Group, 1985). In the Ontario case the environmental functions of concern were primarily those supporting production of agricultural products. Information was obtained from test sites within each of the data units showing supply/response for key agricultural crops to changes in precipitation and soil chemistry. This initial step permitted modeling of a number of scenarios over these hybrid biophysical/political units. Some applications included:

Satisfaction of future demands for particular products

Several runs were carried out to evaluate different future demands for agricultural production to be placed on the resource base and of changes in the constraints (energy costs, climatic changes). For example, the model was run to test the impact of different urban expansion scenarios, each of which would subtract different quantities of land with known characteristics from each of the data units. The questions asked were:

i)   are the present production quantities for key products feasible under each of the urban expansion scenarios?

ii)   what are the major spatial shifts required in key products to satisfy these scenarios?

iii)   are there any specific spatial units where the pressures will intensify to the point where the alterations are not practical?

The impact of policies on resource supply

A second approach examined the impact of particular policies on land use options. For example, if it was decided to reduce the impact of farming practices on soil and water quality, a buffer strip along all water courses could be created (Land Evaluation Group, 1983b). What would be the impact of this on the satisfaction of a number of demands? Would it still be possible to achieve feed grain self-sufficiency without sacrificing other major products under this scenario? What would be the impact of continued erosion over 25 years?

The impact of climatic change

A third type of application of the model was the evaluation of the ability to meet production goals under different climate change scenarios. Each climatic scenario was run in terms of its projected impact on physical attributes of the land base. The changes in climate would therefore enhance or reduce productivity for particular crops on each type of soil within each unit. The questions focused on the need to change location of crops and the impact on overall production. The model trials showed where production would be most important to the achievement of future scenarios. The model was also run to discover whether adjustments in location of cultivation of eight crops could be done to maintain production or to optimize production of priority products. Typical "answers" included "no", only one of the two production goals could be satisfied under a particular scenario, or "yes", but only if over 80 per cent of the land with particular biophysical characteristics in a particular data unit was put into one crop. No further work was done on implications for other sectors such as forest products or habitat protection, as essential productivity-response relationships were not available for these sectors.

While this type of modeling procedure has proven relatively effective for scenario testing for agriculture, it is clear that there are many more capabilities which could be exploited. For example, the addition of productivity-response information for tree crops, biomass production and waterfowl production would permit the extrapolation of these approaches across other sectors.

One element in the further development of this type of modeling that would enhance its utility for the assessment of environmental futures would be the development of integrated biophysical baseline data on biophysically defined units. If biophysically homogeneous units at an appropriate scale were defined, then productivity response, in the face of alterations in key climatic or other physical variables, could be estimated for the most important environmental functions. This holds particularly for those functions that are viewed as most strategic to the achievement of government strategies or defined goals (Conservation Strategies). This would entail the development of productivity-response models for the major food crops, for forest products and wildlife as initial steps. In addition, the definition of relationships between the utility of particular parts of the environment to serve recreational functions, buffering, habitat or any other functions which are particularly sensitive to alterations in biophysical variables would be valuable. In some cases the environmental functions such as many forms of built environment are remarkably insensitive to biophysical change, depending more upon attributes of site and relative location than physical attributes of the resource base. These kinds of functions can be factored into this type of model primarily as subtractions from the available environmental resource at the outset of each run or can be separately modeled in terms of demand functions and used to develop scenarios which are then tested through the primarily biophysical supply/constraint model. Some types of scenarios amenable to this type of approach and pertinent with reference to the attainment of the goal of sustainable development are highlighted in Table 22.3.

Specific demand scenarios can be tested against the ability of the land base to respond. The objective of such model runs would be to answer the following kind of questions:

i)   can we maintain desired levels of supply of key functions given hypothesized disruptions or competing demands?

ii)   do current trends in use or abuse of aspects of the land base threaten our future desired uses?

iii)   do current (or proposed) policies create satisfactory outcomes?

iv)   where are the pressure points? Are particular uses or levels of productivity critical, for particular areas in many scenarios?

v)   can we plan to reduce fragility of the system (sensitivity to surprises) or overdependence on specific parts of the resource?

vi)   can all our goals be simultaneously satisfied by the resource base, or must we plan now to make adjustments/trade-offs between our goals?

If feasible scenarios from a land supply/constraint approach can be identified, the final step would be to evaluate these in terms of satisfaction of goals. In some cases (for example, agricultural production), economic analyses of the results would be valuable because a significant subset of physically feasible scenarios are not economically viable. Other means of evaluation of political and social desirability could also be applied where appropriate. One way of viewing this entire approach is to see the supply/constraint modeling as one side of the overall strategy and the scenario building or demand-evaluating side interfacing with it through the operator. The role of the operator, through as many iterations as is required, is to try to optimize both sides of the model. The objective of sustainable development is therefore to modify demand for, or manage supply of, environmental resources in an anticipatory fashion so that unacceptable outcomes are made less probable.

Table 22.3. Typical scenarios to be modeled.

Production capability under changed land supply
- reduced land base due to urban growth
- changed land capability due to climatic change

Impact of land use practices
- changed capability due to erosion, chemical use
- changed productivity response due to new crops, practices
- altered environmental quality due to use trends (toxification)

Changes in the demand for the land base
- need to achieve self-sufficiency on land base
- need to consecrate land areas to habitat/recreation
- major impact constraints for particular products

Social impact
- changed demand for natural areas
- changes in social acceptance of particular land use practices

Policy impact
- changed regulations on fertilizer use, pesticides
- changed policy: use regulations (agricultural land protection, buffer strips)
- trade agreements, barriers
- immigration/emigration scenarios
- changes in capital depletion changes (forest stumpage, mineral royalties)

Catastrophe
- environmental change scenarios/sensitivities (for example, climate change)
- drought/flood scenarios (short-term)
- environmental disaster scenarios (X% of base sterilized by accident)
- crop damage scenarios (agriculture, forestry)
- civil disruptions/economic surprises

22.9. Conclusion

The objective of the testing of scenarios is to identify those points of intervention where undesirable futures can be averted or at the very least the risk of unacceptable outcomes can be reduced.   Such approaches can also identify those "hot spots" which are particularly fragile or sensitive under a range of demand scenarios. The conceptual approach and modeling procedure described in this Chapter have very significant information requirements. Like all models, the utility is related directly to the amount of work necessary to create it. One of the most serious problems will be the selection of appropriate scales. Clearly, the more detailed the output the more directly applicable to solving problems at a regional scale. In contrast, the greater the detail required, the more expensive and complicated will be any modeling procedure.

The crux of the practical problem in defining an overall program to amass information and to integrate it in a way that permits the assessment of future land use options is the selection of an appropriate level of generalization. The ideal is impossibly complex and expensive. The practical is probably inappropriately general. Yet, the overall problem of creating something usable and something which will, in fact, be able to identify sensitivities, points of intervention and the need for pre-planning, lies in the selection of a small number of clearly definable biophysical variables and a limited number of socio-economic factors which modify supply. These must correlate well with the productivity response for the most important environmental functions. The next step is the selection of moderately-sized data units which are preferably biophysically based to permit better synthesis. The use of geographical information systems is recommended so that output can be obtained not only for biophysical units but also geopolitical ones. The adoption of a supply/demand modeling framework, while it would not address all environmental questions for Europe's future, will permit the testing of many of the most probable scenarios in terms of potential land use impacts. The spatial perspective will permit not only the broader scale identification of supply/demand problems, but also the regionalization of these to the point where specific interventions can be possible in order to reduce sensitivity or to avert probable/possible future difficulties.

This Chapter has been developed as a contribution to the definition of a program which will allow the examination of the determinants of land use for Europe in order that Europe's environmental future can be addressed. In a sense it is an attempt to see the unseeable. Surprise events will continue to be just that - surprises. By casting our net broadly and by understanding the response of those environmental functions upon which society is so dependent, we are at least better armed to deal with surprises. We are also able to identify the weakest links in our system, and perhaps plan in advance to reduce situations of particular fragility or sensitivity, especially those which threaten the sustainability of the system. At the very least these types of approaches help us to understand, at a continental scale, what is occurring and what are the implications for the sustainable development of the land base.

The author would like to express his appreciation for the input, comments and critiques of the following individuals whose help was invaluable in the preparation of this Chapter: P.D. Bircham, W.K. Bond, H.C. Bruneau, R. Gélinas, C. Lisiecki, J.D. McCuaig, L.C. Munn, I. Reiss, W. Simpson-Lewis.

REFERENCES

Adamus, P.R. and Stockwell, L.T., 1983, *A Method for Wetland Functional Assessment.*   Federal   Highway   Administration, Washington.

Audet, R. and Le Hénaff, A., 1984, *The Land Planning Framework of Canada: An Overview.* Working Paper 28.   Lands Directorate, Environment Canada, Ottawa.

Bardecki, M., Bond, W.K. and Manning, E.W., 1988, *Wetland Evaluation and the Decision-making Process.* Ryerson Polytechnical Institute, Toronto, Canada.

Beattie, K.G., Bond, W.K. and Manning, E.W., 1981, *The Agricultural Use of Marginal Lands: A Review and Bibliography.* Working Paper 13. Lands Directorate, Environment Canada, Ottawa.

Bond, W.K., Manning, E.W., Bircham, P.D., McCuaig, J.D. and McKechnie, R.M., 1982, *The Identification of Impacts of Federal Programs on Land Use: A Manual for Program Managers.* Interdepartmental Committee on Land, Environment Canada, Ottawa.

Bond, W.K., Bruneau, H.C. and Bircham, P.D., 1986, *Federal Programs with the Potential to Significantly Affect Canada's Land Resource.* Lands Directorate, Environment Canada, Ottawa.

Brundtland, G., 1985, *Mandate for Change: Key Issues, Strategy and Workplan.* World Commission on Environment and Development. Geneva, Switzerland.

Chorley, Lord R., 1987, *Handling Geographic Information.* Report of the Committee of Enquiry, United Kingdom.   Department of the Environment. HMSO, London.

Clark, W.C. and Munn, R.E. (eds.), 1986, *Sustainable Development of the Biosphere.* Cambridge University Press, Cambridge.

Coote, D. and Shields, J., 1990, SOTER/GLASOD Manual. Agriculture Canada/International Soil Reference and Information Center, Wageningen, The Netherlands.

Crain, I.K. and MacDonald, C.L., 1984, *From Land Inventory to Land Management: The Evolution of an Operational Geographic Information System.* Proceedings, Autocarto Six. Ottawa. **1**, 41-50.

Environment Canada, 1970, CLI: Objectives, scope and organization. *Canada Land Inventory Report 1.* Lands Directorate, Environment Canada.

Environment Canada, 1976, Land capability for agriculture: A preliminary report. *Canada Land Inventory Report 10.* Lands Directorate, Environment Canada, Ottawa.

Franklin, H., 1969, *The European Peasantry.* Methuen, London.

Galeski, B. and Wilkening, E., 1987, *Family Farming in Europe and America.* Westview Press, London.

Gélinas, R., 1988, Development and Application of "Environment Units": Hybrid Map Units Designed to Integrate Environmental and Socio-economic Data for Land Modelling. In R. Gélinas, D. Bond and B. Smit (eds.), *Perspectives on Land Modelling,* Polyscience Publications Inc., Montreal, Canada. 69-85.

Gosselink, J.G. and Lee, L.C., 1987, *Cumulative Impact Assessment in Bottomland Hardwood Forests.* University of Georgia and US Environmental Protection Agency, Washington D.C.

Government of Canada and Government of the United States of America, 1985, *North American Waterfowl Management Plan: Draft.* Canadian Wildlife Service, Environment Canada/Fish and Wildlife Service, US Department of the Interior.

de Groot, R.S., 1986, *A Functional Ecosystem Evaluation Method as a Tool in Environmental Planning and Decision-making.* Nature Conservation Department, Agricultural University, Wageningen, The Netherlands.

de Groot, R.S., 1987, Environmental functions as a unifying concept for ecology and economics. *The Environmentalist,* **7**(2), 105-109.

Harte, J., 1979, The risks of energy strategies: an ecological perspective. In G.T. Goodman and W.D. Rowe (eds.), *Energy Risk Management.* Academic Press, London. 43-59.

Heckland, J.M., 1984, *Land use Budgeting: Models for Land Use Evaluation and Budgeting.* Nordic Consulting Group. Asker, Norway.

Hoekstra, T.W. and Joyce, L.A., 1988, An overview of the southern United States multiresource modelling. In R. Gélinas, D. Bond and B. Smit (eds.), *Perspectives on Land Modelling.* Polyscience Publications Inc., Montreal, Canada. 17-25.

Holling, C.S., 1986, The resilience of terrestrial ecosystems: Local surprise and global change. In W.C. Clark and R.E. Munn (eds.), *Sustainable Development of the Biosphere.* Cambridge University Press, Cambridge. 292-317.

Jacobs, P. and Munro, D.A., 1987, *Conservation with Equity: Strategies for Sustainable Development.* IUCN, Cambridge, U.K.

Jackson, W. and Berry, W. (eds.), 1984, *Meeting the Expectations of the Land: Essays in Sustainable Agriculture and Stewardship.* North Point Press, San Francisco.

Land Evaluation Group, 1983a, *Potential Applications of a National Land Evaluation System.* Guelph University School of Rural Planning and Development, Guelph, Ontario.

Land Evaluation Group, 1983b, *Land Evaluation under Alternative Erosion Control Policies.* Guelph University School of Rural Planning and Development, Guelph, Ontario.

Land Evaluation Group, 1985, *Applications of the Prototype Canadian Land Evaluation Systems to Selected Issues.* Guelph University School of Rural Planning and Development, Guelph, Ontario.

Land Evaluation Project, 1982, Feasibility of a national level land evaluation system. *Report No. 5 (81-83).* University School of Rural Planning and Development, University of Guelph, Guelph, Ontario.

Mandale, M., 1984, Marginal land utilization and potential: Kent County, New Brunswick. *Working Paper 31,* Lands Directorate, Environment Canada, Ottawa.

Manning, E.W., 1972, *The Changing Land Use of a Diversified Agricultural Region: The Heretaunga Plains of New Zealand.* Victoria University of Wellington, Wellington, New Zealand.

Manning, E.W., 1985, Canada's land: Relating research to reality. *Zeitschrift der Gesellschaft für Kanada-Studien,* 1, 61-84.

Manning, E.W., 1986a, Towards sustainable land use: a strategy. Canada-China Bilateral Symposium on Territorial Development and Management, Beijing. *Working Paper 47.* Lands Directorate, Environment Canada, Ottawa.

Manning, E.W., 1986b, Planning Canada's land resource base for sustainable development. In I. Knell and J. English (eds.), *Canadian Agriculture in Global Context.* Centre on Foreign Policy and Federalism, University of Waterloo, Waterloo, Ontario.

Manning, E.W., 1987a, Land Data Requirements for National Land Policy and Program Formulation. In *Proceedings of the Monitoring for Change Workshops.* Lands Directorate, Environment Canada, Ottawa.

Manning, E.W., 1987b, Models and the decision-maker. In R. Gélinas, D. Bond and B. Smit (eds.), *Perspectives on Land Modelling.* Polyscience Publications Inc., Montreal, Canada. 3-7.

Manning, E.W. and McCuaig, J.D., 1984, Planning operational research: a conceptual approach and applications to land research. *The Operational Geographer,* 5.

McCuaig, J.D. and Manning, E.W., 1982, Agricultural land use change in Canada: process and consequences. *Land Use in Canada Series, Volume 21.* Lands Directorate, Environment Canada, Ottawa.

Munn, L.C., 1986, The Canada Land Inventory. In *Proceedings of the NATO Seminar on Land and Its Uses.* Institute of Terrestrial Ecology, Midlothian, Scotland.

Munn, R.E., 1987, Environmental Prospects for the Next Century: Implications for Long-term Policy and Research Strategies. *RR-87-15*, International Institute for Applied Systems Analysis, Laxenburg, Austria.

Munton, R., 1987, The conflict between conservation and food production in Great Britain. In C. Cocklin *et al.* (eds.), *Demands on Rural Lands: Planning for Resource Use.* Westview Press, London.

Peterson, E.B., 1987, *Cumulative Effects Assessment in Canada: An Agenda for Action and Research.* Canadian Environmental Assessment Advisory Council, Ottawa.

Smit, B., 1981, Procedures for the long-term evaluation of rural land. *CRD Publication 105.* Land Evaluation Project Team, University School of Rural Planning, University of Guelph, Guelph, Ontario.

Smit, B. and Brklacich, M., 1985, Feasibility of constructing a multisector land evaluation system: The New Brunswick pilot study. *Working Paper 42*, Lands Directorate, Environment Canada, Ottawa.

Ward, G.N., 1985, Heritage conservation - the built environment. *Working Paper 44*, Lands Directorate, Environment Canada, Ottawa.

Index

The GeoJournal Library

1. B. Currey and G. Hugo (eds.): *Famine as Geographical Phenomenon*. 1984
   ISBN 90-277-1762-1
2. S. H. U. Bowie, F.R.S. and I. Thornton (eds.): *Environmental Geochemistry and Health*. Report of the Royal Society's British National Committee for Problems of the Environment. 1985   ISBN 90-277-1879-2
3. L. A. Kosiński and K. M. Elahi (eds.): *Population Redistribution and Development in South Asia*. 1985   ISBN 90-277-1938-1
4. Y. Gradus (ed.): *Desert Development*. Man and Technology in Sparselands. 1985   ISBN 90-277-2043-6
5. F. J. Calzonetti and B. D. Solomon (eds.): *Geographical Dimensions of Energy*. 1985   ISBN 90-277-2061-4
6. J. Lundqvist, U. Lohm and M. Falkenmark (eds.): *Strategies for River Basin Management*. Environmental Integration of Land and Water in River Basin. 1985   ISBN 90-277-2111-4
7. A. Rogers and F. J. Willekens (eds.): *Migration and Settlement*. A Multi-regional Comparative Study. 1986   ISBN 90-277-2119-X
8. R. Laulajainen: *Spatial Strategies in Retailing*. 1987   ISBN 90-277-2595-0
9. T. H. Lee, H. R. Linden, D. A. Dreyfus and T. Vasko (eds.): *The Methane Age*. 1988   ISBN 90-277-2745-7
10. H. J. Walker (ed.): *Artificial Structures and Shorelines*. 1988
    ISBN 90-277-2746-5
11. A. Kellerman: *Time, Space, and Society*. Geographical Societal Perspectives. 1989   ISBN 0-7923-0123-4
12. P. Fabbri (ed.): *Recreational Uses of Coastal Areas*. A Research Project of the Commission on the Coastal Environment, International Geographical Union. 1990   ISBN 0-7923-0279-6
13. L. M. Brush, M. G. Wolman and Huang Bing-Wei (eds.): *Taming the Yellow River: Silt and Floods*. Proceedings of a Bilateral Seminar on Problems in the Lower Reaches of the Yellow River, China. 1989   ISBN 0-7923-0416-0
14. J. Stillwell and H. J. Scholten (eds.): *Contemporary Research in Population Geography*. A Comparison of the United Kingdom and the Netherlands. 1990
    ISBN 0-7923-0431-4
15. M. S. Kenzer (ed.): *Applied Geography*. Issues, Questions, and Concerns. 1989   ISBN 0-7923-0438-1
16. D. Nir: *Region as a Socio-environmental System*. An Introduction to a Systemic Regional Geography. 1990   ISBN 0-7923-0516-7
17. H. J. Scholten and J. C. H. Stillwell (eds.): *Geographical Information Systems for Urban and Regional Planning*. 1990   ISBN 0-7923-0793-3
18. F. M. Brouwer, A. J. Thomas and M. J. Chadwick (eds.): *Land Use Changes in Europe*. Processes of Change, Environmental Transformations and Future Patterns. 1991   ISBN 0-7923-1099-3

KLUWER ACADEMIC PUBLISHERS – DORDRECHT / BOSTON / LONDON